THE CONIFER MANUAL

Forestry Sciences

VOLUME 34

The titles published in this series are listed at the end of this volume.

HUMPHREY J. WELCH

The Conifer Manual

Volume 1

This book represents a full revision
and up-dating of the following works

Manual of Cultivated Conifers in the Cold and Warm-Temperate Zone
by P. den Ouden and Dr. B. K. Boom (1965)

and

Manual of Dwarf Conifers
by Humphrey J. Welch (1979)

KLUWER ACADEMIC PUBLISHERS
DORDRECHT / BOSTON / LONDON

Library of Congress Cataloging-in-Publication Data

Welch, H. J. (Humphrey James)
The conifer manual / by Humphrey J. Welch.
p. cm. -- (Forestry sciences ; v. 34)
"Represents a full revision and up-dating of ... Manual of cultivated conifers in the cold-and warm-temperate zone, by P. den Ouden and Dr. B.K. Boom ... and Manual of dwarf conifers, by Humphrey J. Welch."
Includes bibliographical references.
ISBN 0-7923-0616-3
1. Ornamental conifers. 2. Conifers. 3. Ornamental conifers--Varieties. 4. Conifers--Varieties. I. Ouden, P. den (Pieter den), 1874-1964. Manual of cultivated conifers. II. Welch, H. J. (Humphrey James). Manual of dwarf conifers. III. Title. IV. Series.
SB428.W39 1990
634.9'75--dc20 89-48627

ISBN 0-7923-0616-3

Published by Kluwer Academic Publishers,
P.O. Box 17, 3300 AA Dordrecht, The Netherlands.

Kluwer Academic Publishers incorporates the publishing programmes of
D. Reidel, Martinus Nijhoff, Dr W. Junk and MTP Press.

Sold and distributed in the U.S.A. and Canada
by Kluwer Academic Publishers,
101 Philip Drive, Norwell, MA 02061, U.S.A.

In all other countries, sold and distributed
by Kluwer Academic Publishers Group,
P.O. Box 322, 3300 AH Dordrecht, The Netherlands.

Cover photo: A beautiful Pungens Spruce.
(Reproduced with permission, © M. Nimmo).

Printed on acid-free paper

Printed in the Netherlands

TABLE OF CONTENTS

Volume One

TABLE OF CONTENTS

Volume Two

Author's Introduction

The research that has culminated in the publication of this Manual was in the first place undertaken for a new edition of the book *Manual of Cultivated Conifers* by P. den Ouden and Dr B. K. Boom (both now deceased), published in 1965.

The claim in that book to have included every cultivar published since 1753 was not entirely realised, and the stated objective, i.e. that the book should serve as a basis for the international register that was even at that date under consideration required a lot of space to be devoted to quite archaic information.

Each of the authors of the *Manual of Cultivated Conifers* was well qualified for the task. P. den Ouden had been systematically collecting conifer data throughout his long life as a nurseryman and had already produced his *Naamlijst van Coniferen* (published in 1937 and adopted as a standard for varietal names at the International Horticultural Congress in Berlin in 1939) and also his book *Coniferen Ginkgo en Ephedra* (in Dutch) published in 1949. Dr B. K. Boom, as the government taxonomist in Holland, had been recording data regarding new conifers as part of his official duties. So the *Manual of Cultivated Conifers* could be readily compiled at that time from available information.

My own book, *Dwarf Conifers – A Complete Guide*, published the following year, owed a great deal to help from Dr Boom, who spent several days with me at the Wansdyke Nursery at the time the book was in preparation.

Unfortunately this particular contribution by Dr Boom does not seem to have continued after his retirement and since then little if any systematic recording of new introductions was done in this field (despite the flood of new conifer cultivars being introduced worldwide) save by one or two private conifer enthusiasts. My own efforts at recording data at Devizes[1] made possible the appearance in 1979 of a larger book: *Manual of Dwarf Conifers*. That book was clearly in effect a partial up-dating of the den Ouden and Boom book under revision, so the decision was made to produce an entirely new work, with the present title and with the following objectives.

1. To bring the taxonomy into line with current practice.
2. To extend the species descriptions and make them (by the use of side headings) more easy to use.
3. To supply some form of 'keys'.
4. To release space for new material by relegating several hundred obsolete names to an Appendix.
5. To bring the coverage of cultivars up to date so far as possible, incorporating material from *Manual of Dwarf Conifers*.

The volume of material was now too great to be contained within a single book of manageable size so the new work has had to be produced in two volumes. Unfortunately this has necessitated the division of the big, descriptive chapter between the two volumes. The genera *Abies* to *Phyllocladus* will be found in this volume and the genera *Picea* to *Widdringtonia* in the volume which it is intended shall follow shortly.

There remains to me only a duty to acknowledge the great help received from other people – always a difficult task and (in the case of a major com-

[1] Since developed as the World Conifer Data Pool. See p. 56.

pilation of this kind) an impossible one because there have been so many contributions, large and small, direct or indirect, from so many sources that individual acknowledgement would be quite impossible. I here mention just a few; other debts of gratitude are acknowledged throughout the book but my sincere thanks go to everyone able to trace his or her own contribution anywhere in the book.

J. H. Lewis, my successor as International Conifer Registrar (and for many years my mentor in all matters botanical) has been most helpful. My thanks go to A. G. Mitchell, K. D. Rushforth and A. D. Silva, amongst others, for their critical help on technical matters, to Dr John Mills for permission to use the unpublished work of the late Laurence Gough, to Adrian Bloom, Robert Fincham, D. M. van Gelderen and many another plantsmen for help with details of new cultivars and to Don Mitchell for reading through the entire MS.

I am spared having to draft the usual encomium to a long-suffering secretary since this is a luxury I have had to do without. But in its place I think an apology is due to the word-processor I have had to make do with instead of a secretary for all the abuse and unparliamentary language that has been hurled at it since we first met.

And lastly, since I have been doing the work at home ever since an event that I still refer to as my retirement, my thanks go to my wife for putting up with the loss of the best room in the house for use as a library, for accepting the loss (for whatever that might have been worth) of many hours of my society, and – most of all – for her maddening, but steady encouragement to peg away and complete a job which had turned out to be far more arduous than I had bargained for when I took it on.

Humphrey J. Welch

Picture Credits

(1) Bailey, *The Cultivated Evergreens* (1923)
(2) Beissner, *Handbuch der Nadelgehölzer* Ed. 3 (1930)
(3) Engler, *Naturliche Pflanze-familien* (1926)
(4) *Flora of China* Vol. 7 (1978)
(5) Forestry Commission of Great Britain, *Conifers in the British Isles* (1972), drawings by Mrs. K. Darter.
(6) Mrs. Bente King.
(7) Mulligan, Brian, Kirkland, WA 98035, U.S.A.
(8) Loudon, *Arboretum et Fruticetum Britannicum* (1838)
(9) Nicholson, *The Gardener's Dictionary* (1884)
(10) Nimmo, M., Haverfordwest, Dyfed, U.K.
(11) Pallas, *Flora Rossica* (1878)
(12) Silva Tarouca, *Unserer Frieland Nadelhölzer* Ed. 2 (1923)
(13) Sudworth, *Forest Trees of the Pacific Slope* (1908)
(14) Veitch, *Manual of the Coniferae* Ed. 2 (1900)
(15) Mariosa, Peter, Lynton, N. Devon, U.K.
Unnumbered photographs supplied by the author.

A Conifer Bibliography

In addition to the following works consulted during the preparation of this book, much information relating to conifers can be found in general dendrological books, technical periodicals and in regional and local floras.

Antoine, F.
1840–47 Die Coniferen nach Lambert, Loudon et anderen.
Bailey, L. H.
1925 The Cultivated Evergreens.
1933 The Cultivated Conifers in North America.
Baker, R. T. & Smith, H. G.
1910 A Research into the Pines of Australia.
Beissner, L.
1887 (Apr.) Systematische Eintheilung der Coniferen.
1887 (Oct.) Handbuch der Coniferen-Benennung.
1891 Handbuch der Nadelholzkunde.
1891 1st Supplement to Handbuch der Coniferen-Benennung.
1892 2nd Supplement to Handbuch der Coniferen-Benennung.
1909 Handbuch der Nadelholzkunde. Ed. 2.
1930 Handbuch der Nadelholzkunde. Ed. 3. (Revised J. Fitschen).
Bloom, Adrian
1975 Conifers in your Garden.
Brotero, F. de A.
1827 Historia Natural dos Pinheiros, Larices et Abetos.
Browicz, K.
1982 Chlorology of Trees and Shrubs in S.W. Asia. Vol. 1.
Callen, G.
1976–7 Les Conifères cultivés en Europe.
Carrière, E-A
1855 Traité Générale des Conifères.
1867 Ditto. Ed. 2.
Chadwick, L. C.
1976 A Study of the genus Taxus.
Chaudun, V.
1956 Ornamental Conifers. (Translation from the French by Higgins, V.)
Cheng, W. C. et al.
1978 Flora Reipublica Popularis Sinicae. Vol. 7. 'Gymnospermae'.
Chittenden, F. J. See Royal Horticultural Society.
Clinton-Baker, H.
1909–1935 Illustrations of Conifers. 4 Vols.
Coltman-Rogers, C.
1920 Conifers and their Characteristics.
Cope, E. A.
1986 Native and Cultivated Conifers of Northeastern N. America.
Coulter, J. M. and Chamberlain, C. J.
1910 Morphology of Gymnosperms.
Critchfield, W. B. and Little, E. L.
1966 Geographic Distribution of the Pines of the World.
Dallimore, W.
1908 Holly, Yew and Box.
Dallimore, W. and Jackson, A. B.
1923 Handbook of Coniferae including Ginkgoaceae.
1931 Ditto. Ed. 2.
1948 Ditto. Ed. 3.
1960 Ditto. Ed. 4 (Revised by Harrison, F. G.)
Debazac, E. F.
1964 Manuel des Conifères.

De Laubenfels, D. J.
1972 Flora de la Nouvelle Caledonie et Dependances. No. 4. 'Gymnospermes'.
Den Ouden, P.
1937 Naamlijst van Coniferen.
1949 Coniferen, Ephedra en Ginkgo.
Den Ouden, P. and Boom, B. K.
1965 Manual of Cultivated Conifers.
De Vos, C.
1867 Beredeneerd Woordenboek der Voornamste Heesters en Coniferen in Nederland gekweekt.
1887 Handboek tot de Boomen, Heesters en Coniferen.
Eichler, A. W.
1899 "Coniferae" in Engler A. & Prantl, K. Die naturlichen Pflanzenfamilien.
Eiselt, M. G. & Schröder R.
1960 Nadelgehölze.
1976 Ditto. Ed. 4.
Elwes, H. J. and Henry, A.
1906–15 Trees of Great Britain and Ireland. 6 vols.
Endlicher, S.
1847 Synopsis Coniferarum.
Engelmann, Geo.
1880 Revision of the genus Pinus.
Farjon, A.
1984 Pines: Drawings and Descriptions of the Species.
Fitschen, W. See Beissner, 1930
Fitzpatrick, H. M.
1929 Coniferae: Keys to the Genera and Species.
1965 Ditto. Ed. 2.
Florin, R.
1955 The systematics of the Gymnosperms.
Forbes, J.
1839 Pinetum Woburnense.
Franco, J. do A.
1950 Abietos.
Fraser, H.
1875 Handy Book of Ornamental Conifers.
Gaussen, H.
1944 on Les Gymnospermes actuelles et fossiles.
Gordon, G. and Glendinning R.
1858 The Pinetum.
1862 Supplement to The Pinetum.
1875 The Pinetum. Ed. 2.
1880 Ditto. Ed. 3. (As Ed. 2, save for 2 additional pages).
Harrison, C. R.
1975 Ornamental Conifers.
Henkel, J. B. & Hochstetter, W.
1865 Synopsis der Nadelgehölzer.
Hillier, H. G.
1964 Dwarf Conifers.
Hochstetter, Wm.
1882 Coniferen oder Nadelgehölzer.
Hoopes, J.
1868 The Book of Evergreens.
Hornibrook, M.
1923 Dwarf and slow-growing Conifers.
1939 Ditto. Ed. 2.
Iwata, T. and Kusaka, M.
1954 Coniferae Japonica Illustratae.
Jackson, A. B.
1946 The Identification of Conifers.
Jäger, H.
1865 Die Ziergehölze.
Jäger, H. & Beissner, L.
1884 Die Ziergehölze. Ed. 2.
1889 Ditto. Ed. 3.
Jay, B. A.
1952 Conifers in Britain - A Guide to Identification.
Jenkins, C. F.
1939–51 Hemlock Arboretum Bulletin.
Kent, A. H. See Veitch, 1900.
Kirwan, C. de
1867 Les Conifères indigènes et exotique.
Klika, J. et al.
1953 Jelicnaté.
Knight and Perry
1850 Synopsis of the Coniferous Plants grown in Gt. Britain.
Kobayashi, Y. et al.
1975 Matsu-zukan.
Krüssmann, G.
1955 Die Nadelgehölze.
1960 Ditto. Ed. 2.
1972 Handbuch der Nadelgehölze.
1979 Die Nadelgehölze. Ed. 3.
1983 Handbuch der Nadelgehölze, Ed. 2. (Revised Hans-Dieter Warda).
1985 Ditto. (English translation by Epp, Michael E.).
Kumlien, L. L.
1936 Hill's Book of Evergreens.
1946 The Friendly Evergreens.
Lambert, A. B.

1803–07 A Description of the Genus Pinus. Vol. 1, 4to.
1824 Ditto. with an Appendix by D. Don. Vol. 2.
1828 Ditto. In 2 vols. 4to.
1832 Ditto. Ed. 3.
1837 Ditto. Ed. 1. Vol. 3.

Lawson, P. and Son
1851 List of Plants of the Fir Tribe.

Lemmon, J. G.
1898 Conifers of the Pacific Slope.

Little, E. L.
1953 Check List of Native and Naturalised Trees of the United States.

Little, E. L. and Critchfield, W. B.
1969 Subdivisions of the Genus Pinus.

Liu, T-S.
1971 A Monograph of the Genus Abies.

Loudon, J. C.
1838 Arboretum et Fruticetum Brittanicum. (8 vols.)

Magrini, G.
1952 Le Conifere. (Italian.)

Makino, T.
1966 New Illustrated Flora of Japan.

Masters, M. T.
1881–1904 Numerous papers in Journals of the Linnean and Royal Horticultural Societies.

Mayr, H.
1890 Monographie der Abietineen des Japanische Reiches.

Meyer, F.
1952 Die Nadelhölzer, including Ginkgo.

Mirov, N. T.
1967 The Genus Pinus.

Mitchell, A. F.
1972 Conifers in the British Isles, published by H. M. S. O. Mitchell also contributed largely to the Report of the Conifer Conference, 1970, published under the same title ('Conifers in the British Isles') by the Royal Horticultural Society, also in 1972.

Morgenthal, J.
1950 Die Nadelgehölze.
1952 Ditto. Ed. 2.
1954 Ditto. Ed. 3.
1964 Ditto. Ed. 4.

Mottet, S.
1902 Conifères et Taxacées.

Murray, A
1863 Pines and Firs of Japan.

Nelson, J. (Senilis)
1866 Pinaceae.

Parde, L.
1937 Les Conifères.

Parlatore, P.
1868 'Conifereas' in DC., Prodomus.

Pilat, A.
1964 Jelicnaté

Pilger, R.
1926 'Gymnospermae' in Engler & Prantl, Die natürlichen Planzenfamilien. Ed. 2. Vol. 13.

Ravenscroft, E. J.
1863–84 The Pinetum Britannicum. 3 Vols. Large folio.

Rehder, A.
1927 Manual of Cultivated Trees and Shrubs.
1940 Ditto. Ed. 2.
1949 Bibliography of Cultivated Trees and Shrubs.

Richard, L. C.
1826 Commentatio Botanica de Conifereis et Cycadeis.

Royal Botanical Gardens, Kew, U. K.
1896 Handlist of Coniferae. Further editions issued in 1903, 1925, 1928, and 1961.
1964 Catalogue of living Plants. Part 1.

Royal Horticultural Society.
1892 Report of Conifer Conference held in 1891. (Vol. 14 of the Society's Journal.)
1932 'Conifers in Cultivation'. Report of the Conifer Conference held in 1931. (F. J. Chittenden, Editor.)
1972 'Conifers in the British Isles'. Report of the Conifer Conference held in 1970. (Elspeth Napier, Editor.)
See also under Mitchell, A.F.

Rushforth, Keith.
1987 Conifers.

Salomon, C.
1884 Deutschlands winterharte Bäume und Sträucher.

Schelle, E.
1909 De Winterharten Nadelhölzer Middeleuropas.

Sénéclause, A.
1868 Les Conifères.

Shaw, G. R.

1909 The Pines of Mexico,
1914 The Genus Pinus.
Siebold & Zuccarini.
1842 Flora Japonica.
Silba, J.
1984 An International Census of the Coniferae.
1986 Encyclopaedia Coniferae.
Silva Tarouca, E. G. and Schneider, C.
1913 Unsere Frieland-Nadelhölzer.
1923 Ditto. Ed. 2.
Slavin, A. D. See Royal Horticultural Society.
Smith, Richard of Worcester.
1861 Plants of the Fir Tribe. (No copy known to exist.)
1864 Ditto. Ed. 2. 'Enlarged and revised edition'.
1867 Ditto. Ed. 3.
1872 Ditto. Ed. 4.
1874 Ditto. Ed. 5. (No copy known to exist.)
Smith, J. R. P. van Hoey. See under Van Gelderen.
Sudworth, G. B.
1908 Forest Trees of the Pacific Slope.
Swartley, G. B.
1985 The Cultivated Hemlocks.
Testu, Charlotte
1970 Conifères de nos Jardins.
Valavanis, W. N.
1976 Japanese Five-needle Pinè.
Van Gelderen, D. M. and van Hoey-Smith, J.R.P.
1986 Conifers.
Van Melle, P. J.
1947 Review of *Juniperus chinensis* et al.
Veitch, J.
1881 Manual of Conifers.
1900 Ditto. Ed. 2. (Edited by Kent, A. H.)
Von Tuebeuf, C. F.
1897 Die Nadelhölzer.
Webster, A. D.
1896 Hardy Coniferous trees.
Welch, Humphrey J.
1966 Dwarf Conifers - A complete Guide.
1979 Manual of Dwarf Conifers.
Wilson, E. H.
1916 The Conifers and Taxads of Japan.
Wolf, C. B. & Wagener, W. E.
1948 The New World Cypresses.

1. General Introduction

It would seem to be an inauspicious commencement to a book of this calibre for an admission to be immediately required that the key word in its title might be in any way unsatisfactory, but that is the case, because the word 'conifer', although widely used in horticultural literature and commonplace in gardeners' conversation turns out, on examination, to be one of those words which, despite their widespread use, almost defy satisfactory definition.

THE WORD 'CONIFER'

The term includes the true cone bearers such as the Firs, Pines, Spruces, Larches and Hemlocks, but as generally understood it includes the Junipers (where the fruit is more like a berry than a cone), the Yews, whose juicy scarlet aril is well known, and also several genera such as *Ginkgo* and *Podocarpus* in which the fruit makes no pretence of looking like a cone or even a modification of one.

The word *Gymnosperm* is no improvement. As a botanical term (*gymno* Greek, naked; *sperma* Greek, seed) it is at the same time too narrow and too exclusive, and the word 'Coniferae', itself having a precise botanical use, is unsatisfactory for use in any looser and wider sense. The phrase 'Narrow-leafed evergreens' fails to include the conifers which are deciduous and a few whose leaves are not notably narrow.

But if, as we are told, the sole value of words is as a means of communicating ideas, the word conifer[1] has undoubtedly established its claim to acceptance over many centuries and so quite justifies its use in our title.

THE ORIGIN OF CONIFERS

The conifers, as the term is thus understood, have played a big part in the development, over many millenia, of the earth and its flora into the world as we know it today. A few prehistoric genera (notably *Ginkgo* and *Metasequoia*) have survived the hazards of the ages and remain with us today, but a wide range of prehistoric genera are only known from their fossilised remains. These are found from Paleozoic times, and from their number and distribution it is evident that at one time or another, most of the earth's surface was covered with conifer forest.

This widespread distribution resulted from the spread of the conifers by a slow process of colonisation of new territory, from an origin probably somewhere in Asia. In the northern hemisphere this colonisation was assisted by the existence, at the time, of large land masses that have since slid apart to form the continents as we have them today. Any similar spread into fresh fields in the southern hemisphere was hindered by the absence of such land masses, whilst the heat and other tropical hazards seem to have presented an almost complete barrier, since surprisingly few genera succeeded in crossing the equator.

The spread of the conifers by this adventurous colonisation has had interesting effects. The different populations, as they slowly spread this way and that, lost touch with one another. Each advancing group adapted to meet the new conditions

1. The first known appearance of the term appears in the work by a French botanist Pierre Belon, published in Paris in 1553 with the title 'De arboribus coniferis, resiniferis, aliis quoque nonullis sempiterna fronde virentibus, cum earundem icomibus ad vinum expressis'.

Figure 1. Masterly use of one dwarf conifer in an expanse of lawn.

it was encountering as it felt best, thereby slowly developing the generic and specific characteristics that we know today. But sometimes the direction taken by this evolutionary process in one group would be the same as that taken by another pioneering group several continents away. The result is that a genus will not infrequently contain species with quite remarkable similarities despite the fact that one is found in western North America (that being the limit of conquest of the group that set off westwards), another in eastern North America, another in Mexico or Japan, where, in the one case the equatorial barrier, and in the other open sea, called a halt for further colonisation.

THE MORPHOLOGY OF CONIFERS

Each and every part of a conifer provides a fascinating study for somebody, and much has been committed to print.

The roots

These have several more or less obvious functions. They are the means whereby the tree draws moisture and nutrients from the soil and they also form its anchorage. These functions and their relationship to soil conditions and the uptake of nutrients and the reaction to the local climate are of prime interest to foresters and soil chemists and landscape architects. They should be of equal interest to the gardening public, whose attitude 'Out of sight - out of mind' is regrettable. More on this subject will be found in Chapter 4 'Cultivation'.

The trunk

Most conifer species start life as a seedling with a single stem which, barring accidental damage, develops into a forest tree with a single, erect trunk often of massive size, the resulting timber from which is so valuable to mankind. The bark on young trees is usually smooth, but with age it thickens and gets rougher, often becoming furrowed, cracked, or otherwise irregular in a way that is characteristic of the species. This makes the bark a valuable aid to identification, more so because the specific pattern seems little affected by environmental factors.

The branches

The branching pattern differs between the families comprising the conifers. Also, some genera and species and sometimes even individual trees deviate from the family pattern, these last being the fastigiate, pendulous, globose and other habit deviants that (for that very reason) have been selected and brought into garden cultivation.

Only general observations can therefore be made. In most of the *Pinaceae* (broadly, 'the Pine family') the year's growth on the leading shoot in a young plant ends with the development of an apical bud and a cluster of smaller buds around it. In the Spring these develop respectively into a vigorous extension of the leader and into a whorl of much shorter, more or less horizontally held side shoots destined to become the main branches of the future tree. This situation is repeated each subsequent year except that each side shoot now carries its own terminal bud and cluster of sub terminal buds. To make the most of the light, only those at the side of these horizontal shoots break into growth so the resulting 'spray' is more or less flat. One result of this is that quite a large side branch cut from a Spruce of Fir will lie quite flat on the ground. Another, associated with the much greater extension, year by year, of the main stem is the development of the characteristic upright-conical outline and regular branching of the traditional 'Fir tree' which makes it so attractive and spectacular as a single specimen or in a small group, but which, repeated *ad nauseum* in acre upon acre of afforestation, results in the monotony and the depressing effect that set the minds of the environmentalists against the mass planting of conifers.

This character is not maintained throughout the life of the tree: the vigour of the different shoots developing from a whorl is often uneven and many accidental factors tend to break up the early symmetry. Because the leaves are retained for several years the foliage on a young tree is very dense: the heavy shade so produced gradually smothers the interior foliage (and also, eventually, the lower branches) so that as the tree matures the foliage is only borne on the outside. And as the tree approaches maturity the leader loses dominance, growth being finally restricted to the topmost branches. The tree develops a tall, clear trunk and a broad, rounded or flat crown and

Figure 2. An attractive group of *Chamaecyparis lawsoniana* cultivars in the Valley gardens, Windsor Great Park, U.K.

matures to a beauty that not even an environmentalist could dispute.

In several other families, notably within the *Cupressaceae* (such as the Cypresses, Thujas and some Junipers) no true buds are formed. The foliage consists of either awl-like (juvenile) leaves held in pairs or in whorls of three, or of scale-like (adult) leaves in decussate pairs, and the new growth arises from the leaf axil, in the latter case from a leaf of the lateral pair. (The meaning of the terms 'facing pair' and 'lateral pair' become obvious if a tiny piece of foliage be laid flat on the table, and an 'axil' is the angle above the leaf between it and the shoot.)

Shoots develop only at intervals that vary with the vigour of the shoot and its position on the plant (rarely from successive axils) and usually on alternate sides of the shoot. This gives the flat, pinnate foliage that is characteristic of the scale-like foliage and which may or may not be repeated the following year (according to the species). This method of producing new growth seriously restricts the ability of species in this group to (in gardeners' language) 'break from old wood', so heavy pruning should be avoided. In *Chamaecyparis lawsoniana* the inability seems complete, making this species a bad subject for hedging.

In a plantation of *Chamaecyparis lawsoniana* a number of trees usually will have formed multiple leaders. If the growing tip of such a tree is examined the uppermost lateral to have formed will be found to be almost as vigorous as the true leader – an obvious rival. A similar situation can be found on young plants in the nursery: it should therefore be possible to avoid this type of tree by selection when planting.

The buds

Most conifers outside the Families *Araucariaceae* and *Cupressaceae* prepare for Winter by developing true winter buds and giving them the protection of short leaves held forward in a cluster around the bud or of leaves completely modified into bud scales. This is often supplemented by heavy coatings of resin. The buds can also be considered as the tree's provision for the following Spring, for each contains the primodials of all the leaves that will develop then. When each bud starts into growth it either pushes the bud-scales off or grows through them (in which case they are described as 'persistent' and are an aid to identification of the species) and develops leaves which are often of a delicate pale shade, attractively contrasting

Figure 3. 'The Whidley Dragon'. An ancient *Juniperus scopulorum* plant on sand dunes on Whidley Island, Washington, U.S.A.

with the sombre colour of the old foliage. In the case of the pines the bud lengthens considerably, along with all the other buds in the cluster to form the vertical 'candles' that are such an attractive feature at this season. The shoot rapidly extends to its full length and the leaves to their full size, and by the Autumn it will have developed, in its turn, another terminal bud or bud cluster. But if the candle be snapped off before the leaves have begun to grow, this cluster of buds will develop at the tip of the remaining stump. This is a technique well known to the growers of Bonsai trees. Applying it to a large specimen in the open is rather tedious, but systematically carried out for several years, where such an effort is justified, an exceedingly dense growth can be induced.

These conifers do not seem to have any great ability to forecast weather a year ahead, nor any means of anticipating being transplanted, so this need to pre-determine next year's growth must bring its problems. A forecast made by the tree during a particularly auspicious year followed by an unusually difficult growing season finds the tree in a situation to which it can only adapt by reducing the length of the shoot and leaves, since their number has been pre-determined. And of course, *vice versa*. So, provided a newly transplanted conifer can be persuaded to stay alive and by the autumn has formed winter buds, the fact that it has made hardly any noticeable growth is no grounds for discouragement.

The leaves

A seedling conifer greets the light above ground with rudimentary seed-leaves (cotyledons). These vary in number, size and shape between the species, so the differences must be welcome to the forestry nurseryman as confirmation of the purity of the seed he has sown. These are followed by what are usually described as 'juvenile' leaves, and foliage characteristic of the species (the 'adult' phase) only appears later, some species in several Families continuing to produce only the juvenile growth for many years.

Notably within the Family *Cupressaceae* all kinds of anomalies occur in this 'growing up' process. Several species of *Juniperus* retain the awl-like juvenile leaf-form throughout life: other species never abandon it entirely and continue to produce patches of juvenile foliage on even a mature tree. Within the Genus *Chamaecyparis* a third stage of leaf development can occur, intermediate between the long, awl-like leaves and irregular branching of the juvenile phase and the small, scale-like leaves in decussate pairs and the flat leaf-spray of the adult foliage. For such growth the descriptive but uninspired term 'Semi-juvenile' is used. Moreover, individuals occur that shew no tendency ever to reach maturity. This is not unknown elsewhere, being far from rare in the Genus *Homo sapiens*, but in plants it is known as 'juvenile fixation'. Since this individual peculiarity is usually (but not invariably) associated with a much reduced rate of growth it has been a fruitful source of some of our most popular 'dwarf conifers'.

Adult conifer leaves are (with the inevitable few exceptions) linear, i. e. much longer than wide or thick, ranging from the needles of the pines and cedars to the wider, flat leaves of *Cephalotaxus* or *Podocarpus*, and they are always produced in one of two ways. In the **Spiral** system the leaves emerge one at a time along an imaginary line drawn in a long spiral around the shoot. In the **Cyclic** system the leaves emerge at intervals along the shoot in opposite pairs or in whorls of three (rarely four or five). (In decussate pairs, each pair is at right angles to the adjacent pairs.) The spiral arrangement is much more general than is often supposed, due to the way that some widespread genera (e.g. *Abies*, *Picea*) twist each leaf-stem so that all the leaves lie flat in a comb-like (pectinate) arrangement, especially on horizontal branches, to make the most use of the light reaching the plant from the sky.

Flowers and cones

The reproduction of conifers by means of seed, involving the sex mechanism, follows that process as it has existed in both plants and animals ever since life on earth began – the fusion of a specially developed female (egg) cell with a similar male (sperm) cell, neither capable of sustaining life indefinitely alone, to form a new individual. To secure this happy event was fairly simple. All that was needed was a mechanism to produce these peculiar cells at a propitious time of year and synchronise their production, to produce enormous quantities of lightweight pollen (male cells)

Figure 4. Low, spreading forms of *Juniperus scopulorum* at McLeod, Sweet Grass Co., S.W. Montana, U.S.A., at 5200 ft. (7)

and leave the rest to the wind. That this haphazard arrangement only occasionally resulted in a good seed crop was not important to species with a life expectancy of a hundred years or more, and it proved adequate to enable conifers to dominate the vegetation over most of the world for many millenia.

But in the slow changes that are inevitable in all forms or life and which have resulted in the world as we know it today, conifers failed to move with the times. Becoming monoecious was a good move, making the process less dependent on the wind and saving isolated trees from lifelong celibacy, but other competitors for space realised the possibilities of overcoming their own immobility by encouraging birds and insects to distribute the pollen on their behalf and so they developed flowers that were easily seen from a distance, often (in one way or another) attractively scented and/or a source of food. So arose the flowering plants of our day, and since(even for plant-life) the wheels of time grind slowly, but they grind exceeding small the one-time world dominance of the conifers had gradually to give way.

The flowering parts of conifers are therefore, as might be expected, simple and often quite inconspicuous. But that is not to say that they are uniform throughout the group. On the contrary, they differ from genus to genus and from species to species in size, structure, distribution on the tree and every imaginable way to such an extent as to virtually defy any simple classification. All that is invariable is that always there will be found male organs (technically the microsporangiate strobili) which produce the pollen and separate female 'flowers' (the megosporangiate strobili) carrying the ovules, which receive the pollen and develop into the seed-containing mature cones. But a few main groups can be easily recognised. In the true cone-bearers, members of the Family *Pinaceae*, the male flower consists of a group of stamens set in a catkin-like cluster, often brightly coloured and sometimes produced in such large numbers as to colour the whole tree seen from a distance. The very small, lightweight pollen grains are released to the wind in enormous numbers, only a microscopic proportion of them having any hope of fulfilling their mission in life by meeting up with the waiting ovules on the female flowers. These (already a cone-shaped structure that needs to change in size but little in shape) consist of a central core and a number of scales in a spiral pattern, each consisting of a bract scale and one or more exposed ovules. After pollination the scales lengthen to cover the ovule/s which thereafter develop unseen until they are released as ripe seed when cone opens. Very sensibly trees often develop female cones at the top of the tree and the male flowers lower down, thereby increasing the chances of fertilisation from the rising clouds of pollen.

Figure 5. *Abies amabilis* (left), *Chamaecyparis nootkatensis* (centre) and *Pinus monticola* (right) on the trail to Mt. Constance, Olympic Mts., Washington, U.S.A. (7)

The *Cupressaceae* is an important Family that are true conifers, although the 'cone' is no longer cone-shaped (i. e. conical). In *Chamaecyparis* and *Cupressus* it consists of several pairs of scales like miniature shields which fit together to form an irregular globe (the term used is 'peltate'). Each scale has a central stem which lengthens when ripe to release the seed. Other genera in this family have cones that might be likened to miniature urns and in *Juniperus* the scales, instead of becoming dry and woody, remain soft so that the fruit looks more like a berry than a cone (which structurally it still is).

The term 'conifer' is extended well beyond its original meaning in the Family *Taxaceae* in which the fruit consists of a single seed partly or wholly surrounded by a fleshy aril (the well-known Yew berry) and is stretched to its limit in the case of the ever-nonconforming *Ginkgo* and some species in the Family *Podocarpaceae* in which the fruit no more looks like a cone than does a plum (hence the popular name 'Plum-fruited yew' of *Prumnopitys andina*).

Every detail of the ripe cone, its average size, shape and colour; its occurrence singly, paired or in clusters; the way it holds itself on the tree and the degree of reluctance toopen and shed the seeds are all generic or specific characters that are of value in identification, and will be found in the descriptions in Chapter 7.

2. The Uses and Value of Conifers

CONIFERS IN NATURE

The role of conifers in afforestation is outside the scope of this book, but for amenity planting much can be said on the subject of this chapter. Conifers are (with a very few exceptions) evergreen, hardy, tolerant of soil conditions and also (within limits) of aspect and situation: they come in all sizes and shapes and in a wide variety of colour, habit and foliage texture and are largely free of pests and diseases. It would be difficult to find a group of plants of greater landscape value, nor (with their relatively slow growth and long life-span) one with a higher ratio of garden value to cost, whether measured in expenditure of money or as time spent in care and maintenance.

Nature has used conifers for some of her most successful plantings, and as always we can learn from her. She always chooses species that do well in the area, often effectively using the narrow, spire-like ones *en masse* in tier upon tier on steep mountain slopes or singly or in small groups in confined ravines, and preferring the broader, round-topped trees for the broad valleys and open plains.

USING CONIFERS

But although we may learn, let us also be warned. Nature always makes use of plenty of space, and plenty of space in so many cases (especially in the small gardens of today) is just what we do not have. All the principal conifer species available to us are forest trees somewhere in the world and as such they will behave, despite their attractiveness as a small plant in a five-gallon container at the garden centre!

The main contribution made by plantings of mature conifers to a stately home or other open landscape is the sense of restfulness and serenity that only they can produce. We often forget that these magnificent trees looked very inadequate when first planted by the grandfather of the present owner. In the confines of a garden (especially a small one) we cannot hope to reproduce quite the same effect, but by using smaller-growing species or by choosing compact or dwarf cultivars we can make use of the hardy, dense, evergreen foliage and rich dark colours of conifers to give us shelter, privacy, background and an air of permanence and solidity. But let us think of them as the dark tones in the garden picture or (if imagination carries us that far) as the bass voices in the garden choir, to be employed with restraint. A garden over-planted with conifers can be dull and even depressing.

In the making of any new garden larger than a pocket handkerchief, having a planting plan is essential, if mistakes are to be avoided. The use of squared paper brings the making of a plan roughly to scale within the capacity of anyone. It should show the outlines of the house and any permanent features, such as existing trees, that are to remain. The exact position of windows should be shewn, to stimulate thought as to what 'vistas' can be created and warn of the folly of plantings that in time will block the light. If the area available is so large that it will have to be tackled a bit at a time, an overall planting plan before starting work at all is even more important. If the site slopes, levels taken by a professional surveyor and shewn on the plan as 'spot-levels' or contour lines may be well worth while. But even on a dead level site learn to think in three dimensions and force yourself to visualise the scene when your plantings

Figure 6. *Juniperus communis*. Bush forms of varying habit are widespread on chalk hills in Britain. (10)

mature. Always, and whatever the scale, simple plans are the most effective. Draw a circle on your plan shewing the area each tree will cover in 25 (or 50) years' time and mentally picture its height by then and the effect it will have on the skyline and the shade it will throw on the bedroom windows. By so doing you might become the first planter in history not to have planted too many trees for the space available, set far too close together. If your courage fails you when the planned planting is done, leave your permanent trees where you have set them however pathetic they look at present and interplant with 'nurse' plants, to be removed as the others grow.

Except in a garden so small that each plant is almost a member of the family an 'assorted trees' effect should be avoided. As a broad rule, alternating conifers with deciduous trees is undesirable: the effect is patchy and the latter smother the foliage of the conifers, which then become unsightly in winter. A group that has grown together into a 'clump' seldom looks well if it contains trees of different colours or forms, so if the plan includes 'clumps' they should be restricted to a single species or cultivar. Cultivars chosen for their shape, colour or other form of individuality should be given adequate space to allow them to remain individuals.

A SOLITARY SPECIMEN

The potential 'architectural' value of a single plant is considerable. A fastigiate Juniper (a columnar Yew would grow more slowly but last longer) sited well away from competition can be quite spectacular, as also can a large spreading or a pendulous specimen well out in the open on a large lawn. Especially in the immediate vicinity of buildings a single conifer can often be more effective than a group.

HEDGES AND SHELTER BELTS: WOODLAND

Conifers can make excellent hedges. Badly chosen they can also create an excellent disaster, as many planters of × *Cupressocyparis leylandii* hedges can testify. A hedge is usually urgently needed for social reasons so the temptation is to demand something that will quickly reach a height of six or eight feet. Unfortunately there is no conifer known that, having reached the householder's target, will stop growing. Since × *Cupressocyparis leylandii* has a growth potential of 30 m or more and will not take 'No!' for an answer, planting it as a hedge is bound to mean trouble. The answer is to decide on the height to which you want the hedge to grow, choose a compact conifer cultivar that will be controllable at that height, and be patient while it grows. If immediate privacy is essential it must be provided by the use of a deciduous species or the erection of some sort of fence. Only species known to stand clipping and to be willing to 'break from old wood' should be chosen for a hedge, so all forms of *Chamaecyparis lawsoniana* should be avoided. The hedge conifer *par excellence* is the common Yew, *Taxus baccata*, which amongst its endearing qualifications for this particular use includes tolerance of alkaline soil and shade. It is also an excellent subject for topiary.

Shelter belts are but hedges on a larger scale: their functions are basically the same. But because much more space is normally available the choice of species is wider. Especially in a coastal area species known to be salt-tolerant should be chosen. These include several species of Pine, but unless there is space to plant 'in depth', with several distinct rows of differing subjects, pines should be avoided since after serving well for some years they develop bare trunks and let in draught at ground level, which is arguably worse than unbroken wind. Shelter belts should not be too densely planted and may include deciduous subjects. The objective is to break up the wind, rather than to present it with an impenetrable barrier.

Except for securing wind shelter, conifers should be used sparingly in rough woodland. In a large area an occasional specimen can be acceptable (provided a cleared area around it is maintained), but conifers throw a dense shadow, and ground-cover planting amongst deciduous trees (we are told) provides much better habitat for wild life. (Our garden birds, however, seem to prefer conifers for nesting sites, especially the dense foliage of the *Cupressaceae*.)

DWARF CONIFERS

Thus far we have been considering the use of conifers as specimen trees in the landscape and their value in gardens for shelter belts, backgrounds

Figure 7. Group of dwarf conifers used to form a 'feature' at Glasnevin Botanical Gardens, Dublin.

and hedges, all aids to the forming of the backdrop to a stage upon which the players will be the flowers and other garden plants. But with the dwarf forms we come to a group in which they themselves are being increasingly used as players.

In Chapter One I have commented on the difficulty in defining 'conifer', and a satisfactory definition of 'dwarf conifer' is no less elusive. In many cases the description 'slow-growing' is much nearer the truth.

To avoid any misunderstanding, perhaps I should explain straight away that throughout this book the term dwarf conifer relates to plants that are naturally dwarf, whether they be species, variety or cultivar and I have nothing to say about Bonsai trees; i.e., artificially dwarfed plants (they may or may not be conifers) which owe their restricted size to root pruning or other cultural practices. The plants I am dealing with are all natural dwarfs – healthy garden plants in their own right. A dwarf conifer may either be of a species which (although truly a conifer and even perhaps tree-like in habit) reaches its maturity when only a metre or so high, or it may be a specimen of a normally tree-like species which as a result of one of Nature's pranks (of which more below) shews much less than the normal rate of growth and so can only achieve, even in maturity, a much smaller plant than the species normally produces.

The existence of such forms has long been known. Loudon, writing in 1838 mentions 10 forms; Carrière in 1867 records about 40 and Beissner in 1891 still more. They had quite a vogue in the middle years of the last century, in the days when horticultural keeping up with the Joneses meant the possession of a better arboretum than Lord So-and-so's in the next county. The few such plantings of those expansive days that still survive, remain as proof (or otherwise) of their claim to be dwarf forms, but with the decline in the number and spending power of those large establishments the demand for the dwarf forms virtually disappeared, and many recorded varieties seem to have been lost to cultivation by the end of the century.

With the rise in the vogue for alpine gardening and of rock garden construction based on the aim of simulating nature as closely as possible, a new and altogether happier use for these conifer microforms was found and the demand increased. Apart, however, from occasional references in gardening journals and in general textbooks on conifers there was no literature on the dwarf forms and so when the late Murray Hornibrook published the first edition of his work *Dwarf and Slow-Growing Conifers* in 1923 he was breaking new ground.

In that book he described and listed about 250 varieties. In 1939 he was able to increase the number to about 530, some of which have since been lost to cultivation. On the other hand new cultivars were being introduced, so in my first book *Dwarf Conifers* in 1966 I listed about 500 cultivars, increasing the number to nearly 900 in *Manual of Dwarf Conifers* in 1979. Since then new cultivars have proliferated in uncontrolled profusion.

Apart from a very small number of species that (at least in cultivation) are never larger than a shrub (applying the description 'dwarf' to these would be an insult) and a number of vigorous but prostrate species where doing so is ridiculous even if customary, dwarf forms can arise in the following ways.

ALPINE FORMS

Several species of conifer whose range includes an area of high mountains, although normally of tree-like habit, have developed stunted, shrubby or even prostrate forms at or above the treeline, many of which make excellent garden plants. As these are established in the wild and have a natural habitat they have in some cases been given varietal rank by the botanists so they will have names of Latin form such as *Abies balsamea* var. *hudsonia*, even although they have been brought into garden cultivation. Because they are botanical varieties we find them more or less variable and in a few cases an outstanding clone has been given a cultivar name.

Because of their origin these alpine or mountain forms (the word alpine is commonly used with a wider meaning than its true sense of 'from the Alps') are indestructibly hardy and within the life-span of a single plant seldom revert to the lowland habit of the species and so they are a very useful group. They seldom, if ever, produce cones.

SEEDLING MUTATIONS

From time to time a seedling will appear showing some distinct variation from the rest. This may be a peculiarity in vigour (in this way arise giant and dwarf forms), in colour (presenting us with golden and glaucous forms), in habit (fastigiate, weeping and globose forms) or in foliage (thread-like, congested and monstrous forms). These seedling mutations are very uncommon. It is possible that dwarf seedling forms arise more frequently than we suppose but the chances of survival in the wild or in a forestry nursery (where the culling of the runts is normal practice) are very small indeed, and the probability of raising a new dwarf form from a packet of conifer seed is of the order of millions to one against. These seedling forms as a class are stable and show no tendency to revert to the typical arboreal form, although there is a tendency for any peculiarity to become less pronounced with age; i.e., golden forms become greener, congestion of foliage gets less noticeable, dwarf forms develop more vigour and so forth.

Examples of seedling mutations are – most cultivars of *Chamaecyparis lawsoniana* and *Ch. obtusa*, *Ch. pisifera* 'Nana', *Picea abies* 'Clanbrassiliana' (probably) and most if not all the cultivars of *Thuja occidentalis*. Unfortunately for us, a trick Nature has done once she can do again, so at any time a seedling may turn up showing the same peculiarity or combination of peculiarities as an already established cultivar, in anything from a broadly similar to an indistinguishably identical manner. At either end of the range the situation is clear enough. Where there are marked differences from all existing cultivars and the new seedling is felt after prolonged trial to have some distinctive garden value it can be propagated and given a new cultivar name, as has been done in the case of all the various globose forms of *Thuja occidentalis* that have been named. At the other end of the range a seedling so like an existing cultivar as to be indistinguishable from it even in old age may sometimes get propagated in its place and no harm will have been done. In any intermediate situation on the other hand, in cases where small differences from existing clones are discernible (or are found to develop as the new seedling ages) the possibility of confusion arises in two ways. The plant may stray into cultivation under an existing cultivar name and so give it a reputation for being variable (this is the probable explanation in the *Chamaecyparis lawsoniana* forms 'Forsteckensis' and 'Tamariscifolia', where in both cases I believe there are several clones in cultivation) or it may be enthusiastically named by its excited and impatient raiser, only to turn out in the end to be, for all practical purposes, identical with an existing cultivar. In every case the exercise of caution and restraint on the part of the raiser is necessary and no attempt should ever be made to introduce a new cultivar until the mother plant has been allowed to grow freely for some years so as to have outgrown any merely juvenile tendencies, and even then the opinion of some authority should be sought on whether the new form is worth introducing. I return to this subject below in Chapter Five.

The true weeping forms are of the seedling mutation class and so this peculiarity is stable. Such a tree has a normal trunk and main branch system and only its lower order branchlets and terminals are pendulous, an example being the species *Picea breweriana*. Forms that must be grafted on to a stem as is done with standard roses or be trained up to a cane in their nursery days unless they are to remain for ever a sprawling victim of their own congenital inability to lift their branches off the ground are more accurately described as prostrate. Allowed to grow thus they are usually more curious than beautiful, but where a bold rock face is to be clothed they can be very effective. Plants in this group are *Picea abies* 'Inversa' and 'Reflexa' and *Cedrus libani* 'Sargentii'.

JUVENILE FIXATIONS

All seedling conifers commence life with a type of foliage quite unlike the adult foliage which the plant will normally develop later. In the *Cupressaceae* the juvenile leaves are acicular (awl-like) and the branching open and irregular. Several species of Juniper retain this foliage throughout life. All species of *Chamaecyparis* normally change with time to the small scale-like leaves and the more regular branching of the adult foliage. *Ch. pisifera* passes through a stage of producing what is descriptively called semi-juvenile or intermediate foliage. In this genus (and also in *Platycladus orientalis*) several clones are in cultivation in which the foliage has remained juvenile and others in which it has never progressed beyond the semi-juvenile stage, the former being described as squarrose and the latter as plumose. These fixed juvenile forms may grow into trees and may even set seed on juvenile foliage, but in the majority of cases they also show much reduced vigour, so are amongst us as dwarf forms. Several of them are excellent garden plants.

These longer and more freely held leaves were once thought to give a plant something of the appearance of heather. As a result, many of these forms have been given the cultivar name 'Ericoides' so it is not suprising that there is much uncertainty in gardens as to the true identity of these forms. At one time two botanists named Siebold and Zuccarini supposed that they were all species of a distinct genus which they named *'Retinospora'*, but the occasional development of adult foliage dispelled this notion and these juvenile forms have been distributed, not without difficulty, amongst the species to which they each belong. The name *'Retinospora'* survives on many old labels and, incredibly enough, in a few nurseries (where often it is not even spelt correctly) although it has been officially abandoned by botanists for many years. As might be expected in the circumstances these forms are often variable and sometimes unstable. A few (e.g., *Chamaecyparis pisifera* 'Squarrosa Intermedia') may suddenly throw up strong adult growth which soon takes over if not removed. Others may show plumose growth on squarrose forms and *vice versa*, and even the most stable of the juvenile forms may occasionally throw a small spray of the typical, adult foliage. This, by revealing to us the specific loyalty of our plant, is very interesting, but any tendency of the plant to extend this habit should be cut away immediately.

BUD MUTATIONS

This is the usual origin of variegated forms. These have arisen as sporting branches shewing this peculiarity on a normal green plant. It can be argued that any departure from the normal in plant life must represent in some degree a loss of perfect health and this may be the reason why these variegated forms are often lacking in vigour although apparently healthy in other respects. As this loss of vigour makes them slow-growing it can be an advantage from the gardener's point of view. Such mutations are often unstable, so a watch must be kept for growth reverting to the normal colour which, unless removed, is apt to take over.

WITCHES' BROOMS

Several genera of conifer are prone to sudden and so far unexplained mutations of quite a different kind. These can occur within a growth bud or at any point along a shoot and they invariably result in a change in vigour. Local stands of trees of exceptionally vigorous growth, of great interest to foresters, may possibly have originated in a mutation which resulted in an increased rate of growth, but the phenomenon is usually only taken notice of where a decrease in vigour results in a dense mass of foliage popularly referred to as a witch's broom. Forms of abnormal and congested growth are found on many kinds of plants, but whereas these are always otherwise associated with parasites or other pathological causes, within the conifers often no causal mechanism can be found. Frequently the reduction in rate of growth and leaf size is not accompanied by a reduced number of buds and where this occurs the broom may get so dense that its interior becomes a mass of dead tissue and where the rest of the tree has continued to grow normally the whole broom will eventually perish through lack of light.

Except for this the growth is quite healthy and it has been found that cuttings or grafts taken from the broom develop into dwarf plants with a quaint charm all of their own and many of our most-sought-after dwarf conifers have arisen in this way, *Picea abies* 'Pygmaea' being a typical example.

Although in most cases the broom growth never develops male strobili or sets cones, these latter have been recorded on several genera including Spruce, Pine and Douglas fir. Invariably the broom cones have been smaller than normal, in scale with the broom growth on which they appear and they carry very small seeds. These are, however, often fertile and accounts of the progeny resulting from the sowing of such seeds have been published from time to time by several writers, references being given below.[1] The general finding has been that whilst about half of the seedlings turned out to be quite normal in habit and rate of growth, the others shewed definite genetic dwarfism in various degrees and gave rise to compact or prostrate little plants of which ouite a number are now available as named cultivars. Time will tell, but so far these genetic dwarfs shew less tendency to develop further mutations or reversion than clones propagated vegetatively from witches' brooms.

If seeds collected from brooms in cultivation are viable, dwarf plants developed from naturally sown seed must be expected to have occurred as well. Because of their slowness of growth the survival rate of any such seedlings in nature must be very low, but it is highly probable that many of the dwarf plants found in the wild have originated in this way, and where a small and very local population of similarly dwarfed plants has been found (as in the well known cases of *Picea glauca* 'Albertiana Conica' and *Tsuga canadensis* 'Minuta') this almost certainly was the case.

Whether or not all variations turning up in the seed beds were due in the first place to a bud mutation on the mother plant is doubtless an interesting speculation for geneticists.

1. 1927 *Hexenbesen: ihre Morphologie, Anatomie und Entstehung*. Liernur, A. G. M., Rotterdam, Nijgh and Van Ditmar, 57 pp. Contains a full Bibliography. (German and Dutch).
1933 'Das Problem der Hexenbesen'. Tubeuf, C. von, Z. *PflKrankh, PflPath. Pflschutz.* **43**: 194-242. (German.)
1933 'Verebung der Hexenbesenbildung bei der Kiefer.' Liese, J., Z. *Forst- u. Jagdw.* **65**: 541-4. (German.)
1936 'Unexplained brooming of Douglas Fir and other conifers in British Columbia and Alberta'. Buckland, D. C. and Kulit, J., *For. Sci.* **3**: 236-242.
1965 'Dwarf Seedlings from Witches' Brooms in Jack Pine'. Johnson, A. G. *et al.* in *Sci. Journ. Ser. Paper No. 5618, Univ. of Minn. Agr. Exp. Sta.,* St. Paul 1, Minnesota. (2 pp) 15th Jan. 1965, and *Paper No. 5758* (2 pp.) 15th April 1965.
1966 'Dwarf seedling from Broomed Douglas Fir'. Duffield, J. W. and Wheat, J. G., *Silvaé Genetica.* **4**: 129-133
1967 'Dwarf Conifers from Witches' Brooms'. Fordham, A. J., *Arnoldia.* **27**: 29-50.
1967 'Witches' Brooms and Dwarf Conifers'. Corley, R. S., *Gar. Chron.* **161**, 20.
1975 'Witches'-brooms. Sources of New and Interesting Dwarf Forms of Picea, Pinus and Tsuga Species'. Waxman, S., *Acta Horticulturae.* **54**. 25.

CULTIVARIANTS

This is a word which I coined in 1966 to describe two cases where a plant that differs sufficiently in habit from the mother plant to fill a distinct place in our planting schemes (and consequently to be called for by the gardening public) can be produced at will by particular selection of cutting wood or scion material. One case was first drawn attention to by the late Murray Hornibrook in his classic work on the dwarf conifers where in his remarks on *Chamaecyparis lawsoniana* 'Fletcheri' he points out the possibility of securing diminutive and often bun-shaped plants from this and certain other upright varieties by the selection of cutting wood from weak-growing shoots low down on the plant. It is a phenomenon particularly characteristic of the genus *Chamaecyparis* and seems to be directly related to each species' low ability to 'break from old wood', which in *Ch. lawsoniana* is very low. It is probable that this phenomenon is far more widespread than is realised and that it can operate in both directions, since tempting but ill-advised selection of strong-growing propagating material of many dwarf cultivars can wholly destroy the original dwarf character in two or three cutting-generations.

The other case consists of low-growing specimens of normally tree-like species which have been produced by appropriate selection of scionwood, such as in former times would have been given the name *prostrata*.

If an Abies be examined (Picea and several other genera are the same) it will be found to carry two types of shoot, leader growth on which the foliage is radial, and side growth on which it is not. The nurseryman knows that the best trees are always produced by the use of leader growth as grafts, but unfortunately there is never enough of this and so he has to use side growth as well, in spite of the additional trouble he knows he will have to take with the resulting plants by careful staking and other professsional knowhow to induce it to form the symmetrical, upright plant which most of his customers demand. Left to itself a graft from side wood tends to maintain this horizontal habit however large the plant grows. A large specimen so produced of any of the good forms of the beautiful blue spruce *Picea pungens*, or of any other of the more spectacular Abies or Picea, (consisting, as it does, of the lowest and only branch of what would otherwise have been a tall tree), tumbling down over the rocks for a dozen or more feet from where it was planted can be a magnificent sight. It is amusing to recollect that such a plant may well have started life as a graft intended for sale in the normal upright form which was accidentally left (nurserymen are busy folk) until it had become too late to take its training in hand.

Any attempt to bring this sort of phenomenon within the scope of nomenclature is difficult. The intention of a clause (Art. 11 (e)) introduced into the 1980 edition of the *Cultivar Code* is obscure, since each of the examples given was already a recognised cultivar under previous editions. A cultivariant (certainly one in the second case) is in no sense a clone, it is an individual plant. A closer analogy might be to a 'standard rose' or a 'fan-trained peach', i.e. a 'description' of its state.

USING DWARF CONIFERS

The garden uses of dwarf conifers are many. There is, however, one golden rule for the use of dwarf conifers (along with miniature plants of any kind, no doubt) and that is, ALWAYS PLANT THEM IN THE OPEN AWAY FROM OTHER LARGE PLANT MASSES. The dwarf conifers are serious, dignified little plants and they resent being made to look ridiculous. Our object must be to use dwarf conifers to set the scale of the miniature garden picture and not let them be put out of scale by the overtowering presence of larger plants.

The key word in the use of dwarf conifers is *scale* and from this arises another useful principle, which is to plant dwarf conifers wherever possible together in groups. This is more of a recommendation than a rule as there are clear exceptions, but in a general way it can be said that dwarf conifers usually look best together. Planted in this way they are in scale with each other and so if your space (or pocket) indicates the purchase of any particular number of trees – be it only half a dozen – I would say without hesitation that you would probably get more landscape value and pleasure from your new treasures by grouping them together as a feature than by dotting them about here and there all over the garden where they can only be viewed one at a time.

Such a feature could well be made on a lawn.

On a large lawn it could look well a little to one side of the main area so as not to appear lost, yet have mown grass all round it. The bed should be raised up very slightly towards the middle, near which one conifer larger than the rest and of a sturdy upright habit will look well dominating the lower varieties around it. In a large planting, other dominant specimens will need to be used to form fresh centres of interest at other spots and it is not at all a bad plan in planting to start with these dominant plants, adding the lower planting around them. The number of trees to plant in a given area will depend upon the varieties chosen and upon how patient you can rely upon being whilst waiting for the trees to grow to give a mature effect, but an allowance of one plant per square metre is a reasonable compromise. Do not worry if you do not know the plants you buy and their habit and growth potential; set them out to form a pleasing group as they are, and re-arrange them as requisite in a year or two.

On a smaller scale the same principles can be followed in grouping a few dwarf conifers to form a definite garden picture. If a good composition is achieved it is at all times pleasing to the eye and since many of the dwarf conifers change colour two or even three times a year the tone values of the group are constantly altering. Some of our plants put on their richest colouring during the winter and what could be expected to give more pleasure than such a group planted where it can be seen from a principal living-room window from whence its winter beauty could be enjoyed in comfort whatever the weather happened to be outside.

Since, in addition to the scenic effect they create, your dwarf conifers will demand close examination and admiration as individual specimens there should always be sufficient paths (or even stepping stones) so that no tree is further from a path or other convenient view-point than, say, two metres.

In every case background is important and needs careful consideration. In the large plantings referred to the group will be large enough to hold one's attention from whatever direction one approaches it and isolation itself may be all the background that is necessary, but in smaller plantings it will generally be the case that our garden picture will mainly or always be viewed from one direction and here the background must be chosen (as a picture frame is chosen) to set off the picture and isolate it from its surroundings. A background of tall trees – at a respectful distance – is excellent and where space is limited an evergreen hedge may be the answer; or the group may be placed so that a fence or even a building will serve the purpose.

The use of dwarf conifers in rockeries and alpine gardens is, of course, well established and many would way that this is the best place for them. As well as suggesting an alpine atmosphere their evergreen sturdiness of itself sets off the usually low-growing flora around them and any who have attempted alpine gardening without the use of dwarf conifers will be surprised at the improvement brought about by the introduction of a few well chosen speciemens.

For this class of work the selection of a suitable form of tree is especially important. It should always be remembered that one is seeking to reproduce a scene from nature in one of her wilder moods, so anything savouring of smugness is to be avoided and plants should be selected for each situation of a form which would be expected in just that situation in the wild. For this reason the globose forms and others of a neat and formal outline are in a general way the least useful. Wholly prostrate forms of the *Juniperus horizontalis* type should be used sparingly, for they almost all spread eventually over a large area that could be more usefully used for other alpines, but in this class both *Juniperus conferta* and *Juniperus procumbens* 'Nana' are useful plants because of their willingness to grow downhill. They are consequently valuable for covering banks of old tree roots, and the former will even hang down and cover vertical rock faces like an aubrieta.

The dwarf Picea are a very useful group. They range from low-growing forms which never reach much over 30–40 cm high but spread slowly outwards (examples are *Picea abies* 'Nidiformis', 'Repens' and 'Tabuliformis') through less prostrate cushion and bun forms ('Pumila Nigra' and 'Gregoryana' and its various forms) to globose plants ('Clanbrassiliana') and pyramidal forms ('Remontii'). In this latter class comes the popular *Picea glauca* 'Albertiana Conica'. This plant grows into a neat green cone which with a little attention may, if you wish, be made to look as though it had been turned on a lathe. A single specimen of this variety always catches the eye, and a group (for preference of varying heights) can be most effective, but always – because of its formal symmetry – it should be planted somewhere on the low ground and well out in the open.

Low-growing varieties of irregular outline such as *Juniperus sabina*, or plants of the *Juniperus* × *media* 'Pfitzeriana' group are very useful and this variety itself can be utilised in a large alpine garden. It will in time become a large plant but it stands being kept to size by heavy pruning better than most of us do.

Many of the real pygmies will be quite at home in suitable nooks and crannies amongst the rocks and there are several of the slow-growing and bushlike forms of Juniper which if planted young and deliberately left unpruned will grow into a gawky, one-sided plant which will look most attractive at the foot of a rock cliff, where it will appear as though the plant had so grown in an attempt to reach up and outwards to the light. There is quite a colour range of these, giving us scope for selections either toning or contrasting with the colour of the rock cliff behind. *Juniperus* × *media* 'Blaauw' is a most attractive blue-grey. 'Kaizuka' is a rich mid-green. 'Plumosa' (green) and 'Plumosa Aurea' (a rich golden-yellow for most of the year) and the two variegated forms, if left to themselves, form lower-growing and more compact plants and may need some training to give us the ruggedness of outline we want just here, but the rich colour of 'Plumosa Aurea' is worth the effort.

It is, however, on the higher parts of the rock garden, at the top of the rocky outcrops, where the greatest care in the choice of suitable varieties is so important. At a first glance these more prominent positions would seem just fine for showing off the beauty and symmetry of some dwarf tree of upright or pyramidal habit and so the tendency is to plant such a tree where in nature it would have been either blown out of the ground in a gale many years ago or at best have been so knocked about by the wind over the years as to have lost all pretension to symmetry of outline. We may not be able to analyse our failure but the eye is instinctively disturbed by such incongruity, whereas a dwarf pine in the same spot (planted with its stem well out of the upright and with its lower branches and all the foliage below main the branchlets removed so as to expose the bare bones of the poor windswept veteran) will look just inevitably right and so be aesthetically satisfying. Several of the asymmetrical cultivars of the Scots pine, *P. sylvestris* ('Beuvronensis', 'Nana', 'Westonbirt') and the many other dwarf pines now becoming available will do excellently.

For any situation more than three or four feet above the level of the path from which it is to be seen I personally feel that spreading forms should be limited to those with definitely pendulous branch tips, as those with stiff, upturned growing points give to my eye an unpleasant shock-headed effect. Plants of the form I prefer are *Juniperus* × *media* 'Pfitzeriana', *Chamaecyparis lawsoniana* 'Tamariscifolia' and, of course, the dwarf weeping Cedar of Lebanon, *Cedrus libani* 'Sargentii'.

Juniperus recurva var. *coxii* is quite entitled to a paragraph to itself because of its unique charm as a young plant, especially when planted near water and if the foliage is drastically pruned away to expose the trunk and main framework of the tree. But be warned that (inasmuch as the tree is called the Coffin Juniper because of the demand for its timber in Burma for this purpose) you must expect it to get too big in time.

A few dwarf conifers are essential in the heather garden. Here, contrary to the general rule they should be isolated plants relatively few and far between, only enough of them being used to emphasize the open, windswept terrain. The prostrate forms are here quite out of place but the choice may fall on any varieties which grow with a rugged, irregular outline such as the varieties of Juniper already mentioned in this chapter,and here out in the open safer use can be made of the upright to conical forms. The dark blue-green mass of *Chamaecyparis lawsoniana* 'Ellwoodii' or the slender grace of *Juniperus communis* 'Suecica' or 'Hibernica' can make a delightful contrast with the low masses of the surrounding heathers and I know of no better combination than the winter-flowering *Erica carnea* and the loosely conical shape of *Thuja occidentalis* 'Ellwangeriana Aurea' with its golden foliage. This is at its best when the Erica are in flower and it seems to revel in the hardest winter weather that comes.

In the United States, where large numbers of houses are of the prefabricated type erected on low walls of concrete, the development of what is there called foundation planting has been necessary. This consists of evergreens planted close to the house to hide the concrete foundations, and for this conifers are widely used. In this country where it is more usual to carry the facing brickwork down the ground level such planting is unnecessary but it is still quite usual for a path to follow the outline of the house at up to a metre or more away. A border so formed is always a problem. Its nearness to the house makes it architecturally important but it is always dry and frequently very draughty, both being conditions in which few plants are happy. Hence the sound advice given in books on garden layout to make it as wide as possible.

Provided that there is sufficient width (one metre being perhaps a fair minimum) several of the dwarf conifers can be used in such a border with good effect. Care should be taken to avoid overplanting. This tendency is inevitable because the need to cover unsightly foundations as soon as possible can easily lead to the use of varieties much too vigorous for such a restricted spot, but in this country we have no such excuse.

As a group the junipers are probably best able to deal with the dry conditions, and the semi-prostrate forms of *Juniperus* × *media* and *Juniperus sabina* are particularly useful. Where there is ample room for it to develop, a plant of *Juniperus* × *media* 'Pfitzeriana' planted right at the main corner of the house can be very impressive, but in most gardens today it would be well to choose the compact form 'Pfitzeriana Compacta'. Low-growing, conical and upright forms can be used to give a point of emphasis where needed, but these should not (for both practical and architectural reasons) be planted in front of windows. In this sort of situation, where the wall of the house forms the background, regard must always be had to its colour and tone in order to secure a good contrast, and in the choice of planting the amount of sunshine the plants will get should be borne in mind.

Figure 8. Dwarf conifers used to provide interest near the main door of a house.

Figure 9. Another example of 'front-door' planting of dwarf conifers.

The entrance door to the house needs special attention. In a very simple small forecourt garden a pair of Irish yews (the golden form where the house walls are grey stone or dark red brickwork and the dark green form where the walls are of any light colour) may be all that is necessary to give simple emphasis to the doorway. Where there is greater room more elaborate planting can be attempted and where there is plenty of space a grouping of conifers to flank the entrance can be made quite a feature, especially if the ground rises towards the house, necessitating a low flight of steps. These large groups should be planned not only for the effect architecturally as one approaches the building but so as to give a picture from the doorway itself, from whence it will most frequently be viewed both by our visitors and ourselves. In all but the simplest schemes it is usually better to secure balance than to attempt symmetry.

In all these suggestions I do not want to give the impression that dwarf conifers should be used alone. Although, as I have said, they are usually most effective used in groups they readily and happily associate with other plants, provided these latter are not too exuberant. The evergreen azaleas and the smaller daphnes, for example, even when not in flower, make a good combination with dwarf conifers by virtue of their contrasting foliage texture, and doubtless the same can be said of most of the slow-growing broadleafed shrubs. Carpeting plants can be allowed to grow right to the trunks of the trees and thereby save much weeding, providing that care be taken to avoid the more rampant and invasive species and varieties. Many dwarf bulbs look beautiful when in flower (if planted in fairly bold groups rather than individually dotted about under the dwarf conifers) but the garden varieties usually look out of scale.

I am frequently asked whether any of the dwarf conifers are suitable as indoor plants and my invariable reply is to suggest that two or preferably three litte trees should be established in pots so that they can be brought indoors in turn. By this system of horticultural shift work they should all continue in excellent health and a small tree or two can give a lot of pleasure to an invalid or to one whose own green-finger days are over, and they would need far less constant attention than do Bonsai plants. The trees should be changed around every month or two. During their indoor shift they should have the coolest and lightest position available and they should be plunged to the rims in the garden during their off duty shifts.

In any garden a few selected dwarf conifers established in pots can be very useful in many ways. They can be used as ornaments in their own right in a cold greenhouse or on the terrace and they may be plunged here and there in the alpine garden or border wherever and whenever a note of evergreen is needed to emphasize the floral picture of the moment. It should be borne in mind that all conifers in pots will need care in the matter of regular watering and they will enjoy frequent heavy syringeings on summer evenings (all conifers like this), especially in the dusty atmosphere of large cities.

Amongst the miscellaneous uses to which dwarf conifers can be effectively put is the planting of graves. Provided that care be taken to select varieties that can be relied upon to remain dwarf and that full advantage is taken of the range of available colours and habits of growth, a simple group of dwarf conifers can be quite as effective as any attempt to grow flowers in such a restricted area, and of course will be far less trouble to maintain. It is a particularly useful solution to the problem set by those graves that are cared about but which are impossible to visit frequently.

3. Choice of Species and Varieties

The range of habit, rate of growth, type and colour of foliage and the many other characteristics of conifers, especially among the large number of garden cultivars is very great. Not only so, but the production of many of the newer and rarer introductions is in the hands of specialist nurseries, so the range of what will be available in local nurseries and garden centres will vary from place to place (and indeed, from country to country). This makes the preparation of listings that will have any real value very difficult. Local enquiry as to what is known to succeed in the district and is locally available is usually the best, often in remoter areas the only, avenue of choice.

CLIMATIC ZONES AND HARDINESS RATINGS

One factor that must influence selection is hardiness.

The (approximately 600) recognised species of conifer, spread amongst about 60 genera are widely distributed throughout the temperate areas, especially in the northern hemisphere. Some are native in arctic regions or on high mountains so are virtually indestructible but a few genera are tropical or semi-tropical. These are outside the scope of this book, but even among the species in cultivation in the temperate zones there are differing degrees of hardiness.

On pp. 33 to 36 will be found climatic zone maps for Europe, North America and the British Isles. They are for use in conjunction with the hardiness ratings given in the species descriptions in Chapter Seven and in Volume Two. Apart from the limitations imposed by the small scale necessarily used, these maps must be regarded only as very rough guide, because hardiness in plants is a very complex subject.

Presumably all plants are hardy in their natural habitat because of their having adapted themselves over innumerable generations to the local climate. So they are able to tolerate normal extremes of heat in summer and of cold in winter and survive periods of inadequate moisture (drought) and they will only fail when conditions are exceptionally severe.

The ability to survive and prosper of a species you wish to establish in another part of the world largely depends, of course, upon the climate of the region to which you intend to take it, but it is also affected by the inherent adaptability of the species itself. The latter can vary to a surprising extent. It is not uncommon to find of two species which grow happily side by side at home that one will turn out to be 'hardy' in a strange land across the seas whereas the other proves a complete failure. One possible explanation of this apparent inconsistency is that the hardiness of a species having a wide distribution in nature can vary considerably from one part of the habitat to another, and that the hardiness of plants raised from collected seed will be similarly affected by the area in which the seed was collected.

The effect of climate on plant life is itself a complex affair. Not only are the extremes of heat in summer and cold in winter important: vital also is the inter-relationship of climatic factors and the pattern they follow month by month around the year. The monthly incidence of rainfall, of instance, as well as the total annual amount, can vary from place to place throughout the world to an extent almost incredible to anyone who has not had occasion to study the subject.

Although infinitely varied, climate can be roughly classified into the 'continental' type characteristic of the large land masses where hot and often dry summers alternate with very cold winters, and the 'oceanic' type found in islands and coastal areas where the moderating influence on temperature of the ocean or sea produces a much more equable weather-pattern. The transfer of a plant from one type of climate to the other is one very obvious source of stress. Many are the species that would have laughed at a much lower temperature at home which will succumb in winter in an equable climate such as in Britain because they have not had the benefit of the hot dry summer they needed to harden the tissues and ripen their buds.

Another factor that will often cause unreliable behaviour (as regards hardiness) of introduced plants is that the internal changes that occur in plants in preparation for approaching winter occur gradually over a period. Similarly, the plant wakes up slowly as the weather warms up in spring. A relatively mild frost in early Autumn or late Spring can cause more losses than much lower temperatures in mid-winter.

Finally, there is the matter of micro-climate. Any experienced gardener will be able to relate experiences of plants that failed time and again until tried 'in a sheltered spot' or against a south wall. Conversely, so complex is the subject, such a spot would have been the worst choice for one of the Chinese Abies characterised by early Spring growth and consequent vulnerability to frost damage. For such, the coldest and most inhospitable spot available, where bud-break will be most delayed, would be the place to choose.

The complicated picture thus drawn explains the difficulty of producing foolproof hardiness maps. Average minimum winter temperature, although it is only one of the many factors involved, is probably the most important single factor. It is, moreover, something that can be measured and recorded, so hardiness maps based on actual temperatures recorded over many years are therefore the best guide available. But a potential planter of conifers (or of any group of plants) would almost certainly do better by seeking advice (if planting on a commercial scale) from a neighbouring landowner or Department of Forestry or (if developing a garden) from a neighbouring gardener or the local nursery than by putting too much trust in a hardiness map.

Meteorological data can only be used to produce maps of hardiness zones by pre-determining what readings are to be taken as the boundaries between the zones. If these are changed the same data will result in a different map, and unfortunately this is just what has happened in some cases. For instance, a map published by the Arnold Arboretum in Massachusetts, U.S.A. and a map (based presumably on more-or-less identical data) published by the United States Department of Agriculture, (usually referred to as the U.S.D.A. map), do not agree as to where all the steps should come, and wherever they differ the zones on the maps must also disagree. The U.S.D.A. map is now more generally followed and is used in this Manual, but many published works have used the Arnold Arboretum map, so it is important before using published hardiness ratings to check the climatic zonings to which they relate.

The map used for Europe is the hardiness zone map produced by Professor Dr W. Heinze of the Berlin University of Technology and Prof. Dr. D. Schreiber of the Bochum Ruhr University, [published in colour in *Gartengehölze*, A. Bärtels, 2nd edition Ulmer, Stuttgart, (1981), and described in detail in *Mitteilungen der Deutschen Dendrologischen Gesellschaft* Vol. 75 (1984)] and used in a black and white version of Messrs Parey of Paris, with kind permission of Professor Heinze. In preparing this map the authors followed the method used in the U.S.D.A. map, using data collected by European meteorological stations over a period of 30 years. As they also selected the same limits for the zones the two maps are strictly comparable.

The European map being necessarily on a small scale, Prof. Heinze has kindly produced for me a map of the British Isles to a much larger scale and with zones in closer steps. This is based on meteorological data comparable to that used in his hardiness zone map of Central Europe.

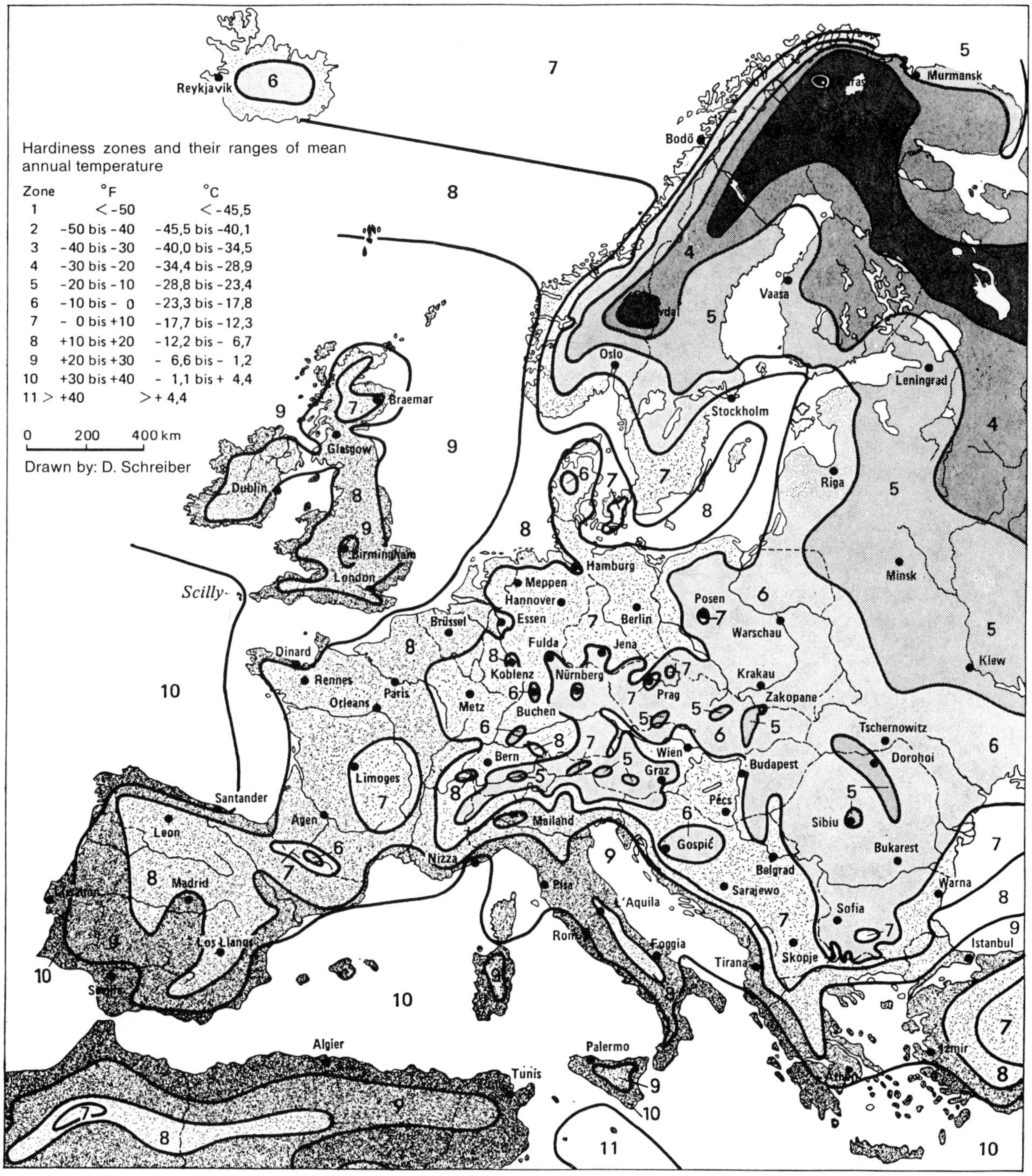

Hardiness zones for trees and shrubs in Europe

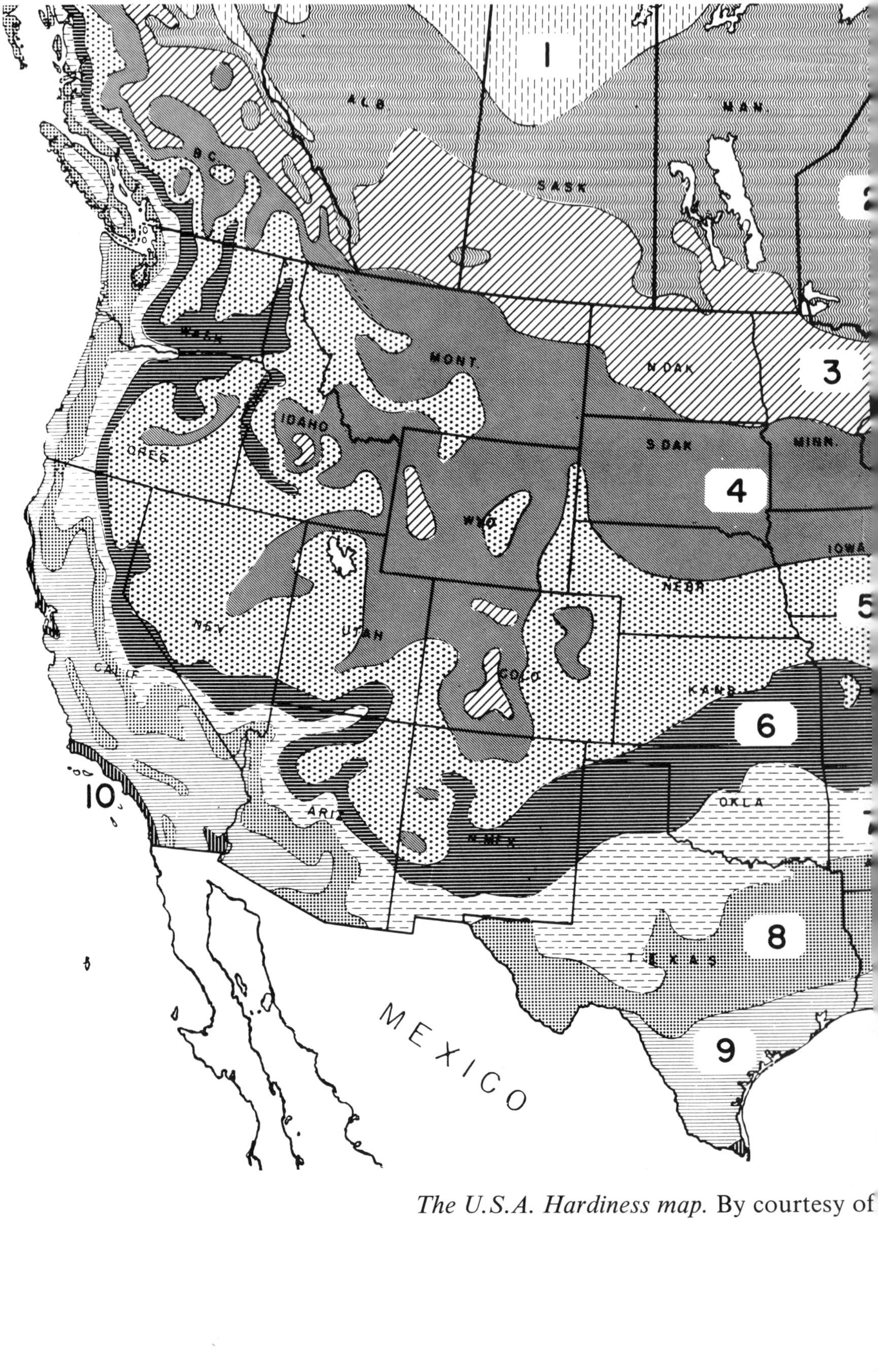

The U.S.A. Hardiness map. By courtesy of

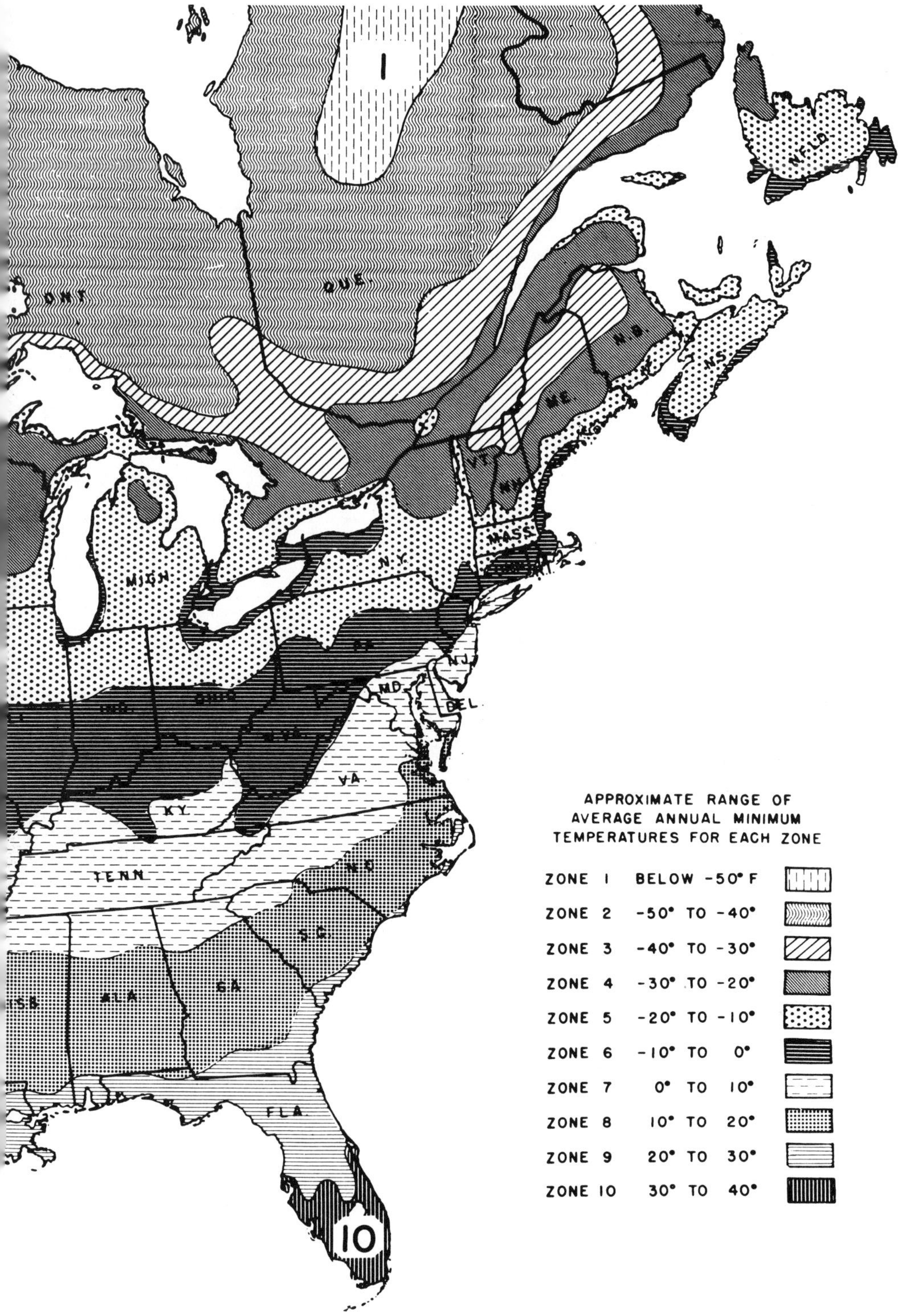

ited States Department of Agriculture.

BN–6914

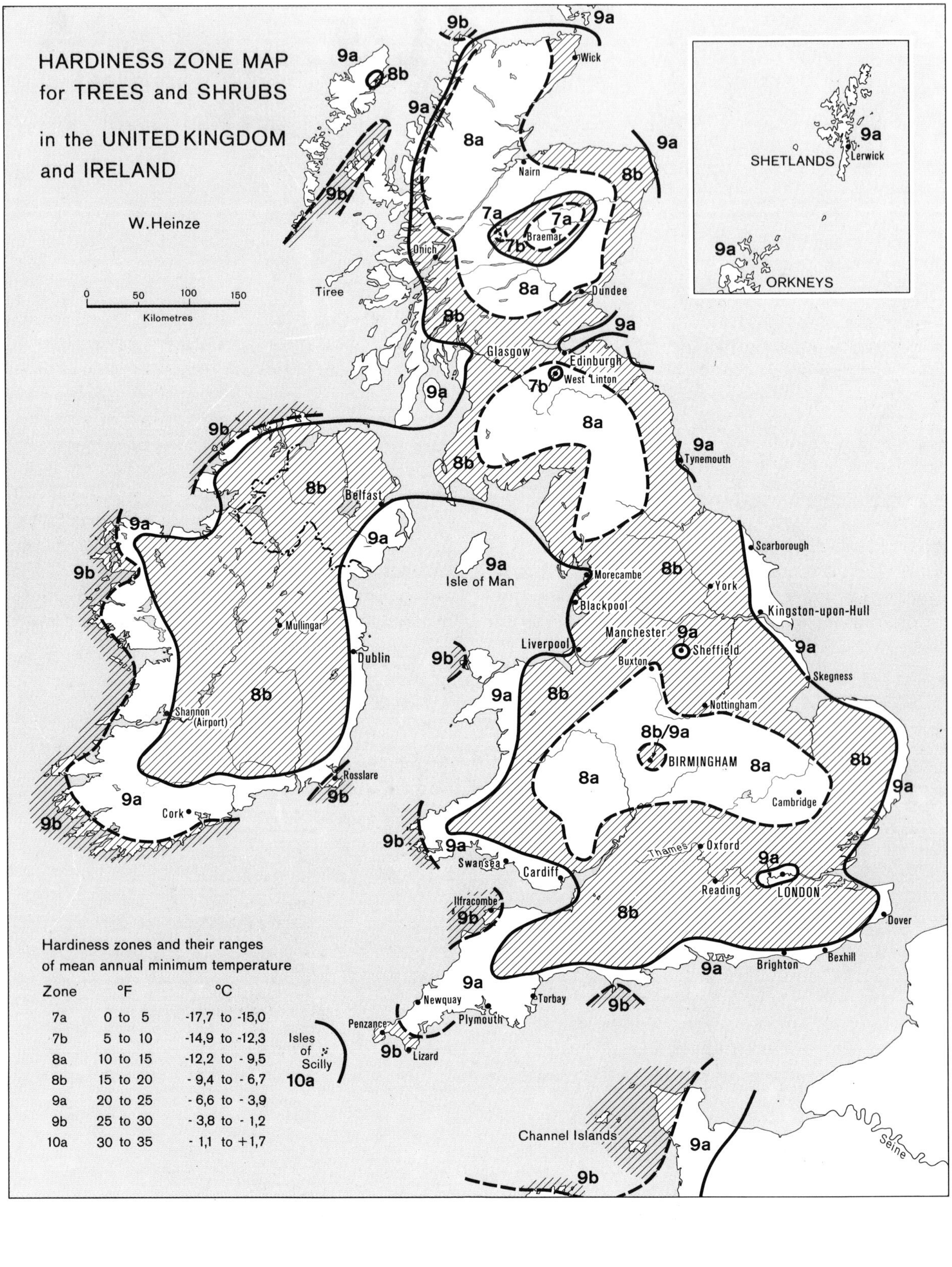
HARDINESS ZONE MAP
for TREES and SHRUBS
in the UNITED KINGDOM
and IRELAND
W.Heinze
0 50 100 150
Kilometres
Hardiness zones and their ranges
of mean annual minimum temperature
Zone °F °C
7a 0 to 5 -17,7 to -15,0
7b 5 to 10 -14,9 to -12,3
8a 10 to 15 -12,2 to - 9,5
8b 15 to 20 - 9,4 to - 6,7
9a 20 to 25 - 6,6 to - 3,9
9b 25 to 30 - 3,8 to - 1,2
10a 30 to 35 - 1,1 to +1,7
SHETLANDS
Lerwick
ORKNEYS
Wick
Nairn
Braemar
Onich
Tiree
Dundee
Glasgow
Edinburgh
West Linton
Tynemouth
Belfast
Isle of Man
Morecambe
Blackpool
Scarborough
York
Kingston-upon-Hull
Mullingar
Dublin
Liverpool
Manchester
Sheffield
Buxton
Skegness
Nottingham
Shannon (Airport)
BIRMINGHAM
Cambridge
Rosslare
Cork
Swansea
Cardiff
Thames
Oxford
Reading
LONDON
Dover
Ilfracombe
Brighton
Bexhill
Newquay
Torbay
Plymouth
Penzance
Lizard
Isles of Scilly
Channel Islands
Seine
7a
7b
8a
8b
8b/9a
9a
9b
10a

4. Cultivation

Little can be said in this chapter about cultivation of conifers that does not equally apply to all trees, and nothing that takes us outside the range of good horticultural practice.

SOIL PREPARATION

In afforestation, the areas involved are usually so large and so exposed to climatic conditions that any amelioration is impracticable and the only course open is the choice of suitable species and the planting of a large number of small trees in the hope that a sufficient proportion of them will survive. In doing this the landowner is but copying Nature's own methods, but with the advantage of not having to scatter his seed to germinate *in situ* and of being able to obtain seed from selected provenances.

In amenity planting conditions are different: the site and its surroundings are usually pre-determined: the range of species has to be wider: the cost of each tree is much greater and the survival rate aimed for correspondingly great. Trees intended to remain as specimens, whether in open parkland

Figure 10. Nature responds to damage by developing a new leader.

or in the confines of a suburban garden, and also shelter belts and hedges – all plantings regarded as being permanent – will in most cases justify expenditure on site preparation, although the cost will usually limit what can be done. This, following normal good horticultural practice, will in the first place, consist of deep digging. A large area could be double-trenched, but the excavation of a deep hole at the site of an individual tree can be disastrous, especially in an impervious clay soil, because it forms a 'sump' in which water will collect. Similarly, attempts to 'improve the drainage' in an area of water-logged soil by dumping gravel or stones at the bottom of the trench will be useless unless true drainage to a pipe drain or to lower ground is also possible. If not, the only resource is to raise the level with permeable soil in which the young trees can be planted or to restrict one's choice to water-tolerant species.

Soil improvement always consists of treating some extreme condition (pure sand, clay, chalk and so on) to bring it nearer to that desirable, balanced commodity known as 'good garden soil'. It usually consists of digging in the obvious ingredient to rectify the existing imbalance. Peat is traditionally a good additive for any soil, but only if moist and well mixed in – large dollops merely dumped into the planting hole only cause settlement later as the peat is absorbed. Raw material such as fresh farmyard manure or unrotted leaf-mould should never be introduced when a tree is being planted; fertilisers should be used sparingly and be limited to slow-acting organic forms of plant-food.

When we come to the cultivation of conifers in gardens, especially the small gardens of today, very different considerations apply. The site is now closely defined and the environment is determined, often much to the disadvantage of plant life, by the neighbouring development. On the other hand the demands we make of our conifers and they on us are more limited, since we are now restricting our thoughts to the dwarf and slow-growing forms, so the objective is to keep them healthy but not to encourage vigorous growth.

In these circumstances some attempt may be desirable to improve the physical condition (texture) of the soil, but anything we are likely to have to contend with that justifies being charitably regarded as garden soil will probably meet fertility requirements. Especially is this so since, because there is never an answer to the question 'How big will it grow?', such plantings must be regarded as being of a temporary nature.

As a generalisation, if conifers do well in your district you can be assured of success. If they are conspicuously absent from the local landscape it is much more likely to be an unsuitable climate than any peculiarity of the soil. In such conditions you may have to provide an immediate environment for your garden conifers that is to their liking if they are to succeed.

Planting in land containing roots of perennial weeds is an invitation to disaster. A decision to devote a year to 'cleaning the ground' would be well repaid, and any treasures you have that are thereby made homeless would lose nothing if well planted and cared for in good cultivated soil elsewhere in the meantime.

PURCHASE OF PLANTS

The selection of suitable varieties for different situations has been briefly dealt with in Chapter Three.

Some garden owners will choose trees they have admired at a Flower Show or seen at a local nursery or garden centre, caring little whether they are labelled correctly, or at all, but the keen plantsman will want to do better than this. There is a good range of conifer species in cultivation and also a very wide selection of named garden cultivars. Plants offered at local centres are usually of reliable, well-tried varieties and (hopefully) are correctly labelled, but even at the largest garden centres the range available will be limited, since the production of garden conifers (and especially the newer cultivars) is still very much in the hands of specialist nurseries. These can usually be found from advertisements in gardening magazines, since they cannot rely wholly on local trade.

Conifers are propagated in several ways. Species are grown from seed and are either (if used in afforestation) planted out when little more than seedlings or are grown on in the nursery for several years until they reach 'planting' size. Most cultivars are grown from cuttings. These are described as 'on their own roots', to distinguish them from plants that have been grafted on a seedling understock. There is actually no inherent objection to grafted plants. *Pinus, Cedrus, Pseudotsuga, Larix*, some of the *Picea* and a few varieties of some other genera are impossible to propagate economically from cuttings so these must be increased by grafting and the resulting plants are healthy and long-lived. But conifers respond to rootstock vigour just as much as do fruit trees. As is well known, an apple variety worked on the dwarfing stock EM IX remains small and suitable for use as a cordon of dwarf bush whereas the same variety worked on a vigorous stock such EM XVI will soon become a large tree. Similarly, a dwarf conifer grafted on to a strong-growing seedling of an arboreal species will make much stronger growth than we want and may even lose character entirely. Unfortunately, producers of general nursery stock whose normal aim is, of course, to produce a good, vigorous and healthy looking plant, tend to overlook the different requirements when it comes to dwarf forms and the unwelcome vigour resulting has brought all grafted plants of dwarf conifers under a cloud of suspicion. It is always possible to detect a grafted plant – the healed-over but still discernible remains of the graft will be found on the trunk an inch or so above the soil, but it is impossible to tell what rootstock has been used. Some estimate of the vigour of the plant can be formed by a careful assessment of the amount of annual growth it has been making in the nursery, but the safest course is to avoid grafted plants whenever possible. This is specially the case if you are buying the real pygmy forms for a trough garden or window box.

But whatever its origin, each plant you buy should be closely examined for quality, especially if you are looking at plants in containers of any kind. The widespread adoption of containerisation of plants is a modern development that, like so many developments of our time, has pernicious results. It cannot be too clearly understood that the objective of containerisation is the convenience of the seller and the buyer, not the benefit of the plant. For shortlived or deciduous plants no great harm may result, but in the case of plants with a permanent hardwooded stem such as conifers, containerisation is aways detrimental, sometimes disastrous. Of this, more below.

The foliage of any of the cypress family should look bright and fresh and should be cool to the touch. A dull, unnatural grey or brown indicates an unhealthy plant, warm foliage a dead one. In the case of the pine family (in the wide sense) the state of the leaves is less important than the condition of the dormant buds. These should be plump and firm. If a grafted tree, the graft should look strong, well healed and not too conspicuous and already be merging into a shapely stem. The shape of a plant should already shew consistence with the shape and character of the future tree and so require the minimum of corrective training or loss from unavoidable hard pruning. Broken, twisted and duplicated leaders should be avoided: bushy varieties should have divided into several stems almost at ground level, and so on. Most of all, avoid plants of all spreading or ground-cover varieties such as *Juniperus* × *media* 'Pfitzeriana' which, instead of having been pruned when small in the nursery to encourage a well-balanced starfishlike plant have been allowed to develop as one long shoot, most of which well at once have to be sacrificed if *ever* a shapely plant is to result.

But the most important part of your prospective purchase – often the least thought of – is the root system. A large tree depends upon its roots both for anchorage and as the means by which it obtains water and food from the soil. The main roots radiate outwards and downwards in the soil, branching again and again into smaller roots and finally into a mass of rootlets, the whole system occupying a bowl-like space (in shallow soils a saucer-like space) in the ground. The main roots must be more or less straight and be distributed reasonably well around the trunk, like the spokes of a wheel, for the tree to be wind-hardy (i.e. resistant to gales coming from any direction) but its anchorage ultimately depends upon the weight of the soil intimately attached to it by the many minor roots and the fine capillary roots whose main function is the extraction of water and nutrients from the soil.

From this picture we may learn a lot – even about choosing plants. The capillary roots appear to be short-lived: they are functional only at their tips and they are being constantly renewed. During transplantation many of these roots are inevitably destroyed, so we aim to preserve as many as possible of the fibrous roots and cosset the tree while it is growing fresh capillaries. Also, we stake our tree until we feel it has taken hold of enough weight of soil to anchor itself. But (and a vital

Figure 11. 'The Three Corpses'. The effect of keeping young conifers too long in small containers.

but the main structural roots are permanent. As we arrange them when we plant a tree, so – for better or for worse – they remain throughout that tree's lifetime.

In former times, standard nursery practice was to hold a tree in the nursery for anything up to ten years, transplanting it every second year, with the result that it reached its final home with a well-balanced root system of main and subsidiary roots of all sizes, all contained in a fairly hefty 'root-ball' neatly tied up in hessian, only needing to be dropped into a hole of appropriate size to give the tree a very high chance of developing well and reaching a good old age.

But nowadays, with advanced technology, a tree of the same size is produced in less than half the time and is offered for sale to the public in a container of (maybe) one quarter of the cubic capacity of the old-style root-ball. Greater understanding of soil chemistry and plant physiology has produced this more profitable rate of growth with a much smaller root system. But what a different system it is! Instead of the radiating, straight, main roots and the mass of subsidiary roots that the tree depends on for survival, it will probably now have one or two roots going round and round the container like an enormous watch-spring. Seedlings may be successfully used in afforestation planted in a slit in the soil opened with a spade and closed with one's heel because they have not yet developed permanent roots and so they can start from scratch, so to speak, but these watch-spring roots in the container are permanent roots that the plant cannot replace. Plant them as they are and in fifty years (or less, if the tree blows down before) you will find them still there. I have seen windfallen × *Cupressocyparis leylandii* trees thirty feet high in which the outline of the container and the malformed roots it contained were still clearly discernible. My next statement may receive no more acceptance than does the remark that horses are much better for the land than tractors, but the brutal fact is that these watchspring roots are worthless and must be removed. So doing compels the tree to develop new permanent roots – the straight anchor roots it must have for anchorage in years to come. Planting them as they are denies it any such chance. The normal reaction to such harsh advice is a belief that you will just kill the tree. With care you probably will do no such thing, but if you do, better to kill it the year it is planted than have it blow out of the ground ten, twenty or thirty years later. Nursing in spare land for a couple of years of a valuable specimen that has had to undergo heavy surgery of this kind may be a fag, but no one, surely, would grudge a little extra trouble taken now to ensure pleasure for his grandson!

THE PLANTING SEASON

After what has been said in the previous section it will be evident that the main advantage claimed for container-grown plants, i.e. that it extends the buying season and facilitates 'instant gardening' does not apply to trees intended to remain permanently.

It is true to the extent that conifers in containers may be acquired at any time throughout the year, but being disturbed at the roots is always a shock, especially so for an evergreen tree and so should only be undertaken at a time of the year when conditions hopefully will be favourable to the tree's recovery. By all means pop your latest acquisitions into the ground to give an immediate effect, but leave them in the containers in which they were bought and keep them watered until the real planting season comes round. Especially in the case of large specimens being moved to a new home, or plants that have become 'pot-bound' in a container and which must therefore be subjected to the humane brutality advocated in the previous section the work should only be taken in hand during the optimum periods.

In broad terms, these are the autumn from August to October and the spring from March to May. During these months the soil is warm and the roots are active and capable of maintaining sufficient flow of sap to prevent the drying out of the foliage. To disturb conifers during the cold winter months is most unwise. The roots then are inactive and incapable of replacing moisture lost by the leaves. Winter sunshine, especially if the ground is frozen, can desiccate a tree beyond recovery in a few hours. Unwelcome arrivals in the depth of winter are best 'heeled-in' in a sheltered spot not reached by the sun and left there until the spring.

The warmer end of each period is the best, provided the water requirements of the plant are understood and met. Conifers lifted early August and kept watered will have a bristling mass of new roots by October, so the former is an ideal time provided that autumn rain or heavy irrigation has penetrated the ground to the full depth of the root system. Never attempt to dig conifers in dry soil, it always results in torn roots.

The later that planting is done in the spring, the more care must be taken to reduce water stress during the hot weather ahead. Although not to be recommended, if sufficient trouble be taken, planting or transplanting can be done successfully at any time during the summer if the tree is provided artificially with total shade and is watered and syringed frequently and regularly for the rest of the season.

PLANTING

Much of what has gone before has a bearing on the actual planting. The verbal picture on p. 39 of a large tree's bowl-shaped root system underground tells you exactly how and why you should take out a hole about twice the diameter you would otherwise be inclined to do, so that you can spread the roots out more or less straight right to their tips, remembering that the tree's chance in life is in your hands. Roots twisted into a small hole will remain like that. Any roots forced to turn upwards will die. Be parsimonious at planting time with fertilisers, especially artificials, and eschew heavy dumping of extraneous materials such as peat. If soil conditions are very poor backfill immediately around the roots with good imported soil. Loosening the subsoil with a fork is good practice, provided that it is thoroughly well firmed before the planting is done, to minimise settlement later. Plant the tree to the level it was growing at before, using a bamboo or lath laid across the hole, but making an intelligent allowance for the inevitable settlement. (Many trees finish up too deep in the ground because this had been forgotten). Backfill slowly, sifting the soil around and between the roots and gently firming as you go. Leave the final stamping (with the full weight of your heel) until the soil has dried out a bit following a thorough soaking either contrived by you or provided by heavy rain.

Staking depends on local circumstances and the size of the tree. The one obvious but often overlooked *sine qua non* is for it to be adequate. If you are going to use a strong post this should be fixed immovably in position in the hole before the tree goes in. Unless a stake is needed for the stem-training of a young tree, it need not be very tall. Its function is to supply temporary anchorage, not to prevent the tree swaying gracefully in the wind, so a stout stake solid in the ground and firmly supporting the tree at one third its height will be far better than a full-length flimsy one uselessly rocking to and fro in the ground in phase with the tree. The ties should be firm, to avoid chafing, but must allow for the increase in diameter of the trunk that will inevitably occur. Good proprietary tree-ties are available and these should be used for tying or some biologically degradable material such as hessian . Never use metal or plastic wire or thread such as binder twine. A large specimen is best anchored with guy-ropes to three pegs in the ground.

AFTERCARE

Regular inspection throughout the tree's life should be an established routine: monthly the first few years until the tree has fully established itself, later annually, and at least quinquennially even in maturity.

Check the stakes for rot at ground level. (A stake held up by its own tree is far-too-common a sight). Replace ties regularly (strangulation by binder-twine is a death no tree deserves and one that ought not to be on anyone's conscience).

WATERING

I am a great believer in the heavy initial soaking mentioned above, even when the ground seems to be wet. Nothing is as good for settling a tree in.

Any subsequent programme of watering depends on local circumstances. A regime just sufficient to keep the tree alive is of course a minimum. Beyond that, a programme of heavy watering frequent enough to keep the tree from suffering any water stress throughout the whole of its first growing season can have quite gratifying results. All conifers of any age or size are damp-climate plants and respond to frequent syringings in hot weather, especially if done during the evening so that the foliage remains wet during the night.

PRUNING

There is nothing mysterious about pruning.

The *objective* in pruning a conifer is to persuade your tree to develop and grow (or in some cases not to grow), in a way that it would not do if left to itself.

The *theory* of pruning is the belief that plants (more precisely, each separate species of plant) will always react to damage in the same characteristic way and that plants cannot differentiate between causes of damage so do not distinguish the work of the pruner from any other disaster that comes their way.

The *practice* of pruning is the application of acquired knowledge of how the particular sort of tree you are dealing with will (hopefully running true to form) react to the damage you are causing it in a way that will secure your objective on this particular occasion.

So, if you are encouraging a young conifer plant to become a symmetrical, conical tree with a single leader, you will eliminate double or multiple leaders and shorten any abnormally vigorous laterals. If your objective is a dome-shaped bush or a shrub thickly furnished to the ground you will cut all lanky growth hard back to encourage a branch-system to develop close to ground level. If the task you are facing is to cut down to size a tree or hedge has outgrown its welcome and seems determined to go on growing, you are up against a battle of wills. Heavy pruning usually stimulates vigorous growth to absorb the unabated vigour of the roots, but species differ as to just how much punishment they can take. A five hundred-year old Yew tree cut to the ground will start again unhesitatingly, but a Lawson cypress never 'breaks from old wood' and, if stripped of green foliage (by you or by circumstances such as bad weather), it will die. So who wins the battle of wills depends on the species you are contending with and whether you have let it go too far before tackling the job.

The tendency of plants, when a branch is severed, is to produce buds and break into growth just below the point of severance. If re-shaping an overgrown plant requires growth to be encouraged at a particular point, have the courage to cut well below to that point, since any less drastic treatment will only produce growth where it is not wanted. So if for instance you have purchased a badly grown lop-sided plant of *Juniperus* × *media* 'Pfitzeriana' consisting of a single long shoot, there is no other way of producing from it a well balanced plant – you must sacrifice most of your plant by cutting it back to within a few inches of the ground. Only so will you encourage the development of a shapely plant. But when such harsh treatment is needed on an old plant it may be wise to do it over a period of two or three years, leaving some shoots unshortened until there is abundant new growth on the branches first taken in hand.

5. Classification and Nomenclature

Because of the impact on mankind of so important a factor in his environment it is not surprising that references to plants and trees are to be found in his folklore and in his earliest attempts at recording his experiences in writing or as pictures on the stone wall of his residence. Such references are also scattered throughout classical and medieval literature. Information slowly accumulated amongst early botanists (and also, presumably, amongst early foresters and early woodworkers) but as in every major plant group, this process was much accelerated by the development in 1753 by the Swedish botanist Carl Linné (Linnaeus) of the binomial system of distinguishing species that bears his name and is in universal use today.

BOTANICAL CLASSIFICATION

Botanists had long had a system of classification based on the observation that Nature herself had

Figure 12. A planting of young conifers at Castlewellan, Saintfield, N.I.

grouped her children into assemblies of individuals which were recognisably similar, usually interfertile and often separated from all other assemblages by genetic or other barriers that prevented or at least restricted interbreeding, and that, despite these distinguishing differences, different such assemblages often shewed similarities that pointed to the existence of larger groups. These are the two groups we now know by the names **Genus** and **Species**. To the former the early botanists gave names *(Pinus, Juniperus* etc.) all nouns in the Latin language they were accustomed to using. The latter they distinguished from one another by adding a written description (diagnosis) again, of course, in Latin. This system was unwieldy and inconvenient and the Linnaeus breakthrough came from his realisation that the descriptions could be replaced without loss of precision by the use of a single adjective or a noun used in a similar way. So was born the binomial system in universal use today.

Nature having provided us, as is her wont, with anomalies and exceptions of all kinds, no cast-iron definition of either term of universal application has ever been or ever will be contrived. Moreover, since every attempt at classification (of any kind) depends, as it must do, to some extent on judgment and opinion and be based, as it must be, on the best knowledge available *at the time*, finality in botanical nomenclature can never be hoped for. But by laying down that the name of each species shall consist of a 'two-tier' binomial in Latin form the problem was reduced by Linnaeus to more or less manageable proportions.

Fortunately for horticulturists this two-tier system provides all they need to draw from the botanical sytem as a basis on which to found their own nomenclature. But the number of species recognised by science biologists is enormous and some further botanical classification at higher ranks was found to be essential. To many gardeners the terms used in these higher of botanical classification are complicated and perhaps a little confusing, so a digression at this point might be helpful to some readers.

The system has a hierarchical structure of ranks arranged rather like the various courses of stone in a Egyptian pyramid. Starting at the top, *Regnum Vegetabile*, the Vegetable Kingdom (i.e. life excluding animals) is broken down, course by course, into an ever enlarging inventory of smaller units, each with its own rank and name, the bottom course representing the category *Species.* (In this course each separate known species can be likened to a single stone, the concept *species* being therefore the unit upon which the whole pyramidal system is erected.) Conversely, starting at this lowest rank and working up the pyramid the units become grouped, course by course, into fewer but larger units.

To render the system manageable there is a convention that all names in the categories above *Genus* shall end in a particular way (i.e. be formed by the use of a prescribed suffix or word ending). This serves rather like the pips on the shoulder or the stripes on the sleeve in a military or similar uniform which make the rank of the wearer evident on sight.

In descending order the ranks are:

Rank	*Suffix*
Division *(Divisio)*	–phyta
Class *(Classis)*	–opsida
Order *(Ordo)*	–ales
Family *(Familia)*	–aceae
Tribe *(Tribus)*	–eae
Genus	No prescribed suffix.
Species	No prescribed suffix.

Note that the plural of *Genus* is *Genera. Species* is spelled the same in both the singular and the plural.

Should even this hierarchical choice or ranks be insufficient to meet the needs of a particular group, further intermediate ranks may be created by the use of the prefix *Sub-* and these, too, have their prescribed endings. Fortunately these are little if ever used in the conifer world, but always the rule is that the relative order of the ranks must remain the same. There must be no jumping rank!

In practice, since the system had to follow the vagaries of nature it has produced what would be pictorially represented by a most lop-sided pyramid. Of the two Classes in the **Pinophyta** with which we are concerned, **Ginkgoopsida** is represented only by a single order *Ginkgoales*, whereas **Pinopsida** is divided into two Orders. Of these, *Taxales* contains only one Family, *Taxaceae*, which is responsible for five genera, whereas *Coniferales* divides into no less than six families, containing between them over 100 genera.

TABLE I

The botanical classification of the Coniferae.

Division		PINOPHYTA						
Class	*GINKGOOPSIDA*	*CONIFEROPSIDA*					*TAXOPSIDA*	
Order	Ginkgoales	Coniferales					Taxales	
Family	*Ginkgoaceae*	*Pinaceae*	*Taxodiaceae*	*Cupressaceae*	*Podocarpaceae*	*Araucariaceae*	*Cephalotaxaceae*	*Taxaceae*
Genus	Ginkgo 1	Abies 39 Cathaya 1 Cedrus 2 Keteleeria 2 Larix 9 Picea 33 Pinus 93 Pseudolarix 1 Pseudotsuga 4 Tsuga 10	Athrotaxis 3 Cryptomeria 1 Cunninghamia 2 Glyptostrobus 1 Metasequoia 1 Sciadopitys 1 Sequoia 1 Sequoiadendron 1 Taiwania 1 Taxodium 2	Actinostrobus 2 Austrocedrus 1 Callitris 14 Calocedrus 3 Chamaecyparis 7 × Cupressocyparis 3 Cupressus 13 Diselma 1 Fitzroya 1 Fokienia 1 Juniperus 50 Libocedrus 8 Microbiota 1 Neocalllitropsis 1 Papuacedrus 2 Pilgerodendron 1 Platycladus 1 Tetraclinis 1 Thuja 5 Thujopsis 1 Widdringtonia 3	Acmopyle 2 Dacrycarpus 9 Dacrydium 25 Decussocarpus 12 Falcatifolium 5 Halocarpus 3 Microcachrys 1 Microstrobus 1 Parasitaxus 1 Phyllocladus 5 Podocarpus 94 Prumnopitys 10 Saxegothea 1	Agathis 19 Araucaria 20	Amentotaxus 2 Cephalotaxus 4 Torreya 5	Austrotaxus 1 Pseudotaxus 1 Taxus 7
								Total 559

In Table One on p. 46, in which is set out the hierarchy as now generally accepted, the number of species in each genus is that currently recognised by many botanists, but in the present state of incomplete knowledge of some genera, botanists are not unanimous, and so different figures may be found in other listings.

The botanical system, as well as extending upwards into the hierarchy described above extends also downwards, so to speak, in the infra-specific categories *sub-species, varietas* and *forma*. Reference to these will be found later in this Chapter.

HISTORICAL REVIEW

Before it reached the present level of stability, conifer nomenclature probably as long and as chequered a history, measured in terms of the human life-span as the trees themselves had, measured on the geological time-scale.

The fluctuations, and even the oscillations, of botanical opinion during the shorter period referred to have had one important effect for anyone concerned to ascertain the 'correct name' of any conifer: it is unsafe to rely on the nomenclature in any of the older conifer books. This may surprise those who know about the 'Priority Rule' and who therefore might not unreasonably expect the older a book the more reliable it should be. The reason is that often either a revision in classification resulting from fuller study or the retrospective application of the *Botanical Code* will have required the acceptance of a change of name. Additionally, the emergence of a *Cultivar Code* in 1953 has resulted in extensive changes in the naming of cultivated conifers. So conifer names from older books should be used with caution unless verified from present-day authorities.

Although to some readers it may be mere 'water under the bridge', a record of the convulsions conifer nomenclature has suffered in reaching its present stability is of value. The following account of early writers by Joseph Hooker is reprinted from the Preface to the First (1896) Edition of the *Hand-list of Coniferae grown in the Royal Gardens* (generally referred to as the Kew Hand-list).

The following historic sketch of the various essays of Botanists to limit the genera and species of *Abietineae*, shows how divergent have been their views. I have endeavoured to make it more instructive, by giving (where there are materials for doing so) under each attempt the number of species known to its author, thus further showing the progressive discovery of species during the last hundred and eighty years or thereabouts.

Tournefort, in 1717, indicated the three genera *Abies*, *Pinus*, and *Larix* (including *Cedrus*); and his *Cedrus* is *Juniperus*.

Linnaeus, in 1753, included all under *Pinus*, of which he enumerated 10 species.

In 1731 the first edition of *Miller's Gardener's Dictionary* appeared, where he has *Pinus*, *Cedrus*, *Larix* and *Abies*, with 16 species in all.

In 1789 A. L. de Jussieu, in his *Genera Plantarum* has only two genera. *Pinus* and *Abies*.

In the same year the first edition of Aiton's *Hortus Kewensis* appeared, wherein all the *Abietineae* (19 species) are included under *Pinus*.

In 1803 was published the first edition of Lambert's *Genus Pinus*, containing about 35 species of *Abietineae*, all included under *Pinus*.

In 1813 the second edition of the *Hortus Kewensis* appeared, in which, as before all the *Abietineae*, (increased to 29) are included under *Pinus*.

In 1826 Louis Claude Richards's *Mémoire sur les Conifères* was edited by his son. This classical work is the earliest that dealt scientifically with the order *Coniferae*, and in it the Tribe *Abietineae* was first established, to include *Pinus* (under which were placed Spruces and Silver Firs), and *Larix* (which included the cedars), to these were added, *Cunninghamia*, *Agathis (Dammara)*, and *Araucaria*, which now form a distinct Tribe.

In 1827 Link, in the *Journal of the Academy of Sciences of Berlin*, proposed the five genera: *Pinus*, *Picea* (for Spruces), *Abies* (for Silver Firs), *Larix*, and *Cedrus*.

In 1832 the second edition of Lambert's *Genus Pinus* appeared, wherein the *Abietineae* were all, as before, included under *Pinus*; which however was broken up by David Don (Lambert's Herbarium keeper) into six sections, *Pinus, Abies* (for the Spruces), *Picea* (for the Silver Firs and Tsugas), *Larix*, and *Cedrus*. About 48 species in all are described.

In 1838 Loudon's great work, the *Arboretum et Fruticetum Britannicum* appeared. The Tribe *Abietineae* is here adopted as defined by Richard, with the genera *Pinus, Abies, Picea, Larix*, and *Cedrus*: and here that confusion of *Abies* and *Picea* which so distresses British foresters and planters had its origin. At p. 2105 there is a clavis of the genera, where the genus *Abies* is attributed to Link, but the characters given to it are those of Link's *Picea*; and in like manner *Picea* is atttributed to Link, though the characters given are those of Link's *Abies*, and they are followed by *D. Don* in brackets. Turning now to the full account of the genera, at p. 2293, we find "*Abies*, D. Don in Lamb. Pin. vol. iii." with the addition of '*Picea*, Link" as a synonym, followed by descriptions of the Spruces. And at p. 2329 we have '*Picea*, D. Don in Lamb. Pin. vol. iii," with *Abies*, Link, as a synonym; followed by descriptions of the Silver Firs. Now not only was the edition of Lambert's work subsequent to Link's establishment of *Picea* for Spruces and *Abies* for Silver Firs, but Don nowhere in Lambert's *Genus Pinus* proposed these names as generic. Disregarding this blundering, which probably originated in the elaboration of the conifers being a composite work of Don and Loudon working apart, there remains but one argument for preferring Don's nomenclature, which is, that Linnaeus called the Silver Fir *Pinus Picea* and the Spruce *P. Abies*, assuming that he was following the practice of the ancients, in which he is known to have been mistaken.

Owing to the great merit and utility of Loudon's *Arboretum*, his nomenclature has hitherto been universally adopted in the United Kingdom. It contains descriptions of about 70 species of *Abietineae.*

In 1841 Link (in Lennaea, vol. xv., p. 481) reviewed the whole Tribe of *Abietineae*, retaining, as before, *Pinus, Picea, Abies, Larix*, and *Cedrus*, and enumerating 52 species.

In 1841–46, Antoine's *Die Coniferen* appeared, in which all the *Abietineae* are referred to *Pinus*, with sections of *Pinus* proper, *Larix, Cedrus, Picea*, Don, *Abies*, Don, and *Tsuga*; he describes 90 species.

In 1842 Spach, in his *Histoire Naturelle des Végétaux*, adopted *Pinus, Abies* (including under it as sections *Picea*, Link, *Tsuga*, and *Pseudotsuga), Cedrus*, and *Larix.*

Endlicher, in the same year, in his *Genera Plantarum* reverted to Linnaeus' practice of including all under *Pinus*, but made four sections-*Pinus* proper, *Picea*, Link, *Abies*, Link, and *Larix*, including *Cedrus.*

In the same author's *Synopsis Coniferarum*, published two years later, he maintains the genus *Pinus* entire, as before, but subdivides it into 11 sections, with 109 species.

In 1850 Lindley and Gordon published (in the *Journal of the Horticultural Society*, vol. v.) "An "Enumeration of *Coniferae* cultivated in Great Britain," where two genera only are adopted, *Pinus* and *Abies*, and the latter is subdivided into Firs (including Spruces), Larches, and Cedars, with 119 species.

Meisner in 1836–43 published his *Genera Plantarum Vascularum*, wherein he includes all *Abietineae* under *Pinus*, with, as sections, *Pinus* proper, *Picea*, Link, *Abies*, Link, and *Larix*, including *Cedrus.*

In 1855 Carrière's Traité général des Conifères appeared; he keeps up *Pinus, Abies, Link, Picea*, Link, *Larix, Cedrus*, and *Tsuga*, which includes *Pseudotsuga*. This work enumerates 133 species, many of which are now ranked as varieties.

In 1858 appeared the first edition of *Gordon's Pinetum*, in which he maintains *Pinus*, Abies. Don (including *Tsuga* and *Pseudotsuga*), *Picea*, Don, *Larix, Cedrus*, and *Pseudolarix*. He enumerates 131 species, of which many are imperfectly known.

In 1865 Henkel and Hochstetter published their *Die Coniferen*, with the following arrangement, *Pinus, Abies* (including *Picea*, Link, *Abies*, Link, *Tsuga*, and *Pseudotsuga*), *Cedrus, Larix*, and the new genus *Pseudolarix*, Gord. They describe 130 species, which should be reduced much as Carrière's should be.

In 1867 the second edition of Carrière's *Traité* was published, in which all the genera of the first edition are retained together with *Pseudotsuga* and *Keteleeria*, and 153 species are descriged, excluding many dubious ones.

In 1868 Parlatore's monograph of the order appeared in the 16th volume of De Candolle's *Prodromus*. The genus *Pinus* is divided into two sub-genera *Pinus* proper for the Pines, and *Sapinus*; which latter includes as sectious only *Picea*, Link, *Abies*, Link, *Cedrus, Larix, Pseudolarix, Tsuga*, and *Pseudotsuga*. The species described are 162.

In 1875 the second edition of Gordon's *Pinetum* was published, in which the same genera are retained as in the 1858 edition, with 150 species.

Lastly, in the *Genera Plantarum* (1882), Mr. Bentham and I have distinguished *Picea*, Link, *Abies*, Link, *Tsuga, Pseudotsuga, Larix*, and *Cedrus*, to which must now be added *Pseudolarix* and probably *Keteleeria*, both at that time very imperfectly known.

In 1886 Dr. M. T. Masters published (*Journal of the Linnean Society*, xxii pp. 169–212, tt. 2–10, with 32 woodcut figures in the text) some "Contributions to the History of certain Conifers." This was followed in 1890 by a "Review of some points in the comparative Morphology, Anatomy, and Life History of the Coniferae" (*Journ. Linn.* Soc, xxvii., pp. 226–332, with 29 woodcut figures in the text), and in 1895 by "Notes on the Genera of *Taxaceae* and *Coniferae*" (*Journ. Linn.* Soc., xxx., pp. 1–42).

In 1889, two years after his death, Dr. A. W. Eichler's elaboration of the Coniferae appeared in Engler and Prantl's *Die Natürlichen Pflanzenfamilien*. He classifies the genera under two primary divisions: *Pinoideae* and *Taxoideae*. The former he again divides into *Abietineae* and *Cupressineae*, with several ternary divisions, and the latter into *Podocarpeae* and *Taxeae*. The genera are limited, with few exceptions, as they are by Bentham and Hooker; but he retains *Chamaecyparis, Thuiopsis* and *Pseudolarix*, and sinks *Pseudotsuga* in *Tsuga* and *Prumnopitys* in *Podocarpus*.

In 1891 L. Beissner published a *Handbuch der Nadelholzkunde*. This was preceded by a *Handbuch der Coniferen-Benennung* (1887), followed by two supplements (1891–2) on *Einheitliche Coniferen-Benennung*. Beissner deals critically with the numerous natural and garden varieties and hybrids. He has also paid much attention to nomenclature, though he accepts familiar names rather than revive obscure ones on the rule of priority.

In 1891 "Conifer Conference," and the papers read thereat form the fourteenth volume of the *Journal* of the Society. This is one of the most interesting volumes issued by the Society, containing a great variety of valuable information concerning *Coniferae*, including a general introduction and a list of the species cultivated in Great Britain and Ireland, by Dr. M. T. Masters, F.R.S.

In 1894 H. Baillon (*Histoire des Plantes*, xii. pp. 1–45) published an account of the order, with a synopsis of the genera, which he reduced to twenty-four, including *Casuarina*. He divides the order into eight "series," namely: *Taxées, Cupressées, Junipérées, Athrotaxées, Nagéiées, Araucariées, Pinées*, and *Casuarinées*. The inclusion of the last is inexplicable, and some changes in nomenclature will not meet with general approval *Belis, Salisb.* is revived for *Cunninghamia*, R. Br.;

Podocarpus, Labill. for *Phyllocladus*, Rich.; *Agathis*, Salisb. for *Dammara*, Lamk., and *Nageia*, Gaertn. for *Podocarpus*, L'Herit."

The arrangement of genera in the Hand-list itself followed that in *Genera Plantarum* with the following exceptions. *Prumnopitys* is again separated from *Podocarpus*, and *Keteleeria, Pseudolarix, Tetraclinis* and *Widdringtonia* become new genera. *Cupressus* and *Chamaecyparis* are distinguished at the Sectional level and *Thujopsis borealis* becomes *Thuja nootkatenssis*.

The same arrangement was followed in the Second and Third editions, issued in 1903 and 1925,

Figure 13. A young *Gingko biloba* in winter. This genus, which is quite unlike all other conifers in its foliage, together with *Larix* (widespread) and *Glyptostrobus, Metasequoia, Pseudolarix* and *Taxodium* (all much more rarely seen), are the only deciduous (i.e. bare of leaves in winter) conifers (10).

respectively, so that by the time the Second Conifer Conference was held by the Royal Horticultural Society in 1931, nomenclature at the generic level and been stable for hearly 40 years. But in 1930 a final decision had been made at the International Botanic Congress held at Cambridge regarding the application of the 'Priority Rule' at the specific level. A Check-list prepared for the Conference by Mr. W. Dallimore incorporated the new rule and pointed the way out of what he described as the 'chaotic' state of conifer nomenclature prevalent at that time.

In an article *Reference List of Conifers and Conifer Allies Grown Out of Doors in the British Isles* prepared for the Third Conifer Conference, held by the Royal Horticultural Society in 1970, Dr D. R. Hunt of the Royal Botanical Gardens, Kew was able to state "Since Dallimore's list was published both conifer taxonomy and conifer nomenclature have remained relatively stable." This is certainly a situation for which we should be most grateful, signifying as it does that after nearly sixty years of stability we have little to fear from irritating upheavals in the future.

Nevertheless some important changes have been made that we must now take into account. These have mostly been the result of further exploration in those areas, such as Mexico and South East Asia, among the tropical and semi-tropical genera previously inadequately understood. Other techniques, such as chemo-taxonomy, embryological studies and cultural behaviour have been developed as methods of diagnosis, supplementing rather than displacing the visual study of morphology upon which classification has had to rely exclusively in the past. It is interesting that such studies largely confirm the previous classification: nevertheless these fresh approaches have cleared up a few 'conundrums', especially amongst hybrids and some popular garden conifers resulting from juvenile fixations.

The present century has produced many important publications that have greatly increased our knowledge of conifer taxonomy. The early 1900's received a good start in the works of G. R. Shaw on the pines and this led on to a comprehensive worldwide study of that genus by N. T. Mirov. In the 1940s and 1950s the works of R. Florin and J. T. Buchholz working on *Cupressaceae* and *Podocarpaceae* led to a more detailed knowledge of those families, and the works of M. Martinez of Mexico City and of E. L. Little of the U.S. Forestry Service in Washington DC led to a much better understanding of the distribution of North American conifers. From the early 1950's to the present day the works of H. Gaussen of Toulouse in France and of D. J. Laubenfels of Syracuse University have greatly enhanced the morphological and geographical data available on a number of conifer species. T. Liu, Head of the Department of Forestry in Taiwan has added much to our knowledge of *Abies* and *Picea* in his recent monographic studies of these genera.

In 1978 W. Cheng and L. K. Fu revised the nomenclature of Chinese conifers, providing a descriptive and illustrated account in *Flora Reipublica Popularis Sinicae*. The late 1970s and the 1980s have seen the publication of the detailed work of R. P. Adams and T. Zanzoni on the genus *Juniperus* and, finally, John Silba of New York has collected and summarised the findings of all these workers in his *International Census of the Coniferae* (1984), and in *Encyclopaedia Coniferae* (1986) he has collected detailed morphological data on individual species on the basis of more available herbarium material and much field work. These two publications are taken as the basis of the present work.

The findings in a still more recent book (*Conifers* by Keith Rushforth, Christopher Helm, London) and an excellent presentation of the subject in a very readable style have also been taken into account in the present work where indicated.

NOMENCLATURE OF CULTIVATED PLANTS

But whilst the botanists have been busy about their self-appointed task of CLASSIFYING plants (this means all plants – past, present and future) horticulturists ever since the days of Pharaoh have been working on the same plants in quite a different manner. They, passing over all the other individuals that failed to interest them in the species in question, have been SELECTING particular plants that would make 'good garden forms' or be for some reason or other of special interest of value *seen through horticultural eyes*.

This essential difference (i.e. between classification and selection) was clearly pointed out by Professor Alfred Rehder in his *Bibliography of*

Cultivated Trees and Shrubs published in 1949, in which (p. xi) he states "The purpose of botanical nomenclature is ...classificatory, whilst the purpose of horticultural (nomenclature) is selective".

But the wholly mistaken idea that the place of origin is the criterion lingers on and still causes much confusion. Whether a plant now in cultivation and needing a name for horticultural reasons first saw the light of day in a spot under cultivation or was found somewhere 'in the wild' is quite immaterial. If, in either case, a findling turns out to be something new to science and of sufficient interest some botanist will have to add an extra item to the universally accepted botanical CLASSIFICATION by 'validly publishing' a name for it in an appropriate category under the *International Code of Botanical Nomenclature* referred to below. If, on the other hand, it was a case of SELECTION FROM WITHIN THE SPECIES, to that species it still belongs and the finder may propagate it and propose a name for it as a new cultivar under the *International Code of Nomenclature of Cultivated Plants*. Each case that arises must fall into one or other category, never into both.

THE CODES OF NOMENCLATURE

There are two of these. They are the *International Code of Botanical Nomenclature* issued by the International Association for Plant Taxonomy (pop ularly referred to as 'The Botanical Code' or as ICBN) and *The International Code of Nomenclature of Cultivated Plants* issued by the International Commission for the Nomenclature of Cultivated Plants of the I.U.B.S. (popularly known as 'The Cultivated Code' or ICNCP).

The *Botanical Code* came first, the current edition dated 1988 being the latest of a series of revisions of a Code first used in the middle of the last century. The *Cultivated Code* came very much later: although the need for a separate Code had long been realised, one did not appear until 1953. This also has since developed through several revisions, the current edition being dated 1980. The uncertainty during this long period when only a *Botanical Code* was available as to how garden plants should be named or classified has, alas, left us a big legacy of trouble. Of this, more below.

The *Botanical Code* is of course based firmly on the Linnaean system and by elaborating rules as to priority, valid publication, the avoidance of any repetition of a name that might cause confusion and other similar matters it establishes a basis for stability. It does not control botanical classification: it concerns itself not with 'what' a botanist may or may not name but only with 'how' he must do it. A botanist studying a particular group of plants may put forward whatever proposals for classification or re-classification within the group he thinks fit (sobered only by the knowledge that he must satisfy the botanical world that he is making an acceptable contribution to scientific knowledge). But having made up his mind as to what the new classification is to be he is absolutely bound by the *Botanical Code* as to which existing names he must use and which he may not use and in the formulation of any new names that his new classification requires him to invent (or as the Code puts it 'find').

The *Cultivated Code* has a similar objective (i.e to establish a basis of stability) within its own sphere. The relationship between the two Codes is quite simple and definite but is often misunderstood. Ignoring complications such as hybrids and unresolved arguments between botanists, botanical classification has established that every plant must needs belong to one particular species having its own scientific name. This always consists of two Latin (or Latin-form) words, and traditionally they are usually printed in italics. e.g. *Juniperus communis*.

The *Cultivated Code* accepts and uses this binomial botanical classification unaltered, but for horticultural purposes distinguishes any selected individuals (known as cultivars) by adding a third element (the cultivar name, or epithet.). This, since 1st January, 1959 has had be a 'fancy' (i.e. NON Latin-form) name of from one to three words, and it is printed in such a way as to make the distinction clear at a glance. eg. *Juniperus communis* 'Hornibrook's Gold' or *Juniperus communis* cv. Hornibrook's Gold.

To this extent the horticultural system 'rides on the back' of the botanical system, but in no sense does this mean that the use of cultivar names is an extension of botanical nomenclature. It is no such thing. The three infra-specific ranks, Subspecies, *varietas*, and *forma* are a true extension of the botanical system into (botanically speaking) minor matters, but a cultivar name, and cultivar

Figure 14. A corner of the conifer collection at the U.S. National Arboretum, Washington, D.C., U.S.A.

groupings of any kind are not.

The *Cultivated Code* (Article 10, Note 3) states that the concept of cultivar is essentially different from the concept of *varietas*. It could well have added the words 'and of any other botanical category', since any such must always be (in everyday language) a 'kind' of plant, distinguished from other 'kinds' as the result of botanical classification, whereas a cultivar, whether it consists of a single clone or a thousand, is always a single entity looked at through horticultural eyes (the result of horticultural selection). It follows from this that the circumscription of a cultivar, however closely in any particular case at any particular time it may appear so to be can therefore (because of this essential difference) never be the circumscription of a *forma* (nor of any botanical taxon) because, on the one hand, the latter concept must include individuals either existing or which may be found in the future which do not or will not fall within the circumscription of the cultivar and because, on the other hand, a fresh selection giving rise to a new cultivar may at any time be made from within the *forma* (or as the case may be). A familiar case is *Juniperus squamata* var. *meyeri* which was recognised botanically on the obvious assumption that a population somewhere in the wild might one day be discovered. But it is now usually considered to be a cultivar, *Juniperus squamata* 'Meyeri' and has given rise by mutation to several new cultivars such as 'Blue Star' and 'Blue Carpet.' Where a cultivar consists of "a group of similar clones" the distinction becomes difficult to follow, but always the element of selection must be present (otherwise the concept cultivar is lost and the botanical name is adequate). The statement that "a cultivar may, on occasion, be co-extensive with the botanical category under which it is classified" is a conceptual absurdity, and legalising an absurdity does not make it any less absurd.

In the case of long-lived trees such as conifers the situation is simplified by the fact that cultivars usually consist of a single clone and a GREAT ADVANTAGE FOLLOWS FROM KEEPING IT THAT WAY.

The *Cultivated Code*, like the *Botanical Code* before it, is mainly based on the two concepts, 'priority' and 'valid publication' of a name. The detailed definitions of each term should be studied in the *Code* itself, but in broad terms, covering most situations, priority means that the chap that 'got in first' with his name wins and valid publication usually boils down to 'found its way into print'.

LATIN-FORM CULTIVAR NAMES

Before the advent of a *Cultivated Code* it was traditional to name garden conifers, not by the use of a fancy name such as was usual even then amongst many other garden plants, but by adding on one, two or even three Latin-form names to the specific name. Although this is no longer permitted, the *Code* states that any such Latin-form name validly in use before 1st January 1959 for a plant subsequently considered to be a cultivar is normally to be retained as its cultivar name. These names can, however, sometimes cause misunderstanding and confusion to the unwary in the following ways.

Firstly, gardeners who know that all botanical names are in Latin form and that cultivar names nowadays must be 'fancy' ones may not recognise that many of these older garden forms with their Latin-form names are cultivars and that the names are correctly so used, provided that the new typographical rules are followed. (Thus, in the case of Lord Clanbrassil's spruce, *Picea abies* 'Clanbrassiliana' and *Picea abies* cv. Clanbrassiliana are both correct, *Picea abies clanbrassiliana* and *Picea abies* var. *clanbrassiliana* are not.)

Secondly, a study of the work of many early writers who gave names to conifers makes it clear that they were introducing the plants as cultivars and that their intentions were horticultural rather than botanical and that when they used the terms 'var.' and 'f.' they were using them as abbreviations of the English words variety and form (or their equivalents in several European languages) rather than of the Latin words *varietas* and *forma* as used by botanists. To cite a few cases – Murray Hornibrook was undoubtedly using the term 'var.' in this way. The French writer Carrière heads his lists of variants – 'Varietés horticoles'. The German writer Beissner makes the same distinction clear by using the abbreviations only for botanical taxa.

Thirdly, with only a *Botanical Code* available at the time, attempts were made to regulate these names within this now-seen-to-be-inappropriate Code and this gave them a spurious botanical status. One writer to do this was Professor Rehder of the Arnold Arboretum of Harvard University, Jamaica Plain, Massachusetts, U.S.A. Making the obvious choice amongst the botanical categories available to him, the Professor sought to reduce the confusion in what he called the "clonal garden forms" by listing them all, uniformly, as *formae*.

With hindsight, now that we have a *Cultivated Code*, this is now generally regarded as having been unfortunate. For instance, the *Kew Handlist of Coniferae*, 5th Edition, 1951, says of the coniferae: "The general practice was to bestow Latin, rather than fancy names on the conifers, so that under species such as *Chamaecyparis obtusa* and *Juniperus chinensis* quite considerable numbers of varieties are listed which prove on examination to be cultivars and not variations of botanical standing... cultivars should... no longer be allowed to masquerade as botanical entities."

Botanists will say 'yes, it is unfortunate, but these combinations were published so are part of botanical history and so must remain with us'. I have, however, elsewhere (*Taxon* **27** (2/3): 187–190, May 1978) drawn attention to the fact that in many cases the requirements of the *Botanical Code* regarding valid publication of these *formae* were not met, being supported only by a reference to a description of a cultivar (a cultivar by definition is not a taxon) which lacked the necessary botanical intention on the part of the original author. We can fall back on the logical principle that as cultivars these plants began, cultivars they are now considered to be, CULTIVARS THEY HAVE BEEN ALL THE TIME.

The simple wording of the clause in the

Cultivated Code quoted above about a plant "subsequently considered to be a cultivar" tends to obscure the fact that such 'consideration' involves a decision based on a botanical judgment of exactly the same kind (and with equally important effects) as the earlier decision to admit the plant to a place in the botanical system. It involves a reversal of that now-seen-to-be-regrettable decision. Nomenclaturally, as with all changes, it reduces that name to synonymy. Both names cannot acceptably co-exist. This need for a decision is the crux of the problem because it requires each case to be examined separately. Fortunately, the majority of these Latin-form names relate to plants long since lost to cultivation – often to a solitary tree, unlikely ever to have been propagated – and in such circumstances the exercise is merely academic, but there remains a considerable number over which each writer since 1st January 1959 has had to make up his own mind. The first of these writers was Krüssmann who in the third edition of *Die Nadegehölze* (1960) listed a large number as being cultivars. This lead was followed by den Ouden and Boom in *Manual of Cultivated Conifers* five years later and by most writers on conifers since.

A large number of obsolete and obsolescent names in this Manual are transferred to Appendix One in Volume Two. Each remaining Latin-form epithet has been examined with a view to ascertaining the intention of its author, whether this was horticultural or whether it was botanical. Lacking facilities for research in primary documents and into the academic histories I have had to choose a practical rather than an academic path. Where a name was either derived from a recorded incident or from a personal or place name such as 'Maxwellii' and 'Beuvronensis' or – in contrast – was recorded as appearing sporadically or in some particular locality no problem arises, but if it consists of a descriptive such as *pendula, fastigiata* or *nana* I have had to consider, case by case, whether the description (although the name was given to a clone) ALSO fits a 'kind' of variant that is of sufficient botanical interest to justify continued recognition as *forma*. When in my opinion in the situation there are now two entities, I must also retain the name for the clone to which it was first given. For example, under *Abies alba* I must list 'f. *pendula*' and also the cultivar 'Pendula', i.e. the clone named by Carrière in 1867. Another and quite legitimate use of such Latin adjectives is in listings of cultivars for special horticultural purposes.

AUTHORITIES

To secure accuracy and precision the *Botanical Code* requires the citation of a botanical name to be followed by the name of the author responsible. (e.g. *Abies mariesii* Masters.) The *Cultivated Code* does not mention this, neither requiring nor forbidding it, but since precision is at least as important in horticulture as it is in botany (and when commercial considerations arise, possibly more so) the citation of authors' names is becoming quite usual with horticultural writers.

TYPES AND PROTOLOGUES

The use of authors' names (often referred to as 'authorities') permanently ties the name given to a plant to the name of the coiner, inventor or finder who gave it. There remains a further problem – how to determine, in years to come, which particular plant owns the name.

In an ideal world this could be done by an undying specimen plant cared for by an undying curator who used indestructible labels. In a real world, the botanical system relies on the use of types. A 'type' in this connection is an actual piece (sometimes a tiny scrap) of what was once the living plant used by the author when raising a name for it. Type was perhaps an unfortunate word to have chosen because there is no connection between the word 'type' in this context and 'typical' in its everyday meaning. The 'type-specimen' in the herbarium may or may not be typical, in the sense of being characteristic, of the species (or of whatever other category the publisher of the name considered appropriate at the time.

The type is irrevocably associated with the name. Species do not have types: it is names that have types, so whatever changes in botanical classification may later occur, the author, the name he coined and the type stay together through thick and thin.

Unfortunately horticulture cannot rely on this system. Frequently the distinctions between cultivars (habit, hardiness, resistence to disease and so on) would not show up in a herbarium type

Figure 15. A group of *Chamaecyparis lawsoniana* 'Pottenii' at one time in the Westonbirt Arboretum in Gloucester, U.K. Attractive when young, with a fastigiate habit and soft, mid-green foliage, this cultivar loses shape (often as the effect of snow) and is seldom seen in maturity (10).

specimen. The nearest approach to a fool-proof (if that is an appropriate word to use) system is what is called a *Protologue* (From Greek, *protos*, first, and *logos* word or discourse). This term is defined as "Everything associated with the name at its first publication, i.e. diagnosis, description, illustrations, references, synonymy, geographical data, citation of specimens, discussions and comments".

Anything not clearly stated in the Protologue (or charitably to be inferred therefrom) must be suspect. However well intentioned, anything added at a later date must be open to argument.

It does not need much intelligence to see that failure here in the past is the source of all the confusion in the nomenclature of garden conifers that plagues us today. The protologue for *Cryptomeria japonica* 'Kusari-sugi', for instance, consists solely of the words in Dallimore and Jackson, "A dwarf form with clustered branches", and of 'Husari-sugi', all we are told is – "Branches curiously bent". It is true that to achieve valid publication the *Cultivated Code* requires a 'description', but this need only be perfunctory. Frequently it consists of nothing more than the description in the introducing nurseryman's catalogue, savouring (as such publications tend to do) much more of an advertisement than of a factual description.

Nothing less than a system that will give authoritative answers to BOTH the following questions will prevent confusion continuing indefinitely.

(1) We have a plant: what is its correct name?
(2) We know the name: have we the right plant?

The authors of the *Manual of Cultivated Conifers* (published in 1965) claimed although not with complete success) to have included and documented all the names of cultivars published since 1753, making the book a very useful working tool for everyone concerned with this important group of plants.

Each of the authors of the manual referred to was well equipped for the task. Peter den Ouden had been systematically collecting conifer data throughout his long life as a nurseryman, his efforts having already resulted in his *Naamlijst van Coniferen* published in 1937 (which was adopted as a standard for varietal names at the International Horticultural Congress in Berlin in 1939) and also his book *Coniferen Ginkgo en Ephedra* (in Dutch), published in 1949. Dr Boom, as the government taxonomist in Holland, had been recording data regarding new conifers as part of his official activities. So the book they were working on could at that time be readily compiled from available information. Dr Boom did however, comment even then on the number of novelties being introduced, often inadequately documented, and since 1965 the number of new introductions has proliferated in every part of the world where conifers are grown. Names have been given to novelties with scant regard to their real horticultural value and often with no regard at all to legitimacy or the need for documentation, and names have been pirated, changed and re-used without permission.

So little wonder that widespread confusion has developed. And, of course, it will continue unless and until we mend our ways.

THE WORLD CONIFER DATA POOL

Unfortunately this particular contribution made by Dr Boom was not continued after his retirement and since then there has been little if any systematic work done in this field save by one or two private conifer enthusiasts.

In 1963, following a visit from Dr Boom to my Wansdyke Nursery at Devizes, Wiltshire, U.K. I started systematically collecting conifer data. At first this was limited to the dwarf forms but as time went by the need for someone to undertake a complete coverage became so evident that I founded the World Conifer Data Pool (WCDP). This is a project having as its objective the collection and handling of data relating to ornamental conifers used in horticulture throughout the world.

These data are now scattered in books, periodicals, nurserymen's catalogues, human memories and a variety of other sources in many countries. WCDP is intended to become an on-going source of reference material available to everyone. Such an objective can only be attained by the 'pooling' of knowledge by a very large panel of Contributors. So the title 'Data Pool' was preferred to the usual 'Databank', and with a view to spreading the data-capturing net as widely as possible, membership is free and is open to botanical and horticultural institutions and members of their staff, including the librarians, to societies and associations interested in conifers, to nurserymen, amateurs and anyone else, old or young, interested in ornamental conifers and their nomenclature. The only qualification is an interest in what is being done and a willingness to help the good work.

The data are most conveniently collected by the use of the form reproduced on pp. 60 and 61. No copyright is claimed for this form: it may be freely reproduced. Because the form is intended to cover all cases it looks more complicated than is really the case, since only such information as is available in each case can be recorded. The remainder of the form remains blank unless and until further information comes to light.

It is hoped that there will be a very widespread response to this appeal for collaboration. If nurserymen or others who have introduced new cultivars in the past would take this opportunity of putting the facts on permanent record, and if anyone introducing a new one would take advantage of the scheme a new era would slowly dawn. Communications should be sent to me at the address given on the form. Information as to new introductions received by WCDP will be passed on to the International Registration Authority for registration of the name.

CONIFER NAMES REGISTRATION

In March 1963, a number of eminent botanists

met at my request to consider and advise me on the detailed application of the (then recently revised) *International Code of Nomenclature for Cultivated Plants* to the particular problems of garden conifers. The meetings took place at the British Museum and were chaired by the late Mr W. Gilmour, Director at the Cambridge University Botanic Garden and at that time Chairman of the International Commission for the Nomenclature of Cultivated Plants of the International Union of Biological Sciences.

At one meeting it was suggested that the appointment of an International Registration Authority was desirable for this group of plants and Mr Gilmour was asked to ascertain the views of the Royal Horticultural Society and in due course (in 1964) the Society was appointed International Registration Authority for Ornamental Conifers. Dr Boom submitted to the Committee a draft Register based on the names and data to appear in the *Manual of Cultivated Conifers*, at that time in preparation.

In 1972, at my suggestion, Dr Boom's list was circulated for comment to a large number of bodies and individuals concerned with conifer nomenclature and a start was made on the extraction of data from existing reference books, but the list itself was never published, so the book *Manual of Cultivated Conifers* when it appeared was, and for many years has remained, the only comprehensive check-list of conifer names available in the English language. And in the absence of any action to publicise the availability and to stress the importance of the registration of new cultivar names this has never 'caught on' in the conifer world. As a result, throughout nearly a quarter of a century new conifer cultivars have been introduced and names have proliferated without any guidance or control, or even (in a large number of cases) any regard to the requirements of the

Figure 16. A corner of the Pygmy Pinetum at Devizes, Wilts., U.K.

International Code of Nomenclature for Cultivated Plants.

Following my retirement from business in 1979 the Society enquired as to my willingness to help and in 1981 appointed me Registrar. All my records were made available and progress was stepped up, but in 1983 I resigned and handed the task of preparing a Register over to the present Registrar, John Lewis, a professional botanist for many years at the British Museum. I understand that the Register is to be issued in eight or ten parts and that it is hoped to complete the work before the end of the century.

Part One, covering *Abies* to *Austrotaxus* is marked by a very high standard of scholarhsip with perhaps a somewhat botanical bias. It will be a very useful tool for research workers in this field, but the arrangement of the names alphabetically within each genus (necessary for the special purposes of a Register) renders it difficult to adapt for general horticultural purposes, which require cultivar names to be listed alphabetically within each species.

I understand that the Registration Authority have made the following rulings.

(1) The cultivar class (See Art. 50 of the *Cultivated Code*) within which a name shall not be used for more than one cultivar shall, in the case of three genera *Chamaecyparis,* × *Cupressocyparis* and *Cupressus*, be that group and in every other case shall be the genus.
(2) The starting point (Art. 44) for the valid publication of cultivar names shall be *Miller's Dictionary*, Ed. 6, (1752).
(3) The re-use of names within a cultivar class will not normally be permitted. (Art. 48).
(4) Names contrary to the Articles of the *Cultivated Code* will not be accepted, and names contrary to the Recommendations will be strongly disfavoured.

Registration, unlike a grant of plant patent or plant variety rights, does not confer on the registrant any commercial advantage. Unless and until registration of new conifer names becomes normal practice in the horticultural trade the only advantages of registration (and this only provided it is done before commercial distribution is commenced), would seem to be that the applicant knows that the name he proposes is not already registered within same cultivar class and that he has the assurance that his name is safe from being registered by anyone else for a different clone.

Here it might be opportune to issue a warning against using an old name taken from a book for what is a new selection. (This is technically known as 're-typification'.) In most cases names consisting of descriptive Latin adjectives such as *pendula*, *fastigiata* or *nana*, given in the days when such trinomials were allowed, were no less clonal in character than personal or place names such as 'Clanbrassiliana' and 'Beissneri'. Some such names have, regrettably, over the years come into use loosely to cover all clones embraced by the adjective in question and have thereby lost all value save as a collective name. But it is highly undesirable for this practice to be extended. With long-lived trees such as conifers the ideal is to restrict the concept 'cultivar' to a single clone. This means validly publishing a new name for each new selection that is to be introduced. And the time to do this is BEFORE ANY DISTRIBUTION IN THE TRADE IS COMMENCED. An objection that is sometimes raised to this suggestion is that some introductions fail to become popular and are 'dropped', leaving an abortive name on the list to cause confusion later. But any such case can be suitably endorsed in the listings, and the practice here advocated would prevent any development of the much greater potential for confusion that rears its head as soon as any new introduction is distributed (private circulation for evaluation purposes is here excluded) without a properly documented name.

Figure 17. *Picea abies* 'Pygmaea' and *Tsuga canadensis* 'Pendula'. A nice foliage contrast.

Figure 18. A young planting of prostrate *Picea abies* cultivars that soon needed re-planting as the plants grew.

RECORD OF CONIFER DATA ★ PART ONE

Please supply answers to whatever questions you can.

The WORLD CONIFER DATA POOL

1. NAME OF PERSON OR INSTITUTION SUPPLYING THE DATA

2. ADDRESS

Signature | Date

3. GENUS | 4. SPECIES [1]

5. CULTIVAR NAME [2] | 6. Has any different name ever been suggested or used?

7. ORIGINATION OF THE PLANT [3]

"How" — Originated as:- Seedling [4] | Mutation | W/broom | Other (?)

"Where" — Location or address [5]

"When" — Full date, or at least the year, if known. 19

"By whom" — Name and address of raiser/finder

8. FIRST VALID PUBLICATION OF THE NAME (i.e. the first appearance "in print" of the name with a description.)

"Where" — Name of book, periodical or Nurseryman's catalogue [6]

"When" — Supply page no. | and date 19

"By whom" — Name of author, editor or owner (as the case may be)

9. INTRODUCTION TO THE TRADE / OFFICIAL DISTRIBUTION [7]

"By whom" — Name and address of the nursery or Institution

"When" — Earliest catalogue appearance or known distribution [8] | Page no. | Date 19

10. DESCRIPTION The description need not be technical but should be factual and in some detail. State habit, vigour, leaf size and colour, shape of buds etc, and add any known details of origin, personal anecdotes etc. worth recording for future interest. Important would be how it can be distinguished from related cultivars. Continue on plain paper.

NOTES

1. If a botanical "sub-species", "var." or "f." is involved, please include this information.
2. Any new name must comply with the requirements of the *Cultivar Code.* Any name raised since 1st Jan. 1959 not so complying is illegitimate and MUST be replaced.
3. Questions should be answered as fully as the information that is available (or can be ferreted out) permits.
4. If raised from seed, the parentage or the origin of the seed to be recorded, if known.
5. Full postal addresses should be given.
6. If the answer to Question 8 ("Where") is "Nursery catalogue", the answers to Question 9 become vital.
7. The earliest known commercial use of the name is usually the one to be recorded.
8. Please supply a copy, if available, for the WCDP archives, or a photocopy of the cover and relevant page/s.

Until a regional organisation has been developed in each country outside U.K., please send completed forms to the Co-ordinator, "Treetops", Buzzacott Lane, Combe Martin, Devon, EX34 0NL, U.K.

COMPUTERISATION OF DATA ★ PART TWO

Note. It is of vital importance NOT to complete this Part unless identity of the plant being described with the name in Part One can be established beyond all reasonable doubt and argumentation.

Because of the absence (and, now, inaccessibility) of data relating to obsolete names, no comprehensive data-base is possible. But for conifers in current production it should be possible BY A DETERMINED CO-OPERATIVE EFFORT WORLDWIDE to assemble the requisite data. Please signify your willingness to help us in this task by putting a tick here   or by writing to WCDP at the address given overleaf.

In computer jargon, any such database would have "fields" to receive the data in Part One (scientific name, author, date of valid publication, origination, etc) and also the following "fields", for a coded description of the plant and its capabilities.

Please fill in the fields using the symbols from the following codes. (See example at foot of page).

A-D	E-M	N-U	V-X	Z

1. Vigour	2. Habit	3. Foliage-colour	4. Foliage-type	5. Hardiness	
A Pygmy	E Prostrate, procumbent	N White or Cream	V Juvenile	Z,1	Hardy in Zone 1
B Dwarf	F Spreading, wider than high	O Grey	W Intermediate	Z,2	do. Zone 2
C Compact	G Squat or globose	P Green	X Adult	Z,3	do. Zone 3
D Species norm.	H Upwards spreading	Q Blue	Y Abnormal	Z,4,	do. Zone 4
	J Broadly conical	R Red		Z,5	do. Zone 5
	K Narrowly upright	S Brown		Z,6	do. Zone 6
	L Pendulous (true leader)	T Yellow		Z,7	do. Zone 7
	M Pseudo-pendulous (sprawling)	U Purple or mauve		Z,8	do. Zone 8
				Z,9	do. Zone 9
			(per U.S.Dept. Agric. map.)	Z,10	do. Zone 10

Age of plant

01 A young plant
02 A mature plant

Seasonal State

03 Spring
04 Summer
05 Autumn
06 Winter
07 Throughout the year (or remainder of year).

Colour shade

10 Pale
11 Medium
12 Dark
13 Variegated
14 Mottled
15 Striped, streaky

Colour intensity

20 Bright
21 Dull

Surface texture

30 Matt
31 Slight sheen
32 Glossy

Frequency, or amount present

60 None: never
61 Little or sparsely: rarely
62 Medium: sometimes
63 Predominantly: frequently
64 Entirely: always

A-D	E-M	N-U	V-Y	Z
B,01 C,02	F,01 J,02	P,03,11,20,32 P,07,12,21,30	X,64	Z,9

This would be the coded entry for a cultivar that as a young plant is dwarf, spreading wider than high, which becomes more upright in maturity (but still only ranking as "compact" for size); and carries wholly adult foliage of a medium green colour, bright and glossy in Spring, soon darkening and becoming dull green with a matt surface for the rest of the year. It is reliably hardy in Zone 9.

6. Identification

KEYS

The intention of a key is to enable its user to establish the identity (genus, species, etc.) of plants covered by the key.

The basic idea is disarmingly simple. Keys always take the same form: the user is presented with a series of selections to make, one by one, each choice leading him to the next. Older keys often required a choice to be made from amongst several alternatives: in modern keys (as befits the days of computers and binary notation) the choice is always between two possibilities. Such keys are termed dichotymous: they may be simpler but they are indisputably longer.

The process might be likened to a small animal climbing from the ground to his nest at the top of a tree. He must make a choice each time he comes to a fork, and should the poor little fellow make even a single wrong choice he will obviously never reach home. Using a key carries the same hazard and this is the reason that keys are of very little use to the inexperienced beginner but increase in value as the standard of knowledge of the user rises.

The compilation of a key is normally quite straightforward, but in the case of the cultivated conifers it is extremely difficult as the result of several factors. Firstly, botanical classification is primarily based on the reproductive part of plants (the flowers and fruit). In the case of conifers these can be either absent, (this is always the case in young trees), or they may occur only at the top of older trees so be inaccessible from the ground. Deprived in this way of his principal weapon, to produce any kind of usable key the compiler has to fall back on variations in the foliage and often must rely on other considerations that may well cross botanical boundaries.

Even so, because foliage characteristics vary with the age and vigour of the tree, (the latter often due to climatic conditions and nutritional state) and also always differ considerably at different levels within the same tree (this can readily be confirmed by looking over a fallen monster) it is sometimes virtually impossible to give an unequivocal answer to questions posed in a key. Moreover, in a garden or arboretum there is always the possibility that the tree under examination is a hybrid or an example of an infra-specific variant or a cultivar. These last are by definition variants selected and cultivated BECAUSE of that variation, which can be far more noticeable than the differences between species.

From practical considerations no key can be exhaustive, covering every variation, and any one of the above factors can render any key virtually useless.

Because of the difficulties referred to above, any attempt to produce a workable key of such a large group as the conifers must result in a lengthy, cumbersome and often bewildering affair, largely duplicating the information in the descriptions supplied elsewhere. So, instead of keys, simple 'Guides to Identification of Species' will be found on pp. 63–89, the intention being that these should provide the searcher, stage by stage, with pointers to one or more possibilities, progressively narrowing the search and leading him on to the full species descriptions given later on in this volume and in Volume 2, saving him on the one hand from wild guesses and on the other from a tedious and possibly unrewarding search.

But whether 'keys' or 'guides', the assistance of someone with a knowledge of conifers greater than that possessed by the searcher is greatly recommended. A key alone should never be relied on in any situation where an authoritative identification is important.

GUIDES TO IDENTIFICATION OF SPECIES

The following guides are based as far as possible on the habit of the tree under examination and its trunk and branch-system and on the foliar parts on side branches available within easy reach. Identification is always much easier and more certain if the flowering parts and/or cones are available, and in some cases it is impossible in their absence. Because these guides on the one hand and botanical classification on the other involve quite different criteria, apparent inconsistencies will appear, some names turning up in more than one Group. This is especially true of genera containing much diversity, such as *Cupressus, Juniperus* and *Podocarpus.*

Method of using the guides

The guides are to be used to narrow down the field of search progressively, stage by stage, as follows.

Stage 1. From a quick inspection of the tree and its foliage provisionally select the most promising Main Group from p. 64.

Stage 2. Turn to 'Go to' Listing on the page indicated and from the genera or species there listed reject the alternatives that are clearly inapplicable to the tree being identified.

Stage 3. Finalise the search by checking out the remaining 'possibles' one by one from the full descriptions of the genera and species, making use of the 'Guides to Identification of Species' there provided for several of the larger genera.

If no clear identification results, look again at the Main Groups. The descriptions given are necessarily brief and to some extent arbitrary, so anomalies occur and border-line cases are frequent.

It must be borne in mind that plants in garden cultivation may very well be cultivars, that is to say, variants from the norm and selected on that account. Variation within a species can often be greater than the normal differences between similar species. It is therefore impossible to 'key out' such plants.

The twelve main groups

Main Group One. Trees, conical when young, more or less broad-crowned in maturity. Leaves needle-like in clusters of 2, 3, or 5 (rarely 6 or 7), segments together making a circular cluster. Go to p. 65

Main Group Two. Trees, narrowly conical when young, broad crowned or flat-topped in maturity. 2nd year shoots woody. Leaves thin or needle-like, arranged in two ways, (1) Singly, emerging spirally and widely spaced on long shoots, and (2) in clusters of 10–30 on short spur-shoots.
Go to p. 65

Main Group Three. Large trees, narrowly conical when young, broadly so in maturity. 2nd year shoots are woody. Leaves more or less strap-shaped (parallel sided), emerging singly from the shoot in a spiral arrangement (but on side-branches misleadingly twisted so as to appear to be in opposite ranks, like a fish-bone); leaving scars on the shoots when they fall. Go to p. 66

Main Group Four. Trees or large bushes. Leaves all as in Main Group Three, but 2nd year shoots remain green. Leaves hard or leathery.
Go to p. 69

Main Group Five. Large or medium-size trees. 2nd year shoots green. Leaves emerging singly, spirally arranged on shoot; decurrent at the base; free-standing but loosely covering the shoot; often incurved and four-sided or triangular in cross-section, and noticeably tapering from a broad base towards a pointed apex. Go to p. 62

Main Group Six. Large or medium trees or large shrubs. 2nd year shoots green. Leaves large to very large, lanceolate, ovate-lanceolate or ovate.
Go to p. 74

Main Group Seven. Large to medium size trees, erect when young, abundantly branched, with dense foliage, the branchlets usually lying in flat sprays. Leaves scale-like in opposite, decussate pairs, the facial pair flattened, the lateral pair folded, lying smoothly against the shoot with only the tips free; the buds being completely hidden.
Go to p. 75

Main Group Eight. As Main Group Seven, but branchlets spreading around the main shoot, i.e. not in flat sprays. The distinction between this Main Group and Main Group Seven is not absolute; border-line cases occur. Go to p. 80

Main Group Nine. Unusual-looking deciduous tree with wedge-shaped, fan-veined, leathery and mainly two-lobed leaves on long stalks. 'The Maidenhair Tree'. Go to p. 83

Main Group Ten. Large to medium trees or spreading to prostrate bushes. Foliage in whorls of three or four. Go to p. 83

Main Group Eleven. Trees or bushes bearing cladodes (modified branchlets masquerading as leaves, the true leaves being reduced to scale-like dimensions). Non-woody second year.
Go to p. 85

Main Group Twelve. Prostrate, procumbent, spreading or rounded shrubs. Leaves in pairs or threes, either scale-like and decurrent, covering the buds, or awl-like (a persistent juvenile state). In some species, both are often present in the same plant. It must be borne in mind that a young plant or dwarf cultivar of any arboreal species may well appear to fall into this-group. Go to p. 86

'GO-TO' Listing for Main Group One

"Trees, conical when young, more or less broad-crowned in maturity. Leaves needle-like, in clusters of 2, 3 or 5, (rarely 6 or 7), segments together making a circular cluster."

Note that the drawings differ widely in scale

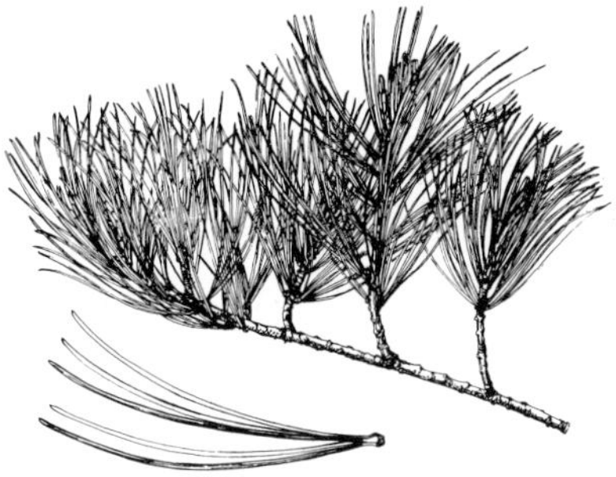

5 needle pine (*P. parviflora*). (5)

2 needle pine (*P. sylvestris*). (5)

1.1. This group consists of a single genus, recognised readily from its well-known 'pine needles'.
See Vol. 2. ***PINUS***

'GO-TO' Listing for Main Group Two

"Trees, narrowly conical when young, broad-crowned or flat-topped in maturity. 2nd year shoots woody. Leaves thin or needle-like, arranged in two ways, (1). Singly, emerging spirally and widely spaced on long shoots, and (2) in clusters of 10–30 on short spur-shoots."

2.1 Leaves thin, rigid and sharp; evergreen. Spur shoots roughened. Cones large, erect, falling at maturity.
Turn to p. 178. ***CEDRUS***

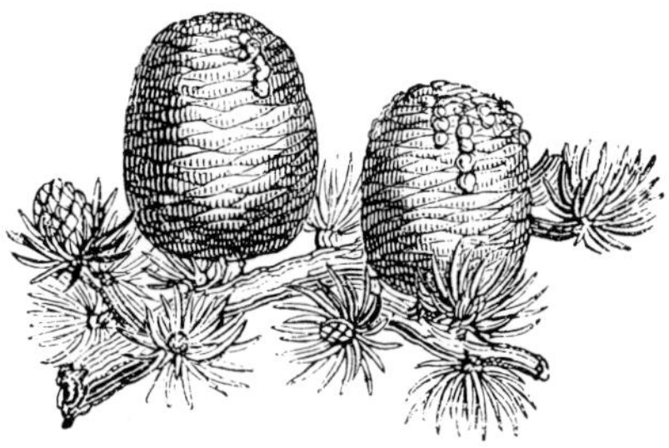

Cedrus (*C. libani*). (8)

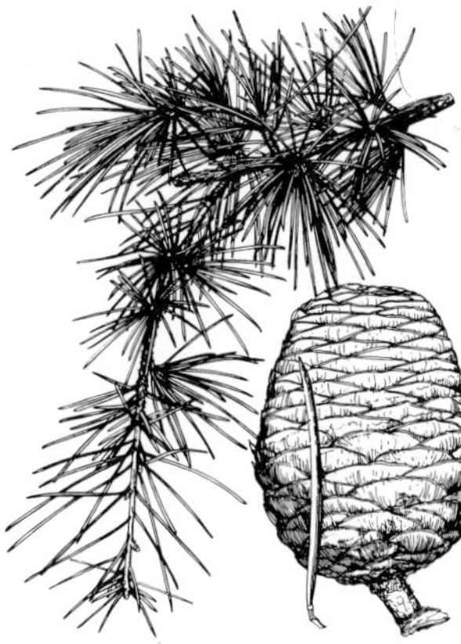

Cedrus (*C. deodara*). (5)

2.2 Leaves thin, narrow and flat; soft to the touch; deciduous. Cones sub-globose and much smaller. Male flowers solitary.
Turn to p. 405. ***LARIX***

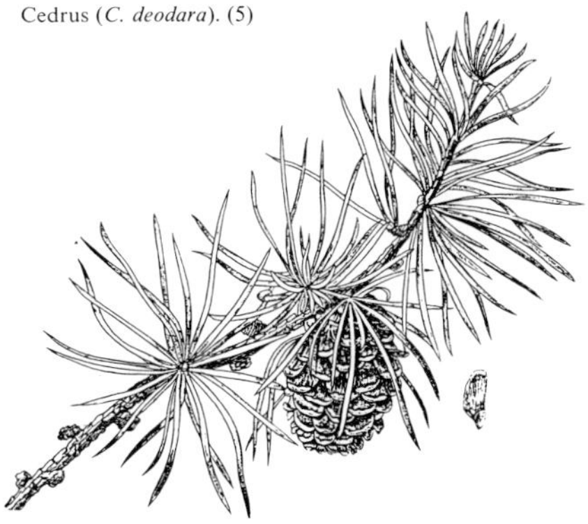

Larix (*L. leptolepis*). (2)

Note that the drawings differ widely in scale

Pseudolarix. (5)

2.3 Deciduous. leaves tufted as in *Larix* but relatively broader, borne on long, curved spurs with distinct annual rings. Cones ovoid, with woody scales, ripening and disintegrating the first year, only seen on mature trees. Male flowers borne in clusters.
See Vol. 2. ***PSEUDOLARIX***

'GO-TO' Listing for Main Group Three

"Large trees, narrowly conical when young, broadly so in maturity, 2nd year shoots woody. Leaves more or less strap-shaped (parallel-sided), emerging singly from the shoot in a spiral arrangement (but on side-branches misleadingly twisted so as to appear to be in opposite ranks, like a fish-bone); leaving scars on the shoots when they fall."

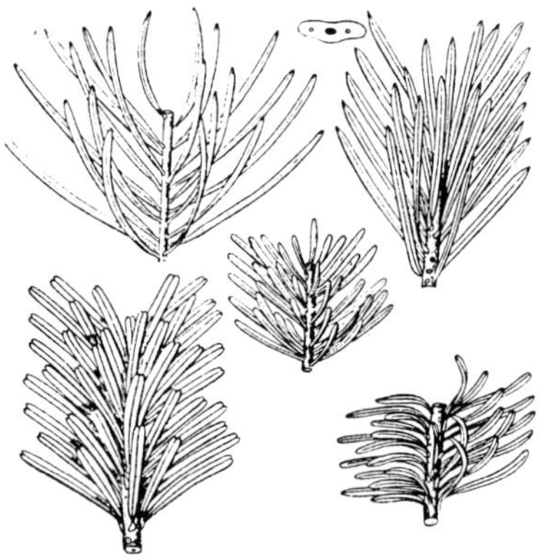

Abies (*Various species*). (12)

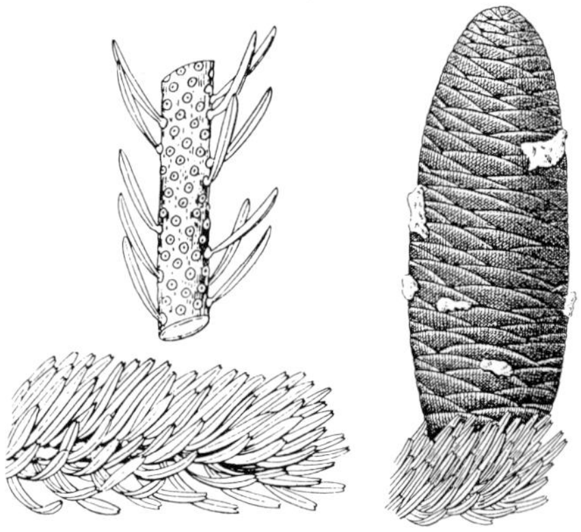

Abies (*Various species*). (12)

3.1 Branches whorled. Leaves with distinct resin canals, keeled below only; leaf scars circular. Cones held stiffly erect, disintegrating in early Autumn as soon as ripe. Leaving the central core on the tree.
Turn to p. 92. ***ABIES***

Keteleeria. (5)

3.2 Foliage similar to *Abies* (but less hardy), with the same circular leaf scars, but with a raised mid-rib on both sides of the leaf. Also the cones differ by falling to the ground whole when ripe. It is very rare in cultivation.
Turn to p. 404. ***KETELEERIA***

Note that the drawings differ widely in scale

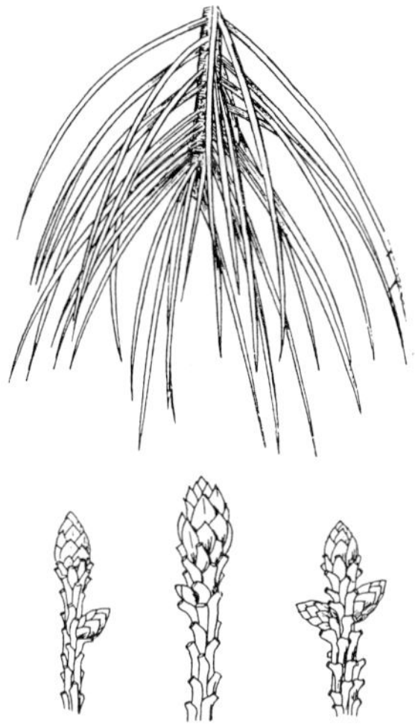

Picea (*Various species*). (12)

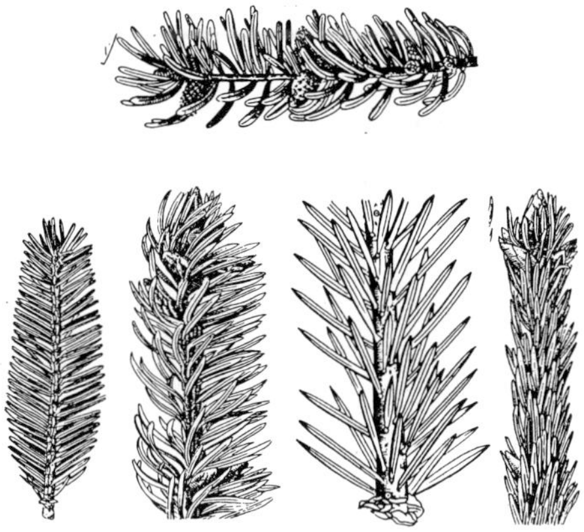

Picea (*Various species*). (12)

3.3 Leaves flattened or four-sided (according to the species), arising from peg-like projections which remain on the shoot at leaf-fall. The characteristic arrangement of the leaf-spray is similar to *Abies* but the leaf-tip is sharply pointed, never rounded or notched.
See Vol. 2. ***PICEA***

Pseudotsuga. (12)

3.4 Habit and foliage similar to *Abies* but leaf-scars are less noticeable, oval and slightly raised. Leaves very soft, aromatic when bruised. Buds reminiscent of the beech tree, *Fagus sylvatica*, sharply pointed, glossy brown. Cones hang downwards with bracts projecting, ripening in one year.
See Vol. 2. ***PSEUDOTSUGA***

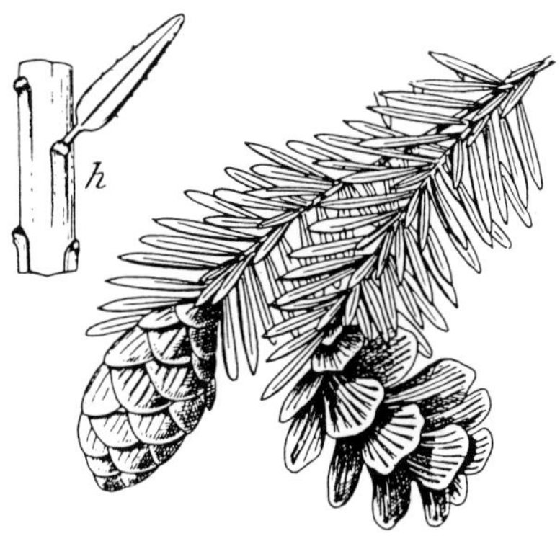

Tsuga (*Ts. heterophylla*). (12)

3.5 Main branches whorled, (save in *T. mertensiana*). Leaves short, flat, blunt-ended and of different sizes on the same shoot, held on a slender stalk which lies pressed against the shoot. Cones very small, hanging from the branch-tips.
See Vol. 2. ***TSUGA***

Note that the drawings differ widely in scale

Cunninghamia. (2)

3.6 An unusual-looking tree with leaves held in more-or-less two ranks but of triangular shape. Listed in Main Group Six. Turn to p. 279.
CUNNINGHAMIA

Metasequoia. (10)

3.7 Deciduous tree with branchlets of two kinds as in *Taxodium* (See 3.9 below). Deciduous leaves light green in opposite pairs (i.e. not alternate), sessile; with two stomatic bands above; perennial branches becoming brown and corky the 2nd year. Cones pendulous on long stalks.
Turn to p. 421. ***METASEQUOIA***

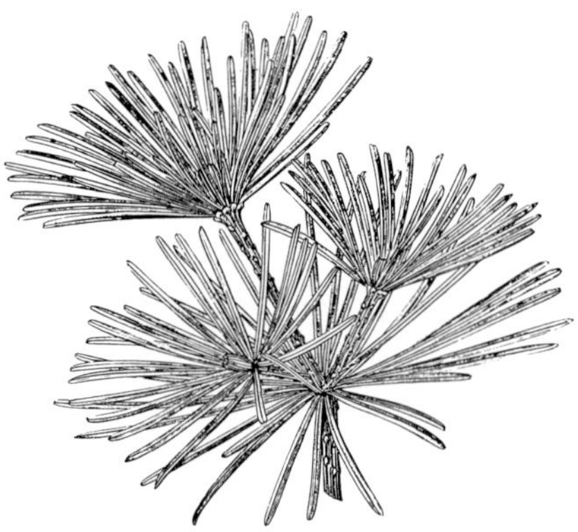

Sciadopitys. (14)

3.8 Curious tree with paralled-sided 'leaves' radiating like the spokes of an umbrella. Here listed in Main Group Eleven.
See Vol. 2. ***SCIADOPITYS***

Note that the drawings differ widely in scale

Taxodium. (2)

3.9 Deciduous. Branchlets of two kinds. Leaves on annual branchlets (which fall with the leaves in autumn) lie in an apparently pectinate and alternate arrangement; stomata inconspicuous. Perennial branches turn brown and corky the 2nd year. Old trees may develop 'knees', root-like growths above ground.
See Vol. 2. ***TAXODIUM***

'GO-TO' Listing for Main Group Four

"Trees or large bushes. Leaves more-or-less strap-shaped (parallel-sided), emerging singly from the shoot in a spiral arrangement (but on side--branches misleadingly twisted so as to appear to be in opposite ranks, like a fish-bone); leaving scars on the shoots when they fall. Leaves hard or leathery. 2nd year shoots remain green."

Cephalotaxus. (6)

4.1 Large shrub or small tree with branchlets opposite and large (up to 75 mm long) leaves lying flat on either side of the shoot or forming a V-groove and apparently 'opposite'. The upper side has a prominent midrib sunk in a depression, and 2 broad bands of stomata beneath. Bud scales remain after the branchlet has developed. The shoots continue a bright green for 3 years. The fruit is a large drupe, hence the name 'Plum Yew'.
Turn to p. 191. ***CEPHALOTAXUS***

Glyptostrobus. (4)

4.2 Upright deciduoud tree with branchlets of two kinds as in *Taxodium* (with which it is frequently confused) but the perennial branchlets remain green for several years and are marked by small dots.
Turn to p. 321. ***GLYPTOSTROBUS***

Note that the drawings differ widely in scale

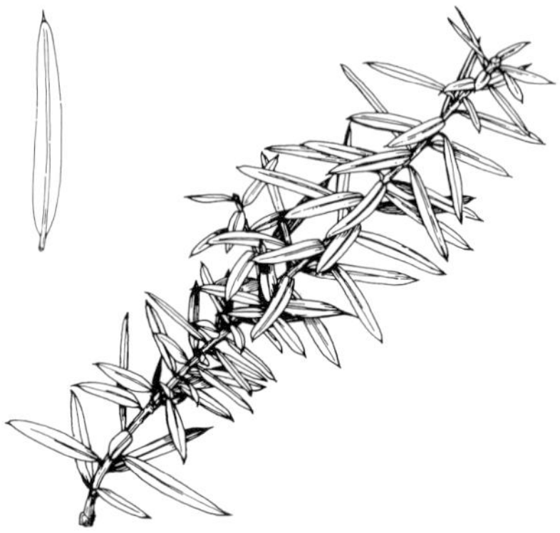

Podocarpus. (*P. totara*). (5)

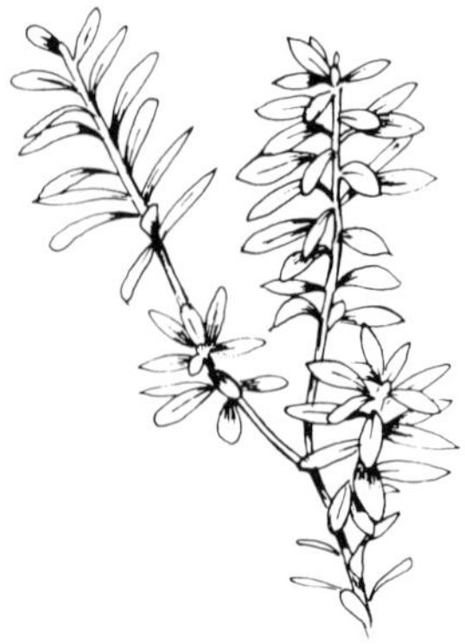

(*P. nivalis*). (6)

4.3 Small trees or shrubs (in cultivation) with branches alternate; leaves on short stalks; usually parallel-sided and sharp-pointed; soft or of leathery texture. From a large and very diverse genus only a few species are reliably hardly in cool temperate areas. (See also Main Groups Five, Six and Twelve).
See Vol. 2. ***PODOCARPUS*** (*in part*)

Prumnopitys. (6)

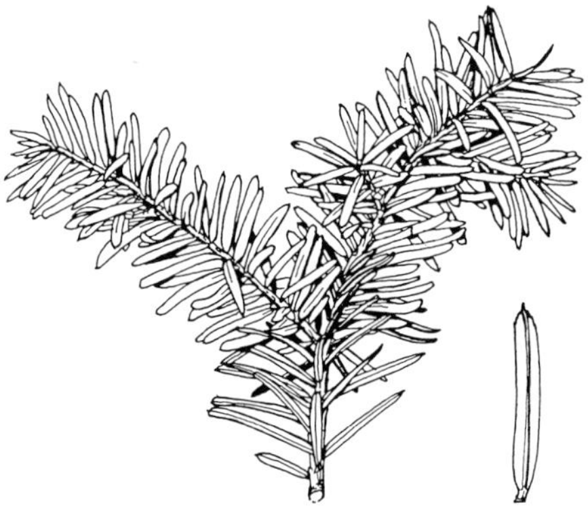

Prumnopitys. (5)

4.4 Small tree or large shrub in cultivation, very similar to *Podocarpus*, under which it was for many years and by some authorities still is listed. Only one species is hardy in Britain. The branches are alternate; the leaves are short-stalked, irregularly arranged and widely spaced; convex and dull bluish-green above and weith two whitish bands of stomata below. The fruit is yellowish-white, shaped like a plum (hence the name 'Plum-fruited Yew').
See Vol. 2. ***PRUMNOPITYS***

Saxegothea. (5)

4.5 An uncommon tree resembling *Taxus*. Branches opposite or in whorls. Leaves short, indistinctly two-ranked; dark green above, paler, with 2 rows stomata beneath, slightly twisted; midrib prominent above and beneath.
See Vol. 2. ***SAXEGOTHEA***

Note that the drawings differ widely in scale

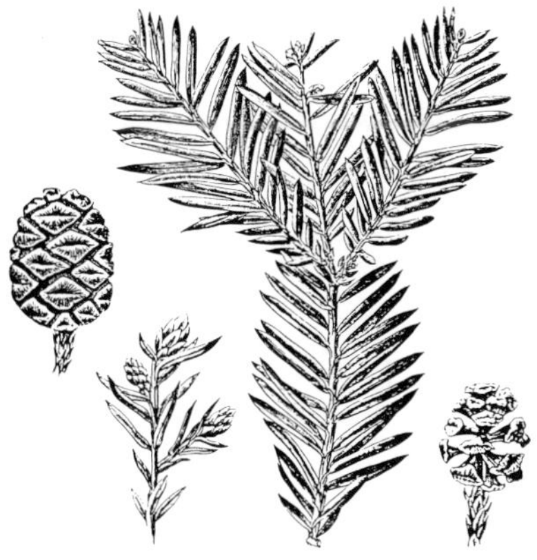

Sequioa. (2)

4.6 Tree with thick, spongy bark. Leaves dimorphic; on laterals they are more or less 2 ranked, decurrent, shortest at the base of the shoot; upper surface convex with central groove, dark green; lower surface with green midrib and two conspicuous stomatic bands. On main or flowering shoots the leaves are scale-like, arranged in several ranks, loosely appressed. Winter buds scaly; cones small, ripening the 1st year. It is commonly seen in Britain. See Vol. 2. ***SEQUOIA***

Taxus. (*T. baccata*). (12)

4.7 Large or medium-size tree, much planted in churchyards. Shoot ribbed; leaves distinctly two ranked, dark green and convex above with a prominent midrib; yellowish green beneath with two indistinct stomatic bands of stomata beneath. The buds are globular, small. Fruit is a scarlet glutinous aril open at the top, the well-known 'Yew Berry'. See Vol. 2. ***TAXUS***

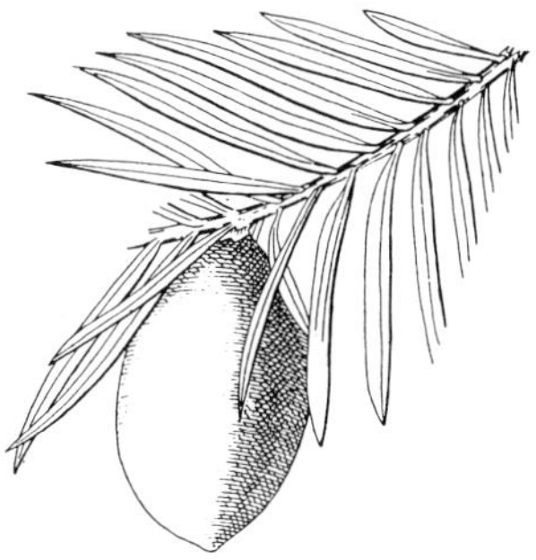

Torreya. (*T. californica*). (12)

4.8 Shrub or small tree with branches opposite or sub-opposite and with luxuriant foliage similar to *Cephalotaxus*, but the leaves are hard to the touch and are tipped with a fine, hard point. They are dark green and convex above and have 2 sunken, white, stomatic bands and a green midrib beneath. The fruit takes two years to ripen. See Vol. 2. ***TORREYA***

'GO-TO' Listing for Main Group Five

"Large or medium-size trees. 2nd year shoots green. Leaves emerging singly, spirally arranged on shoot; decurrent at the base; free-standing but loosely covering the shoot; often incurved and four-sided or triangular in cross-section and noticeably tapering from a broad base towards a pointed apex."

Note that the drawings differ widely in scale

Athrotaxis. (*A. cupressoides*). (5)

(*A. laxifolia*). (5)

5.1 Small tree with spirally arranged, small, imbricate leaves spreading and incurved or loosely or closely appressed. Uncommon outside large collections.
Turn to p. 171. ***ATHROTAXIS***

(*A. selaginoides*). (5)

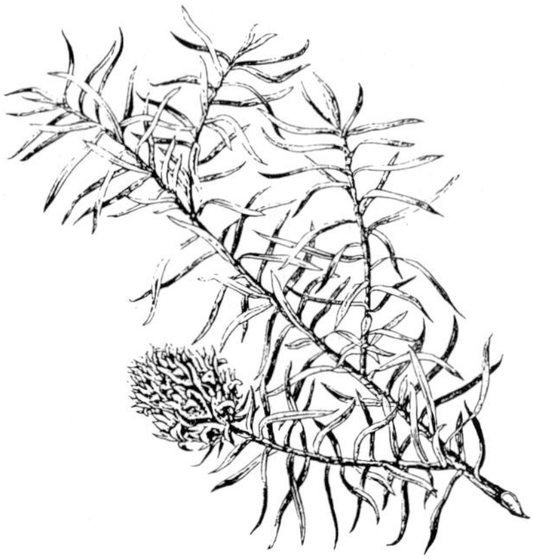

Cryptomeria. Juvenile foliage. (2)

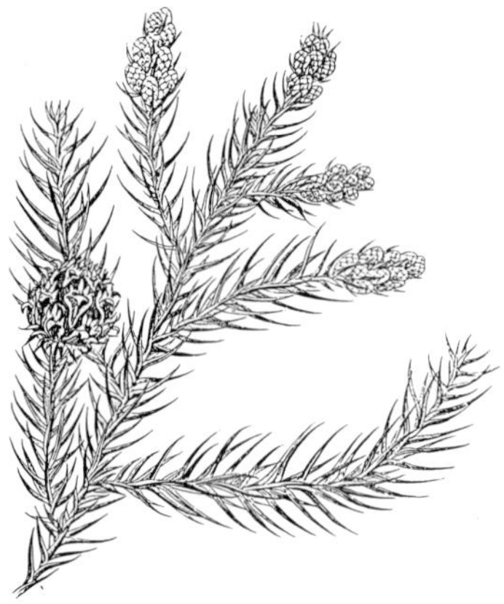

Cryptomeria. Adult foliage. (2)

5.2 Large tree with grass-green foliage. Leaves are four-sided with stomata on all faces; broad at base, narrowing to a sharp-pointed tip; at first spreading widely but strongly incurved. The spent cones frequently remain on the tree for several years.
Turn to p. 263. ***CRYPTOMERIA***

Note that the drawings differ widely in scale

Dacrydium. (*D. franklinii*). (5)

5.3 A diversified genus of trees and shrubs from the southern hemisphere, very few of which are hardy in cold temperate areas. The foliage is '*Cupressus*'-like, but the leaves are in threes.
Turn to p. 315. ***DACRYDIUM***

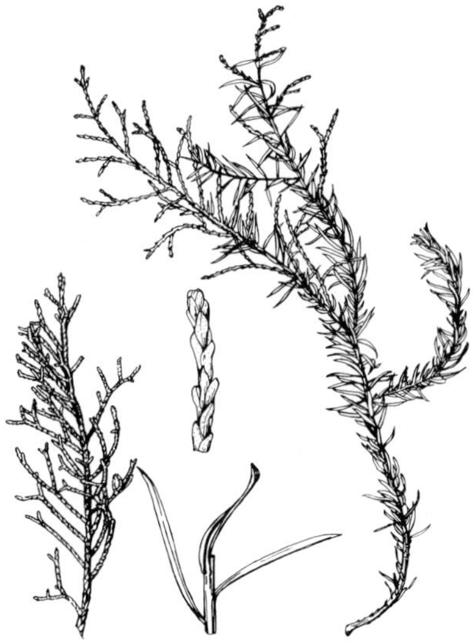

Halocarpus. (*H. bidwillii*). (5)

5.4 A genus of shrubs recently separated from *Dacrydium* on minor botanical grounds.
Turn to p. 322. ***HALOCARPUS***

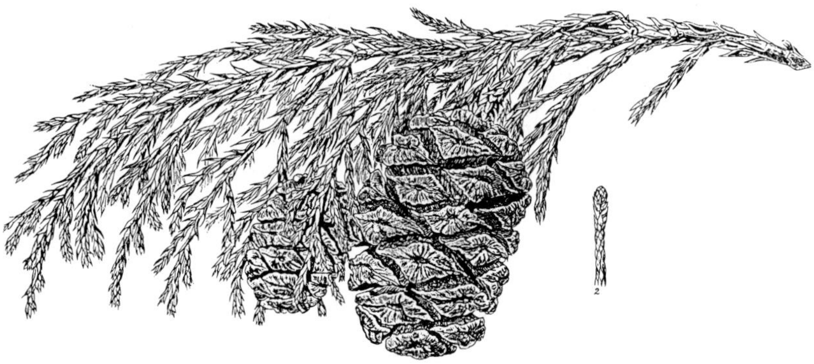

Sequoiadendron. (2)

5.5 The well-known 'Big Tree' of California, frequently seen as a large tree even in cultivation, with leaves 4–7 mm long, incurved and held forward along the shoot, retained for several years and giving a ropey look (and feel) to the foliage, which Mitchell tells us smells readily of aniseed if bruised. Distinctive is the very thick, corky bark and the cones persisting on the tree for several years.
See Vol. 2. ***SEQUOIADENDRON***

Note that the drawings differ widely in scale

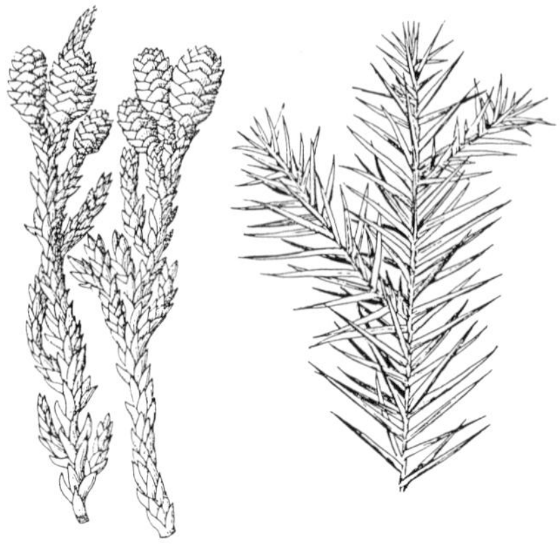

Taiwania. (2)

5.6 A monotypic genus resembling the much commoner *Cryptomeria*. The leaves are triangular in cross section with conspicuous stomatic hollows on all surfaces, broad at the base, spreading but strongly incurved towards the sharply pointed tip.
See Vol. 2. ***TAIWANIA***

'GO-TO' Listing for Main Group Six

"Large or medium trees or large shrubs. 2nd year shoots green. Leaves large to very large, lanceolate, ovate-lanceolate or ovate."

Agathis. (*A. australis*). (1)

6.1 Leaves large, flat, ovate-lanceolate, 25 to 75 mm long by 6 to 12 mm wide, finely striate above and beneath, no visible stomata. Only one species is hardy, and that only in the most favoured areas in Britain.
Turn to p. 167. ***AGATHIS***

Araucaria. (5)

6.2 Curious tall tree with a rounded crown, frequently bare of lower branches. Bark thick, ringed at the base, suggestive of an elephant's foot. Foliage very dense, closely covering the branches, the leaves triangular, broad at the base, lying forwards but curving outwards to a very sharp point. The formidable appearance that results explains the popular name 'Monkey-puzzle Tree'.
Turn to p. 167. ***ARAUCARIA***

Note that the drawings differ widely in scale

Cunninghamia. (2)

6.3 Unusual tree with large, 70–80 mm long by 5 mm board, glossy, long, dark-green leaves narrowed at the base and decurrent on the twig, stiff, narrowly lanceolate, sharply pointed, emerging spirally but lying in two ranks or twisted all ways, giving an untidy appearance. The upper surface is concave with slightly raised margins and midrib and two very narrow bands of stomata; the lower surface s convex, with a green midrib and two broad stomatic bands, the edge being finely toothed. The bark is fibrous, stringy and corky on the surface but hard beneath.
Turn to p. 279. ***CUNNINGHAMIA***

6.4 Two species of ***Podocarpus*** have unusually large leaves.
See Vol. 2. ***PODOCARPUS***

Podocarpus. (*P. macrophylla*). (6)

(*P. salignusa*). (5)

'GO-TO' Listing for Main Group Seven

"Large to medium size trees, erect when young, abundantly branched, with dense foliage, the branchlets usually lying in flat sprays. Leaves scale-like, in opposite, decussate pairs, the facial pair flattened, the lateral pair folded, held closely against the shoot with never more than the tips free. The buds are completely hidden. Rarely, some awl-like, juvenile leaves may be present".

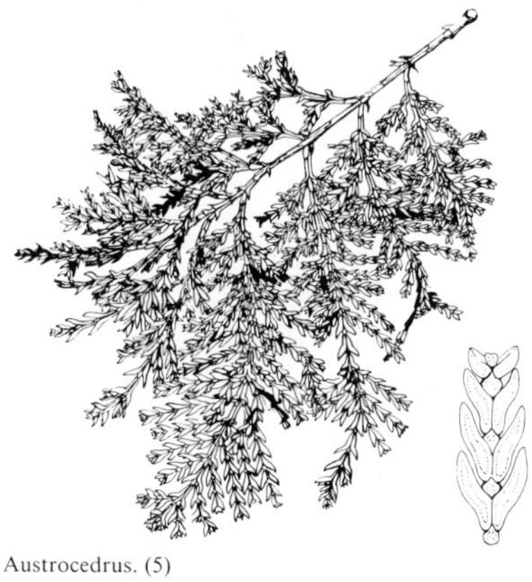

Austrocedrus. (5)

Note: This group cannot in some cases be identified with certainty in the absence of cones or (as the case may be) juniper berries.

7.1 Usually seen as a small tree. Foliage as in *Thujopsis*, but with white bands above and below. The facing pair of leaves are bluntly rhombic or ovate, and very much shorter than the lateral pair, which are incurved at the tip. A rare tree.
Turn to p. 174. ***AUSTROCEDRUS***

Note that the drawings differ widely in scale

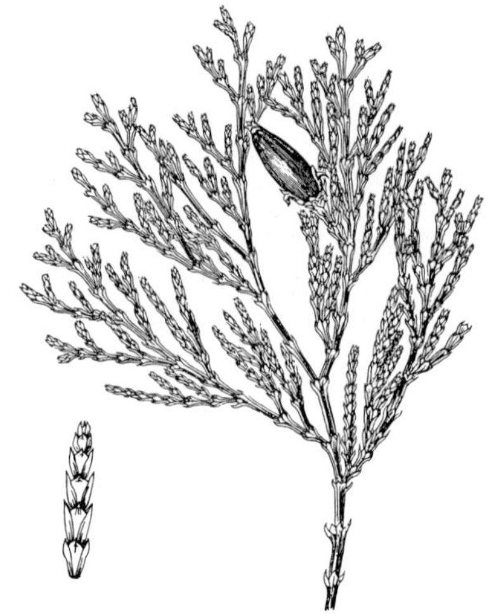

Calocedrus. (*C. decurrens*). (2)

7.2 Tall, usually narrow, often a monumental tree with scale-like foliage in much-flattened fan-like sprays, the facing and lateral pairs of equal length. Foliage smells of turpentine when bruised. Cones small; leathery. For many years known as *Libocedrus*.
Turn to p. 174. ***CALOCEDRUS***

Chamaecyparis. (*Ch. formosensis*). (5)

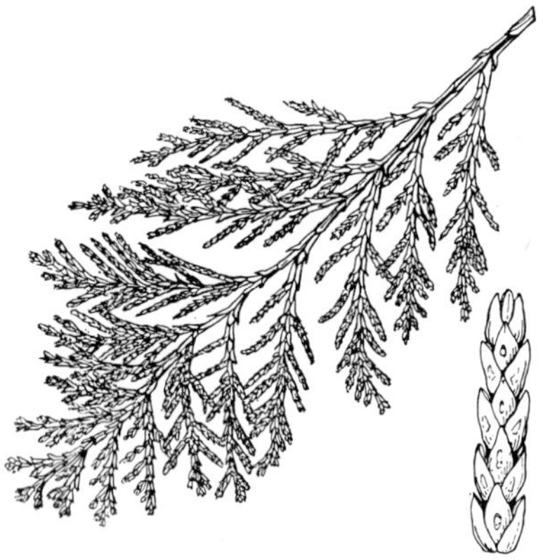

(*Ch. lawsoniana*). (5)

7.3 The False Cypresses. Usually tall, conical or round-topped trees with foliage (save in ***Ch. thyoides***) in flat sprays. The leaves are very small, scale-like and hard to the touch, the facing leaves flattened, with tips rounded or triangular, the lateral leaves folded. Cones globular, **small**, seldom over 8 mm diam, consisting of 8 to 10 scales, each holding two seeds, in most species ripening the first year. The numerous cultivars can vary from the norm in any respect, making them sometimes difficult to identify.
Turn to p. 194. ***CHAMAECYPARIS***

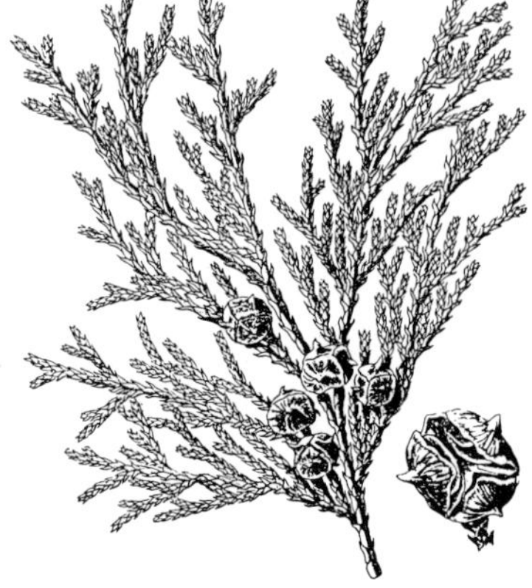

(*Ch. nootkatensis*). (2)

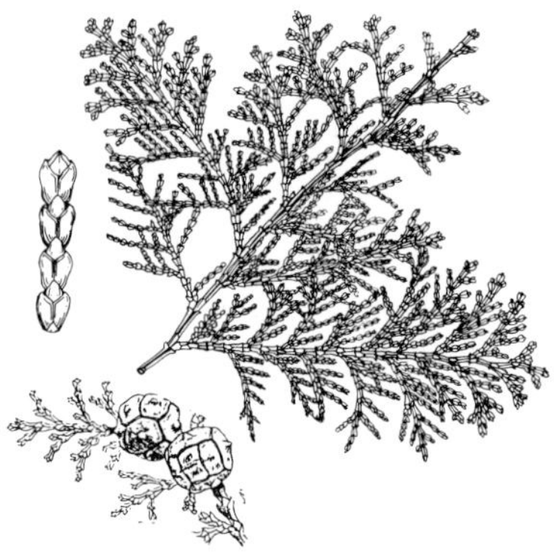

(*Ch. obtusa*). (5)

Note that the drawings differ widely in scale

7.3 ***CHAMAECYPARIS*** (Cont.)

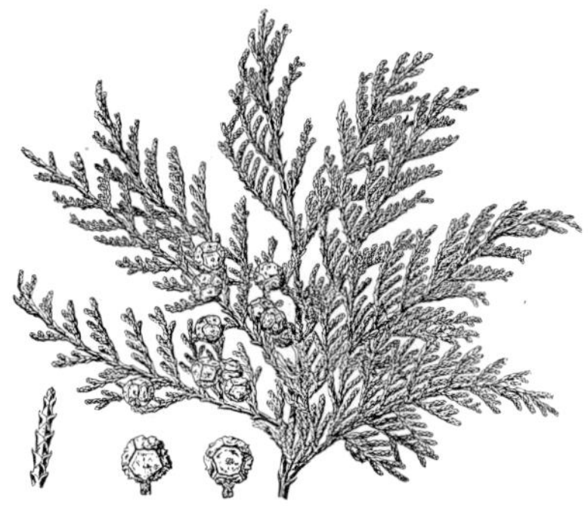

(*Ch. pisifera*). (2)

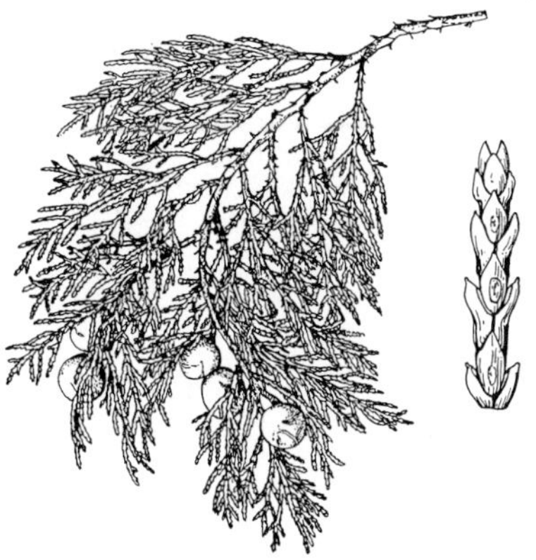

(*Ch. thyoides*). (5)

(*Ch. pisifera* 'Plumosa'). (2)

(*Ch. pisifera* 'Squarrosa'). (2)

(× *Cupressocyparis leylandii*). (5)

7.4 A bigeneric hybrid genus with foliage intermediate between the parent genera, *Chamaecyparis* and *Cupressus*, one or other parent usually predominating. Of several such hybrids on record, only this one species has been widely planted. (*Ch. nootkatensis* × *C. macrocarpa*).

In this Main Group would come the cultivar 'Leighton Green' and other cultivars tending towards *Ch. nootkatensis*. See also Main Group Eight. Turn to p. 282.

× ***CUPRESSOCYPARIS leylandii***
(In part)

Note that the drawings differ widely in scale

Cupressus. (*C. lusitanica*). (5)

7.5 Of the True Cypresses only the following species hold their foliage in flat, pinnate sprays. (For other species see Main Group Eight.)

CUPRESSUS lusitanica
C. 1. var. ***benthamii***
C. macnabiana
C. torulosa
C. t. var. ***cashmeriana***

(*Dacrydium franklinii*). (5)

7.6 Bush or medium (in cultivation) tree with pendulous habit and at first sight cupressus-like foliage, but the small, scale-like leaves emerge spirally in threes, not in opposite pairs. See Main Group Five.
Turn to p. 315. ***DACRYDIUM***
(Part.)

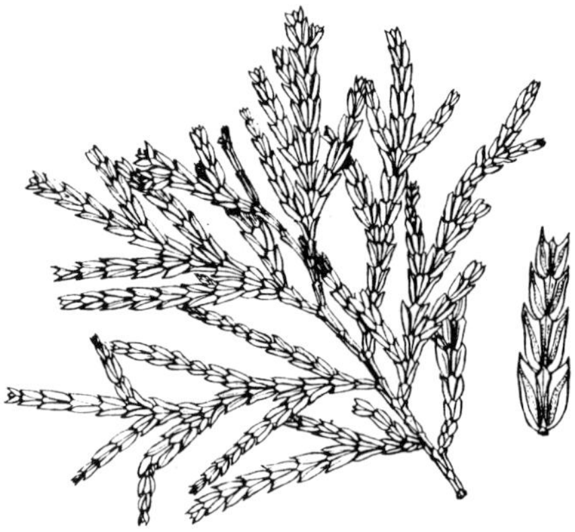

Fokienia. (5)

7.7 A small tree or large shrub, very rare in Britain, being hardy only in the most favoured areas. Foliage similar to *Thujopsis*, but even more flattened, and glistening white beneath.
Turn to p. 317. ***FOKIENIA***

Note that the drawings differ widely in scale

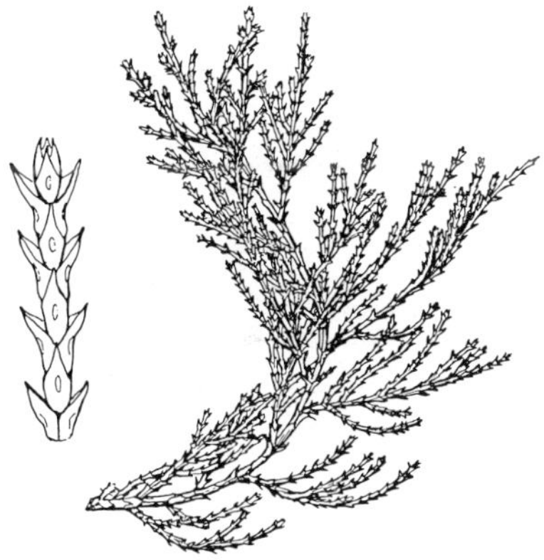

Juniperus. (*J. thurifera*). (5)

7.8 In this large genus, only ***J. thurifera*** carries foliage in flat sprays. It is at once identifiable by the musty smell given off by the plant.
Turn to p. 399. ***JUNIPERUS*** (Part.)

Platycladus. (Formerly *Thuja orientalis*). (5)

7.9 Round-topped tree with branchlets held erect in vertical sprays. Cones erect on shoot, at first fleshy but ripening dry and woody, with three pairs of scales, each with a prominently hooked projection. (Formerly listed under *Thuja* or at one time under *Biota*.)
See Vol. 2. ***PLATYCLADUS***

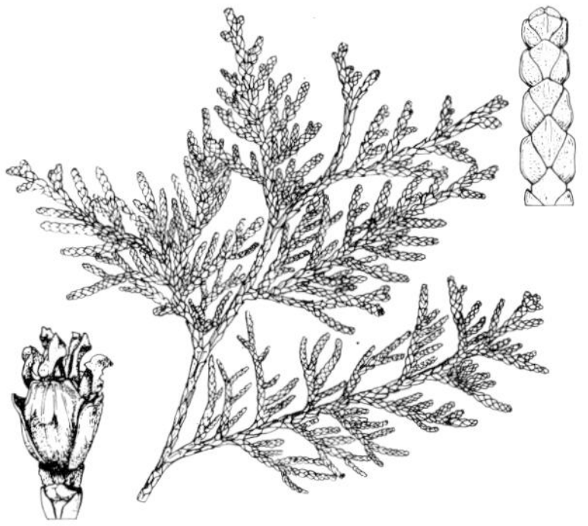

Thuja. (*Th. koraiensis*). (5)

(*Th. occidentalis*). (5)

7.10 Large tree with an erect leading shoot. Final growth in flattened sprays. Foliage in most species aromatic when bruised. Cones solitary; scales 8 to 12.
See Vol. 2. ***THUJA***

Note that the drawings differ widely in scale

7.10 ***THUJA*** (Cont.)

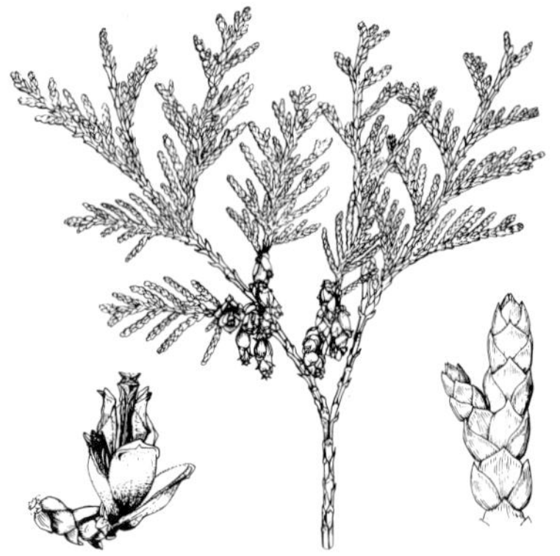

(*Th. plicata*). (5)

(*Th. standishii*). (5)

7.11 Tree with *Thuja*-like foliage much flattened, as though having passed through the rollers of a mangle. Glossy green above, white-striped beneath. Cones sub-globose; conescales 6 to 8.
See Vol. 2. ***THUJOPSIS***

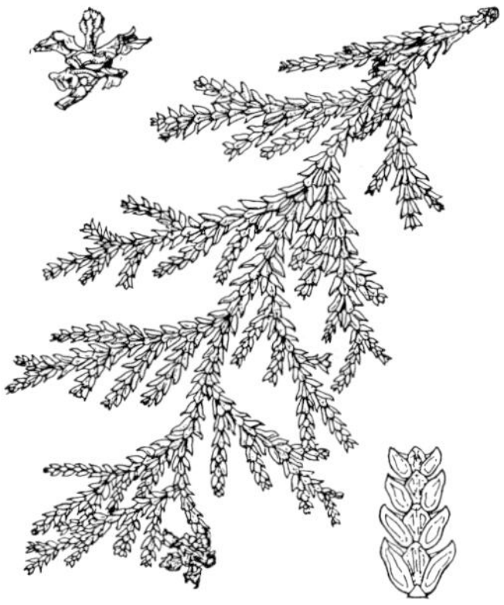

Thujopsis. (5)

'GO-TO' Listing for Main Group Eight

"Large to medium size trees, erect when young, abundantly branched, with dense foliage, the branchlets around the main shoot, i.e. not in flat sprays. Leaves scale-like in opposite, decussate pairs, the facial pair flattened, the lateral pair folded, lying smoothly against the shoot with only the tips free; the buds being completely hidden."

Note: The distinction between this Main Group and Main Group Seven is not absolute: border-line cases occur.

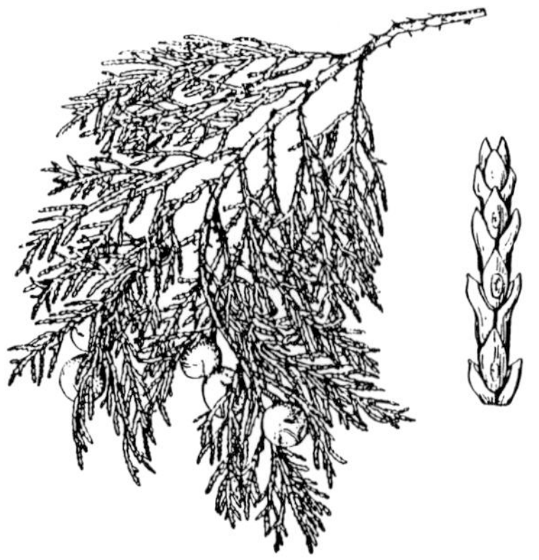

(*Ch. thyoides*). (5)

8.1 In this important genus, this species alone carries its branchlets in irregular, short angular sprays, not in the flattened sprays characteristic of the genus. Turn to p. 260.
CHAMAECYPARIS thyoides

Note that the drawings differ widely in scale

(× *Cupressocyparis leylandii*). (5)

8.2 See also Main Group Seven.
In this Main Group come 'Haggerston Grey' and other plumose clones of Leyland Cypress that incline towards *C. macrocarpa*. Turn to p. 282.
× ***CUPRESSOCYPARIS leylandii***
(Part)

Cupressus. (*C. macrocarpa*). (5)

8.3 The True Cypresses. A large and difficult genus, with small, scale-like leaves, in most species closely appressed, the ultimate shoots arising all round the shoot, not in flat, pinnate sprays. The cones are **large**, globular 12 mm or more in diam, each scale holding six or more seeds, ripening the second year and hanging on the tree for several years.
Turn to p. 288. ***CUPRESSUS***

Diselma. (15)

8.4 A species rare in cultivation. Ultimate branchlets tetragonal, all leaves of equal size. Leaves occasionally in threes.
Turn to p. 316. ***DISELMA***

Note that the drawings differ widely in scale

8.5 Shrub or small trees bearing both small, scale-like adult leaves, closely appressed as in *Cupressus* and/or awl-like, juvenile leaves. The latter are characteristic of young plants; adult leaves predominate on mature trees, but small areas of juvenile foliage are usually present. Of the species falling into this Main Group, many are seldom if ever seen in cultivation, and the following are usually represented by named cultivars, which may or may be uncharacteristic in one or more respects. ***JUNIPERUS (Part)***

Juniperus. (*J. chinensis*). (6)

(i) Juvenile leaves in 2s or 3s, at the base or tip of the shoot; adult leaves closely appressed, blunt-tipped. Fruit large, ripening 1st year.
Turn to p. 326. ***J. chinensis***

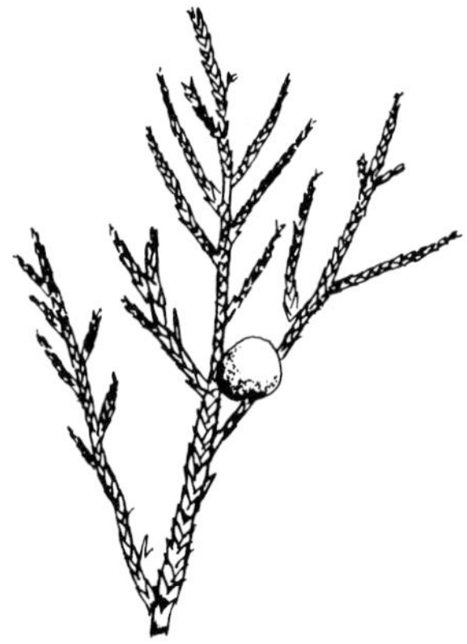

(*J. scopulorum*). (6)

(ii) Adult leaves in 2s, widely spaced, sharply pointed. Fruits globose, 15–20 mm, ripening 2nd year to black, whitened by a powdery bloom.
Turn to p. 390. ***J. scopulorum***

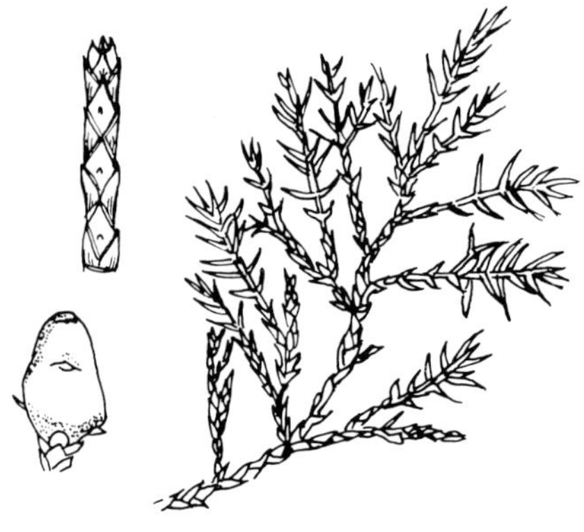

(*J. virginiana*). (15)

(iii) Juvenile leaves in 2s at base or tip of shoot; adult leaves in 2s; overlapping, sharply ponted. Fruit ovoid, ripening 1st year to glaucous blue-black.
Turn to p. 400. ***J. virginiana***

Note that the drawings differ widely in scale

'GO-TO' Listing for Main Group Nine

"Unusual-looking tree with wedge-shaped, fan-veined, leathery and mainly two-lobed leaves on long stalks."

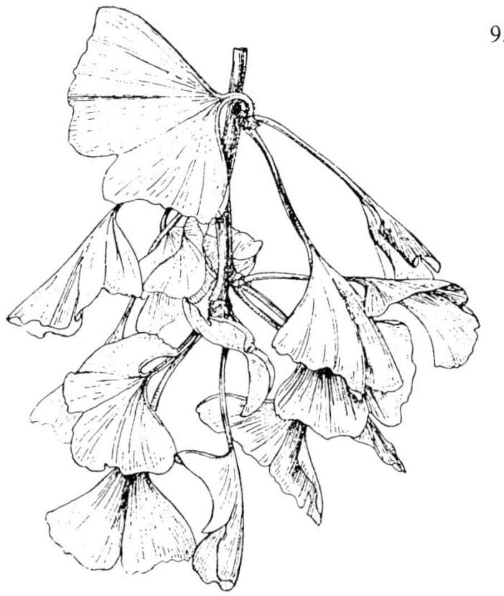

Ginkgo. (5)

9.1 The Maidenhair Tree. Quite unlike all other conifers and difficult to recognise as one.
Turn to p. 317. ***GINKGO***

'GO-TO' Listing for Main Group Ten

"Large or small trees, or spreading to prostrate shrubs with foliage in whorls of three or four."

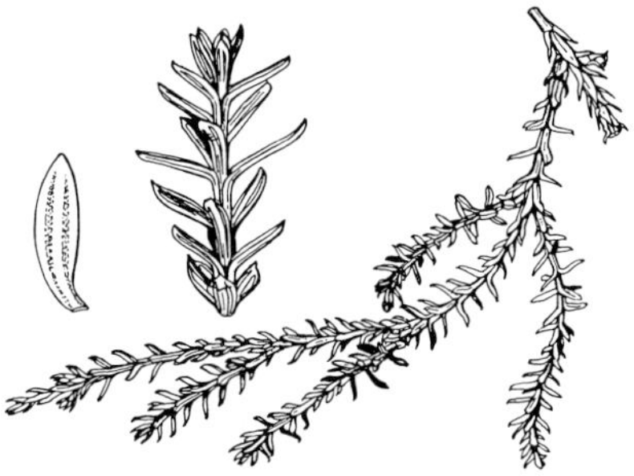

Fitzroya. (5)

10.1 A genus of trees or shrubs with very slender foliage and scale-like leaves held in whorls of four. Not hardy in Britain. ***CALLITRIS***

10.2 Tree with foliage bearing a vague resemblance to *Cupressus* but the dark blue-green leaves are borne in threes. The foliage is very dense, the leaves hard to the touch; thick, 3–4 mm long, curving out but with the blunt tips incurved. 2 stomatic bands.
Turn to p. 317. ***FITZROYA***

10.3 A group of junipers readily distinguishable by the narrow, sharply pointed leaves in whorls of three. (See also Main Group Twelve).
JUNIPERUS (Part)

Note that the drawings differ widely in scale

10.3 ***JUNIPERUS*** (Cont.)

(i) Leaves large (15–25 mm long, 2–4 mm wide), decurrent. ***J. drupacea***

Juniperus. (*J. drupacea*). (12)

(ii) Leaves not decurrent to the (usually three-sided) shoots but standing off at a steep angle from a swollen, foot-like base.

J. Communis
J. conferta
J. oxycedrus
J. rigida
J. taxifolia

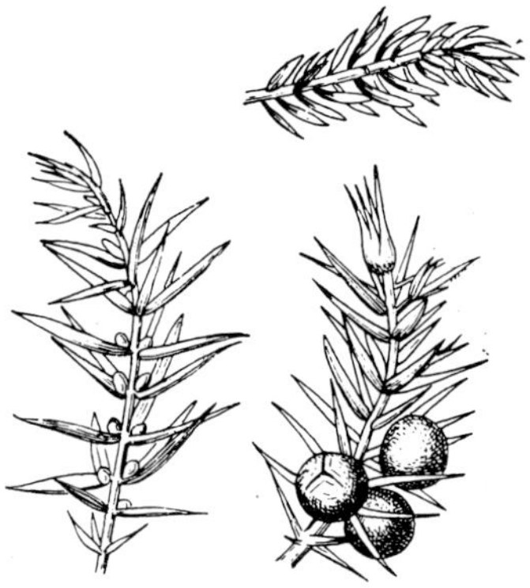

(*J. communis*). (12)

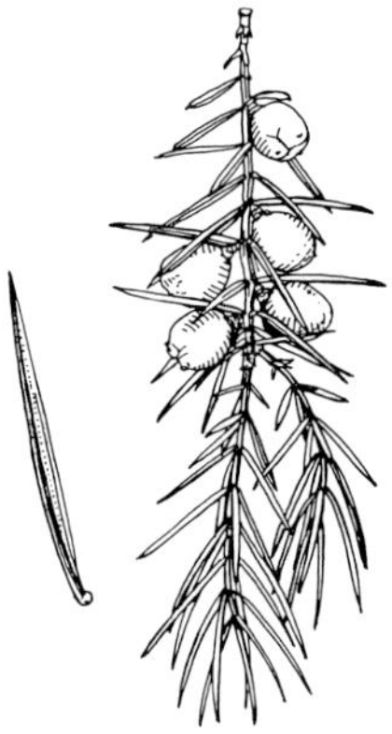

(*J. rigida*). (5)

(iii) Leaves long-tapered, finely pointed; pointing forwards but not decurrent to the shoot.

J. coxii
J. morrisonicola
J. procumbens
J. recurva
J. squamata

(*J. recurva*). (14)

(*J. squamata*). (15)

Note that the drawings differ widely in scale

Pilgerodendron. (15)

10.4 Small tree or shrub with thick, triangular leaves in whorls of four, overlapping, adpressed at base, then spreading widely, bluntly pointed; grass green with a broad white band above. It is at once distinguished from *Fitzroya* by the colour.
See Vol. 2. ***PILGERODENDRON***

'GO-TO' Listing for Main Group Eleven

"Curious-looking trees or bushes bearing cladodes (modified branchlets masquerading as leaves, the true leaves being reduced to scale-like dimensions. Non-woody second year."

Phyllocladus. (12)

11.1 Shrub. Leaves and cladodes emerge spirally, the cladodes are flat, rhomboidal, arising each from the axil of a true leaf (which are tiny) and looking as though made by a small child out of modelling clay. One species only is hardy.
Turn to p. 425. ***PHYLLOCLADUS***

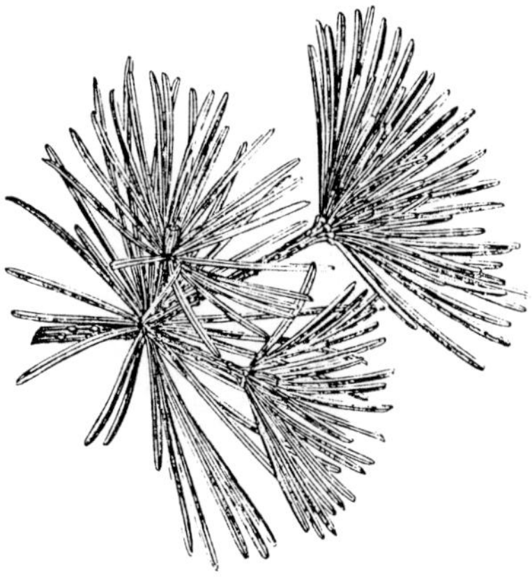

Sciadopitys. (12)

11.2 An unusual-looking large tree. The 'leaves' are joined together in pairs, are 50 to 100 mm long and are arranged in whorls of between 10 and 30, like the spokes of an umbrella (hence the popular name 'Umbrella Pine'.)
See Vol. 2. ***SCIADOPITYS***

Note that the drawings differ widely in scale

'GO-TO' Listing for Main Group Twelve

"Prostrate, procumbent, spreading, flat-topped or rounded shrubs. Leaves in pairs or threes, either scale-like and decurrent, covering the buds, or awl-like (a persistent juvenile state). In some species, both are often present in the same plant."

It must be borne in mind that a young plant of a prostrate or spreading cultivar or form of any species may well appear to fall into this group.

(*Dacrydium laxifolium*). (15)

12.1 The smallest conifer. A low, sprawling plant with dark green leaves. Female plants cone when only a few cm across. Turn to p. 316.
DACRYDIUM laxifolium

12.2 As well as the species listed in (i) to (x) below which come within this main group, clones individually fitting the present description are in garden cultivation in the following species.

JUNIPERUS (Part)
J. chinensis
J. communis
J. scopulorum
J. recurva
J. rigida
J. taxifolia
J. virginiana

(i) A prostrate plant with bright mid-green leaves in 3-s (as in *J. communis*)
Turn to p. 353. ***J. conferta***

(ii) A species from Central Asia imperfectly known and probably not in cultivation. Several clones often listed under this species belong to *J. chinensis*
Turn to p. 355. ***J. davurica***

(*J. horizontalis*). (15)

(iii) A more or less completely prostrate species with either acicular leaves in 2's or 3's, or adult foliage in four ranks; pointed or with rounded tips and with a conspicuous gland on the back. Numerous widely diverse selections from the wild have been named as cultivars and are widely planted for ground cover.
Turn to p. 362. ***J. horizontalis***

Note that the drawings differ widely in scale

(J. × *media*). (15)

(iv) A hybrid species (*J. sabina* × *J. chinensis*) not universally accepted, but of practical value to cover the ever-increasing number of cultivars developed from the well-known 'Pfitzer Juniper'. Main branches rise at 45° and arch over; side branches numerous, all pointing forwards. (Note: authorities not accepting *J.* × *media* list these cultivars under *J. chinensis*.) Turn to p. 370. ***J.* × media**

(*J. pingii*. Formerly *J. squamata* var. *wilsonii*). (15)

(v) A species that includes a group of cultivars still widely grown under the loose name *J. squamata* var. *wilsonii*. Turn to p. 378. ***J. pingii***

(*J. procumbens*). (15)

(vi) A prostrate species that forms a large, dense mat, mounding at the centre. Foliage is wholly in 3's, decurrent but tips free, sharply pointed; fresh grass-green; shoots stout, with tips very slightly ascending. Turn to p. 380. ***J. procumbens***

(*J. sabina*). (15)

(vii) A widely planted species consisting of low, spreading shrubs. It is at once identifiable by the unpleasant smell of the crushed foliage. Turn to p. 384. ***J. sabina***

Note that the drawings differ widely in scale

(*J. sargentii*). (15)

(viii) A ground-covering plant with long, horizontal branches and numerous ascending laterals, clothed with a strongly incurved foliage. The crushed foliage smells of camphor.
Turn to p. 389. ***J. sargentii***

(*J. squamata*). (15)

(ix) A shrubby species, occasionally a small tree, more usually seen in cultivation as one of the extremely blue clones.
Turn to p. 396. ***J. squamata***

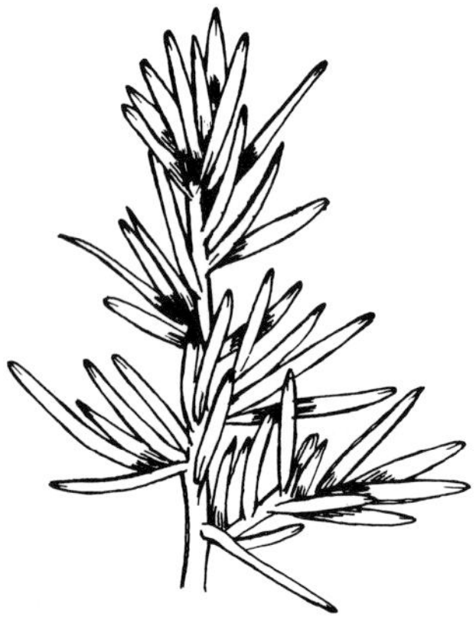

(*J. taxifolia*). (15)

(x) Probably only known in cultivation in the completely prostrate var. *luchuensis* (Koid.) Satake. It resembles *J. conferta*, but the needles are broader, flatter and somewhat boat-shaped. The tips of the leaves are sharply pointed but not prickly.
Turn to p. 399. ***J. taxifolia***

12.3 A dense, mat-forming plant with Cupressus- or Thuja-like foliage.
Turn to p. 423. ***MICROBIOTA***

Note that the drawings differ widely in scale

Microcachrys. (15)

12.4 Small, spreading shrub with dark green foliage with scale-like leaves set spirally in four ranks. The fruits are fleshy, ripening scarlet.
Turn to p. 424. ***MICROCACHRYS***

Microstrobus. (15).

12.5 Much branched shrubs with scale-like overlapping leaves spirally arranged, incurved, yellow-green with white stomata on the inner face.
Turn to p. 424. ***MICROSTROBUS***

Podocarpus. (*P. alpinus*). (15)

12.6 A low spreading, multi-branched shrub with dark green leaves and crimson-red berries.
See Vol. 2. ***PODOCARPUS*** (part).

P. alpina
P. nivalis

7. Varieties Currently Available

In this chapter will be found descriptions of conifers sufficiently hardy to survive in the open ground in the temperate zone. Coverage of the species, varieties and cultivars obtainable in the trade (and, consequently, commonly to be met with in parks and gardens) is fairly complete, but also included are conifers that will only be found in an arboretum or in a very comprehensive or specialist conifer collection. A few listed here may not yet be in cultivation. Species that are reliably hardy only in the more favourable areas covered by this book are so described: tender genera are mentioned briefly for the sake of completeness.

In the *Manual of Cultivated Conifers*, the claim was made to have mentioned all the names of cultivars published since 1753, whether in cultivation or not, so as to provide a basis upon which an International Register could be built. Whilst this was a most commendable aim, the claim to completeness was not fully met, and a large part of the book was taken up in recording archaic data at the expense of information of current interest. In order to release more space for the latter in the present book, a large number of names known (or reasonably assumed) to be obsolete are transferred to Appendix One (See Volume Two). The cultivars in question may indeed be extinct (many names were raised – quite illegitimately – for a single plant that was never propagated) or may be no longer identifiable, or are not known to be in current production.

Nevertheless, because conifers, like their names, are long-lived this list of relegations must be regarded as provisional, since in a few cases exemplars in excellent health may well exist, somewhere. Of these, some may in maturity have shewn qualities that would merit re-introduction to the trade. Readers are invited to draw my attention, through the publishers, to any names in Appendix One that are not yet due for honourable retirement. In the meantime they will be included in the Index to facilitate reference to the information recorded in the earlier book in which they were described.

CITATION OF 'AUTHORITIES'

Since botanists all over the world have been publishing names for plants (taxa) for a hundred and fifty years or more it is too much to expect that none of them will ever have invented different names for the same taxon or ever have used the same name for different ones. As a step towards securing accuracy and precision in this situation, the *Botanical Code* requires the name of the author to be printed following the name of the taxon. This helps, but it is about the equivalent of citing 'Moses' or 'St Paul' for a Biblical quotation. If the botanist in question has been known to have rushed into print more than once the particular year will usually have been added, but often the student, whether botanical or Biblical, needs more – he needs full 'chapter and verse'. Since a botanical student may also need directing to information sources for several synonyms, these citations could take up a lot of space for each name, so a system of abbreviations has been developed. This will be found set out in detail in the *Botanical Code.*

That this can result in a string of letters and numbers that may convey nothing at first sight to the casual enquirer is less serious than would appear, since the references given are frequently to rare books or obscure publications only to be found in a good botanical library where the librarian would have no difficulty in understanding the cryptic information put before him. So citations for botanical names are here given in this abbreviated form.

The use of author citations in the case of horticultural names is a relatively recent development. The *Cultivated Code* does not mention 'authorities', neither requiring them nor forbidding their use. But precision is no less important in the case of cultivars than it is in botanical circles (and, in view of the commercial considerations sometimes involved it may be much greater) so the practice is becoming general with horticultural writers of using the botanical system suitably adapted for its new use.

CITATION OF LATIN-FORM CULTIVAR NAMES

The principle I have argued in Chapter Five (i.e. that the pre-1st January 1959 garden forms that received Latin-form names and are now considered to be cultivars must have been such all the time) simplifies our task in this chapter, since we are thereby entitled to disregard all the attempts to give them pseudo-botanical status in the intervening period. This unfortunate embroilment of cultivars with the botanical system before *Cultivated Code* was available often did not stop with a mere temporary admission. There are many cases where generic and/or specific changes were made during this period that affected the cultivars, and others where 'var.' became changed to 'f.' or *vice versa*. (In botanical jargon these are, respectively, 'changes of position', and 'changes of rank', and each such change results in a 'new combination'.

Take a case at random: the popular conifer *PICEA abies* 'Clanbrassiliana'. This plant (apart from suffering several alternative spellings) appears 'in the literature' in the following different combinations.

Pinus clanbrassiliana	Loddiges	(1820)
Picea clanbrassilliana	Lawson	(1836)
Abies excelsa var. *clanbrassiliana*	Loudon	(1838)
Pinus abies clanbrassiliana	Antoine	(1842)
Pinus picea clanbrassiliana	Endlicher	(1847)
Picea excelsa clanbrassiliana	Carrière	(1855)
Picea abies var. *clanbrassiliana*	Th. Fries	(1890)
Picea abies f. *Clanbrassiliana*	Rehder	(1949).

The cultivar status of this particular clone is beyond dispute, since the mother plant is with us today, so we may here disregard all of these combinations. Details of all these unfortunate attempts to find a home for this plant within the botanical system are on record in the World Conifer Data Pool files, but in the listings in this chapter the entry (in the section relating to *PICEA abies*) can be simplified to: –

-- **'Clanbrassiliana'** Th. Fries, 259 (1890); under *Pinus* Loddiges 31 (1820).

Here the first citation is to the first use of the basionym in its correct botanical 'position'. The second entry (following the semicolon) refers to the first known use of the basionym. (Further entries given in any other case relate to synonymy or draw attention to other matters of interest in the case of that particular cultivar.)

ABIES

Pinaceae

The Silver firs (often – more briefly – the Firs) form a genus of about 50 species, many with attractive foliage, from temperate regions of the northern hemisphere. Some species are among the tallest and most monumental of all trees, but these (save for their value in forestry) have been mainly planted in arboreta and very large gardens and have given rise to few cultivars. A number of foliage and habit variants were recorded during the latter part of the nineteenth century but these never became widely distributed and, in many cases, can no longer be traced in cultivation. A few of the smaller-growing species such as *A. koreana* are more suitable for the smaller gardens of today, and several species, including *A. balsamea* and *A. lasiocarpa* have developed slow-growing or mountain forms that serve well as 'dwarf conifers' because of their neat habit and attractive foliage.

The Firs have had more than their fair share of trouble from changes of name. They were at first included, along with the cedars, hemlocks, larches and some other genera in the genus ***Pinus***, but this large conglomeration was later split up, the Firs becoming ***Abies*** and the Spruces, ***Picea***. Unfortunately, the distinction was not at first universally accepted and – worse still – several nineteenth century writers on conifers reversed these two names, so many of the older conifer books must be used with caution. Each of these factors has given rise to a number of now-rejected names, and a further source of confusion and oftentimes exasperation is that the working of the inflexible law of priority in botanical nomenclature has saddled the Norway Spruce (which has given birth to innumerable garden cultivars) with the unfortunate name *PICEA abies*. So as a generic name ***ABIES*** refers to one group of trees: used as a specific epithet it refers to quite a different group. It would be difficult to imagine anything more confusing.

As a guide through this particular nomenclatural labyrinth a listing of the principal authorities and the names they used will be found in Appendix Two.

Abies Miller, Gard. Dict., Abridged ed, 4, v. 1 (1754); *Pinus* Linnaeus, Sp. Plant. 1000 (1753)

Figure 19a. *Abies alba*. Silver fir. Bark on main trunk. (10)

Figure 19b. *Abies cilicica*. Cilician fir. Bark on main trunk. (10)

Figure 20a. *Abies firma*. Momi fir. Bark on main trunk. (10)

Figure 20b. *Abies grandis*. Grand fir. Bark on main trunk. (10)

Figure 20c. *Abies holophylla*. Manchurian fir. Bark on main trunk. (10)

Figure 20d. *Abies homolepsis*. Nikko fir. Bark on main trunk. (10)

Figure 20e. *Abies pinsapo.* Spanish fir. Bark on main trunk. (10)

Figure 20f. *Abies procera.* Noble fir. Bark on main trunk. (10)

in part; *Peuce* L. C. Richard in Ann. Mus. Hist. Nat. Paris, **16**: 298 (1810) in part; *Picea* D. Don ex Loudon, Arb. Frut. Brit. 4: 2329 (1838). Silver Fir (Gt.Brit.), Fir (Amer.)

Size and habit Monoecious evergreen trees, all (save a few species) of forest size; habit symmetrically conical or narrow-conical when young, broader with age.

Trunk Simple, rarely forked; *Bark*: usually smooth, thin on young trees, often thick and furrowed at the base on old trees.

Branches Whorled, horizontal, in tiers.

Branchlets Smooth, grooved in a few species; woody the second year, with circular scars where leaves have fallen.

Winter Buds Globose or ovoid, often in clusters, usually resinous.

Leaves Firm in texture, leathery, emerging spirally but twisted (on side branches) and spreading in two ranks (pectinate); linear or linear-lanceolate, entire, sessile, narrowing near the base, usually flattened and grooved above, in most species having 2 white or pale stomatic bands beneath, keeled below, rarely 4-sided and with stomata on all 4 faces; rounded and variously notched or pointed at the apex. On cone-bearing shoots the leaves are held more erect and are arranged radially. At leaf-fall, the scars left on the shoot are circular. This at once distinguishes *Abies* from *Pseudotsuga* (where the scars are oval) and from *Picea* in which genus small peg-like protuberances mark where a leaf has fallen.

Flowers Male flowers in short catkins from the leaf axils on the underside of the branchlets which, on falling, leave gall-like scars. Female flowers in erect cones composed of numerous scales, each with two ovules adaxially at the base and subtended by a narrow exserted or included bract.

Mature cones Cylindrical or barrel shaped, usually confined to the topmost branches of mature

Figure 21a. *Abies alba.* Silver fir. A specimen 40 m high. (10)

Figure 21b. *Abies concolor.* 'Violacea'. A selection with conspicuously silvery-white foliage.

Figure 21c. *Abies firma.* Momi fir.

Figure 21d. *Abies grandis.* Grand fir.

Figure 22a. *Abies magnifica.* (10)

Figure 22b. *Abies pinsapo.* Spanish fir. A young specimen at Grayswood Hill, Surrey, U.K.

Figure 22c. *Abies pinsapo.* 'Pyramidata' at Arboretum Les Barres, France.

Figure 22d. *Abies procera.* Noble fir. (10)

trees, held erect, with closely imbricated, fan-shaped scales and hidden or projecting (Exserted) bracts; ripening and falling to pieces in the first autumn, leaving the axis standing.

Seeds Ovoid or oblong, two on each scale, with resin vesicles; each with one large and thin wing.

Habitat The genus is widely distributed throughout the northern hemisphere, reaching as far south as Guatemala in America, to northern Africa and the Himalayas.

Cultivation Silver firs will grow well on any moist soil. They will stand shade very well and are wind resistant. The Asian species break bud early so are prone to damage from frost. They should be planted in a cold situation where bud-break will be retarded. Propagation of the species is from seed; the cultivars and dwarf forms are grafted, *Abies alba* being used for the small leaved species and *Abies nordmanniana* for the large leaved (although wide compatibility seems to exist throughout the genus).

Scions should be of leading growth, those from lateral shoots seldom make good plants.

This genus is one of the most difficult for identification from the foliar parts alone and the soon-disintegrating cones are seldom available. A very complete Key is given by Mitchell (1972). The 'Guides' on pp. 98–100 must be used with caution because of the inherent variability of several of the species. The taxonomy follows the *International Conifer Register*, Part One (1987), with the exceptions noted.

A. abies (L.) Borbas (1902). Now known as *PICEA abies* (L.) Karst.
A. acicularis Hort. Now known as *PICEA alcockiana* Carrière.
A. acutissima Knight & Perry (1850) (pt.) Now known as *PICEA glauca* (Moench) Voss
A. ajanensis Lindley & Gordon (1850). Now known as *PICEA jezoensis* (S. & Z.) Carrière.
A. alba (1) Michaud (1810). Now known as *PICEA glauca* Moench (Voss)

A. alba (2) Miller (1768)). Formerly *A. pectinata* Lamarke & De Candolle (1815). European Silver Fir (Gt. Brit.); Silver Fir (Amer.)

Size and habit Tree, 30–50 m, conical when young, in old age irregular and round-topped.

Trunk Straight and tapering. *Bark*: greyish, smooth, scaly on old trees. See Illus. No. 19a.

Branches Horizontal, often long and heavy, sweeping upwards at the tips, or a few metres out from the bole turning to vertical.

Branchlets Greyish-green with short, blackish pubescence; leaf scars almost circular.

Winter Buds More or less ovoid, blunt, brown, rarely slightly resinous; bud scales rounded at apex.

Leaves Pectinate or in two ranks forming a V-shaped valley, flat or with edges slightly down-curled, linear, 2–3 cm long, the upper ranks shortest, 2–2.3 mm broad, slightly notched at apex, on fruiting branches upright and pointed at apex; grooved and shining dark green above, with 2 silvery-white stomatic bands and a green mid-rib beneath.

Mature cones Short-stalked, cylindrical or slightly tapering; 10-16 cm long, 3-5 cm wide, brownish green when young, ripening to red-brown. *Scales*: large and broadly rounded above, few in number, bracts narrowly spoon-shaped with an oval or rounded blade, tapering above and below, apex long-pointed, exserted and reflexed. Cones only borne high on the tree.

Seeds Subtriangular, 7–9 mm long; wing twice as long as wide, reddish, wedge-shaped, truncate at apex.

Habitat Mountains of Central and East Europe.

Cultivation This species flushes very early so is prone to damage from Spring frosts. It is also prone to insect attack, so is useless for forestry and is now seldom planted for ornament.

A number of variants were found and given names in Europe during the last century, but

(text continues on p.102)

GUIDES TO IDENTIFICATION OF ABIES SPECIES

Characteristics shew variation with the age of the tree and even within a single tree

	SHOOTS: Pubescent (Use a hand lens)	Glabrous or nearly so	Grooved, furrowed or ridged	Smooth (i.e. not furrowed)	LEAVES: Tips notched (bifid). Note that inconsistencies occur.	Tips pointed or rounded (not notched)	Length (mm.)	STOMATA: On top of the leaf as well as below	On the underside of the leaf only	LEAF SHAPE (See definitions on p. 100)	Remarks
alba	X			X	X		15–30		X	F	Foliage sparse; dark glossy green above
amabilis	X				X		20–30		X	D	Foliage dense, covering shoot; dark green; gives off odour of tangerines
balsamea	X			X	X		15–25	X		F	Dark green; white patch near tip
borisii-regis	X					X	25–30	X			Foliage dense; stomata at tip only
bornmuelleriana		X		X	X		25–40	X			Leaves form a V groove; scattered stomata above
bracteata		X		X		X	30–50		X	B	Hard; sharp tip; buds like *Fagus sylvatica*
cephalonica		X		X		X	20–30		X	A	Leaves spring from large, sucker-like bases
var. graeca (=equi-trojanii)		X		X	X		15–30		X		Leaves form V groove above and are parted below, with vivid silver bands
chensiensis	X		X		X		25–40		X	F	Leaves widely parted; two hard, sharp tips
cilicica	X					X	25–40		X	D	Lax, forming V groove above; shiny dark green
concolor		X		X	X		40–60	X		C	Foliage glaucous grey or bluish green; leathery, sparse
var. lowiana		X		X	X		30–75		X	C	Foliage dark green, varying from pectinate to forming a wide groove
delavayi		X	X		X		15–35		X	F	Leaves curved; margins strongly down-curved
var. smithii (=forrestii)	X						30–35		X		Leaves form a V groove; dark, dull green with two silvery white bands below
fargesii	X		X			X	20–45				Leaves below shoot spreading; coarse glossy; deep yellow-green
firma	X				X		20–50		X	F	Leaves spread widely on very thin stalks, shewing light between; yellow-green above and beneath
fraseri	X				X		15–20				Very dark green; noticeably pale-tipped; broadest above the centre
grandis	X				X		20–50		X	F	Leaves in ranks, varying in length; they give off odour of oranges

GUIDES TO IDENTIFICATION OF ABIES SPECIES

Characteristics shew variation with the age of the tree and even within a single tree

	SHOOTS: Pubescent (Use a hand lens)	Glabrous or nearly so	Grooved, furrowed or ridged	Smooth (i.e. not furrowed)	LEAVES: Tips notched (bifid). Note that inconsistencies occur.	Tips pointed or rounded (not notched)	Length (mm.)	STOMATA: On top of the leaf as well as below	On the underside of the leaf only	LEAF SHAPE (See definitions on p. 100)	Remarks
holophylla		X		X		X	20–45				Light-green; held well above the shoot
homolepis		X	X		X		20–30				Leaves parted above and below shoot and give off minty aroma when crushed
kawakamii	X		X		X		15–30		X	F	Leaves narrow. Bark on trunk corky
koreana	X		X		X		10–20		X	G	Leaves radial, standing out stiffly; blackish-green; white patch at tip; white beneath
lasiocarpa	X		X		X		15–30	X		C	Leaves held forwards, forming narrow V groove. Rare in cultivation in UK
var. arizonica	X		X		X		10–25		X	D	Leaves silvery-grey; bark thick, corky
magnifica	X		X			X	25–40	X		E	Foliage less crowded than in *A. procera*; each leaf circular in cross section
mariesii	X			X	X		15–30		X	D	Leaves blunt, grooved; shorter than in *A. magnifica*
nebrodensis	X			X		X	15–20		X	H	Leaves held upwards, dark glossy green
nephrolepsis	X		X		X		10–25		X	F	Leaves spreading; yellow-green
nordmanniana		X	X		X		20–30		X	D	Foliage dense, rich green; old leaves persist for many years; slightly aromatic
numidica		X	X			X	15–20		X	G	Foliage variable in colour
pindrow				X	X		30–70		X	H	Leaves sparse, long and thin; two sharp tips
pinsapo		X	X			X	10–20	X		A	Leaves stiff, standing straight out; blunt, often curved inwards
procera	X			X	X		10–35	X		E	Shoot orange; foliage dense above shoot leaves adpressed close to shoot
recurvata		X				X	20–30		X	C	Shiny yellowish-green above and below
religiosa	X		X			X	20–35		X	D	Upper leaves form a narrow groove lower leaves hang down
sachalinensis	X		X		X		30–35		X	D	Leaves held radially, light green
sibirica	X			X	X						Soft, shiny, light green
spectabilis	X		X		X		30–60		X	E	Deep glossy green leaves form a wide V groove; apex of leaf squared off

GUIDES TO IDENTIFICATION OF ABIES SPECIES

Characteristics shew variation with the age of the tree and even within a single tree

Species	SHOOTS: Pubescent (Use a hand lens)	Glabrous or nearly so	Grooved, furrowed or ridged	Smooth (i.e. not furrowed)	LEAVES: Tips notched (bifid). Note that inconsistencies occur.	Tips pointed or rounded (not notched)	Length (mm.)	STOMATA: On top of the leaf as well as below	On the underside of the leaf only	LEAF SHAPE (See definitions below)	Remarks
squamata		X		X		X	15–25	X		C	Leaf length reduces towards terminal bud
veitchii	X			X	X		15–25		X	D	Leaves point forwards; glossy dark green 2 silvery bands below
vejari		X		X		X	20–25	X			Scattered rows of stomata above; leaves curved

Arrangement of leaves in side shoots

A. Leaves arranged radially around the shoot.
B. Leaves arranged pectinately, the two lateral sets lying in a single plane.
C. Leaves irregularly arranged, those on the lower side of the shoot not being truly pectinate.
D. Leaves densely crowded, those in the ranks nearest the central line imbricate, directed forwards, their bases not appressed to the shoot. The leaves below the shoot are in two lateral sets, loosely held.
E. Leaves in the ranks nearest the central line covering and at first appressed to the shoot, then curved upwards. The leaves below the shoot are in two sets, as in D.
F. Leaves partly arranged pectinately, the lower ranks lying in a single plane, the upper ranks directed upwards and outwards so as to form a V-shaped valley.
G. Leaves all held upwards or outwards vertically or nearly so.
H. Leaves irregularly arranged, all directed more or less forward; those above imperfectly covering the shoot; those below mostly pectinate but with some leaves directed downwards and outwards.

Figure 23. *Abies alba*. Detail of foliage from a side branch.

Figure 24. *Abies amabilis.* Detail of foliage from a side branch.

most of these are probably no longer in cultivation. See Appendix One in Volume Two, Zone 6. (See Illus. Nos. 21a, 23)

-- **'Aurea'** (Nelson) Mattfeld in Mitt. d. d. Ges. **35**: 22 (1925); under *Picea pectinata* Nelson 42 (1866); *A. pectinata auricoma* Carrière, 281

Figure 25. *Abies alba*. Silver fir. A ripe cone (10)

(1867). A variegated form with some constant golden-yellow leaves, otherwise green; uncommon in cultivation.

-- **'Columnaris'** (Carrière) Rehder in J. A. A. **1**: 55 (1919); under *A. pectinata* Carrière in Rev. Hort. 39 (1859). Tall and slender; branches short, equal in length, leaves somewhat shorter than usual. Originally found in a forest on Mount Pila. Probably more than one clone is now in cultivation under this name.

-- **'Compacta'** (Parsons) Rehder in J. A. A. **6**; under *Picea pectinata* Parsons Nurs. Cat. 65 (1887). Dwarf, globular, densely branched, broader than tall; leaves dark green and more glossy than usual. Raised by Parsons and Sons, Flushing, NY, USA, 1885.

-- **'Fastigiata'** Den Ouden & Boom, 2 (1965); under *A. pectinata* Carrière 207 (1855); *A. pectinata stricta* Carrière 281 (1867). Branches and branchlets crowded, erect, long, slender, the full-grown tree shaped like the Lombardy Poplar; leaves shorter and much thinner than usual frequently curved upward. Found by Reynius of Avignon in the forests of the Grande Chartreuse, France and put into cultivation by A. Sénéclauze in 1846.

-- **'Green Spiral'** Welch, Manual Dwarf Conifers, 141 (1979). Two trees planted in the Secrest Arboretum in 1916 have developed into trees 10m. high, each with a single, more or less erect but curiously twisted trunk and relatively few, short and pendulous branches which also tend to be twisted and incurved. They are bare of foliage in places, but elsewhere densely clothed. They were grown for many years under the name 'Tortuosa' under which they were received. The origin is unknown, but is certainly not the European plant given that name in 1838 and no longer to be found in cultivation. See also Virgata.

-- **'Microphylla'** (Carrière) Hornibrook. 80 (1932); under *A. pectinata* Carrière 283 (1867). Dwarf; branches crowded, short, thin; winter buds small, red, resinous; leaves narrow, thin. Rare.

-- **'Nana'** (Knight) Zederb. 1934 (1907); under *Picea pectinata* Knight & Perry, Syn. Con. 38 (1850), name only; Knight ex Gordon 151 (1858). Shrub, bushy, 2m. high. Uncommon in the trade.

-- **'Pendula'** (Carrière) Ascherson & Graebner. 191 (1897); under *A. pectinata* Carrière 207 (1855). Pendulous, 10–15 m. high; branches pendulous along the trunk. Raised by Godefroy, nurseryman at Ville d'Avray, France in 1835. Mr Mitchell informs me that there are two specimens at 'Endsleigh', Devon, U.K., one with a horizontal branch 6m. long.

-- **'Pendula Gracilis'** Sénéclauze 14 (1868). As the preceding cv. but branches much more elongated and long-pendulous. Of this form there is a good specimen at 'Thorn House', Wembury, near Plymouth, U.K.

-- **'Pyramidalis'** (Carrière) Voss 1237 (1896); under *A. pectinata* Carrière 207 (1855). Sentinel Silver Fir. Fastigiate when young, becoming conical, compact, 7–10 m; branches ascending; branchlets crowded, short; leaves short, lustrous dark green. Cultivated in England in 1851.

-- **'Virgata'** (Caspary) Zederbaur. 1934 (1907); under *A. pectinata* Caspary in Hemp., Oestr. Forstz. 43 (1883). Branches long and slender, growing downwards, densely set with leaves with scarcely any branchlets. Found in a forest in Czechoslovakia, 1879.

A. albertiana A. Murray (1863). Now known as *TSUGA heterophylla*.

A. alcockiana Veitch (1861). Now known as *PICEA bicolor*.

A. alcocquiana Parlatore (1868). Now known as *PICEA bicolor*.

A. alpestris Bruegger. Now known as *PICEA abies*.

A. amabilis (1) Pardé (1906). Now known as *A. concolor* var. *lowiana*.

A. amabilis (2) (Loudon) Forbes 125 (1839); *Picea amabilis* Douglas ex Loudon 2342 (1838); *Abies grandis* Franco 162 (1949) not Lindley (1833). Red Silver Fir (UK.), Pacific Silver Fir (Amer.).

Size and habit Conical tree to 75 m in the wild, less in cultivation, often furnished to the ground with branches, with either a spire-like or rounded crown.

Trunk Straight and massive. *Bark*: Dull grey and smooth for many years but silvery grey and grooved at the base on old trees.

Branches Spreading in tiers, lower branches pendent.

Branchlets Numerous and densely set, grey or orange-brown; lightly grooved, densely pubescent.

Winter Buds Small, globose, 6–9 mm diam, very white, resinous in winter.

Leaves Leaves pointing forward above the shoot, covering the shoot, spreading at the sides and pectinate beneath, about 2.5 cm long, 1.8 mm broad, often broadest above the middle, truncate or notched at apex, shining dark green above, with broad white bands beneath. Often glaucous above in first year on young trees.

Mature cones Oblong, smooth, 9–15 cm long, 5–6 cm wide, greyish-purple, ripening to brown. *Scales*: 12–18 mm broad.

Bracts: hidden.

Seeds Pale brown; wing obliquely wedge-shaped, as long as broad or even longer.

Figure 26. *Abies amabilis* 'Spreading Star'.

Habitat South Alaska to Western Oregon, U.S.A.

Cultivation This species forms an attractive specimen, but is rare in Britain. David Douglas sent seeds to the Hort. Soc. of London in 1830, and it was re-introduced by W. Lobb in 1851. The crushed foliage smells strongly of tangerines, which makes identification very simple. Growth very rapid when young in cool, humid area of Scotland and Ireland. Zone 6. (See Illus Nos. 24, 25)

-- **'Spreading Star'** Den Ouden and Boom, 5 (1965); *A. amabilis* 'Compacta' Meyer, Plant Explorations **2**: 69 (1963), NON Hornibrook 17 (1923). Low-growing and wide-spreading; stem rather thick, short, with numerous horizontal branches. Named for a plant in the Blijdenstein Pinetum, Hilversum, Netherlands. This magnificent specimen is undoubtedly a cultivariant, the result of a grafted side-branch, but this particular clone has been widely distributed under this name so the retention of this cultivar name is justified. Where there is plenty of space for it to develop it makes a spectacular plant. (See Illus No. 26)

A. americana (1) K. Koch (1873). Now known as *PICEA mariana*.

A. americana (2) Miller (1768). Now known as *TSUGA canadensis*.

A. araragi Siebold. (1830). Now known as *TSUGA sieboldii*.

A. araucana Poiret (1804). Now known as *ARAUCARIA araucana*.

A. arctica Cunningham ex Gordon. Now known as *PICEA mariana*.

A. arizonica Merriam (1896). Now known as *A. lasiocarpa var. arizonica*.

A. × arnoldia Nitzelius (1969). A hybrid *(A. koreana × A. veitchii)* raised in the Botanic Garden at Gothenberg, Sweden.

A. atlantica Lindley ex Gordon (1858). Now known as *CEDRUS atlantica*.

A. baborensis Cosson (1861). Now known as *A. numidica*.

A. balsamea (L.) Miller (1768); *Pinus balsamea* (L.) 1753. Balsam Fir.

Size and habit A symmetrical, acutely conical tree, 15–25 m.

Trunk Slender. *Bark*: greyish-brown, smooth but with many resin bumps, heavily scored on old trees.

Branches Spreading or ascending, stout.

Branchlets Spreading, dull grey with short, dark grey pubescence.

Winter Buds Small, ovoid to globular, reddish, very resinous in winter, seemingly varnished.

Scales: 9–10 mm wide, pubescent.

Leaves Spreading upwards at 90° to the shoot in two rows, parted to form a V-shaped valley; on the underside of the shoot usually pectinate; usually 15–25 mm (often to 30 mm) long, 1.25–1.5 mm wide, distinctly notched at apex, flat, grooved, lustrous dark green above, always with a patch of white stomata at the tip; with two white bands below, each consisting of 5–6 lines of stomata. The bruised leaves have a strong scent of balsam, hence the specific name.

Mature cones Fascicled, ovoid-cylindrical with rounded top, 6–9 cm long, 2.5–3.5 cm broad, dark violet at first, but maturing to various shades of blue, grey and green, usually resinous. *Scales*: stalked, about 15 mm long and wide, rounded above. *Bracts*: hidden or but slightly shewing, with a rounded blade and an awl-shapd apex.

Seeds 5 mm long, triangular wedge-shaped, brown; the wing 10 mm long, broad, thinly membraneous.

Habitat Eastern North America, from Labrador to Alberta and Pennsylvania.

Cultivation This species does not mature well in U.K., but makes an attractive tree when young and is popular for its habit of coning freely even when but a few feet high. Clones ripening to different colours are available from some nurseries in U.S.A. It is recorded as having been in cultivation by Bishop Compton as long ago as 1696. Zone 3.

-- **'Andover'** Wyman, Arnoldia **26**: 14 (1966). Named for a prostrate tree with normal foliage found wild in the woods near Andover, New Hampshire, U.S.A. in 1957. It was almost certainly the result of a tree having suffered accidental damage in early life. Although the name was registered, it does not seem ever to have been propagated, in which case it is not, strictly speaking, a cultivar, since by definition (ICNCP, Art. 10) a cultivar must be "An assemblage of plants...".

-- **'Coerulea'** Carrière 294 (1867) Conical shrub; branches very short; leaves arranged in 2 ranks, spreading, dark green above, silvery-white beneath. This name, and also *A. balsamea coerulescens* Senecl. 6 (1868) and *A. balsamea glauca* Beissn. in Mitt. d.d.d.Ges. **8**: 121 (1899) relate to selections made for the bluish tint to the foliage that cannot now be identified. They should therefore not be used. Any new selection made for its foliage colour should receive a new cultivar name, as in the case of 'Cree's Blue', below.

-- **'Cree's Blue'**. New name. This originated as a seedling of outstanding colour in the nursery of Leighton G. Cree of Colebrook, New Hudson, U.S.A., in a plantation of Christmas trees.

-- var. *fraseri* Nuttall (1818). See *A. fraseri*.

-- **'Globosa'**, According to den Ouden in Naamlijst Conif. 48 (1937), this is a synonym of 'Nana'. The names 'Hemispherica', 'Globosa Nana' and 'Nana Globosa' also can be found in lists but have no authority or useful value. Probably none of these identifiable in cultivation.

-- f. ***hudsonia*** (Jacques) Fernaby and Weatherby in Rhodora **34**: 190 (1932); *A. hudsonia* Jacques in Journ. Soc. Agron. Prat. 199 (1899); *A. balsamea nana* (Nelson) Carrière 294 (1867); *Picea balsamea* var. *nana* Nelson 37 (1866); *A. balsamea globosa* Jäger u. Beissner Ziergehölze **2**: 430 (1884). I here follow the treatment suggested in *Dwarf Conifers*, 142 (1966) and recognise this single botanical *forma* to cover all the low-growing mountain forms of this species, continuing the use of the descriptive names in use for separating distinguishable clones (or, it may be, groups of clones) in current circulation. They are all attractive, low-growing shrubs densely clothed with dark green, glossy leaves usually much shorter than the species norm and, because of their origin, indestructibly hardy.

-- **'Hudsonia'** Welch 107 (1966). This is the clone (or more probably a group of clones) the most common form in European cultivation, with a flattened globose or cushion-forming habit and semi-radial foliage. The difference between 'Hudsonia' and 'Nana' can be clearly seen if a twig of the current year's growth be snapped off and looked at from the broken end.

-- **'Nana'** (Nelson) Carrière 294 1867; *Picea balsamea* var. *nana* Nelson 37 (1866). This clone forms a somewhat more globose or rounded bush, seldom reaching one metre, and carries its (usually very short) leaves more or less radially on the stem.

-- var. *phanerolepis* Fernald in Rhodora 11: 203 (1909). Bracted Balsam Fir. Cones a little smaller; bracts ordinarily as long as, and occasionally a little longer than its subtended scale, with divergent exserted awns (in typical *A. balsamea* the bract, including the erect awn, is ordinarily about 2/3 as long as the lamina of its subtended scale). Also considered as a variety of *A. fraseri*. In this southern species, however, the cones are more ovoid and the bracts are very much longer than the scales and have strongly recurved broad tips. In Loudon, Enc. of Trees and Shrubs (fig. 1953) there is a fairly characteristic plate of the cone of var. *phanerolepis*, given as an illustration for *A. balsamea*. It is possibly a natural hybrid.

-- **'Prostrata'** (Gordon) Carrière 218 (1855); *Picea balsamea prostrata* Hort ex Knight 39 (1850) is an uncommon but quite distinctive clone with a prostrate habit and a completely pectinate leaf arrangement.

-- **'Verkade's Prostrate'**. A spreading form that originated from a witch's broom. It has annual growth of 2.5–5cm.

A. batavorum Siebold. Now known as *CUNNINGHAMIA lanceolata*.

A. beissneriana Rehder and Wilson in Sargent (1914). Now *A. chensiensis* var. *ernestii*.

A. beshanuensis M. H. Wu in Acta Phytotax. Sin. **14**: (2): 19 (1976). A recently described species from Chekiang province, China, shewing affinities with *A. firma* and *A. homolespis*. It is not yet in European cultivation. Zone 7.

A. bicolor Maxim. (1866). Now known as *PICEA bicolor*.

A. bifida Siebold & Zuccarini (1842). Now known as *A. firma*.

A. bifolia A. Murray (1863). Now known as *A. lasiocarpa*.

A.× borisii-regis Mattfeld in Notizbl. Bot. Gart. Mus. Berlin, **9**: 235 (1925); *A. alba* var. *acutifolia* Turrill, Kew Bull. **1**: 34 (1925). King Boris Fir.

Size and habit Probably a natural hybrid, *A. alba* Miller × *A. cephalonica* Loud., forming a tree to 30 m with a broad, rounded-conic crown of strong, parallel raised branches until in maturity.

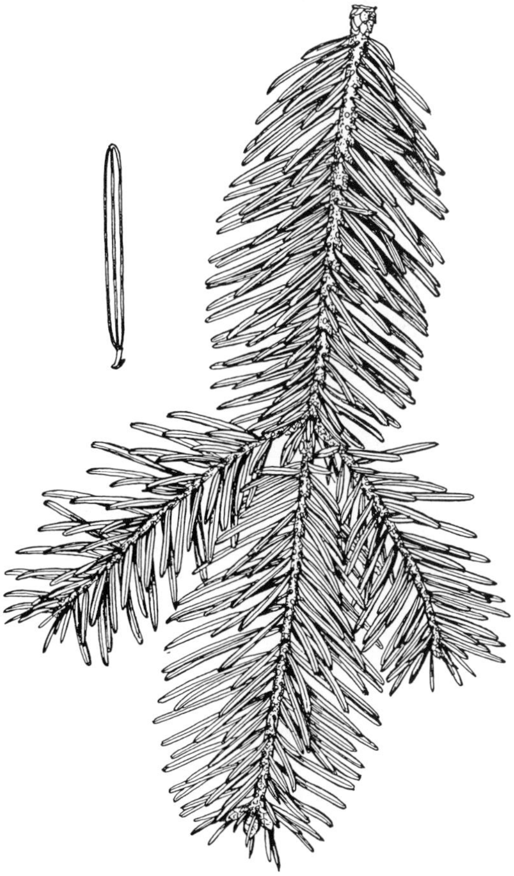

Figure 27. *Abies borisii-regis*. King Boris fir. Detail of foliage from a side branch. A seldom recognised, vigorous tree. Note the narrow leaf and densely pubescent shoot. (5)

Figure 28. *Abies bracteata.*. Santa Lucia fir. Detail of foliage from a side branch. The long, tapered buds and sharply pointed leaves are characteristic. (5)

Trunk Straight, stout, somewhat devoid of branches near the ground. *Bark*: Very dark grey, texture rough.

Branches Ascending.

Branchlets Spreading, short, with dense, yellowish or blackish pubescence when young.

Winter Buds Winter buds moderately resinous.

Leaves Slender, densely set in flattened ranks, on the upper side of the shoot pointing forward; 3–3.5 cm long, 1.5–2 mm broad, often acute and bluntly pointed, sometimes rounded or slightly notched; grooved; rich shining dark green, often without stomata above (or with a few near the leaf-tip) and with two white bands each of 6–10 stomatal lines beneath.

Mature cones Conical-cylindric, narrowing upward, up to 15 cm long, 4–5 cm broad; *Scales*: stalked, 7–10 mm long, often wedge-shaped, margin serrate or undulate, rounded at the apex, the uncovered part shortly tomentose; *Bracts*: projecting and reflexed, with a linear stripe 12–14 mm long and an elliptical blade 5–6 mm long.

Seeds 10–12 mm long; wing 15 mm long, 12–14 mm wide, notched at apex.

Habitat Mountains in South Bulgaria, on the peninsula of Athos, the isle of Thasos, Greek-Macedonia, northern and central Greece.

Cultivation This hybrid is intermediate between its parents and, being variable, is by some authorities considered a variety of *A. alba*.

It was introduced into cultivation in 1883. It was named in honour of King Boris of Bulgaria. It is rare in cultivation, but makes a magnificent specimen tree. Zone 6. (See Illus No. 27)

A. × ***bornmuelleriana*** Mattfeld in Notizbl. Bot. Gard. Berlin, **9**: 239 (1925). Bornmueller Fir.

Size and habit A natural hybrid, *A. cephalonica* Loudon × *A. nordmanniana* (Steven) Spach. which forms a tall tree, furnished to the ground with branches, in habit resembling *A. nordmanniana*.

Trunk Bark smooth, fissured at base of old trees only; black.

Branches At first spreading, becoming horizontal

Branchlets Stout, glabrous, shining green, becoming brown 2nd year year.

Winter Buds Ovoid, but very pointed at apex, 8–10 mm, resinous. *Scales*: Visible through the coating of resin, green with brown tips.

Figure 29. *Abies bornmuelleriana*. Bornmueller fir. Detail of foliage from a side branch.

Figure 30. *Abies fargesii.*. Farges fir. Detail of foliage from a side branch. Note the deeply bifid tip and the purple or red shoot colour. The long, glossy leaves and the absence of any blue tinge distinguishes this species from the *Abies delavayi* group.

Leaves Very densely set short leaves above the shoot held stiffly upwards and forwards to hide the shoot, spreading and pectinate on the lower side of the shoot; 30–40 mm long, about 3 mm broad, rounded or notched and (on fruiting branchlets) pointed at apex; dark green, often with noticeable stomatal lines above, not extending the full length of the leaf; with two broad bands of stomata beneath.

Mature cones Cylindrical to 15 cm long and 6 cm in diam.; scales stalked, about 30 mm long and up to 40 mm wide, abruptly broadening at the base; bracts exserted and reflexed, up to 1 cm long, abruptly contracting to a long pointed apex.

Seeds Up to 15 mm long, 5 mm broad.

Habitat Found at altitudes of 1100–1700 m on mountains along the southern coast of the Black Sea in N W Asiatic Turkey, occasionally forming forests. It was named in honour of Prof. J. Bornmueller, a German botanical explorer who introduced the tree. Zone 5 (See Illus No. 29)

A. brachyphylla Maxim. (1866). Now known as *A. homolepis*.

A. brachytyla Franchet (1899) Now known as *PICEA brachytyla*.

A. bracteata (D. Don) Poiteau in Rev. Hort. ser. 2 IV. 7 (1945). Santa Lucia fir, Bristlecone fir.

Size and habit Slender, broadly conical tree 30–60 m with a narrow crown.

Trunk 0.6–1.5 m in diam; *Bark*: Smooth, brown or grey, fissured at the base of old trees and around dead branches.

Figure 31. *Abies bracteata.* Santa Lucia fir. A bud, foliage and a ripe cone.

Figure 32. *Abies cephalonica.* Detail of foliage from a side branch. The spiky, radiating leaves are pointed, unlike the similar but blunt leaves of *Abies pinsapo.* (5)

Branches Crowded, lower branches pendulous; upper branches shorter and ascending.

Branchlets Stout, glabrous and polished, greenish or red-brown to purple.

Winter Buds Very large, elongated, more than 2.5 cm long, 6–8 mm broad, spindle-shaped, pointed at the apex, pale yellowish, not resinous.

Leaves Very large and loosely set, pectinate save for a few leaves pointing forwards along the shoot; very stiff, linear to linear-lanceolate, 3.5–5 cm long, 2–3.5 mm broad, rigid, sharply spine-pointed, shining dark green above, somewhat grooved towards the base, with broad white bands, each of 10 stomatal lines beneath. Persistent on the tree for several years.

Mature cones Ovoid to nearly cylindrical, 7–10 cm long, 4–5.5 mm broad, shining green, ripening to purplish brown, resinous. *Scales*: kid-

Figure 33. *Abies cilicica.* Cilician fir. Detail of foliage from a side branch. Note the stout shoot, and the stomata, near the tip of the leaf above and in broad bands below. (5)

ney-shaped to roundish from a wedge-shaped base, rounded and notched at margin. *Bracts*: Visible, ending in rigid decurved spines 2.5–5 cm long which give the cones a bristly appearance that is quite distinctive, hence the popular name 'Bristlecone fir'.

Seeds Oblong-wedge shaped 9 mm long, dark reddish-brown, thin, shining light brown.

Habitat S. California, but only on the Santa Lucia mountains at an altitude of 1000–2000 m in cool and damp places.

Cultivation A beautiful and outstandingly distinctive fir, succeeding on chalky soils. Introduced into Britain by the collector W. J. Lobb, who sent seeds to the Veitch nurseries in 1854. Zone 7. (See Illus No. 000)

A. bridgesii. Kellog (1863). Now known as *TSUGA heterophylla.*
A. brunoniana. Knight & Perry (1850). Now known as *TSUGA dumosa.*
A. californica Hort. ex Steudel (1840). Now known as *PSEUDOTSUGA menziesii.*
A. campylocarpa A. Murray (1860). Now known as *A. magnifica.*
A. canadensis (1) Michaux (1803). Now known as *TSUGA canadensis.*
A. canadensis (2) Miller (1768). Now known as *PICEA glauca.*
A. caroliniana (Engelmann) Chapman (1883). Now known as *TSUGA caroliniana.*
A. cedrus Poiret (1805). Now known as *CEDRUS libani.*

A. cephalonica Loudon 2325 (1838). Greek Fir.

Size and habit Broadly conical tree to 20–30 m, usually with a rounded crown.

Trunk Up to 5 m in girth. *Bark*: Greyish- to pink-brown, smooth on young trees, dark grey and fissured into oblong loose plates on old trees.

Branches Long, spreading horizontally in regular tiers.

Branchlets Striped, glabrous, shining red-brown.

Winter Buds Ovoid to ovoid conical, pointed or slightly blunt at apex. reddish-purple; very resinous. Bud-scales protrude through terminal leaves.

Leaves Densely set on thin stalks springing from large bases, radial, spreading at nearly 90°; leaves on upper side of branchlets shorter and directed forward; linear, thin, tapering and sharply pointed at the apex, 15–25 mm long × 2 mm broad, stiff, lustrous bright green, occasionally grooved, usually with some stomatal lines above near the apex and even down to the middle; with two white bands each of 5–6 stomatal lines beneath.

Mature cones Subsessile, narrowly cylindrical, 12–18 cm long, 3–5 cm broad, greenish-brown when young, ripening to brown or reddish-violet, resinous. *Scales*: fan-shaped, rounded, entire. *Bracts*: Projecting and reflexed.

Seeds Triangular-ovoid, 7 mm long, light brown; with wing 16 mm long.

Habitat Greece and southern Yugoslavia.

Cultivation Introduced into Britain from Greece as seeds by General Sir Charles Napier in 1824. The earliest fir to flush, this species is prone to damage from Spring frosts, especially in the young state. Zone 6. (See Illus No. 32)

-- var. *apollinis* (Link) Beissn. (April 1887). Now known as *A. cephalonica* var. *graeca.*

-- **'Aurea'** Carrière (1867). A selection with branchlets and leaves golden yellow when young. Raised at Bourg-Argental, France and probably no longer in cultivation. Indistinguishable from 'Aureovariegata' Beissner 443 (1891).

-- var. ***graeca*** (Fraas) Liu 222 (1971). A sporadic variant, differing from the norm only by the leaves having rounded tips and being pointed forward and held rather densely above the shoot.

-- **'Meyer's Dwarf'** Den Ouden and Boom, 10 (1965); *A. cephalonica* 'Nana' Meyer Plant Explorations **2**: 69 (1983); Hillier Dw. Con. 10 (1964) and Man. 458 (1971). Forms a slow-growing, flat-topped, spreading plant with rigid branchlets and leaves shorter than normal. It is a very attractive form which should be more freely planted.

-- 'Nana' Hort. A mistake for 'Meyer's Dwarf'.

-- **'Robusta'** Carr. (1867) 285. Crown spirally shaped; branches erect, long, stout, decurving below, moderately spirally and widely borne;

branchlets crowded, thick; leaves very densely set, thick, stiff, sharp and blunt pointed, on young shoots narrowed abruptly near the top, blunt or notched at the apex, bluish-white beneath. Originated in France, 1845. Rare.

A. chayuensis. Cheng et Fu (1978). A little-known species; possibly a sub-species of *ABIES forrestii*. (per Rushforth 1987).

A. chensiensis Van Tieghem in Bull. Soc. Bot. France **38**: 413 (1891). Shensi Fir.

Size and habit Conical tree to 40 m with rounded crown, often flattened in old trees.

Trunk Usually straight. *Bark*: Smooth dark-grey to light brown; fissured in older trees.

Branches Short and massive.

Branchlets Glabrous, yellowish-grey, becoming dark grey with age, sometimes grooved.

Winter Buds Ovoid, grey-brown, 8–11 mm diam, slightly resinous in winter.

Leaves Spreading, more or less 2-ranked, forming a V-shaped valley but incurved, linear, broader above the middle; 2.5–3.5 (to 7) cm long, 2 mm wide, grooved, rounded and often notched at the apex, shining yellow-green above, with bands of greyish-green or often blue stomata beneath. Leaves of young plants resemble those of *Keteleeria davidiana*.

Mature cones Sub-sessile, ovoid-oblong with an elongate apex, 8–10 cm long, 2–4 cm broad; green when young, ripening to reddish-brown. *Scales*: About 3 cm wide, jagged margins, pubescent outside. *Bracts*: hidden.

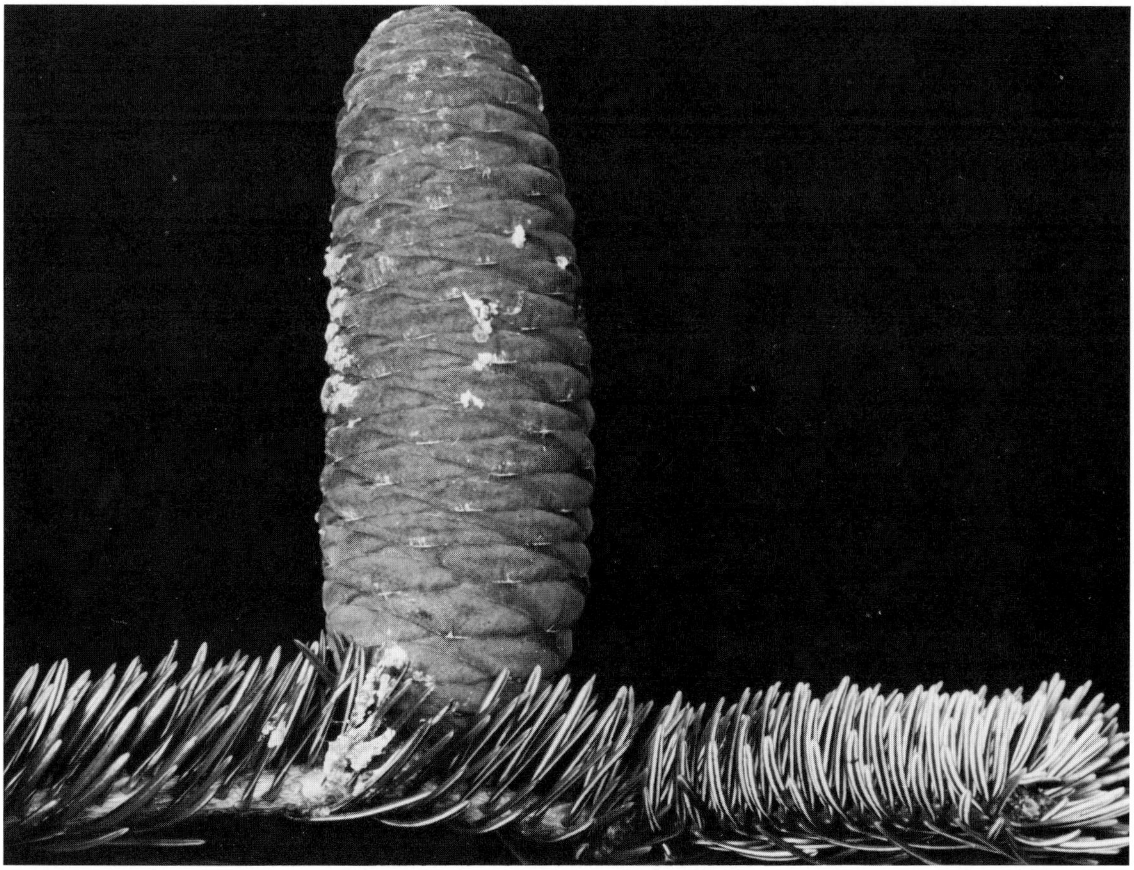

Figure 34. *Abies cilicica*. A ripe cone. (10)

Figure 35. *Abies concolor* var. *concolor.* Concolor White spruce. Detail of foliage from a side branch. The thick, upturned leaves have stomata on both sides. (5)

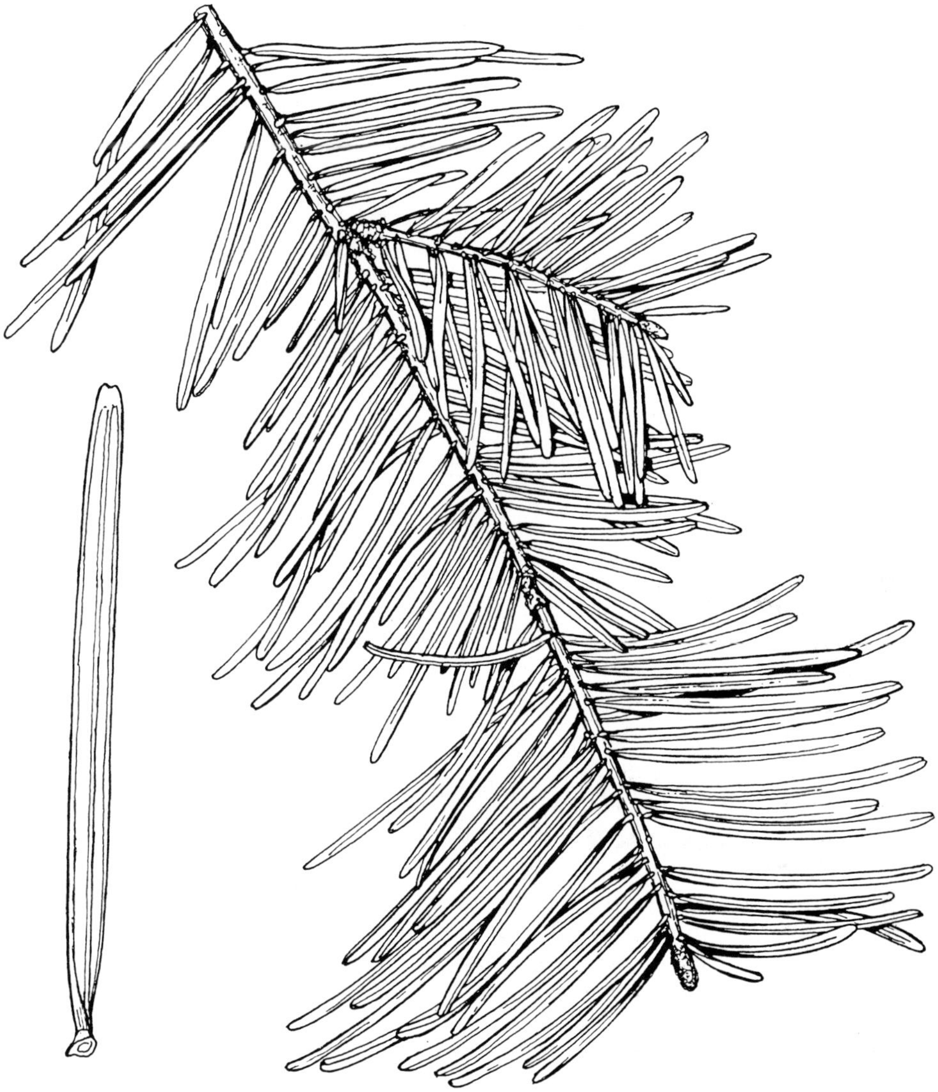

Figure 36. *Abies concolor.* var. *lowiana.* Californian White spruce. Detail of foliage from a side branch. In this var. the leaves are less curved and lack stomata on upper surface. (5)

Seeds Obovoid, 6–7 mm, wing 1.3 cm long, pale brown.

Habitat Yunnan and West Szechuan, Central China.

Cultivation Introduced to Britain in 1907 by the collector E. H. Wilson who sent seeds to the Veitch nurseries, but it is still very rare in cultivation. Zone 6.

-- var. ***ernestii*** (Rehder) Liu 135 (1971); *A. ernestii* Rehder in Journ. Arn. Arb. **20**: 85 (1939); *A. beissneriana* Rehder & Wilson in Sargent, Pl. Wils. **2**: 46 (1914). Widely distributed; vigorous but not markedly different from the species norm. Liu, 137 (1971) sets out the defferences in foliage and cone scales which he considers enough to justify treating it as a *varietas*.

A. chiloensis Carrière, 299 (1867). Now known as *A. spectabilis*.

A. chilrowensis Henkel & Hochstetter, (1865). Probably now included in *A. firma*.

A. chinensis Franchet, 259 (1899). Now known as *TSUGA chinensis.*

A. cilicica (Antoine & Kotschy) Carrière 229 (1855); *Pinus cilicica* Ant. & Kotschy in Oestr. Bot. Wochenbl. **3**: 409 (1853). Cilician Fir.

Size and habit Narrowly conical tree to 20–30 m high. Crown usually somewhat sparse.

Bark Ashy-grey, deeply grooved or breaking up into scaly plates, especially around dead knots, elsewhere smooth, but eventually finely plated.

Branches Crowded, spreading in tiers, the lower branches pendent, the upper branches ascending.

Branchlets Smooth, with scattered pubescence but soon becoming glabrous, yellowish-green or greyish-brown.

Winter Buds Ovoid, 3–4 cm long, chestnut-brown, sometimes resinous in winter. *Scales*: Keeled, free at tips. 1–1.5 mm across.

Leaves On the upper sides of shoots - directed forwards and curving upwards, leaving a V-shaped valley between the two rows. 2–3.5 cm long. ± 1.5 mm broad; thin, rounded or notched at apex; dark green, grooved, lacking stomata above (save occasionally, near the tip) but with two separate, narrow, bluish-white bands beneath, each consisting of about 7 stomatic lines.

Mature cones Almost sessile, cylindrical, 16–30 cm long. 4–6 cm broad, reddish-brown. *Scales*: broad, fan-shaped. *Bracts*: hidden, shorter than the scales; spatulate.

Seeds Obovoid, 13–14 mm long, brownish-red with wing 18 mm long; yellowish-red.

Habitat E. Turkey: Antitaurus Mts.; N. Syria; Lebanon, between 1000 and 2000 m. Introduced by Balansa into France (Paris) in 1854.

Cultivation Very rare in Britain. Zone 6. (See Illus. Nos. 19b, 33, 34)

-- var. *leioclada* Gordon (1858) (part). A minor variant. No longer identifiable.

A. clanbrassiliana Lawson 370 (1836). Now known as *PICEA abies* 'Clanbrassiliana'.

A. coahuilensis (Johnston) Martinez (1963). Now known as *A. durangensis* var. *coahuilensis*

A. communis Lawson 365 (1836). Now known as *PICEA abies.*

A. concolor (Gord.) Ldl. ex Hildebrand. Verbr. Con. 261 (1861): *PICEA concolor* Gord. 155 (1858). Colorado Fir; White Fir.

The species is divided into two botanical varieties, var. ***concolor***,[1] the typical or mountain form, and var. ***lowiana***, the coastal form. The distinction may be of climatic origin, since intermediate forms can be found and *A. concolor* var. *lowiana* shews an affinity with *A. grandis* in Oregon at the northern limit of its range, where hybridisation undoubtedly occurs. Zone 3. (See Illus. Nos. 35, 36.)

The typical form var. *concolor*, is here described, the features distinguishing var. *lowiana* being given below.

Size and habit Columnar-conic tree, 40–50 m with a spirelike crown as a young tree, becoming rounded with age; furnished to the ground with branches.

Trunk At first straight, often dividing later, 0.7–1.3 m in diam. *Bark*: Ashy-grey with a trace of brown; smooth on young trees but with resin blisters (the white colour of the inside gives the reason for the popular name, White Fir), with age becoming darker grey and finely fissured and cracked.

[1] When you partition off part of a room it is undeniable that you also partition off the remaining part of that room. The botanical rule is that whenever one part of a species (or more than one part) is recognised as a *varietas* (Variety) the residue shall be automatically regarded as as variety with the same name as the species. Such a name is called an autonym and *A. concolor* var. *concolor* is just such a case. Because of its rump-like character it is not necessary to quote any 'authority' for an autonym.

Branches In tiers, the lower branches drooping, upper ones ascending and finally becoming upright, giving rise to a rounded crown on old trees.

Branchlets Smooth and stout, pointing forwards, bright yellow-green, sometimes with slight reddish pubescence, but usually glabrous.

Winter buds Ovoid with a rounded apex, brown, resinous, somewhat hidden by leaves. *Scales*: large, broadly triangular.

Leaves On sterile branchlets widely spaced and irregularly arranged and noticeably upcurved, 4.5–8 cm long, 2–2.5 mm broad, acute or rounded at apex, dull blue or grey-green, with numerous lines of stomata above, glaucous on both sides.

Mature cones Sessile or short-stalked, cylindrical but tapering, 10–14 cm long, 3.5–5 cm broad, pale green ripening to purple. *Scales*: broadly fan-shaped from a wedge-shaped base, toothed on both sides, 12 mm long, 24 mm wide. *Bracts*: Hidden, 1/3 as long as cone-scales.

Seeds Cuneate, obovoid, acute at base, 8–10 mm, glossy, dull yellow-brown; wing thin, oblong-cuneate, 20 mm, pinkish.

Habitat Rocky Mts. and mountains of New Mexico. Arizona and New Mexico at levels from 600 to 3000 m. Zone 3. See Illus. No. 35.

-- var. ***lowiana*** (Gordon) Lemmon, 64 (1895); *Picea lowiana* Gordon 53 (1862); *A. lowiana* (Gordon) Murray in Proc. Hort. Soc. London **3**: 317 (1863); *A. concolor* var. *lasiocarpa* Beissner 473 (1891); *A. lasiocarpa* Masters in Gard. Chron. 8 (1880/1) not Nuttall, (1849). Pacific White Fir. See Illus. 36.

Differs from the species norm in having darker (brown or grey) bark, on old trees with deep longitudinal fissures exposing the yellow-brown surface beneath. Shoots slender, copper-brown the second year. The leaves are dark green, grooved, more or less straight, arranged pectinately or rising uniformly at an angle to the shoot forming a V; 3–4 cm, long, in two ranks, the upper rank being shorter. The winter buds are smaller: the cones ripen to chestnut-brown, and the bract-scales are rounded.

Habitat Mountains in the South Pacific coastal region of U.S.A.: S. Cascades, Siskiyous and the Sierra Nevada and cross-ranges.

-- 'Albospica'. This and 'Argentea' are names for now unidentifiable clones with white or whitish foliage such as will turn up occasionally in the seed-beds. See the cultivars 'Candicans' and 'Wattezii' below.

-- **'Archer's Dwarf'** forms a multi-stemmed, upright little bush with pale blue, narrow and recurved leaves, the branches tending to curl downwards. The buds are obtuse or globular.

-- **'Aurea'** Beissner in Mitt. d. d. d. Ges. **15**: 144 (1906). Strong-growing, young branchlets golden yellow in May, afterwards silvery grey. Raised by Ansorge, nurseryman, Flottbeck (near Hamburg), Germany.

-- **'Candicans'** Hort. ex Viguie et Gaussen in Bull. Soc. Hist. Nat. Toulouse, **17**: 320 (1929). A strong-growing cultivar with light-coloured leaves, which originated in Les Barres Arboretum, near Paris. It is very effective if propagated from a side branch and allowed to develop as a large, spreading plant, several metres across. In such a case a leader must not be allowed to develop.

-- **'Compacta'** (Beissner) Hornibrook 24 (1939); *A. concolor* f. *violacea compacta* Beissner 476 (1891); *A. concolor* var. *glauca 'Compacta'* Hillier 10 (1964). Dwarf, irregular, compact shrub; leaves stout, stiff, straight or falcate, 2.5–4 cm long, very glaucous. A very good garden plant. (See Illus No. 37)

-- **'Conica'** Slavin 90 (1932). Conical White Fir. Dwarf, conical, low-growing; leaves horizontal, 2–4 cm long; originated in the Pinetum of the Durand Eastman Park, Rochester, N.Y., U.S.A.

-- **'Elkin's Weeper'** Welch, 143 (1979). A slow-

Figure 37. *Abies concolor* 'Compacta'. A deservedly popular dwarf form.

growing and strongly pendulous form with dark green leaves, always much curled. It was raised in U.S.A. by Harry Elkins.

-- **'Fagerhult'**. A vigorous, very pendulous clone found in a seed-bed on the Fagerhult Nursery in Sweden about 1933. It was received from the late Tage Lundell of Helsingborg, Sweden in 1973 and was at first distributed by the Wansdyke Nursery, Devizes, U.K. as 'Pendula Fagerhult', an illegitimate name.

-- **'Gable's Weeper'**. An extremely dwarf, pendulous form that originated on the Joseph B. Gable nursery, Stewartstown, Pennsylvania, U.S.A. Introduced and registered by Don Smith of the Watnong Nursery, Morris Plains, NJ., in 1969. At 18 years, the mother plant was 40 cm high and 50 cm across.

-- 'Glauca Compacta'. Superfluous name for 'Compacta'.

-- 'Green Globe'. Now listed under *A. lasiocarpa.*

-- **'Husky Pup'**. A dwarf or slow-growing form. The original plant at 15 years is broadly conical in outline, 40 cm tall and nearly as wide at the base. The new shoots, which have an irregular pattern of growth quite unlike most true firs, are thick and short, from 2–6 cm long, olive green changing to grey-brown on older wood. The needles, seldom over 2.5 cm long by 2 mm broad are much smaller than those typical of this species. They are dark grey-green above, with two faint blue-grey lines below. The plant was found in a group of seedlings by Mr Nickolas Sizoo of Bothell, Washington, and was registered and introduced by the University of Washington in about 1965. See A. A. B. G. A. Bull. **9** (2): 1975.

-- 'Horstmann's Igel'. See 'Igel'.

-- **'Igel'** Reuter, Amer. Conif. Soc. Bull. **1** (3): 92 (1984). This arose on a nursery in West Germany in a seed-bed of *Abies concolor* var. *lowiana*. It slowly forms a hemispherical plant with an annual growth of 1.502 cm. It is very difficult to propagate. The name 'Horstmann', being already in use in *Abies pinsapo*, is ineligible for re-use within the same genus, so the simpler name is preferable.

-- **'Masonic Broom'** This clone was raised from a witch's broom in U.S.A. It makes a congested little plant similar to 'Pigglemee'.

-- **'Pendula'** (Simon-Louis) ex Beissner in Mitt. d. d. d. Ges. **5**: 64 (1896); Carrière in Rev. Hort. 117 (1890); *A. concolor* var. *pendens* Sudworth 55 (1897); *A. lowiana* var. *pendula* Fitsch in Beissn. 163 (1930). In view of the repeated use of the name 'Pendula' (and the name 'Pendens' associated with it) for several clones, none of which seem to be still in cultivation, it should not now be used other than in a loose, collective sense.

-- **'Piggelmee'** Draijer ex Grootendorst, Dendroflora **8**: 73 (1971). An attractive and very slow-growing form that originated as a witch's broom on a tree of *A. concolor* 'Candicans' in a garden in Geneva. It retains the silvery-blue colouring but has much shorter leaves and an annual growth of only 3–4 cm. It was introduced to the trade by the Draijer Nursery, Heemstede, Netherlands.

-- **'Pyramidalis'** Krüssmann 44 (1979). Described as having a strong, upright habit and a fastigiate branching system, presumably found in Whitnal Park, Milwaukee, U.S.A., but the name would appear to be illegitimate and needing replacement by an acceptable name.

-- **'Violacea'** (Murray) Beissner 45 (April 1887): under *Picea concolor* var. *violacea* Murray in Gard. Chron. ser. 2, **12**: 684 (1875). Seedlings with more or less glaucous-blue foliage colour turn up from time to time. The clone in cultivation under this name is outstanding, so must be propagated vegetatively.

-- **'Wattezii'** Wattez ex de Graaf in Ned. Tuinb. bl. **16**: 277(1900). Leaves pale yellow when developing, becoming yellowish-white, finally assuming a colour like that of a languishing tree. Raised by D. Wattez, nurseryman at Bussum, Netherlands. Rare in cultivation.

-- **'Wattezii Prostrate'** is merely a cultivariant, i.e. a plant propagated from lateral growth and so forming only a low, spreading bush. It makes an attractive garden plant.

-- **'Wintergold'** Krüssman, Nadel. ed. 3, 45 (1979). A form that is a rather insipid yellow-green during the summer, but which turns a brilliant gold during the winter months.

A. davidiana Franchet (1884). Now known as *KETELEERIA davidiana* Beisnner.

A. delavayi (Van Teighem) Franchet in Journ. Bot. **13**: 255 (1899); *A. delavayi* var. *motuensis* Cheng et Fu, Acta Phytotax. Sin. **13**(4): 83 (1975); *A. fabri* (Masters) Craib, Roy. Bot. Edinb. **11**: 278(1919); *A. minensis* Bord. Rey et Gaussen, Trav. Lab. For. Toulouse **1. 4**(15): 10. f.1-12 (1947). Delavay Fir, Yunnan Fir.

This species is one of a group of Asian Firs so closely related that they must have caused many botanical headaches. Each successive investigator seems to have come to a different judgment as to which was a variety of which species and which should be given specific status on its own. In consequence, considerable disagreement between the different conifer books will be found, as evidenced by the Table on p. 129.

I here follow the treatment by Tang Shu Liu in *A Monograph of the Genus Abies*, (1971).

Size and habit Spectacular tree to 40 m, often crowned with foliage.

Trunk Straight, erect, tapering. *Bark*: Smooth, grey or brownish grey, on old trees flaking or with deep longitudinal furrows.

Branches Stout, relatively few.

Branchlets Numerous, stout, shining reddish-brown, orange of brownish grey; slightly grooved; glabrous or nearly so.

Winter buds Ovoid, large 1–1.2 cm, obtuse at

Figure 38. *Abies delavayi.* var. *fabri.* Faber's fir. Detail of foliage from a side branch. The bark is dark brown. Not distinguished within the species by Liu (1971). See Table on p. 124.

Figure 39. *Abies delavayi* var. *smithii.* Detail of foliage from a side branch. The variety most commonly grown in U.K., for many years listed as var. *forrestii.* The shoot is rough and glabrous.

	Liu (1971)	Silba (1984)	Rushforth (1987)
Delavay Fir	*A. delavayi* Franchet var. *delavayi.*	*A. delavayi* Franchet var. *delavayi.*	*A. delavayi* Franchet.
Forrest Fir	*A. delavayi* Franchet var. *smithii* (Vig. et Gauss.) Liu.	*A. delavayii* Franchet var. *smithii.* var. *forrestii* pro syn.	*A. forrestii* C. Coltman-Rogers. *var. smithii* pro syn.
George Fir	*A. delavayi* Franchet var. *georgei* (Orr) Melville	*A. delavayi* Franchet var. *georgei (Orr) Melville*	*A. georgei* Orr. Seldom grown true in UK.
Faber Fir	Not distinguished within the species	Not distinguished within the species	*A. fabri* (Masters) Craib.
Farges Fir	*A. fargesii* Franchet var. *fargesii.*	*A. fargesii* Franchet.	*A. fargesii* Franchet.
Faxon Fir	*A. fargesii* Franchet var. *faxoniana.* (Rehd. et Wils.) Liu.	Not distinguished within the species.	Not distinguished within the species.
Sutchuensis Fir	*A. fargesii* Franchet var. *sutchuensis. A. sutchuensis* (Franchet) Rehd. et Wils. pro syn.	Not distinguished from *A. fargesii.*	Not distinguished from *A. fargesii.*

apex, resinous, a rich chestnut brown.

Leaves Crowded, held in three or more ranks, shortest in the topmost rank; 1–4 cm long. The most noticeable feature is the strongly revolute (curled downwards) margins, giving the leaf a cross section almost that of a figure 8 on its side, nearly concealing the stomatal bands.

Mature cones Cylindrical or barrel-shape; sessile or on a short stalk, 9 cm by 4 cm, purplish black or blue. Scales closely imbricate. Bracts hidden.

Habitat High mountain areas of N. E. India, N. Burma, S. China, Tibet, N. W. Yunnan, at 2400–4000 m.

Cultivation Named for Father I. M. Delavay, a French missionary who discovered it in 1887. It is still extremely rare in Britain. Zone 8. (See Illus. Nos. 38, 39)

-- var. ***georgei*** (Orr) Melville, Kew Bull. 533 (1958): *A. georgei* Orr, Notes Roy. Bot. Gard. Edinb. **18** (1): 276 (1933). George Fir. Thisand the following variety only differ in minor respects. It forms a smaller tree. The shoots are dark purplish brown, with a dense light reddish-brown pubescence; the leaves are held forwards around the shoot, glistening white on the underside and are narrower and only slightly revolute. The resin on the winter bud is yellow and there are certain differences in the cone and the bracts.

-- '**Headfort**' New name. *A. delavayi* 'Nana' Hillier, Man. 459 (1971). A slow-growing form of the type. Winter buds orange brown. Leaves more or less radially arranged, 1–1.5 cm long, the margins incurved. It originated at Kells, County Meath in Ireland, seat of Lord Headfort, (hence the name chosen). From the picture in *Conifers* (van Gelderen **37**: 1986) I believe 'Headford' to be the same.

-- var. ***smithii*** (Vig. et Gauss.) Liu, 143 (1971); *A. delavayi* var. *forrestii* (Rogers) Jackson ex Chittenden, Conif. Cult. 245 (1932); Forrest Fir. Distinguishing differences in this variety are the orange-brown to red young shoots, the winter buds covered with resin, white in colour, the dark blue-green leaves arranged all around the shoot, only moderately revolute, often twisted so as to expose the glistening white underside. This is the variety usually found being grown as *A. delavayi* in Britain.

A. densa Griffith (1851). Now known as *A. spectabilis* var. *densa*.

A. deodara Standish & Noble (1852). Now known as *CEDRUS deodara*.

A. diversifolia Maxim. (1868). Now known as *TSUGA diversifolia*.

A. douglasii. Lindley (1833). Now known as *PSEUDOTSUGA menziesii*.

A. drummondii Hort. ex Gordon (1858). Now known as *PSEUDOTSUGA menziesii*.

A. dumosa (D. Don) Loudon (1838). Now known as *TSUGA dumosa*.

A. durangensis Martinez, Anal. Inst. Biol. Mexico **13**: 621 (1942). Durango Fir.

Size and habit Tree to 40 m, narrowly pyramidal when young, often round-topped in maturity.

Trunk Usually straight, massive. *Bark*: thick, greyish-red, deeply fissured on old trees.

Branches Wide-spreading.

Branchlets Erect, opposite; blackish purple-brown, prominently ridged and furrowed, glabrous or sometimes with minute pubescence in the furrows.

Winter Buds Oblong-ovoid, thickly coated with yellow-pink resin.

Leaves Narrowly-linear, more or less pectinately held, 25–35 mm long, occasionally longer, rounded or obtuse at apex, light green above, deeply grooved in the lower part of the leaf, with inconspicuous stomata towards the leaf-tip. Beneath, two hands, each of ten lines of stomata, separated by a thin, raised mid-rib. The leaf scar is round-oval.

Mature cones Sub-sessile, nearly cylindrical, 6–10 cm long by 3.5 cm wide, obtuse at apex; pale brown. *Bracts*: hidden.

Seeds Yellowish.

Habitat States of Chihuaha and Durango in N. Mexico. Zone 7.

-- var. **coahuilensis** (Johnston) Martinez, Los Pin. Mexicanos, ed. 3, 400 (1963); *A. coahuilensis* Johnston in Journ. Arn. Arb. **24**: 332 (1943). A geographical form from northern Mexico, differing in details of foliage and in the shape and size of the cone-scales, bract-scales and seeds.

A. elegans Smith of Ayr ex Knight & Smith (1850). Now known as *PICEA abies* 'Elegans'.

A. engelmannii Parry (1863). Now known as *PICEA engelmannii*.

A. equi-trojanii Aschers et Sint. (1883). Now known as *A. cephalonica* var. *graeca* (Fraas) Liu. 223 (1971).

A. eremita Hort. ex Koch (1873). Now known as *PICEA abies* 'Eremita' Carrière (1855).

A. ernestii Rehder. Now known as *A. chensiensis* var. *ernestii*.

A. excelsa Poiret, Encycl. Math. Bot. **6**: 518 (1805). Now known as *PICEA abies*. (And numerous cultivars.)

A. excelsior Franco (1949). Now known as *A. grandis*.

A. fabri (Masters) Craib. (Faberi). Now merged with *A. delavayi*. See Liu, 139 (1971).

A. falcata Rafinesque (1832). Now known as *PICEA sitchensis*.

A. fargesii Franchet in Journ. Bot. **13**: 256 (1899). Farges Fir.

Size and habit Handsome tree to 40 m, narrowly conical, with luxuriant yellowish-green foliage.

Trunk Stout, with very little taper. *Bark*: smooth, pinkish-grey.

Branches Massive, horizontal, upturned at their tips.

Branchlets Varying in colour, from shining reddish-brown to purple; slightly grooved; glabrous or with minute pubescence in the grooves; turning pale orange-brown the second year.

Winter buds Ovoid, 5–6 mm long, yellowish-brown, very resinous.

Leaves Held more or less pectinately, those of the middle rank directed upward or slightly recurved, those of the lower ranks pectinate; emarginate, 2–4.5 cm long, 2–3 mm broad; slightly tapering; tip on young trees strongly bifid; with leathery texture; shining deep yellow-green above, with a prominent mid-rib and two grey white bands below.

Mature cones More or less sessile, 7–19 cm long, 5–7 cm broad, purplish when young. *Scales*: broad, auriculate at base, entire. *Bracts*: exserted, reflexed.

Seeds 5.5 mm long. *Wing*: about 7 mm long, 6.5 mm wide at top.

Habitat Western China.

Cultivation Named in honour of R. P. Farges, who discovered the tree. It is not one of the *A. delavayii* complex, being quite distinctive in the bright colouring of the one-year shoots and the yellowish-green of the leaves. Zone 5. (See Illus No. 30)

-- var. ***faxoniana*** (Rehder et Wilson) Liu, 151 (1971); *A. faxoniana* Rehder et Wilson in Sarg., Pl. Wils. **II**: 304 (1922); *A. delavayi* var. *faxoniana* A. B. Jackson in Chittenden, Conif. in Cult. Faxon Fir. A widespread and rather variable tree. When first introduced it was given specific rank, but is now regarded as a variety, by many authorities of *A. delavayi*. Liu considers that it is more closely related to *A. fargesii*.

From the species it differs in the young shoots having dense, brown pubescence, small buds and its short linear-ligulate leaves rather irregularly held. It was discovered by E. H. Wilson at a very high altitude in Western Szechuan in Central China and introduced by him to the Arnold Arboretum, U.S.A. in 1910.

-- var. ***sutchuenensis*** Franchet in Journ. Bot. **8**: 256 (1899); *A. sutchuensis* (Franchet) Rehder et Wilson 48 (1914). Szechuan Fir. A geographical variant from Szechuan Province which differs only in having shorter, flatter and more crowded leaves more acute at the apex. Discovered by W. Purdon on the banks of the Tow river in western Kansu. No longer distinguished by some botanists.

A. faxoniana Rehder & Wilson. Now known as *A. fargesii* var. *faxoniana*.

A. *firma* Sieb. et Zucc. 15 (1842). Momi Fir; Japanese Silver Fir.

Size and habit Tree to 50 m at first conical but broadening as the tree ages.

Trunk Stout, straight. *Bark*: On young trees pinkish-grey with resin blisters, flaking; on old trees thick and rough to the touch, dark grey.

Branches On young trees ascending: on older trees horizontal, spreading.

Branchlets Stout; grooved; pale grey to light brown; glabrous or with blackish-brown pubescence when young, pinkish-brown the second year.

Winter buds Rather large, 3.5 cm long, ovoid, more or less resinous. *Scales*: Numerous.

Leaves In two ranks, those towards the middle of upper sides of branchlets gradually shorter, spreading, more or less recurved, leaving between them a wedge-shaped furrow; 15–30 (to 40) mm long, 2.5–3 mm broad, emarginate but sharply notched at apex on young specimens; flat with edges down-curled; hard and leathery to the touch, stiff but held on very thin stalks; slightly grooved, shining yellowish green above, often with inconspicuous stomata near the tip; two greyish bands of stomata beneath, not very conspicuous.

Mature cones Cylindrical or cylindric-conical, 10–15 cm long by 3–5 cm broad, yellowish-green at first, ripening to dark or grey-brown. *Scales*: broadly kidney-shaped, auricled, abruptly narrowing to a stalk. *Bracts*: rhomboid, apex lanceolate, exserted and erect. Cones are only borne high up on large trees.

Seeds Obovoid, wedge-shaped, angular, light brown with dark brown mottling; wing broad and also brown.

Figure 40. *Abies firma.* Detail of foliage from a side branch. Note the 'daylight' visible between the thin stems.

Habitat Japan, principally in the Hondo Province, but also on the islands of Shikoku and Kyushu.

Cultivation Introduced into Europe by Dr. von Siebold of Leiden in 1861, and into Britain by the J. G. Veitch nursery in 1861. *Bifida, holOphylla* and *momi* are names raised for minor variants no longer regarded as sufficiently distinct. Pendulous and variegated selections are reported from Japan, but are probably not in European cultivation. Zone 7. (See Illus. Nos. 20a, 21c and 40)

-- var. *brachyphylla* (Maxim). Bertrand 1871. Now known as *A. homolepis.*

A. finedonensis Hort. ex Gordon (1862). Now known as *PICEA abies* 'Finedonensis'.
A. forrestii See *A. delavayi* var. *smithii.*
A. fortunei Masters (1884). Now known as *KETELEERIA fortunei.*

A. fraseri (Pursh) Poir., Encycl. Meth., Suppl. **5**: 35 (1817); *PINUS fraseri* Pursh, Fl. Am. Sept. **2**: 639 (1814). Fraser Balsam Fir.

Size and habit Small to medium tree, 9–12, (rarely 20) m, not reaching a great age; crown slender conical or columnar.

Trunk Straight and tapering. *Bark*: Thin and smooth; warm-brown with numerous resin-blisters on young trees; older specimens covered with pink and closely appressed scales; eventually grey.

Branches Slender but rigid, in regular whorls; ascending.

Branchlets Smooth, grey or pale yellowish-brown with dense, short, red pubescence, persisting 3–4 years, becoming darker each year.

Winter buds Small; broadly ovoid obtuse at apex; orange-brown, coated with pale yellow resin.

Leaves Crowded, directed forward and outward, 12–20 mm long by 2 mm broad, widest above the middle; acute, rounded or slightly notched at apex; flat, grooved, shining blackish-green above with some stomata in the groove near the apex; below with 2 broad bands each of 8–12 stomatic lines, whiter in *A. koreana.*

Mature cones Ovoid or cylindrical, tapering towards a rounded top; 3–5 cm long by 2–4 cm broad, purplish before maturity. *Scales*: About 15 mm long. *Bracts*: Pale yellowish-green; noticeably protuding and reflexed, nearly covering the scales.

Seeds Obovate-cuneate; the wing oblique with wavy edges.

Habitat Limited to mountains from W. Virginia to N. Carolina and in Tennessee, U.S.A. at altitudes of 1600–2000 m.

Cultivation Introduced into the U.K. by John Fraser, a Scottish collector, who discovered it and sent it to the Lee nursery then at Hammersmith in London. It is closely related to *A. balsamea*, from which it is distinguishable by shorter, broader leaves, glistening white beneath, by the dense reddish pubescence on the first-year shoots, and by its smaller cones, covered by the protruding, strongly reflexed bract-scales. Zone 4.

-- **'Klein's Nest'**. A compact nest-like plant, lacking a leader and broader than high. The growth is very congested, with an annual growth of no more than 10 cm. The main branches ascend, the side branches very numerous. The leaves are short, dark green, very densely borne. Buds are numerous and noticeable: nut-brown.

-- **'Prostrata'** *A. fraseri* f. *prostrata* Rehder in Journ. Arn. Arb. **9**. [Welch 44 (1979)] discusses the case for not recognising this variant botanically. It was first observed in Wyman's Framington nursery, Framington, Mass. amongst seedlings of *A. fraseri* received from a French nursery in about 1912 and it forms a widespreading, flat-topped plants less than 1 m high but several metres across. The name was later registered by Seth L. Kelsey [Woody Plant Registrations **2**: 3 (1950)] for a plant introduced by the Highlands-Kelsey nursery in 1938. This was probably the same clone, but whether or not, both were almost certainly cultivariants, so the name should be used only in a clonal sense.

A. georgei. Now known as *A. delavayi* var. *georgei.*
A. gigantea Smith ex Carrière (1867). Now known as *PICEA abies.*
A. glehnii Fr. Schmidt (1868). Now known as *PICEA glehnii.*
A. gmelinii Ruprecht (1845). Now known as *LARIX gmelinii.*
A. grandis (1) Hooker (1840) and Murray (1863). Now known as *A. amabilis.*

A. grandis (2) (Douglas ex D Don) Ldl. in Penny Cycl. **1**: 30 (1833). *PINUS grandis* Douglas ex Don in Lamb. 144a; (1832) *A. excelsior* Franco 162. (1949) Grand Fir, Vancouver Fir.

Size and habit Monumental tree to 100 m in the wild, often widest about midway to the rather narrow crown. See Illus., Nos 20b, 4, 43.

Trunk Straight and tapering. *Bark*: Thin and smooth on young trees but with many resin-blisters, ashy-brown with pale patches; on older trees becoming thick, greyish-brown, deeply furrowed and broken up into small square plates, especially around scars left by fallen branches.

Branches Long, often somewhat pendulous but upswept towards the tip.

Branchlets Thin, olive-green or slightly brownish; at first with fine greyish-brown pubescence but soon becoming glabrous.

Winter buds Small, globose, dark-brown to purple; scaly; slightly resinous.

Leaves Held very flat on either side of the shoot at 80–90°; very unequal in length, 3–6 cm long by 2–2.5 mm broad, the longer leaves alternating with much shorter leaves; flat, flexible but rigid at apex which is usually notched; shining dark green above with a narrow groove and with 2 narrow white bands beneath. The bruised leaves have a strong smell: to most people disagreeable, but others liken it to tangerines.

Mature cones Cylindrical, narrowing towards a rounded or slightly hollowed-out top; 7–10 cm long (occasionally longer, to 12 cm) by 3–4 cm broad; bright yellowish green or greenish-brown, ripening to greenish-purple. *Scales*: 2.5–3 mm long and broad; border entire. *Bracts*: Hidden; half as long as the scale; edges toothed.

Seeds Wedge-shape; 10 mm long. *Wing*: wedge-shaped, 7 mm long by 9–10 mm wide; lustrous; purple.

Habitat From Vancouver Island to Northern California and eastwards to Montana, U.S.A.

Cultivation Introduced into Britain by D. Douglas, collector for the Horticultural Society of London (now the Royal Horticultural Society) and in 1851 by W. J. Lobb, a collector for the Veitch nursery.

It prefers moist but well-drained situations and lower slopes, chiefly along valleys. Zone 2. (See Illus. Nos. 21d, 41)

-- **'Aurea'** Beissner 479 (1891). A clone with golden-yellow foliage raised on the Hesse Nursery, Weener, Ems, W. Germany.

-- **'Compacta'** Hesse ex Beissner, 479 (1891). This old form has recently turned up and is now again being propagated. Mr Gordon Haddow of Kenwith Castle Nursery, Bideford, Devon, U.K. tells me that it makes a globose bush, very tight and compact on its own roots with an annual growth of 10–15 mm, but of rather looser growth when grafted. It carries very small, brown buds and leaves that are fresh green above with a central midrib and slightly recurved edges, dull beneath, 10–15 mm long. The crushed foliage has the rather disagreeable odour characteristic of *A. grandis*, which plants sometimes offered as 'Nana' do not, these being usually a clone of *A. balsamea.*

-- **'Johnson'** Matthews in Journ. For. (U.S.A.) **40**:

Figure 41. *Abies grandis.* Grand fir. Detail of foliage from a side branch. The leaves vary in length and stand off at a wide angle. The shoot is covered with a fine pubescence. (5)

Figure 42. *Abies holophylla.* Manchurian fir. Detail of foliage from a side branch. The rich green leaves are pale green beneath. (5)

579 (1942). A narrowly conical, slow-growing tree with all but the lowest branches strongly ascending; foliage very dense on the copiously branched twigs, leaves very imperfectly or not at all biseriate, 18–25 mm long, under 2 mm wide, minutely or not at all emarginate at apex, very shiny above, thinly glaucous beneath on either side of the midrib. Cones as normal in the species. Found as a solitary tree by John A. Johnson in 1897 on the lower Columbia River about four miles below Rainier, OR, U.S.A. It makes an excellent small specimen.

-- Nana. See 'Compacta'.

-- '**Pendula**' Späth in Mitt. d. d. d. Ges. **5**: 28 (1896). A clone with strongly weeping branches that originated in the Späth nurseries, Berlin.

Figure 43. *Abies grandis*. A ripe cone. (7)

Probably no longer in circulation.

A. gregoryana Low ex Gordon (1962) Now known as *PICEA abies* 'Gregoryana'.

A. griffithiana Hooker ex Lindley & Gordon (1850). Now known as *LARIX griffithiana*.

A. guatamalensis Rehder in Journ. Arn. Arb. **20** (3): 285 (1939); *A. guatamalensis* var. *jaliscana* Martinez in Anal. Inst. Biol. Mexico. 1 (1948); *A. tacanensis* Lundell in Amer. Midl. Nat. **23**: 175 (1940).Guatemalan Fir.

Size and habit Tree to 25–35 m.

Trunk Stout. *Bark*: Blackish-brown, broken up into plates.

Branches Spreading horizontally.

Branchlets Slender, with pulvinus-like ridges and grooves; purplish-red turning to nearly black; slight grey pubescence.

Winter buds Long ovate to ovoid, 5–6 mm long, resinous. *Scales*: broadly triangular, blunt-tipped, keeled.

Leaves Varying in size between 1.5 and 5 cm long by 1.2–1.4 cm broad, arranged in two ranks, the lower row standing off from the shoot at nearly 90°, the upper row (of shorter leaves) pointing forwards; narrowly linear, slightly reflexed, bifid at apex; the upper surface shining green, grooved throughout its length and carrying occasional stomata; below with two white bands, each consisting of 8–10 rows of stomata, separated by a mid-rib.

Mature cones Sessile or short-stalked, oblong-cylindrical, 8–12 cm long by 4–5.5 cm broad, obtuse or rounded and flattened at the apex; purplish brown. *Scales*: hidden.

Seeds Cuneate-obovoid, 8–9 mm long, light brown with wings 10–13 mm long, as broad as long.

Habitat S. E. Mexico and Guatemala, at altitudes between 2000 and 3500 m, according to the latitude. This is the southernmost representative of the genus. Zone 9.

A. hanbury Hort. ex Gordon (1875). Now known as *TSUGA heterophylla*.

A. heterophylla Rafinesque (1832). Now known as *TSUGA heterophylla*.

A. hickelii Flous & Gaussen in Bull. Soc. Hist. Nat. Toulouse **64**: 24 (1932); *A. oaxacana* Martinez in Anal. Inst. Biol. Mexico **19**: 1 (1948). A species with a very limited range on high mountains in Mexico. A conical tree 20–30 m high with an irregular crown; branches purplish-red or reddish-brown; branchlets ridged and pubescent in the grooves, glabrous on older branches. Leaves nearly 2-ranked, narrowly linear; dark shiny green above, rarely with a few scattered stomata near the apex; glaucous below. Cone: cylindrical-oblong, on stalk; rounded or short-pointed at apex, slightly resinous. It has affinities with *A. religiosa* and *A. guatamalensis* and is possibly not in European cultivation.

A. holophylla Maxim. in Mel. Biol. Acad. Petersb. **6**: 22 (1866). Manchurian Fir. Needle Fir.

Size and habit Tree to 30–45 m. Broadly columnar with the crown at first narrow, broadening with age, flat-topped in maturity.

Trunk Stout. *Bark*: On young trees smooth, grey; becoming furrowed and shredding into thin papery scales; pink, orange or brown, the colouration often patchy.

Branches Thick, spreading horizontally.

Branchlets Lustrous pink-brown or yellowish, shallowly grooved with blackish-brown hairs in the grooves on young trees; older trees glabrous.

Winter buds Large, globular; bud-scales numerous; shiny deep-red or red-brown; sometimes lightly coated with white resin.

Leaves Densely set in 2 ranks, those on the upper side of the shoot standing up stiffly to form a V-shaped valley, 2.5–4.5 cm long gradually shortening towards the tip of the shoot but all about 1.5 mm broad; parallel-sided; either blunt or sharply-pointed, flat, stiff, leathery, shiny grass-green, grooved, occasionally shewing stomata above near the tip with two grey-green bands of stomata beneath, less noticeable on older trees.

Mature cones Short-stalked, cylindrical or oblong-cylindrical, 10–15 cm long by 5 cm broad, green at first, ripening to light brown. *Scales*: 20 mm long 30–50 mm wide. *Bracts*: Just visible, slightly reflexed.

Seeds Obovate to wedge-shaped, 10 mm long. *Wing*: 13 mm long by 13 mm wide.

Habitat Manchuria, Korea.

Cultivation Introduced by Komarov in 1898, but is still rare in Britain. It seems to be a very variable species. It is very near to *A. firma* but is distinctive by its acutely pointed leaves, never bifid on young specimens. Zone 5. (See Illus. Nos. 20c, 42)

A. homolepis Sieb. et Zucc. Fl. Japan **17**: (1842); *A. brachyphylla* Maxim. in Bull. Acad. Sci. St. Petersb. ser. 3, 488 (1866). Nikko Fir.

Size and habit Broadly conical tree to 40 m. with 'spiky' appearance; open at first, later filling out and forming a rounded crown.

Figure 44. *Abies homolepsis.* Nikko fir. Detail of foliage from a side branch. The shoot is glabrous, grooved, pink to yellow. Old specimens often cone freely low on the tree.

Figure 45. *Abies religiosa.* Sacred fir. Detail of foliage from a side branch.

Trunk Stout. *Bark*: grey, scaly and peeling, with distinctive pink to pale orange-pink patches until quite a big tree.

Branches Horizontal and wide spreading.

Branchlets Deeply grooved and ridged, white or brownish-yellow, glabrous.

Winter buds Ovoid or bluntly conical partially hidden by the leaves, chocolate brown, resinous, soon whitened.

Leaves Very densely set, those on the upper side of the shoot, spreading obliquely and sometimes leaving a V-shaped space along the shoot, sometimes covering the shoot; pectinate below, outer ones 20–30 mm long, upper ones 10–20 mm long, all 1.5–2 mm broad, flat, short-pointed at apex, grooved, shining dark green, without stomata above but with 2 chalk-white bands beneath.

Mature cones Sessile, cylindrical ± 8 cm long by 3 cm broad, purplish red, ripening to a purple-brown. *Scales*: very thin, kidney-shaped with rounded apex. *Bracts*: hidden.

Seeds Obovate-wedge-shaped, 7–8 mm long, nearly black. Wing: quadrangular, 9–10 mm long, 8–9 mm wide at the top.

Habitat Central Japan.

Cultivation Introduced by Dr. von Siebold into the Netherlands in 1859. Not widely planted, but in most collections in U.K. up to 25 m. Zone 5. (See Illus. Nos. 20d, 44)

-- **'Prostrata'** Several clones seem to be in cultivation which form low-growing, widespread plants. One, at the Brooklyn Botanic Gardens, was received from a Jerome Lukins of Port Chester, NY. Another was registered as **'Shelter Island'** in 1981 by Fred J. Crowe. All such plants are cultivariants, i.e. plants raised from lateral growth that have not succeeded in establishing a true leader. (See Illus. No. 46)

A. hookeriana A. Murray (1855). Now known as *TSUGA mertensiana*.

A. iidaensis Kusaka (1952). Now listed under *A. firma*.

A. × *insignis* Carrière ex Bailly in Rev. Hort. 230 (1890). This is a hybrid, *A. pinsapo* pollinated by *A. nordmanniana*, first raised by Renault, nurseryman at Bulgnéville in the Vosges,

Figure 46. *Abies homolepsis* 'Prostrata'. A specimen 5 m across. It is probably a cultivariant.

France. In characteristics it is intermediate between the parent species, habit and foliage arrangement being close to *A. nordmanniana* but the leaves themselves suggesting the other parent. Several clones, 'Andreana', 'Beissneriana', 'Kentiana' and 'Mastersiana' were given names but seem never to have been distributed. Another hybrid, ***A.*** × ***insignis*** var. ***speciosa*** between the same species, but with the parentage reversed (*A. pinsapo* now being the pollen producer) was raised by Croux, a nurseryman at Sceaux near Paris in 1871, but this also is lost to cultivation. Zone 6.

A. inverta R. Smith ex Gordon (1862). Now known as *PICEA abies* 'Inversa'.
A. jezoensis (1) Sieb. & Zuccarini. Now known as *PICEA jezoensis*.
A. jezoensis (2) Lindley ex Paxton. Now known as *KETELEERIA fortunei*.
A. kaempferi (1) (Lambert) Lindley (1833). Now known as *LARIX kaempferi*
A. kaempferi (2) Lindley (1854). Now known as *PSEUDOLARIX amabilis*.
A. kamtschatica Ruprecht (1845). Now known as *LARIX gmelinii* var. *japonica*.

A. kawakamii (Hayata) Ito, Encycl. Jap. **2**: 167 (1909). *A. mariesii* var. *kawakamii* Hayata in Journ. Coll. Sci. Tokyo **25**: 223 (1908).

Size and habit Conical tree 15–35 m, with dense, rounded crown.

Figure 47. *Abies kawakamii.* Taiwan fir. Detail of foliage from a side branch. The narrow, dark green leaves are in dense layers each side of a brown, grooved shoot.

Trunk Straight and erect; rough or grooved at the base. *Bark*: Nearly white, grey or yellowish-brown, flaking or shredding off in strips.

Branchlets Yellowish to dark brown; deeply grooved; brown-tomentose in the grooves or sometimes nearly glabrous.

Winter buds Small, ovoid, brown or reddish-brown, very resinous and irregular in shape.

Leaves 1.5–2.5 cm long, narrow; densely set in layers on either side of and not hiding the shoot; directed forwards; narrow-pointed and notched at the tip; grooved and dark green above with a few stomata near the tip and with white bands of stomata below.

Mature cones Cylindric-ovoid, 6–10 cm long by 3.5–4 cm broad; violet-purple. *Scales*: Triangular-ovoid-reniform, upper margin rounded outside, border often auriculate and more or less erose. *Bracts*: Oblong-spatulate, apex pointed, not overtopping the scales.

Seeds 7 mm long, nearly black, with a wing 9–10 mm long by 8 mm wide, slightly rounded or truncate at apex.

Habitat Mt. Morrison, Formosa.

Cultivation *A. kawakamii* is close to *A. mariesii* and differs from the latter mainlyby its less pubescent branchlets, its longer cones and its blackish seeds. It was discovered in 1916 by E. H. Wilson and distributed in 1960. Very rare in cultivation. Zone 6. (See Illus. No. 47)

A. khutro(w) Royle (1839). Now known as *PICEA smithiana*.

Figure 48. *Abies koreana.* Korean fir. Detail of foliage from a side branch. This species cones freely when quite small. There is a dense white pubescence on the underside of the leaf.

A. koreana Wilson in Journ. Arn. Arb. **1**: 188 (1920). Korean Fir.

Size and habit A handsome but often short-lived conical tree to 15 m.

Bark Smooth but with resin-blisters; reddish-brown or nearly black with (on young trees) a purplish bloom; on old trees grey and deeply fissured, broken into irregular plates, exposing a red-brown colour beneath.

Branches Numerous, slightly ascending but with the lower branches sometimes long and drooping.

Branchlets Slightly grooved, yellowish with a purplish bloom, at first with a scattering of hairs but becoming glabrous later.

Winter buds Nearly globular, 4–5 mm diam, chestnut-brown, resinous; pure white by winter.

Leaves Densely set, held pectinately beneath the shoot, 1–2 cm long by 2–2.5 mm wide, widest towards the apex; yellowed at the base; apex slightly notched (on young trees sometimes pointed; with slightly down-turned margins; shining green above, with two whitish bands of stomata beneath.

Mature cones Cylindrical, freely borne on quite young trees, 5–7 cm long by ± 2.5 cm broad, apex flattened or slightly hollowed out, sea-green ripening to violet-purple (or blue or dull-brown, according to the clone). *Bracts*: Projecting, reflexed and somewhat tattered-looking.

Seeds Obovoid, ± 1 cm long, violet-purple. Wing quadrangular, rounded, 6–8 mm long and wide.

Habitat Korea: Chirisan, on high mountains to 1000 m, and Quelpart Is.

Cultivation An alpine species discovered by U. Faurie who sent seeds to M de Vilmorin in Paris in 1905. It is abundant above 900 m, either forming pure forests or associated with birch and other deciduous trees and *Picea jezoensis*. It is popular in gardens because of its habit of producing its spectacular cones when still quite small, and being an alpine form it has produced several low-growing forms, useful in gardens as 'dwarf conifers'. Zone 6.

-- **'Aurea'**. This is a clone which has beautiful golden foliage. It arose in 1956 in a batch of seedlings on the nursery of E. H. Lohbrunner Vancouver Island, Canada. The name is of questionable legitimacy. Seedlings tinged with gold occasionally turn up.

-- **'Blauer Pfiff'** Krüssmann 49 (1979). This plant has unusual, blue-green foliage, the colour resulting from the treatment of the seed with cobalt. It was raised by K. Wiltboldt-Mueller of Verden-Eitze, West Germany. It forms a slow-growing globose plant. Other seedlings of similar origin have made conical plants.

-- 'Blue Hit'. See 'Blauer Pfiff'.

-- **'Blue Standard'** A very slow-growing form with an irregular habit, freely producing cones which ripen to a deep violet colour. It was raised from seed by the Jeddeloh nursery, Oldenberg, W. Germany and was introduced in 1962.

-- **'Compact Dwarf'** Den Ouden and Boom 23 (1965); *A. koreana* 'Nana' hort. NON Meyer, Plant Explor. **2**: 70. (1963). This is a slow-growing form without a leader, usually forming a squat or wide-spreading, flat-topped bush with much interlacing of branches. The short, stiff, bifid leaves, 15 mm by 1.75 mm, standing out stiffy from the creamy-grey shoots give the plant a more open appearance than is actually the case. The buds are light brown and up to 3 mm in diam. It seldom cones, but nevertheless makes a handsome and attractive garden plant.

-- **'Piccolo'** Welch 145 (1979). A selection with unusually small leaves. As a young plant, flat-topped and wide-spreading, it is a most attractive 'dwarf conifer', but in time, unless prevented from so doing, it is act to throw up a leader and loses character completely. It was raised on the Konijn nursery at Ederveen, Netherlands.

Figure 49. *Abies koreana* 'Prostrate Beauty' in the Spingarn collection, then of Baldwin, N.Y., U.S.A.

-- **'Pompon'** New name. A slow-growing form that makes a globose plant with well-branched habit and annual growth about 4–5 cm. Foliage colour is bright green. It originated from a witch's broom discovered in 1980 by W. Wustemeyer of Schermbeck, West Germany. It was introduced [Bull. Amer. Conif. **1** (3): 91 (1984)] as 'Hexenbezen Wustemeyer'.

-- **'Prostrate Beauty'** Den Ouden and Boom 25 (1965). A low-growing, wide-spreading clone with dense foliage, coning freely. It originated in the Arnold Arboretum, Mass., U.S.A. Plants of a similar, low-growing propensity turn up, but for these the name 'Prostrata' should not be used since it is illegitimate in form and uncertain in effect. Any new clone should be registered under a new name. One such is 'Starker's Dwarf', below. (See Illus. No. 49)

-- **'Silberlocke'** This was the first of some interesting seedlings that appeared in several nurseries in W. Germany in the 1970's with the leaves curled inwards so as to expose the white underside, giving a silvery-white appearance to the tree. 'Silberlocke' has the normal habit of the species but is somewhat slower-growing. It should be planted in full sun and in poor soil since otherwise it grows vigorously and loses the curl to the foliage. It sets cones which ripen to a deep blue – very attractive in contrast to the foliage colour. It was at first distributed under the name 'Horstmann's Silberlocke' [Krüssmann 35 (1983)], but the name 'Horstmann' being already in use in *Abies pinsapo* and the re-use of a name within the same genus being now disallowed, the simpler name is preferable. The name is sometimes translated to 'Silver Curls'.

-- **'Silberreif'** Hachmann Nursery, W. Germany Cat. (1982). This originated as a seedling on the Hachmann Nurs. at Barmstedt, W. Germany in 1972. It has the same attractive foliage as 'Silberlocke', but grows somewhat faster and shews the attractive curl of the leaves less on

Figure 50. *Abies lasiocarpa.* Subalpine fir. Detail of foliage from a side branch. Note the swollen shoot bases.

second-year shoots. It also bears beautiful blue cones at an early age.

-- **'Silver Show'** Wittboldt-Mueller in Dendroflora **13–14**: 68 (1977). This is another clone with its leaves much curled inwards so as to show the silvery-white underside. It is much slower-growing than 'Silberlocke', is a brighter colour and it retains its character better as the plant ages. At 14 years the mother plant on the Wittboldt-Mueller Nurs was 1.6 m high. If not stem-trained when young it will form an attractive widespreading, flat-topped bush.

-- **'Starker's Dwarf'** Welch, 145 (1979). This attractive form makes a dense flat-topped rounded plant not unlike the vultivar 'Prostrate Beauty' with the same rich deep glossy green foliage. It is, however, slower growing, with an annual growth of about 7 cm. It was named for Carl Starker of Jennings Lodge, Oregon, U.S.A., who received the plant from an unrecorded source in Canada in 1960.

A. lanceolata Desfontaine (1829) and Poiret (1805). Now known as *CUNNINGHAMIA lanceolata.*

A. larix Poiret (1805). Now known as *LARIX decidua.*

A. lasiocarpa (1) Auct. (1855 etc). Now known as *A. concolor* var. *lowiana.*

A. lasiocarpa (2) Lindley & Gordon (1850). Now known as *A. grandis.*

A. *lasiocarpa* (3) (Hooker) Nuttall 138 (1849): *Pinus lasiocarpa* Hooker, Fl. Bor. Amer. **2**: 163 (1839); *A. subalpina* Engelm. in Amer. Natur.

10 555 (1876). Rocky Mountain Fir. Subalpine Fir.

Size and habit Conical or columnar tree 20–30 m with a narrow crown, sometimes a finely tapered slender spire.

Trunk Stout. *Bark*: Rather thin, silvery-grey on young trees, with resin blisters; ash-grey on mature specimens, shallowly fissured or roughened; ash-grey or grey-brown with red-coloured scales.

Branches Horizontal, the foliage most dense in the lower part of the tree.

Branchlets Lightly grooved ash-grey or grey-brown with short, rust-coloured pubescence persisting for 2 years. Leaf scars purple.

Winter buds Broadly ovoid, 10 mm long, 6 mm across; pale brown; very resinous.

Leaves Irregularly pectinate or bristling upwards, often scattered spreading, 2.5–4.5 cm long by 1–1.5 cm broad, flat but twisted at the base; grooved; apex acute or slightly notched; dark bluish-green, but silvery bands on both sides, each consisting of 6 lines of stomata, give a pale, glaucous appearance.

Mature cones Oblong-ovoid; 8–10 cm long, 3.0–3.5 cm across; dark purplish or green, ripening to dark violet-purple. *Scales*: Nearly circular. *Bracts*: Hidden, one third as long as the scales.

Seeds 6 mm long, broadly wedge-shaped. *Wing*: 12 mm long by 9 mm wide, thinly membraneous, violet-purple.

Habitat Alaska to Oregon and N. Mexico, at 2000–3000 m.

Cultivation Introduced into Britain in 1863 by Dr. Lyall, and in 1874 by Roezl. Zone 2. (See Illus. No. 50)

-- 'Argentea' Hort. See var. *arizonica*.

-- var. ***arizonica*** Since the Corkbark Fir shares the habitat of *A. lasiocarpa* and is even found in pure stands it might not unreasonably expect to be treated as a separate species, but since it differs only in its thick, corky, creamy-white bark and in minor differences in the ripe cones, the consensus of botanical opinion seems to be to treat it as a *varietas*. [See Liu, 309 (1971)].

-- **'Arizonica Compacta'** Den Ouden 9 (1937); Welch, 146 (1979); *A. arizonica* var. *compacta* Hornb. 22 (1939); *A. lasiocarpa* 'Compacta' Welch, 108 (1966). The reduction of the species *A. arizonica* (Merriam) Lemmon, to a variety of *A. lasiocarpa* created a minor nomenclatural problem, since Hornibrook [22 and 27 (1939)] described a "var. *compacta*" under each of the species accepted at the time. Welch [146 (1979)] changed his own previous and different suggestion [108 (1966)] and gave recognition to the cultivar name 'Arizonica Compacta' on the grounds that it would preserve well-established trade usage. There still seems to be some uncertainty, so it might be well to point out that, even if the plant for which the basionym 'Compacta' (See below) was used in 1900 no longer exists, the name is, technically, 'occupied' so is unavailable for the widely-grown clone raised as a seedling in the Jan Boer Gz nursery at Boskoop, Netherlands, about 1927. This is a very slow-growing clone which makes a densely branched, conical small tree with leaves crowded, directed forwards and outwards, slightly curved, 15–25 mm long by 1–1.5 mm wide, intense silvery-blue above and with irregular, very conspicuous bluish-white bands of stomata beneath.

-- **'Conica'** Hornibrook 21 (1923). Dwarf; shrub broadly conical, compact; winter buds about 3 mm long, light brown; leaves very crowded above, very thin and fine, curved, pointing forward, about 12 mm long, scarcely 1 mm broad, greyish-green, with stomatic bands not strongly marked beneath. Raised in the Arnold Arboretum from seeds sent by Dr. C. C. Parry from Colorado in 1873. Welch (146 (1979)) states that this is not in cultivation in U.K. unless the cultivar 'Roger Watson' proves to be the same, in which case that name is redundant.

-- **'Duflon'** New cultivar. This extremely slow-growing miniature, probably the smallest fir on record, originated as a dwarf plant discovered near Lake Cushman on the Olympic Peninsula in the Olympic Mountains, Washington, U.S.A., by Mr and Mrs DuFlon of Seattle. It is diminutive in every way, and was introduced into cultivation by E. H. Lohbrunner, of Victoria B. C., Canada.

-- **'Green Globe'** Welch. 144 (1979) incorrectly under *A. concolor*. A most attractive dwarf plant, appropriately named, introduced by the Verkade Nursery of Wayne, NJ. U.S.A. One of the best introductions of recent years. The original seedling was found growing near a tree of *Abies balsamea*, under which species it was at first listed. It is possibly a hybrid with *A. concolor*.

-- **'Mulligan's Dwarf'** A clone of seedling origin, but forming a dense, compact plant with short dark green foliage.

-- **'Pendula'** Ordnung ex Beissner in Mitt. d. d. d. Ges. **18**: 268 (1909), not *A. lasiocarpa pendula* Carr. (1890), which is *A. concolor 'Pendula'*. The typical variety with strongly drooping branches. Raised by Ordnung, Eisenberg, Czechoslovakia.

-- **'Roger Watson'** Welch, 146 (1979). The name was raised for a plant discovered in the garden of Roger Watson of Taunton in Somerset. It forms a very attractive, slow-growing, broadly conical tree with a horizontal branching system and an annual growth of 75 mm. The leaves are 15 mm long by 1.25 mm broad but are otherwise normal, and the stomata are inconspicuous, giving the foliage a dark grey-green colourthat is very attractive. The mother-plant is now in the Pygmy Pinetum at Devizes, Wiltshire, U.K.

A. ledebourii Ruprecht. Now known as *LARIX russica*.
A. leioclada Stevens ex Gordon (1858). Now known as *A. cilicica* var. *leioclada*.
A. lemoniana Booth ex Gordon (1858). Now known as *PICEA abies*.
A. leptolepis. Siebold & Zuccarini (1842). Now known as *LARIX kaempfery*.
A. likiangensis Franchet (1899). Now known as *PICEA likiangensis*.
A. macrocarpa. Vasey (1876). Now known as *PSEUDOTSUGA menziesii*.

A. *magnifica* Murray in Proc. Hort. Soc. Lond. **3**: 318 (1863). Californian Red Fir.

Size and habit A monumental, narrowly-conical tree, 60–70 m, with noticeably whorled branches and a narrow crown.

Trunk Massive, straight, usually clear of low branches. *Bark*: At first thin, relatively smooth, chalky-white, with resin-blisters; soon becoming rough, thick, and corky; purple and grey; eventually with narrow fissures or curled cracks and with prominent scars where branches have been shed.

Branches Crowded; relatively short; in regular whorls; horizontal at first, then strongly upswept.

Branchlets Dark brown, with short pubescence.

Winter buds Very small, ovoid or roundish, pubescent, hidden by the topmost leaves.

Leaves On upper side of shoot curled upwards to 70–80; the lower leaves more widespread, sharply curved near the base and then curving randomly; 20–40 mm long by about 1 mm broad, spaced apart; dark grey-green; keeled above and below, with well-marked bands of stomata on four faces.

Mature cones Ovoid-cylindrical, 15–22 cm long, 7–9 cm across; violet, ripening to pale yellow-green.

Seeds Triangular; 8 mm long, dark brown, fringed at the sides. *Wing*: Wedge-shape, 24–30 mm long, 13–14 mm wide at the top, violet. *Scales*: Flabelliform, entire. *Bracts*: Hidden.

Habitat U.S.A., S.W. Oregon; N. California to W. Nevada, at 1800 to 2,500 m.

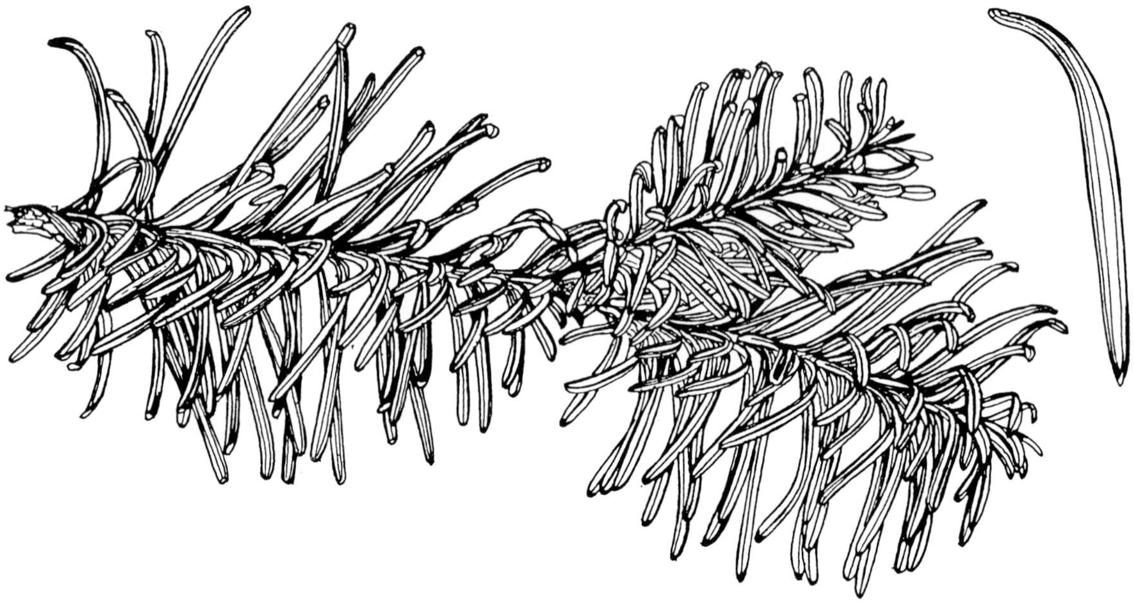

Figure 51. *Abies magnifica*. Magnificent fir. Detail of foliage from a side branch. The leaves are sparser, longer and less upswept than in *Abies procera*. (5)

Cultivation Introduced in 1851 into Britain by Jeffrey (Scottish Oregon Association) and in 1852 by W. J. Lobb (Veitch Nurseries.) Mitchell [32 (1972)] remarks that *A. magnifica* makes a much more regular, open and narrow tree than does *A. procera*, with foliage less dense; also that the leaves are longer and are irregularly curled and (being almost round in cross section) may be rolled between the fingers. Zone 6. (See Illus. Nos. 22a, 51, 52)

-- var. *shastensis* Lemmon in Rep. Calif. State Board For. III, 145 (1890). Now known as *A.* × *shastensis*

A. mariesii Masters in Gard. Chron. New series, **12**: 788 (1879).

Size and habit Tree to about 25 m, with open, narrowly conical crown.

Trunk To 60 cm diam. *Bark* pale grey, almost white on old trees, prominently pocked with small brown lenticels. Fallen branches leave large, dark scars.

Branches Spreading, regularly whorled, stout, the lower branches drooping, upper ones ascending.

Branchlets Reddish-brown, densely pubescent.

Winter buds Small, ovoid, blunt, purplish-brown, resinous.

Leaves Crowded on upper side of the shoot, not pectinate, pointing forwards, gradually longer towards the sides of the shoots, pectinate on the under side; 15–22 mm long by 2 mm broad; bluntly square-ended and notched; flat, grooved, shining dark green without stomata above and with 2 white bands of stomata below.

Mature cones Elliptic-cylindrical, 6–10 cm long by 4–5 cm broad, smooth, violet-purplish. *Scales*: wedge-shaped at base, upper margin broad, rounded, entire. *Bracts*: hidden, 1/3 shorter than the scales.

Seeds Obovate-ovate, 10 mm long by 5 mm wide, pale reddish-yellow.

Wing: hatchet-shaped, 9 mm long by 10 mm wide.

Habitat Japan and Formosa.

Figure 52. *Abies Magnifica*. A ripe cone. (10)

Cultivation Sent to the Veitch nursery in U.K. from Japan in 1878. It differs from *A. amabilis* in the bark, colour of shoot and shorter leaves. Also the smell of the crushed foliage is less suggestive of tangerines. Zone 6.

A. marocana Trabut (1906). Now known as *A. pinsapo* var. *marocana*. (Spelling varies).
A. mayriana Miyabe & Kudo (1920). Now known as *A. sachalinensis* var. *mayriana*.
A. menziesii (2) Engelmann (1862). Now known as *PICEA pungens*.
A. menziesii (3) Franchet & Savatier (1875). Now known as *PICEA jezoensis*.
A. menziesii (1) Lindley ex Loudon (1838). Now known as *PICEA sitchensis*.
A. merkiana Fischen (1959?). Now known as *PICEA sitchensis*.
A. mertensiana (1) (Bongard) Lindley & Gordon (1850). Now known as *TSUGA mertensiana*.
A. mertensiana (2) Gordon (1858). Now known as *TSUGA heterophylla*.
A. mexicana Martinez (1942). Not now distinguished from *A. vejari*.
A. microcarpa. (Poiret) Lindley et Gordon (1850). Now known as *LARIX laricina*.
A. microphylla Rafinesque (1832). Now known as *TSUGA heterophylla*.
A. microsperma Lindley (1861). Now known as *PICEA jezoensis*.
A. momi Sieboldii (1830). No longer distinguished from *A. firma*.
A. montana Nyman (1878?). Now known as *PICEA abies*.
A. morinda (Lawson) Wenderoth (1851). Now known as *PICEA smithiana*.
A. mucronata (1) Rafinesque (1832). Now known as *PSEUDOTSUGA menziesii*.

A. nebrodensis (Lojac.) Mattei in Boll. Orto Bot. Palermo **7**: 59 (1908); *A. pectinata* var. *nebrodensis* Lojac., Fl. Sicul. **2,2**: 401 (1907). Sicilian Fir.

Size and habit Tall, narrow, conical tree, the crown becoming rounded with age.

Trunk Straight, usually free of branches. *Bark*: Smooth at first, later becoming fissured into small, irregular plates.

Branchlets Shoots at first greyish-green, later becoming yellow-brown; longitudinally grooved, usually glabrous (rarely with short, brown pubescence). Leaf scars oval, light red with a white centre.

Winter buds Ovoid, with acute tip; brown; resinous. *Scales*: Triangular-ovate.

Leaves Densely set in 2 ranks, pointing forwards and curved upwards, especially so on fruiting branches. 10–12 mm long, 2–3 mm wide, stiff, flat, blunt or notched at apex; dark green above with a few broken lines of stomata near apex and with two white bands of stomata and a mid-rib (widest at each end) beneath.

Mature cones Short-stalked, cylindrical or elliptical, 7–8 cm long, 3 cm broad. *Scales*: Ovoid-rhomboid, 2 cm long, upper margin reddish-tomentose. *Bracts*: Projecting slightly, linear-spatulate,reflexed, with a long-pointed top.

Habitat Prior to 1800 *A. nebrodensis* was widely distributed on the mountains of North Sicily but it is now nearly extinct in the wild and still very rare in cultivation. Mitchell [33 (1972)] records that all save one of the trees in gardens are grafted specimens. This species is a good case for conservation, but those who plant such trees should remember that they are thereby conserving the clone, not the species. Zone 5.

A. nephrolepis (1) (Trautvetter) Maxim. in Bull. Acad. Sci. St. Petersb. **10**: 486 (1866); *A. sibirica* var. *nephrolepis* Trautv. ex Maxim. in Mém. Div. Sav. Acad. Sci. St. Petersb. 9: 260 (1859); *A. gracilis* Kom. in Acta Hort. Petr. **20**: 203 (1901). Khingan Fir.

Size and habit Narrow and regularly conical tree 25–35 m, with a dense, rounded crown.

Trunk Stout. *Bark*: Whitish grey to light grey-brown; rough to the touch but only shallowly fissured even on old trees.

Branches Short, stiff and horizontally spreading.

Branchlets At first yellow or light brown, shallowly fissured with a little brown or blackish pubescence in the grooves. Second-year shoots are grey, prominent dark-brown scales marking the division.

Winter buds Small, ovate-oblong, reddish-brown, slightly resinous.

Leaves On upper sides of the shoots crowded and pointing forwards, pectinate below; 10–22 (–25) mm long by about 2 mm broad, acute, rounded or sometimes notched at apex, flat but with edges strongly rolled under, narrowing towards the base, grooved and a dull, light grey-green above, rarely with 2 short lines of stomata towards the apex; two narrow grey bands of stomata below.

Mature cones Stalkless; ovate-oblong to cylindrical; 6–7 cm long by 2 cm broad, with a projecting apex; at first reddish, later becoming black-violet, ripening to brown. *Scales*: Kidney-shaped; upwards of 10 mm long (including the stalk), 18–20 mm wide. *Bracts*: Hidden, brown.

Seeds Obovate, 6 mm long, 3.5 mm wide, with wedge-shaped wing, dark purple.

Habitat E. Siberia and Korea, N.E. China.

Cultivation Introduced in 1908. Mitchell [34 (1972)] observes that the crushed foliage of *A. nephrolepis* smells of balsam or paint. He further remarks of this species that it is "like a dull, short-leaved *A. veitchii*, rarely brightly banded white beneath." Leaves broader and less parallel and much less dense than *A. sachalinensis* and shoot not ribbed. Buds, leaves and shoot darker than *A. sibirica*. Zone 3.

A. nephrolepis (2) Nakai (1914). Now known as *A. koreana*.
A. nigra Aiton Miller, Poiret and other early authors. Now known as *PICEA mariana*.
A. nobilis (D. Don) Lindley (1833). Now known as *A. procera*.

A. nordmanniana (Stev.) Spach, Hist. Nat. Vég. Phan. **II**: 418 (1842); *Pinus nordmanniana* Stev. in Bull. Soc. Nat. Mosc. **II**: 45 (1838). Caucasian Fir.

Size and habit Regularly conical tree to 50–60 m, usually furnished to the ground with branches and usually with a narrow-crown.

Trunk Stout, tapering. *Bark*: On young trees smooth, thin, grey; becoming rough, with shallow furrows and scales; blackish-grey on older trees.

Branches Horizontal, in dense regular tiers, upper branches ascending, the lowest branches drooping.

Figure 53. *Abies nordmanniana*. Caucasian fir. A ripe cone. (10)

Branchlets Yellowish-green, grey or greyish-brown, striate (marked with lines), with erect, black pubescence or (often on trees in cultivation) glabrescent.

Winter buds Ovoid, reddish-brown, not resinous. *Scales*: Broadly ovate, obtuse at apex, margins toothed, slightly keeled.

Leaves Very variable, densely set; in the upper ranks short, 1–1.6 cm long, overlapping and pointing directly forward, covering the shoot: longer, 3–4 cm, 2–2.5 mm broad below the shoot; uniform in width; squared-off or notched at apex; flat, grooved; hard and leathery; very shiny dark green without stomata above but with 2 silvery-white bands of stomata beneath.

Mature cones Stalkless or nearly so, oblong-ovoid or oblong-cylindrical, 15–20 mm long by 5 cm broad, greenish at first, ripening to dark brown. *Scales*: Broadly conical, upper margin entire. *Bracts*: Projecting and reflexed, the blade large and nearly quadrangular, truncate at the base, emarginate at the top and with a prolonged apex in the sinus.

Seeds Ovoid, nearly triangular, 10 mm long, the wing 15 mm wide, membraneous, apex straight.

Habitat N.E. Turkey; W. Caucasus; Southern U.S.S.R. at 1000–2000 m, forming forests with *Picea orientalis*. It is not found in the Crimea.

Cultivation It was introduced into Britain (The Lawson nursery at Edinburgh) in 1840 and into Germany (Berlin Botanical Gardens) by von Humboldt, in 1848. It is common in collections, but often suffers from Spring frosts. *A. nordmanniana* resembles *A. amabilis* in its foliage but has non-resinous buds and lacks the

Figure 54. *Abies nordmanniana.* Detail of foliage from a side branch. A variable species with hard, leathery, glossy dark green leaves, always pointing forwards.

tangerine smell from the crushed leaves of that species. Zone 5. (See Illus. Nos. 53, 54)

-- **'Aurea'** Beissner 437 (1891). Leaves golden yellow.

-- **'Aureospica'** Hesse ex Beissner 437 (1891). Hesse ex Beissner 126 (1909). Leaves tipped with golden-yellow; raised by H. A. Hesse, Weener-on-Ems, Germany.

-- 'Glauca' Beissner 437 (1891). Seedlings with more or less glaucous foliage turn up in seedbeds occasionally.

-- **'Golden Spreader'** Den Ouden and Boom, Man. Cult. Conif. 33 (1965). *A. n. aurea nana* Hort. Dwarf; branches spreading, light brown, annual growth 4–5 cm; leaves slightly curved, 12–25 mm long, 1.8–2 mm broad, bright golden-yellow above, pale yellowish-white beneath and with 2 indistinct blue bands. Originated as a seedling in the nursery of S. N. Schoots, Culemborg, Netherlands and introduced into the trade by L. Konijn & Son Reeuwijk, near Boskoop, 1961. It succeeds best in partial shade, but when doing well it is the brightest chrome-yellow of any dwarf conifer.

-- **'Jensen'** Den Ouden and Boom, Man. Cult. Conif. 33 (1965). *A. nordmanniana pendula* Jensen, Cat. **1944**. Pendulous, strong-growing;

leaves shorter (10–18 mm). Raised by Asger M. Jensen, Holmstrup, Denmark. The new name was required because of the prior use of the epithet 'Pendula', q.v.

-- **'Pendula'** (Young) Beissner 437 (1891), under *Picea nordmanniana* Young, Nurs. Cat. **8**: (1874.) Originated in the nursery of Maurice Young, Milford, Surrey, England; a similar clone was also raised by Courtois, nurseryman at Clamart, France in 1869, and seedlings with more or less pendulous habit occur from time to time. See 'Jensen' above.

-- **'Procumbens** '. Browicz & Bugala in Rosznik Arb. Korn. **3**: 73 (1958). Very slow growing, without a trunk and with only a few branchlets trailing over the ground; originated in 1934 in the Körnickie Arboretum in Poland. It is probably a grafted side-branch.

A. novae-angliae Koch (1873). Now known as *PICEA mariana*.

Figure 55. *Abies numidica*. Algerian fir. Detail of foliage from a side branch. The short, thick leaves have broad white bands of stomata beneath. (5)

A. numidica De Lannoy ex Carrière in Rev. Hort. 106 (1886 March): *A. baborensis* Cosson in Rev. Hort. 182 (1886 May): Algerian Fir.

Size and habit Regularly conical tree, 15–20 m with a dense crown.

Trunk Stout, often divided at a low level. *Bark*: Ashy or pinkish-grey, rather smooth on young plants, becoming rougher and cut up into circular plates as the tree matures, and a pale orange-brown.

Branches Crowded, held horizontally in tiers, drooping towards the tips.

Branchlets Glabrous, yellowish-green to light red-brown; lustrous; sometimes with shallow ridges running at an angle across the shoot.

Winter buds Large, ovoid or globose-conic with a pointed apex; pale red to dark brown; not resinous. *Scales*: In 5–6 ranks, closely appressed but with the tips free.

Leaves Densely set; on the upper sides of the shoot - short, 10–15 mm long, bristly, standing out from the shoot at ± 90°; hiding the shoot entirely only on the strongest growths; on the underside of the shoot - pectinate or with some leaves spreading irregularly, 20–25 mm long, about 2.5 mm wide; linear; stiff and leathery; broad, narrowing toward the base, bluntly rounded or slightly notched at apex, (pointed on young plants); narrowly grooved, dark blue-green above, with a few stomata forming a white patch near the apex and, beneath, a broad thickened midrib and two white bands, each consisting of 7–10 stomatic lines.

Mature cones On a short-stalk, ringed around by scales; cylindrical, 12–18 cm long, 4–5 cm broad; smooth, pale green or lilac. *Scales*: 3 cm broad with very broad margins recurved above. *Bracts*: Hidden, much shorter than the cone-scales. Cones are borne only on the highest branches.

Seeds Obovate-ovoid, 12 mm long, the wing 18 mm long.

Habitat Confined to the Babor and Tababor Mountain in E. Algeria

Cultivation It was found in 1861 by Captain de Guibert, but it is still uncommon. *A. numidica* can be distinguished from *A. homolepis* by its short, stout leaves standing out stiffly at 90°, and by its non-resinous buds. In *A. cilicica* the upper ranks of leaves point forward along the shoot; in *A. pinsapo* the leaves are held radially all round the shoot, like a bottle-brush. Zone 6. (See Illus. No. 55)

-- **'Glauca'** Beissner 141 (1909). Leaves shorter, broader, bluish, with fewer stomatic lines above. Found in the Arboretum des Barres, Nogent-sur-Vernisson, France.

-- **'Lawrenceville'** Named for a 24ft tall, dense, conical tree with a very pleasing silvery-blue foliage at Ringwood State Part, Ringwood, N. J. (per Thomas Buchter, 10 Dec. 1973).

A. obliqua Bongard ex Gordon (1862). Now known as *PSEUDOTSUGA menziesii*.

A. obliquata Rafinesque (1830). Now known *PSEUDOTSUGA menziesii*.

A. obovata var. *japonica* Maxim (1866). Now known as *PICEA maximowiczii*.

A. orientalis (L.) Poiret (1805). Now known as *PICEA orientalis*.

A. pardei Gaussen in Bull. Soc. Hist. Nat. Toulouse **57**: 141 (1929). Parde's Fir. A little known fir, at one time considered to be a hybrid, *A. nordmanniana* and *A. pinsapo*, and also sometimes confused with *A. numidica*. It is distinguished by dense black or dark brown pubescence on the young shoots, by resinous winter-buds with keeled scales and by cones with long-extruding bracts held erect. It seems to be only known in cultivation.

A. parlatorei Hort. ex Liu (1971). Now known as *PICEA pungens*.

A. parryana Hort. ex Liu (1971). Now known as *PICEA pungens*.

A. pattoniana Jeffrey ex A. Murray (1853). Now known as *TSUGA mertensiana*.

A. pattonii Jeffrey ex Gordon (1862). Now known as *TSUGA mertensiana*.

A. pectinata (1) Gilibert (1792). Now known as *PICEA abies*.

A. pectinata (2) Poiret (1804). Now known as *TSUGA canadensis*.

A. pectinata (3) De Candolle (1805). Now known as *ABIES alba*.

A. pendula ex Gordon (1862). Now known as *PICEA smithiana*.

A. picea (1) Miller (1768). Now known as *PICEA abies*.

A. picea (2) Lindley (1838). Now known as *ABIES alba*.

A. *pindrow* Royle, Ill. Bot. Himal. t. 86 (1836). Pindrow Fir.

Size and habit Narrowly conical tree to 60 m usually with a narrow, rounded crown.

Trunk Stout and massive. *Bark*: On young stems smooth or merely roughened, silvery-grey; on old trees grey-brown, with flat areas and longitudinal fissures.

Branches Short, often massive.

Branchlets Stout; yellowish-grey or sometimes reddish-brown, glabrous; smooth but with prominent corky ridges.

Winter buds Large, oblong-conical, 8 mm diam; reddish-brown, heavily coated with white resin.

Leaves Sparse and open. On upper side of shoot – short, directed forwards, sometimes leaving medially a narrow space. At the sides and underneath the shoots – longer, 3–6 (–9) cm long, 1.5–2 mm broad, pectinate, spreading out widely and arching over; flat, soft, narrowly linear, abruptly tapering towards the base and gradually so to the rounded (or more usually lopsidedly bifid) apex; shining dark green above with a narrow midrib and recurved edges but without stomata; with a narrow midrib and two greyish-green bands beneath, each of 5–6 stomatic lines.

Mature cones On a short, stout stalk; cylindrical, 10–14 cm long, 5–7 cm broad, rounded at each end; red-brown, ripening to violet purple. *Bracts*: Hidden, reaching only 1/3 length of the scale.

Seeds Blackish-green; 9 mm long with wing 22 mm long, 15 mm wide at the top.

Habitat Afghanistan; Pakistan; N.E. India; Nepal.

Cultivation *A. pindrow* grows together with cedars and spruces as dense forests in the western Himalayas. It was introduced into U.K. by Royle in 1837, but is still very rare. It suffers very much from late frosts save in the coldest areas of Scotland. Two cultivars, **'Brevifolia'** and **'Intermedia'** have been selected, but neither has ever been widely distributed. Zone 7. (See Illus. No. 56)

A. *pinsapo* Boissier in Biol. Univ. Genève **13**: 406 (1838). Spanish fir.

Size and habit A very broad, much branched, conical tree to 30 m with a broadly conical, often a very lopsided crown.

Trunk Tapering. *Bark*: More or less smooth on young trees, flaking off in loose plates, greyish-brown. On older trees the bark is dark grey or black, much puckered around where branches have died. (See Illus. 20e)

Branches Short numerous, in whorls; lower branches drooping, upper branches ascending.

Branchlets Pale reddish-brown, glabrous with raised ridges; darkening to purple the second year.

Winter buds Ovoid, apex blunt, 3–5 mm diam; greenish-brown, ripening to purple; resinous. *Scales*: keeled, the tips free-standing.

Leaves Emerging spirally from sucker-like bases all round the shoot and standing out stiffly at ± 90°, sometimes curved backwards; lozenge-shape in cross-section; 8–20 mm long, 1.5–2.5 mm wide, apex bluntly acute; dull glaucous

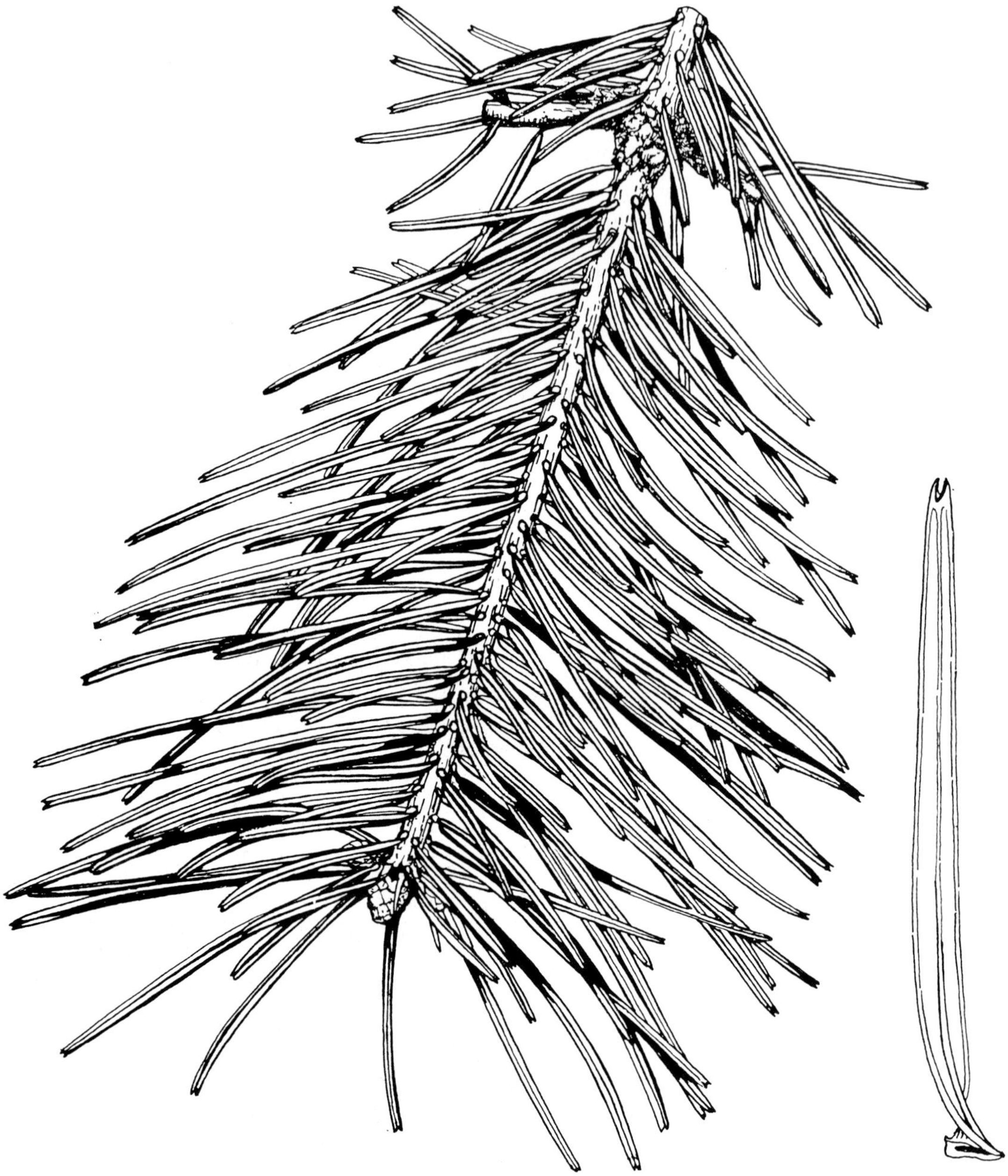

Figure 56. *Abies pindrow*. Pindrow fir. Detail of foliage from a side branch. The long, downward-arching leaves feel hard to the touch.

green, with stomata on all four faces.

Mature cones Sessile, oblong-cylindrical, narrowing sharply to apex, 10–15 cm long, 4–5 cm broad, greenish brown ripening to purplish-brown. *Scales*: nearly triangular, widening from a wedge-shaped base, 2.5 cm wide, hidden.

Seeds Obovate-ovoid, angular, 7 mm long, dark-brown, with the wing 15 mm long, 10 mm wide at the top, thinly membraneous, light brown.

Habitat *A. pinsapo* has a very limited range, occurring only between 1000 and 2000 m on the mountains in southern Spain in the Malaga

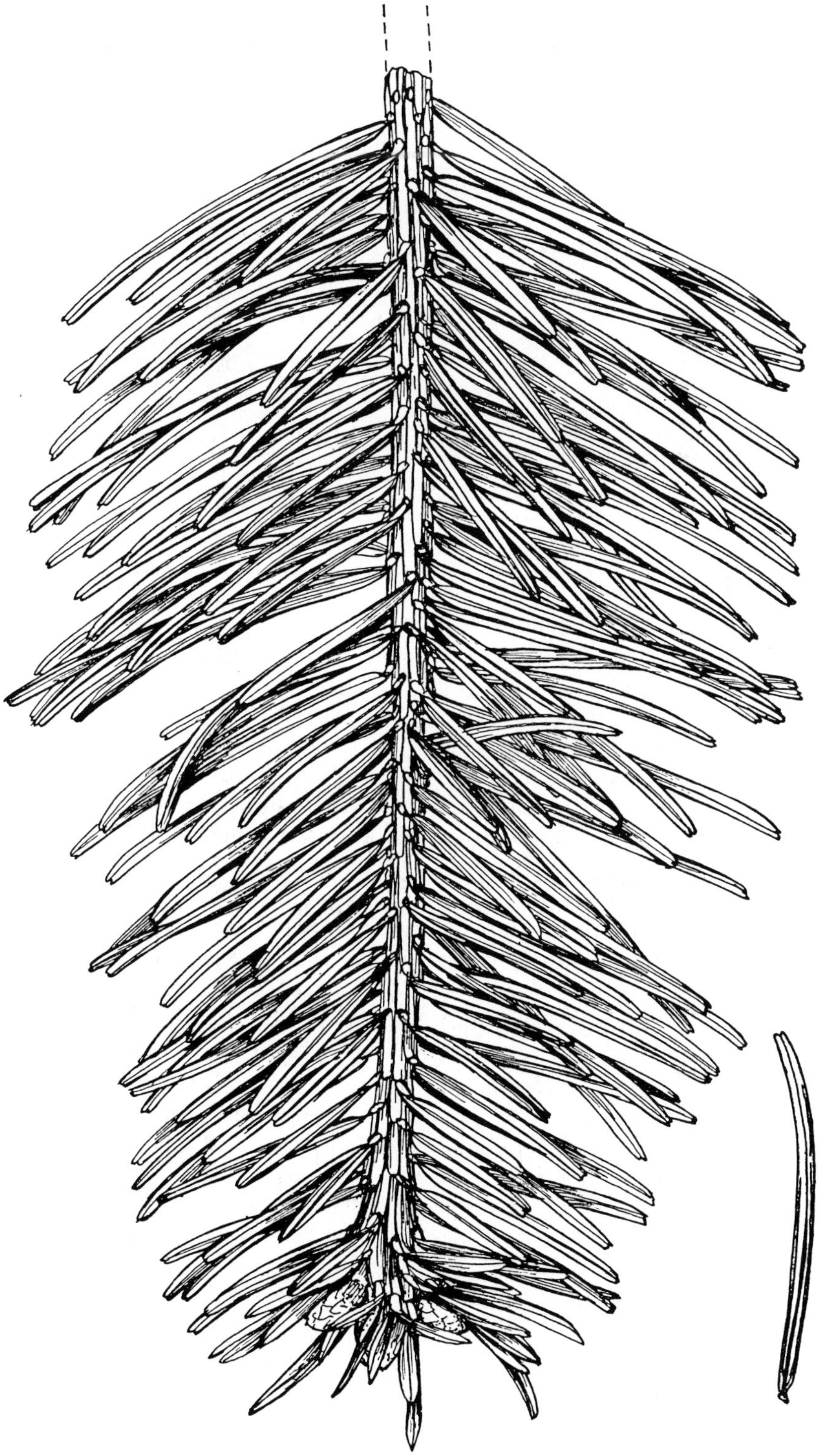

Figure 57. *Abies spectabilis.* Himalayan fir. Detail of foliage from a side branch. Note the stout, furrowed shoot and the pinkish-brown hairs in the furrows. (5)

Figure 58. *Abies pinsapo.* Close-up of foliage.

and Granada provinces. Two populations found in Morocco are recognised botanically. Var. ***marocana*** (Trabut) Ceballos & Bolaños in Serv. For. Invest. Exp. **2**: 18 (1928) occurs only on the mountains of Tetuan, and var. ***tazaotana*** only on Mount Tazaot. Although territorially well separated from the main population in Spain (which we can distinguish as 'var. ***pinsapo***') these African varieties differ therefrom morphologically only in minor respects.

Cultivation It was introduced in 1837 by Boissier who went seeds to the Vilmorin Nurseries in Paris. Despite its limited natural distribution, the species has given rise to a number of variants that have received cultivar names. *A. pinsapo* is one of the easiest species to recognise, being unique in its short, thick, rigid, blunt-ended leaves spreading radially at ± 90° all around the shoot. The leaves of *A. cephalonica* are softer, longer and sharply pointed. Zone 7. (See Illus. Nos. 22b, 22c, 58)

-- '**Aurea**' Sénéclauze. 16 (1868). Branchlets and leaves distinctly golden yellow. Found in the nursery of A. Sénéclauze. Rather common in the trade as a low shrub. It makes an attractive garden plant, seldom developing a leader.

-- '**Clarke**' Clarke, Woody Plant Reg. **1**: (1949). Extremely dwarf-growing. According to W. B. Clarke Company, San Jose, California, a plant from seed was about 20 years old and about 1 m tall in 1948. Save for its slow rate of growth it is normal in every respect.

-- '**Fastigiata**' Sénéclauze, 17 (1868). A dense, columnar clone that arose on the Sénéclauze nursery at Bourg Argental, Loire, France, and which can no longer be identified. The name should not now be used for other variants of similar habit, and any new clone to be brought into cultivation should be registered under an acceptable new cultivar name.

-- '**Glauca**' Carrière, 304 (1867). A strong-growing clone with glaucous and pruinose foliage that was raised by Desfosse and Thuilier, Orleans, France, and is now in wide circulation. In cold areas it suffers from frost, so is difficult to form a good tree. 'Kelleriis' is a much hardier selection.

-- '**Horstmann**' Krüssmann, Nadel. ed. 3, 51 (1979). A cultivar originating from a witch's broom found on the cultivar 'Glauca' that slowly forms a wide-spreading, densely branched, rounded shrub with short, silvery-blue leaves. The name 'Horstmann's Nana' sometimes used is illegitimate in form, and also superfluous.

-- **'Kelleriis'** A seedling (possibly a hybrid between *A. pinsapo* and *A. numidica* raised by D. T. Poulsen of Kelleriis, Denmark, about 1925. It is very robust and hardy and develops into an attractive tree, but it needs stem-training when young. Growth is 40–50 cm annually.

-- 'Robusta' An illegitimate listed name.

-- **'San Diego Creeper'** is a completely procumbent form discovered in 1977 by Fred J. Crowe of San Diego, CA, U.S.A. and registered by him in 1981. No information is given as to the origin: it would appear to be a stable clone of the 'cultivariant' persuasion.

A. procera Rehder in Rhodora **42**: 522 (1940); *A. nobilis* (Douglas ex D. Don) Lindley in Penny Cyclop. **1**: 30 (1833) not of A. Dietr. (1824). Noble Fir.

Size and habit Magnificent, slender, conical tree, 35–50 m with a narrowly conical, rarely a rounded crown.

Trunk Massive; straight. *Bark*: Thin, smooth, grey and with resin-blisters on young trees; thickened and with longitudinal fissures and small cracks on older trees; grey, with plates flaking off to expose the reddish-brown colouring beneath. (See Illus. No. 20f)

Branches Short, in regular whorls, spreading horizontally, the lowest branches drooping.

Branchlets Slender and pendulous; at first reddish-brown, darkening the 2nd year to purplish-red, with a minute, dense, rusty pubescence retained for several years.

Winter buds Small, pointed-ovoid, 3 mm diam, hidden by the leaves; dark red-brown. *Scales*: Elongated, with only the tips free.

Leaves On large, sucker-like bases; appressed for a short distance, then curving upwards or outwards. Leaves above the shoot are short (1.0 mm) and crowded, curving up to ± 90°; the lower ranks spreading and curved outtwards, 2.5–3.5 cm long, 1.5 mm broad, parallel-sided, with a squared-off or slightly notched apex; deeply grooved above; grey-green with two broad stomatic bands at their upper end and two narrower bands beneath.

Mature cones Stalkless or on a very short stalk, very large, oblong-cylindrical, 15–25 cm long, 5–8 cm broad, tapering slightly to a rounded apex; green when young, ripening to purplish-brown. *Scales*: variable, usually bluntly triangular, 3–3.5 cm long and broad, rounded above, entire. *Bracts*: Yellowish; much exposed and reflexed, nearly covering the whole cone; on some cones they turn to the left, on others to the right; margin frayed, with a long apex.

Seeds Wedge-shaped, angular, 10 mm long, with wing wedge-shaped triangular, 14–15 mm long, 10 mm wide at the top, slightly notched.

Habitat Cascade Mountains of Washington and Oregon, also Siskyou Mountains of Clifornia, U.S.A.

Cultivation Introduced into Britain by D. Douglas in 1830. Succeeds best and is most frequently seen in Scottish collections. This species cones freely, often on relatively young trees. Zone 5. (See Illus. Nos. 22d, 59)

-- 'Aurea' Incorrect name for 'Sherwoodii'.

-- **'Blaue Hexe'** Krüssmann 52 (1979). Derived from a witch's broom discovered by Boehlje of Westertede, West Germany in or around 1965. It slowly (Annual growth 25 mm) forms a low, flat-topped or hemispherical plant, very slow-growing and densely clothed with short (10–12 mm), broad, pale-blue leaves. It requires good soil conditions to develop a good colour.

-- **'Glauca'** (Ravenscroft) Rehd. in Rhodora **42**: 523 (1940); under *PICEA nobilis* Ravenscroft, Pinet. Brit. **2**: 183 (1863). Blue Noble Fir. The most glaucous form and a free cone-bearer. To graft this form, take scions from the leaders only. (See below, under 'Prostrata').

Figure 59. *Abies procera*. A ripe cone. (10)

-- **'Jeddeloh'** Krüssmann 52 (1979). Similar to 'Blaue Hexe', but the foliage is more greenish-blue. It was developed from a witch's broom on the Jeddeloh Nursery, Oldenburg, W. Germany, about 1960. It is less attractive than 'Blaue Hexe' when at its best.

-- **'La Graciosa'** A prostrate form raised from seed by Cindy Lou Pease at Evans Farm Nursery about 1980. In seven years the mother-plant was 1.5 m across, 75 cm high. The growth is compact, the foliage colour normal.

-- **'Noble's Dwarf'** New name. This was the first of a group of 'dwarf' cultivars now in cultivation, all rather alike. It originated in the James Noble collection in Golden Gate Park, San Francisco, whence a plant was received by Mr A. H. Nisbet of Gosport, one of the first serious collectors of dwarf conifers in Britain. Nisbet distributed it as 'Noble's form', and it has now apparently in some quarters acquired the name 'Nobel', but this must surely be an orthographic error (which should be corrected) in honour of a great pioneer with the dwarf conifers in America.

-- **'Prostrata'** (Hornibrook) Rehder in Rhodora **42**: 524 (1940); *A. nobilis* var. *prostrata* Hornb. 72 (1932).

Many of the species of fir, in common with some other genera in which the structure of lateral branches is different from true leader growth will, if propagated from the former type of shoot have difficulty in producing a bud that will develop into a true leader and so enable the plant to reassert its arboreal character. So great is this handicap in some cases that such a plant will develop without training into a flat-topped, spreading plant. Since such a plant comprises a single branch of a tree free of the normal competition for light, it will usually produce a well-furnished, densely clothed specimen, and when (as in the case of *A. procera*) the foliage is itself attractive, the result is a very useful and sometimes spectacular garden plant.

This situation is not recognised in the Cultivated Code. An apparent attempt to do so in the latest (1980) edition fails because these plants are not "a clone...having a habit of growth which is clearly distinguishable from the normal habit and which is retained by appropriate methods of propagation" – they are, and remain, individuals produced at will by selection of propagation material taken from **any member** of the species. None of the three examples cited [Article 11, (e)] falls within this description, each being already acceptable as a cultivar, so we must continue for the moment in the use of the word *Cultivariant*, suggested by Welch [23 (1966)] and now quite widely used in this connection.

Nevertheless, these interesting plants have long been in circulation, so understandably have in some cases acquired cultivar names which we can continue to use (but with reservations as to their having any clonal value). All such plants should be watched for strong-growing, upright shoots with radially arranged foliage. These must be cut at once if the prostrate character of the plant is not to be lost. **'Glauca Prostrata'** Hornibrook, 28 (1939) covers cultivariants of the best blue form 'Glauca': it makes one of the most outstanding and attractive garden conifers. The name 'Procumbens' has been used for an intermediate form with foliage that is grey rather than blue. It is however illegitimate, so if this is to be maintained in cultivation an acceptable name will have to be found.

-- **'Sherwoodii'** Sherwood in Amer. Nurs., Dec. 15th (1948) and in Woody Pl. Reg. List **1** (1949). Leaves golden yellow throughout the year. Discovered by the Sherwood Nursery, Portland, Oregon, in 1933. Should be planted more widely.

A. pseudo-menziesii Wenderoth (1851). Now known as *PICEA sitchensis*.

A. pumila Hort. ex Gordon (1858). Now known *PICEA abies* 'Pumila'.

A. pungens Parry (1863). Now known as *PICEA engelmannii*.

A. pygmaea Carrière (1855). Now known as *PICEA abies* 'Pygmaea'.

A. recurvata Masters in Journ. Linn. Soc. Bot. **37**: 423 (1906).

Size and habit Conical tree 30–40 m, with age forming a flat crown.

Figure 60. *Abies recurvata.* Min fir. Detail of foliage from a side branch.

Trunk 1.8–4.5 m in girth. *Bark*: Dark grey or reddish-brown, papery and flaking in strips; frequently with sprouts of foliage.

Branchlets Pale yellowish-grey, lustrous, glabrous, shallowly ridged, increasingly roughened the 2nd and 3rd years.

Winter buds Ovoid, irregular; usually pointed at the apex, very resinous.

Leaves Foliage dense in two ranks; on the upper side of the shoots leaves are strongly recurved, to ±90° (hence the species name); forming a narrow valley; those on the underside pectinate, spreading widely at 80–90°; linear, narrowed at the base; 12–25 mm long, thick, hard; apex acute or pointed, often sharply so; lustrous yellow- or bluish- green above, paler beneath.

Mature cones Borne in small clusters; short-stalked, oblong-ovoid, narrowed at the base, rounded tops; 5–10 cm long, 3–4 cm broad, slightly resinous; violet purple ripening to greyish-brown. *Scales*: Oblique-elliptic, 20 mm long, 10 mm wide, wedge-shaped, narrow at base, *Bracts*: Hidden.

Seeds Ovoid, 6 mm long, with obliquely rounded dark violet wing 5 mm long.

Habitat W. China.

Cultivation Discovered by E. H. Wilson and

introduced by him into U.S.A. in 1910, but still uncommon in U.K. *A. recurvata* is distinguishable from the similar foliage of *A. numidica* by the strongly recurved leaves with acute or pointed tips, often glaucescent or bloomed on the upper side. In *A. numidica* these features are less pronounced and the winter buds are not resinous. Zone 6. (See Illus. No. 60)

A. religiosa (H.B.K.) Schlechtendal in Linnaea **5**: 77 (1830); *PINUS religiosa* H.B.K., Nov. Gen et Spec. **2**: 5 (1817).

Size and habit Handsome tree to 60 m in the wild, less in cultivation, with a narrow crown.

Trunk Stout. *Bark*: At first smooth, greyish-white, becoming rough and deeply broken up into irregular, oblong plates, scaling off.

Branches Slender, spreading or ascending.

Branchlets Glossy light red above later becoming reddish- to purplish-brown, olive-green below; with pulvini-like ridges; at first with minute erect pubescence, later becoming glabrous.

Winter buds Broadly-conic, rounded at tip; shining reddish-brown with green patches; very resinous, strongly scented.

Leaves Densely set in 3–4 ranks, directed forwards; widespreading at ± 10–20° and curving outwards on either side of the shoot; 15–30 (35) mm long, about 1–1.5 mm broad; narrowly linear, gradually tapering to a long-pointed, (rarely somewhat notched) tip: lustrous dark green above, deeply grooved at the lower end, otherwise flattened and with sometimes a few stomata near the tip; two silvery-grey or white bands of stomata beneath.

Mature cones Set on a short, stout stalk; long-cylindrical, 11–17 cm long, 5–8 cm broad, rounded at base, gently tapering towards a rounded tip; dark violet-blue ripening to dark brown. *Scales*: Broad, fan-shaped, 15 mm long, 30 mm broad. *Bracts*: Much projecting with long points, reflexed.

Seeds 15 mm long, obliquely pointed, with obliquely rounded wing, 15 mm long; bright chestnut-brown.

Habitat Mexico and Guamatemala, at altitudes from 1200–2000 m.

Cultivation Discovered by von Humboldt in 1799 and introduced by Hartweg into England (Hort. Soc. of London) in 1838. It is not hardy in cold, temperate regions and is extremely rare in Britain. Mitchell [43 (1972)] comments on a superficial likeness of this species to *Pseudotsuga menziesii* he states "The groove in the leaf, failing before halfway and the outward curved, tapering leaves on shiny brown shoot are diagnostic, but *A. vejari* differs only in minor ways". Zone 8. (See Illus. No. 45)

A. romania Loudon ex Steudel (1841). Now known as *PINUS nigra* var. *caramanica*.

A. rubra Michaud, Poiret and other authors. Now known as *PICEA rubens*.

A. sachalinensis (Schmidt) Mast. in Gard. Chron. **2** 588: (1879); *A. veitchii* var. *sachalinensis* Schmidt in Mem. Acad. Sci. St. Petersb. Ser. 17 **12**: 175 (1868). Sachhalin Fir.

Size and habit Tree to 30 m, with a dense, narrowly conical crown.

Trunk Smooth, circular. *Bark*: On young trees smooth, grey, or white and covered with resin blisters; with age becoming thick, blackish-grey, more or less broken up into irregular plates peeling off to expose the red- or grey-brown surface beneath.

Branches Short, more or less whorled; horizontal or ascending; lower branches drooping.

Branchlets Short, ribbed or grooved, glossy brown or rusty-grey, densely pubescent in the grooves. Both shoot and pubescence are darker in colour the second year.

Winter buds Small; ovoid to globose, 5 mm long, dark shining red to purple, hairy and thickly

Figure 61. *Abies sachalinensis*. Sakhalin fir. Detail of foliage from a side branch. Leaves are pale green, dull beneath. (5)

encrusted with white resin.

Leaves Crowded, directed strongly forward all round the shoot; on the upper side of the shoot held upwards, some shewing the underside; at the sides and below the shoot spreading at ± 45°. Slender, 2.5–4 cm long and not more than 1 mm wide, parallel-sided, rounded or notched at apex, shining green, grooved, without stomata above save at the extreme tip but with 7–8 lines in each of two bands beneath.

Mature cones Sessile, cylindrical, narrowed towards top and base, 7 cm long, 2.5–3 cm broad, olive green when young ripening to dark brown or blackish-purple. *Scales*: Numerous; densely pubescent outside, glabrous within. *Bracts*: Reddish-brown; usually extending but little beyond the scale, narrowing to a tail-like point, not reflexed.

Seeds Wedge-shaped, 5 mm long, greyish violet with broadly wedgeshape wing, 4 mm long, 6 mm wide.

Habitat Northern Japan, Sachalin, Kurile Isles.

Cultivation Introduced by Maries, who sent seeds from Sachalin to the Veitch nursery in London in 1878. This species starts into growth very early so is frequently damaged by Spring frosts, especially as a young plant. Mitchell [44 (1972)] says that the crushed foliage is sticky and smells of cedarwood oil. Two botanical varieties are recorded in the literature. Of these, var. ***nemorensis*** Mayr, Monog. Abiet. Japan, 42 (1890) is considered by Liu [73 (1971)] not to be worthy of recognition in such a variable species, but var. ***mayriana*** Miyabe & Kudo in Trans. Sapporo Nat. Hist. Soc. **7**: 131 (1919) is a variant found in Sakhara: U.S.S.R. and in Hokkaido: Japan (within the general range of the species) that differs in having dark green cones, prominently visible and recurved scales, greenish in tone andin other minor morphological differences. Zone 5. (See Illus. No. 61)

A. sacra David (1875). Now known as *KETELEERIA davidiana*.

A. × ***shastensis*** (Lemmon) emend Liu 319 (1971). Shasta red fir. A naturally occurring hybrid: *A. magnifica* Murray × *A. procera* Rehder, that varies from one plant to another (some specimens deviating only slightly from *A. magnifica*), but always intermadiate between the two parent species. Many specimens grown in gardens as *A. magnifica* turn out, on bearing cones, to be *A. shastensis*.

-- **'Shasta Prostrate'** makes an attractive and completely prostrate plant. It originated in Girard's nursery, Bellingham, Washington, U.S.A., where the mother plant has a good blue colour and has never thrown up a leader. At thirty years it is 2.5 m across and only 45 cm high.

A. schrenkiana. (Fischer) Lindley & Gordon (1850). Now known as *PICEA schrenkiana*.

A. *sibirica* Ledebour. Fl. Alt. **4**: 202 (1833). *A. pichta* Forbes 113 (1839). Siberian Fir.

Size and habit Graceful, conical tree 30–40 m with a dense spire-like crown.

Trunk Straight, tapering. *Bark*: Blackish-grey, with numerous resin blisters.

Branches Closely set, spreading horizontally. Lower branches long, pendent.

Branchlets Numerous; silvery-grey or pale borwn; smooth, not ribbed; with dense minute white pubescence; glabrous and grey-black the second year.

Winter buds Small, globose, brownish-yellow, 2–3 mm long, almost hidden under a dense coating of white resin.

Leaves Foliage dense all round the shoot, directed forward. Leaves on upper side of shoot 10–15 mm long, completely hiding the shoot; at the side and below longer, to 30 mm, upswept or randomly curved, but all uniformly 1.25–1.50 mm broad; bright yellowish-green above, flat but with a narrow groove, rounded or slightly notched at apex, often with 2–3 short lines of stomata near apex; beneath, much paler green with two narrow greyish bands and a broad keel.

Mature cones Set on very short stalk, cylindrical, 5–8 cm long, 3–4 cm broad, rounded at both ends, deep violet-blue ripening to cinnamon brown; densely pubescentinside and out. *Scales*: Broadly wedge-shaped, fringed at the sides. *Bracts*: Hidden, being only half as long as the cone scales.

Seeds Small, wedge-shape, 5–6 mm long with a thin, purple, membraneous wing, 10 mm long, purple.

Habitat North east U.S.S.R.: Siberia to Altai Mountains; North China: Heilungkiang. To 2000 m.

Cultivation Introduced into Britain in 1820, but still rare (possibly not in cultivation at all) in Britain, where it is sometimes misidentified as *A. sachalinensis*. It is subject to injury from Spring frosts. Several selections made around the turn of the century are lost to cultivation. Zones 2/6.

-- var. ***semenovii*** (Fedtschenko) Liu, 188 (1971). A geographical form shewing minor differences in habit and foliar characteristics. It is not accepted as distinctive by some botanists, since the differences may be partly climatic, and intermediate forms occur.

A. sitchensis (1) C. Koch (1878). Now known as *PICEA jezoensis*.

A. sitchensis (2) Lindley ex Gordon (1858). Now known as *PICEA sitchensis*.

A. smithiana Forbes, Lawson and other early authors. Now known as *PICEA smithiana*

A. *spectabilis* (D. Don) Spach. Hist. Nat. Veg. Phan. **11**: 422 (1824); *PINUS spectabilis* D. Don

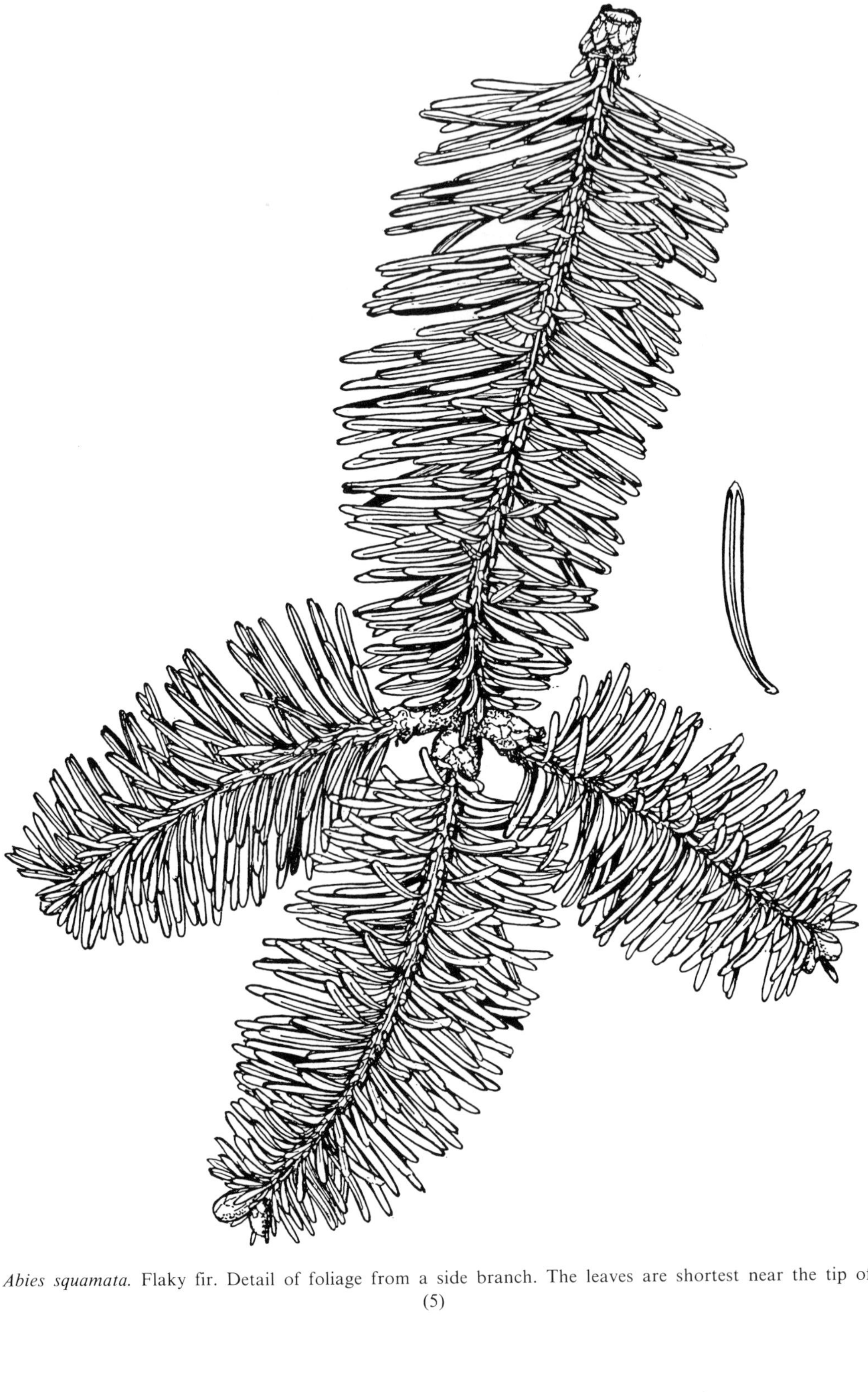

Figure 62. *Abies squamata.* Flaky fir. Detail of foliage from a side branch. The leaves are shortest near the tip of the shoot. (5)

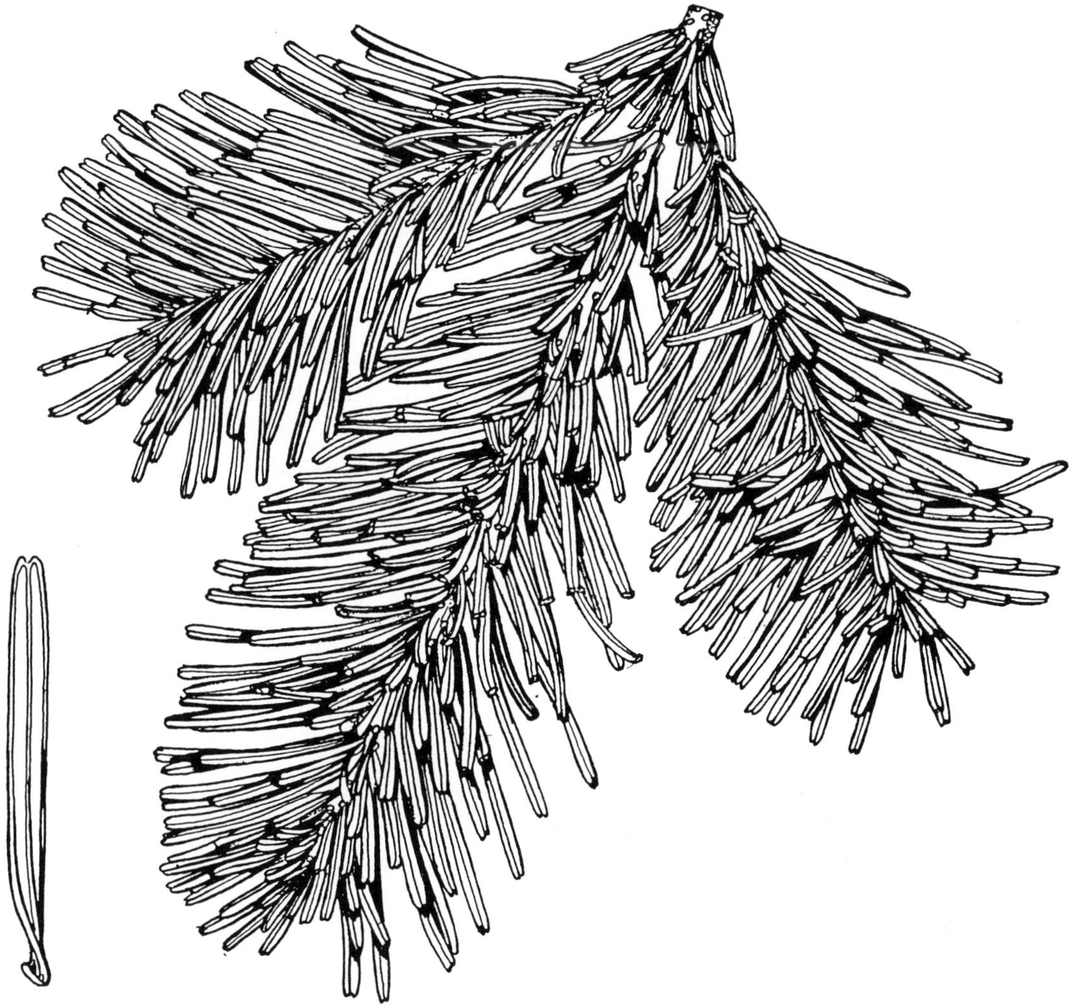

Figure 63. *Abies veitchii.* Veitch fir. Detail of foliage from a side branch. The branches are often upswept. (5)

in Lamb. **3**: (1824); *ABIES webbiana* Lindley in Penny Cycl. **1**: 30 (1833). Himalayan Fir.

Size and habit Striking, broadly conical tree 25–30 m tall, with a dense, umbrella-shaped, eventually an open, flat-topped crown.

Trunk Stout. *Bark*: Smooth, thin, pinkish- or silvery-white on young trees; on old trees dark- or brownish-grey; rough and with long, deep fissures, spiralling round the trunk.

Branches Thick, widespsreading.

Branchlets Stout, noticeably ridged and deeply grooved; grey or reddish-brown with brown pubescence in the grooves; darkening to greyish-black the second year.

Winter buds Large, ovoid-globose; 1 cm diam, yellow-brown; strongly resinous.

Leaves Crowded; more or less pectinate in 3–4 ranks, those in the top rank pointing upwards and outwards forming a valley, exposing the shoot; in the lower ranks spread widely in a flat plane; varying greatly in size (2.5–6 cm long, 2.5–3.5 mm wide); flattened, rigid, linear, rounded or notched at apex with tooth-like tips; stiff leathery, lustrous dark green above, narrowly grooved full length and lacking stomata; beneath with two broad white or pale yellow-green stomatic bands and a glossy, raised mid-rib.

Mature cones Usually on a very short stalk, cylindrical or ovoid-oblong; 15–20 cm long, 3–5 cm broad; violet-purple ripening to brown. *Scales*: 1.5–2 cm long, 4–6 mm broad. *Bracts*: Half as long as the scale: hidden or shewing between the scales.

Habitat North East Afganistan; North India; West Nepal.

Seeds Oblong-ovate, angular, 8 mm long with a broad-ovate wing, 15 mm long, notched.

Cultivation Introduced into Britain (Fulham nursery) by Dr. Wallich in 1879, but still uncommon in English collections, perhaps because this species flushes early so is prone to damage from Spring frosts. Zone 6.

-- var. ***brevifolia*** (Henry) Rehder in J. A. A. **1**: 54 (1919). A geographical form with short leaves, 3–4 cm long, smoother bark and other minor foliar differences and forming a more shapely and conical tree. It succeeds better than the type in Britain.

A. spinulosa Griffith (1847). Now known as *PICEA spinulosa*

A. *squamata* Masters in Gard. Chron. ser. 3, **39**: 299 (1906).

Size and habit Conical tree to 40 m, usually with a narrow crown.

Trunk Stout. *Bark*: Pinkish-brown to dark purple; shaggy; peeling off the underside of the branches in thin papery layers as in *Betula*.

Branchlets Dark brown or purple, with broadened ridges and dense brown pubescence in the grooves, or (sometimes, in cultivation) glabrous.

Winter buds Sub-globose, reddish-brown, conspicuously white resinous.

Leaves Foliage dense; held in distinct ranks; in the top rank the leaves standing vertically, not hiding the shoot; in the lower ranks they are more or less pectinate, stiffly standing off at 80–90°; leaves 12–25 mm long, 2 mm broad, hard, rigid, parallel-sided, with an abrupt point at the apex. Dark greyish- or bluish-green above carrying stomata near the tips; beneath with 2 white stomatic bands. A characteristic feature is the shortening of the leaves towards the tip of the shoot, producing a tapered outline to the year's growth.

Mature cones Stalkless; Oblong-ovoid, rounded both ends; 5–6 cm long, 2–3 cm broad, violet; very resinous. *Scales*: about 15 mm across, wedged-shaped, narrowest at the base. *Bracts*: Showing slightly, with a slender tip, often decurved.

Seeds Oblong-ovoid, 6 mm long, 2 mm broad with a hatchet-shape, brownish-red wing 10 mm long, 14 mm wide.

Habitat West China at an altitude of 4000–4600 m.

Cultivation Discovered by E. H. Wilson and introduced by him to America (Arnold Arboretum) in 1910, but it is still rare in Britain. Zone 5. (See Illus No. 000)

A. standishiana. C. Koch (1878). Now known as *PSEUDOTSUGA menziesii*.

A. subalpina Engelmann, Amer. Natur. **10**: 554 (1876). Now known as *A. lasiocarpa*.

A. subarctica Nyman (1878). Now known as *PICEA abies*.

A. sutchuensis (Franchet) Rehder and Wilson 48 (1914). Now known as *A. fargesii* var. *sutchuenensis*.

A. taxifolia (1) Poiret (1804). Lambert, Presl and other early authors. Now known as *PSEUDOTSUGA menziesii*.

A. taxifolia (2) Jeffery ex Gordon (1858). Now known as *TSUGA heterophylla*.

A. thunbergii Thunberg ex Gordon (1875). Now known as *PICEA smithiana*.

A. torano Siebold (1830). Now known as *PICEA polita*.
A. tsuga Siebold Zuccarini (1842). Now known as *TSUGA sieboldii*.

A. × vasconcellosiana Franco in Port. Acta Biol. II (1/2), 144, t. 1–4 (1946). A natural hybrid (*A. pindrow* × *A. pinsapo*) that was discovered in 1945 by Dr. Franco in the Pena Park at Sintra, Portugal, grown from seed collected from a speciman of *A. pindrow* which stood in the neighbourhood of an *A. pinsapo*. All characteristics incline to one or the other parent or are intermediate.

-- **'Amaral Franco'** Meyer in Baileya **9**: 127 (1961). Differs from other trees of the hybrid growing in the Pena Park at Sintra, Portugal by its conical and more dense habit.

A. veitchii Lindley in Gard. Chron. 23 (1861). Veitch Fir.

Size and habit Broadly conical tree to 30 m with a narrow, pointed crown; often furnished to the ground with branches.

Trunk Stout; deeply fluted. *Bark*: Smooth, dark- or greenish grey, with prominent resin blisters.

Branches Short, slender; lower branches upswept, showing the silvery-white undersides of the leaves.

Branchlets Reddish-brown or green, with curly, dense brown pubescence.

Winter buds Small, nearly globular, 2–3 mm; purplish or reddish, translucently resinous.

Leaves Crowded, in ranks. Leaves in top rank short, directed forward and steeply upward, hiding the shoot; in lower ranks more or less pectinate, spreading horizontally at 45° or curving downwards; linear, 10–25 mm long, not broadening towards the tip; square-ended and notched at apex; tapering toward a large sucker-like base; lustrous dark green above and with chalky-white bands of stomata beneath.

Mature cones Sessile, narrowly cylindrical, 4–6 cm long, 2–2.5 cm broad, apex rounded or with a blunt, oblique tip; bluish purple (or occasionally green) ripening brown. *Scales*: Very densely compressed, 9 mm long, 15 mm broad, entire, pubescent on exposed surfaces. *Bracts*: Three-lobed; shewing slightly, exposing the pointed, often reflexed tip.

Seeds Obovate-ovoid, cuneate-angular, 7 mm long; yellowish, the wing fan-shaped, wider than long, 4 mm long, blackish, notched above.

Habitat Central Japan at high altitudes.

Cultivation Discovered by J. G. Veitch on the Susi San mountain and introduced by him into England in 1861 and by Dr von Siebold into the Netherlands (Leiden) in 1862.

A. veitchii is one of the most successful of the Firs in cultivation and is widely to be found in Britain. Mitchell [47 (1972)] states that it has large pockets under each branch and corresponding ribs between. This at once distinguishes it from *A. amabilis* and *A. nephrolepis* where this peculiarity is absent, or (in *A. nephrolepis* var. *nikkoensis*) present only to a lesser extent. Zone 3. (See Illus. No. 63)

-- var. **sikokiana** (Nakai) Kusaka in Iwata & Kusaka, Conif. Jap. **3**: 212 (1954); *A. sikokiana* Nakai in Bot. Mag. Tokyo, **42**: 452 (1928). Shikoku Fir. This is a geographical variant of uncertain status. At one time given specific rank, and having affinities with *A. koreana*, it differs from typical *A. veitchii* in the young shoots soon becoming glabrous, the leaves broadening out towards their tips and other foliar details. From *A. koreana* it differs by its pectinate lower leaves.

A. vejari Martinez in Anal. Inst. Biol. Mexico **13**: 629 (1942). Vejar Fir.

Size and habit Tree to 30–40 m, with a conical crown.

Trunk Probably stout. *Bark*: Thin, smooth, grey.

Branches In whorls, spreading.

Branchlets Greenish-red or purplish brown, grooved, glabrous; noticeably lustrous beneath.

Winter buds Ovoid-globose, Deep purple-red, with a lumpy coating of white resin.

Leaves Crowded, emerging spirally all round the shoot and held forward in no definite pattern, the white undersides of leaves frequently to be seen. Leaves linear, rigid, nearly straight, 25–30 mm long, 1.5 mm wide, dark green above, slowly tapering to a yellowish point, grooved and with 6–10 broken lines of stomata; glaucous beneath with a prominent raised midrib between two bands of stomata.

Mature cones Sub-sessile, shortly oblong or nearly ovoid, 6–10 cm long, 4–5 cm wide, or occasionally dented at the top. *Scales*: Projecting and recurved.

Seeds Violet-brown, winged.

Habitat Mountains of Mexico.

Cultivation Although this species was described in 1942 it has only recently been introduced into Britain, where it seems to prosper. It is closely related to *A. religiosa*, to a variety of which it may one day find itself demoted. Zone 6.

-- var. *macrocarpa* Martinez, Anal. Inst. Biol. Mexico XIX, 1 (1948). A local population with large fruit; no longer distinguished within the species.

Figure 64. *Abies vilmoriniana.* Close-up of foliage.

-- var. ***mexicana*** (Martinez) Liu, 261 (1971); *A. mexicana* Martinez in Anal. Inst. Biol. Mexico XII 626 (1942). Mexican Fir. A geographical variety with a limited distribution shewing rather minor variations in foliage and fruit. Probably not yet in European cultivation.

-- **'Serpent'** A totally recumbent form registered by F. J. Crowe of San Diego, Cal. U.S.A. in 1981.

A. venusta Koch. Now known as *A. bracteata*.

A.* × *vilmorinii Masters in Journ. Hort. Soc. Lond. **26**: (1901). An artificial hybrid raised by M. de Vilmorin at Verriers, near Paris, in 1858 between *A. pinsapo* and *A. cephalonica*. It makes a vigorous tree, the foliage shewing the influence of each of the parents, but the cones are very close to the pollen-parent *A. cephalonica*. (See Illus. No. 64)

A. vulgaris Wenderoth (1851). Now known as *PICEA abies*.
A. webbiana Lindley (1833). Now known as *A. spectabilis*.
A. williamsonii Newberry (1857). Now known as *TSUGA mertensiana*.
A. withmanniana Hort ex Carrière (1867). Now known as *PICEA orientalis*.
A. yunnanensis Franchet (1899). Now known as *TSUGA yunnanensis*.

ACMOPYLE

Podocarpaceae

Acmopyle Pilger in Engler Pflanzer. **4. 5** (Heft 18): 117 (1903). A small genus allied to *Podocarpus* and *Dacrydium*. A native of New Caledonia, where it forms a forest tree. It is not hardy in Europe. Zone 9.

ACTINOSTROBUS

Cupressaceae

Actinostrobus Miguel. ex Lehmann, Pl. Preis. **1**: 644 (1845). A small genus of evergreen shrubs allied to *Callitris*. Native to Western Australia and not hardy in Europe. Zone 9.

AGATHIS

Araucariaceae

Agathis Salisbury in Trans. Linn. Soc. Bot. **8**: 311 (1807); *Dammara* Link, Enum. Pl. Hort. Berol. **2**: 411 (1822). Kauri Pine: Dammara. A genus of tall trees containing upwards of a dozen species from New Zealand, Australia, Indonesia and the Malay peninsula of which only one is marginally hardy even in the most favoured areas of the British Isles.

A. australis (Lambert) Steudel (1940) is, or at least at one time was an important timber tree in New Zealand, but in Britain it is seen only as a large bush or small tree with flat, leathery leaves in opposite pairs, 6–8 cm long, 1.0–1.5 cm across; quite unlike the narrow leaves generally characteristic of conifers. Zone 9. (See Illus. No. 72)

AMENTOTAXUS

Cephalotaxaceae

Amentotaxus Pilger, Bot. Jahrb. **54**: 41 (1916). A genus of 2 species confined to S.E. Asia. None is hardy in Europe. Zone 9.

ARAUCARIA

Araucariaceae

Araucaria Jussieu. Gen. Plant. 413 (1789). A genus of about 18 species from Australia, Indonesia and S. America, of which only 1 species is reliably hardy in Europe, although what is popularly called the Norfolk Island Pine. *A. heterophylla* is widely grown under greenhouse culture as a decorative plant.

Evergreen trees, 60 m; bark on old trees thick, resinous and ringed with the bases of old leaves or, in some species, rough and falling off in papery scales: branches on young plants symmetrical, on old trees horizontal, usually in whorls: branchlets glabrous when young, rusty coloured, opposite or alternate: winter-buds thick, apex blunt, enclosed by leaves: scales light brown; leaves spirally arranged, awl-shaped or scale-like, stiff, clothing all the branches uniformly, closely imbricated, sharply pointed, persisting for many years, dimorphic in some species: male flowers, terminal cylindrica; female flowers terminal, in ovoid or globose heads: cones globular or ovoid, woody, falling apart at maturity; scales numerous, closely imbricated, elongated, wedge-shaped, ridged, with only one seed; seeds large, adnate to the base of the scale, winged in most species, sometimes edible, cotyledons 2–4.

A. ***angustifolia*** (Bertol.) Kuntze, Rev. Gen. Pl. **3**: 375 (1893); *Colvmbea angustifolia* Bertol., Piante del Bras. 7 (1820); *A. brasiliana* A. Rich. in Dict. Class. Hist. Nat. **1**: 512 (1812); *A. brasiliensis* Auct. Parana Pine: Candelabra Tree.

Tree, 35 m; sparingly branched, head flat; trunk straight; branchlets long and pendent; scales terminal with a recurved spine at apex; seeds up to 5 cm long, 2 cm wide, 8 mm thick, light brown.

Southern Brasil. Introduced by Lee, nurseryman at Hammersmith (London area) from Rio Tanrivo by seed in 1819. Not hardy, much planted in subtropical areas, viz. the Mediterranean. Several clones selected during the last century because of minor differences in leaf shape are no longer in cultivation.

A. ***araucana*** (Molina) K. Koch 206 (1873); *Pinus araucana* Molina, Saggio Storia Nat. Chile 182 (1782); *A. imbricata* Pav. in Mém. Acad. Med. Madrid **1**: 199 (1797). Monkey Puzzle, Chile Pine.

Dioecious tree, 30–35 m, pyramidal; crown often round topped; bark thick, resinous, showing shallow circular undulations; branches very stout, spreading in whorls of 4–8, usually 5, falling off after several years; branchlets few, in opposite pairs, horizontal or pendulous; leaves densely imbricated, ovate-lanceolate, slightly concave above, spiny pointed, 2.5–5 cm long, persisting for many years, on both sides glossy dark green, margin thickened, with longitudinal stomatic lines; cones globular-ovoid, 14–20 cm across, ripe after 2–3 years, falling apart at maturity; seeds oblong, slightly

Figure 65. *Araucaria aruacana.* Detail of base of trunk shewing the characteristic 'elephant's foot'.

Figure 66. *Araucaria araucana.* Detail of foliage and flower.

Figure 67. *Araucaria araucana.* The Monkey-puzzle tree. A specimen at West Cliffe Hall, Hythe, Hampshire, U.K.

Figure 68. *Austrocedrus chilensis* at Bedgebury National Pinetum, Kent, U.K.

Figure 69. *Austrocedrus chilensis*. Detail of foliage and cones.

compressed, 2.5–3.5 cm long, wingless; cotyledons 2.

This is the well-known "Monkeypuzzle tree", probably the most easily recognised of all trees because of its sharply-pointed, stiff, closely set leaves and the fact that because of the interest it created when first introduced into cultivation it was widely planted (and still is being planted, sometimes in most unsuitable situations) so is now quite a common sight. Although quite hardy, being one of the few South American trees to take kindly to our climate, it will never merge with the English landscape and so should only be planted in isolation or where it will not look out of place as it develops. Where the soil and climate are to its liking it will remain furnished to the ground and in such circumstances an isolated tree can be quite striking even if not beautiful, but in dry, shallow soils it usually loses its lower branches, and the sprouts developing from the trunk do little to improve the look of the tree. Several avenues of large trees exist, where the puckerings of the bark at the bases of the trunks are reminiscent of elephants' feet, but even in such a setting the Monkeypuzzle tree has little contribution to make. During the latter part of the nineteenth century, at the height of its popularity, several foliage variants and coloured forms were introduced. It is possible that these were more curious than beautiful, so it is not surprising that none of them is now known in cultivation. Zone 7 (See Illus. Nos. 65, 66, 67)

*A. **bidwillii*** Hook. in London Journ. Bot. **2**: 303 (1843). Bunya Bunya.

Dioecious tree, 50 m, young trees pyramidal, symmetrical, densely branched; bark thick, resinous, scaling off in thin layers; branches in whorls of 10–15, sparingly ramified; branchlets green when young, lateral branchlets long, pendent; leaves spirally arranged, imbricate, lanceolate-ovate on sterile branches, very sharply pointed, to 5 cm long, 5–10 mm broad, thick, stiff, entire, glossy green; on fruiting branches and upper shoots, more closely arranged, often less than 2.5 cm long, curved inward, in both cases with irregular stomatic lines above and with regular lines beneath; cones globose-ovoid, about 30 cm long, 23 cm broad, sometimes 4–5 kg in wight and containing up to 150 seeds; scales numerous, 10 cm long, 7.5 cm broad, terminating in an acute margin; seeds large, pear-shaped, 5–6.5 cm long, over 2.5 cm wide, maturing the third year; wing rudimentary.

Australia, coastal districts of Queensland. Found by Andrew Petric, Superintendent of Government Works at Moretown Bay, named after J. S. Bidwill, who brought it to Kew, where it was distributed by W. J. Hooker in 1843. Not hardy in the temperate zone, rather widely planted in subtropical areas, viz. the Mediterranean.

*A. **cunninghamii*** Sweet, Hort. Brit. 475 (1827). Hoop Pine; Moreton Bay Pine. Tree, 60–70 m: trunk 3 m in girth; bark rough, peeling off in thin layers; branches in whorls of 4–7, the upper ascending, the lower on older plants somewhat depressed; branchlets in dense tufts at the end of the branches; leaves of two kinds, those on young plants and lateral branches spirally arranged, usually lanceolate or triangular, 10–20 mm long, straight or nearly so, spreading, stiff, sharp-pointed, margin entire; on older trees and fruiting branches crowded, shorter, imbricate, short-pointed, incurved, with the midrib decurrent, green or glaucous, with stomata in each surface; cones ovoid-globose, about 10 cm long, 7.5 cm broad; scales terminating in a lanceolate recurved mucro; seeds with a narrow membranous wing on each side. Northern New South Wales and Queensland, Australia.

The most widely distributed of the Australian araucarias and a valuable timber tree. Introduced in 1827. Not hardy in the temperate zone, rather commonly planted in subtropical areas.

A. heterophylla (Salisb.) Franco in Anais Inst. Sup. Agron. **29**: 11 (1952); *Eutacta heterophylla* Salisb. in Trans. Linn. Soc. Lond. **8**: 316 (1807); *A. excelsa* (Lamb.) R. Br. in Ait. 412 (1813); *Dombeya excelsa* Lamb. Decr. Genus. Pinus. ed. 1, **1**: 87 (1807). illeg. name. Norfolk Island Pine.

Tree. 70 m; trunk 15–18 m in diam.; bark scaling off in thin flakes; branches horizontal in whorls of 4-7; branchlets horizontal or pendent; leaves of 2 kinds; on juvenile or lateral branchlets awl-shaped, 8–12 mm lon, soft, incurved, bright green, on older and fertile shoots densely set, imbricate, broadly ovate, incurved, horny-pointed, to 6 mm long, 3–6 mm broad at the base; cones subglobose, 10–15 cm across, broadest at base; scales terminating in an incurved spine; seeds (including wings) 3–6 mm long.

Norfolk Islands, introduced into Britain in 1793.

A. heterophylla (formerly known as *A. excelsa*) is not hardy but grown in pots and tubs as a small house plant. Zone 10.

A. rulei Muell. ex Ldl. in Gard. Chron. 868 **(1861)**. Tree, 15 m, very variable at different ages; branches in whorls of 5–7; branchlets elongate, horizontal or slightly pendent; leafy branchlets very long; leaves very variable, oblong-lanceolate to elliptic, 12–15 mm long, 6–10 mm broad at base, imbricate, concave, stiff, leathery, apex shortly pointed, arched toward the branch and stem, with a prominent dorsal nerve, silvery grey above, glossy green and with minute stomata beneath; cones globose-ovoid, 12–15 cm long, 8–10 cm across; scales 20 mm long, terminating in a spiny apex; seeds about 10 mm long, 6 mm wide; wings narrow. New Caledonia, Papuan Archipelago. Discovered by W. Duncan, botanical collector, who sent it to John Rule of Melbourne, who introduced it into cultivation in 1865. Not hardy but rather widely planted in warm temperate regions.

ATHROTAXIS
Taxodiaceae

Athrotaxis D. Don in Trans. Linn. Soc. **18**: 171 (1839). Tasmanian Cedar. Trees evergreen, densely branched, old trees with poorly-shapen heads; bark slightly furrowed, fibrous, peeling off in longitudinal shreds; branches small, irregularly disposed; branchlets spreading; leaves green, crowded, small, alernate or indistinctly decussate, either short, blunt, scale-like, appressed, or lanceolate and somewhat loosely disposed; male flowers in imbricated spiral catkins; female flowers in spirally imbricated catkins; cones small, resembling those of *Cryptomeria japonica*, maturing the first season: scales woody, wedge-shaped at base, spreading horizontally, with a triangular, spine-like process near the apex; seeds 3–6, small, thin, oblong, with 2 narrow wings, rather similar to those of Sequioa, cotyledons 2.

Three species confined to Tasmania. The name is derived from the Greek athroos = crowded together, and taxis = arrangement, in reference to the position of the cone scales. Propagation by seeds or cuttings, or by grafting onto stocks of *Cryptomeria japonica*.

A. cupressoides D. Don in Trans. Linn. Soc. **18**: 173 (1839); *Cunninghamia cupressoides* Siebold & Zuccarini 9 (1842); *A. imbricata* Maule ex Gordon 16 (1862). Tree, 6–12 m; bark fibrous, slightly furrowed; branches ascending; leaves rhomboid-ovate, 3–5 mm long, 5–12.5 mm broad, obtuse, appressed, thick, keeled, with translucent denticulate margin; cones 7–12 mm across; scales woody, rounded, at the top with a short triangular recurved process. Tasmania, western Ranges and Lake St. Clair district. Zone 8. (Illus. No. 74)

A. laxifolia Hooker, Icon. Pl. t. 573 (1843); *A. donniana* Parker ex Gordon 16 (1862). Tree,

Figure 70. *Athrotaxis laxifolia.* A tree at Bedgebury National Pinetum, Kent, U.K.

Figure 71. *Athrotaxis laxifolia.* Bark on trunk.

Figure 72. *Agathis australis.* A young tree at Bicton, Devon, U.K.

Figure 73. *Athrotaxis laxifolia.* Detail of foliage and fruit.

Figure 74. *Athrotaxis cupressoides*. Detail of foliage.

Figure 75. *Athrotaxis selaginoides*. King William pine. Detail of foliage and fruit.

12 m; leaves slightly spreading, ovate-lanceolate, obtuse or pointed, 4–6 mm long, with an entire translucent margin, upper side with 2 glacous stomatic bands, whitish depressions on the back near the base; cones 18 mm across; scales thickened, with a large acute process on the back. Tasmania, summit of the Western Mountain. Zone 7. (See Illus. Nos. 70, 71, 73)

A. selaginoides D. Don in Trans. Linn. Soc. **18**: 173 1839); *Cunninghamia selaginoides* Siebold & Zuccarini 9 (1842). King William Pine. Tree, 30–35 m; bark fibrous, slightly furrowed; leaves spreading, lanceolate, incurved, acute, rigid and spine-pointed, 6–9 mm long, leathery, keeled on the back with 2 small glaucous depressions and with 2 glaucous bands on the ventral side; cones 12–18 mm long and broad; scales numerous, broad-ovate, ending in a spine-like pricess. (See Illus. No. 75)

Tasmania, Western Mountains at an altitude of 900–1200 m. Introduced about 1857 by W. Archer, Cheshunt, England. Zone 8.

AUSTROCEDRUS

Cupressaceae

Austrocedrus Florin & Boutelje, Acta.Hort. Berg. **17**: 29 (1954). Chilean Cedar. A monotypic genus, rather unconvincingly separated from *Libocedrus* in 1954, but still regarded as of that genus by some botanists.

A. chilensis (D. Don) Florin and Boutelge, loc. cit. An attractive small monoecious tree to 15 m high, with a straight trunk and ascending branches, forming a conical tree, the crown broadening with age. The foliage is most attractive, and is somewhat reminiscent of *Thujopsis*.

The short, much-flattened sprays are densely covered with leaves in decussate pairs, the facing pair being short and obtuse, the lateral pair 2–3 times as long, folded and flattened, with an inward curving tip, deep green, with a narrow white band above and a broad, similarly white band below. The cones, seldom seen in cultivation, are small. It is found on mountain slopes of the Andes, in Chile from northern Valparaiso to Valdivia.

This makes an interesting tree suitable for even a small garden, and with its attractive two-tone foliage, should be more widely planted. Zone 8. (See Illus. Nos. 68, 69)

-- **'Argentea'** described as a seedling found on the Sénéclauze nursery at Bourg Argental, Loire, France, with small, more open and silvery leaves may no longer be in cultivation, but **'Viridis'**, a seedling found in England, wholly lacking the white markings, is still in circulation, but very rare.

AUSTROTAXUS

Taxodiaceae

A monotypic genus, resembling *Podocarpus* in foliage and *Taxus* in fruits, a native of New Caledonia, and therefore not hardy in the temperate zone.

CALLITRIS

Cupressaceae

A genus of 14 species of trees or shrubs with juvenile foliage in whorls of four and adult foliage in whorls of three leaves. All are native to Australia or Tasmania, and none is hardy in the temperate zone, except in the most favoured spots. Extremely rare in cultivation. Zone 9/10.

CALOCEDRUS

Cupressaceae

Calocedrus Kurz in Journ. Bot. **2**: 196 (June 1873). A genus of 3 species allied to *Thuja*, long included in *Libocedrus* but distinct therefrom in the more flattened branchlets, which are more like those of *Thuja*, and in having seeds with one long wing projecting backwards, whereas *Libocedrus* has seeds with a wing each side of the nut. One species only has been widely planted in Britain. From *Thuja plicata* it can readily be distinguished by the very different odour of the crushed foliage and the regular,

Figure 76. *Calocedrus decurrens*. The columnar habit characteristic of trees in U.K. (10)

Figure 77. *Calocedrus decurrens* 'Aurovariegata'. The golden-variegated form.

Figure 78. *Calocedrus decurrens*. Detail of foliage and fruit. (10)

forward pointing final shoots. From *Thujopsis* and *Fokienia* it is distinct by the complete absence of any whitish colouring on the underside of the leaves.

C. decurrens (Torrey) Florin in Taxon **5**: 192 (1956); *Libocedrus decurrens* Torreya in Smithson. Contr. **6**: (1): 7 (1853); *Heyderia decurrens* (Torrey) K. Koch 179 (1873). Incense Cedar.

Size and habit Handsome tree, 30–45 m (much less in cultivation) usually seen as a dense, narrow column scarcely tapering from ground-level to the irregularly rounded crown, but in certain climatic conditions may form a much broader tree.

Trunk Straight, angled, tapering from a broad base. *Bark*: Cinnamon-red, deeply forrowed or broken up even on young trees into large vertical plates or patches which curl outwards.

Branches Short but spreading, in old trees erect.

Branchlets The scaly young shoots soon take on the rich red colouring of the older wood. The final shoots lie in a flattened spray with all pointing forwards at ± 45°.

Winter buds Hidden by the foliage, as is usual in the Cupressaceae.

Leaves The leaves are borne in flattened, fan-shape sprays held erect. They vary in size from 3–5 mm long on lateral growth to twice that length and widely spaced apart on leading shoots; scale-like, closely appressed save for the free-standing pointed tips; dark green on both sides (Slightly paler beneath); oblong-obovate; in opposite pairs in four ranks, all equal in size, the lateral pair boat-shaped, overlapping the facial pair which are minutely glandular.

Flowers Usually borne on separate branches of the same tree. Male flower small, globular, pale yellow, borne on a short stalk with 12–16 scales.

Mature cones Solitary, hanging on stalks, 20 mm long, 10 mm broad, vase-shaped with 4–6 scales, the lowest pair being reflexed, sterile and only half as long as the seed-bearing scales; fleshy at first, becoming leathery, ripening to reddish- or yellow-brown in Autumn, persisting until the Spring.

Seeds 8–12 cm long; 1–2 seeds to each scale, one side very small-winged, the other wing being as long as the scale; with an oblique, blunt tip.

Habitat From Oregon to Lower California, U.S.A.

Cultivation Introduced into Britain in 1853. It is quite hardy, but does best in cool, deep well-drained soil. Zone 5. (See Illus. Nos. 76, 78)

-- **'Aureovariegatá'** (Beissner) Krüssmann 70, (1960); under *Libocedrus* Beissner in Mitt. d. d. d. Ges. **13**: 257 (1904); under *Heyderia* (Beissner) Li in J.A.A. **34**: 17 (1953). With irregular patches of golden yellow foliage.

-- **'Columnaris'** Krüssmann, Nadel. 70. (1960); under *Libocedrus* Beissner, 472 (1884); under *Heyderia* (Beissner) Krüssmann, 124 (1955). This was described as "Regularly columnar with short crowded branches", but since this description would cover most of the trees in cultivation in the U.K. the name is rather pointless.

-- **'Compacta'** Krüssmann, 70 (1960); under *Libocedrus* Beissner, 30 (1891); under *Heyderia* (Beissner) Li in J.A.A. **34**: 17 (1953). Compact in habit and densely branched.

-- **'Depressa'** Welch, 148 (1979); under *Libocedrus* (Scott) Gordon 426 (1875). This is a very dense, compact and globose form, growing as wide as high, which originated in the Scott Nursery at Merriott in Somerset, where the mother plant at 10 years was 1 m × 1 m. I now consider that the specimen at the Westonbirt Arboretum in Gloucestershire, near the Head Forester's office referred to in my earlier book [Welch, Dw. Con. 217. 1966] can be identified with this form, not only because of the relatively close association geographically of the two places but because a propagation of the Westonbirt tree, after 10 years growth in Devizes, corresponds exactly to Gordon's description of the mother plant of the same age at Merriott.

-- **'Glauca'** (Beissner) Krüssmann, 70 (1960); under *Libocedrus* Beissner, 472 (1887); under *Heyderia* (Beissner) Li in J.A.A. **34**: 17 (1953). Leaves conspicuously glaucous.

-- **'Intricata'** Den Ouden & Boom 221 (1965); under *Libocedrus* Hillier, 42 (1964). This is a form raised from seed by James R. Noble of San Francisco, California, U.S.A., in 1938 and introduced to this country by the late Mr. A. H. Nisbet of Gosport. It forms a dense and very slow-growing but distinctly columnar plant. Unfortunately, this and the Somerset plant have become mixed in the trade. From foliar characteristics alone they are difficult to separate, 'Intricata' being only marginally smaller and denser in the spray, but in winter it bronzes much more than does 'Depressa' and the difference in habit as the plants develop is unmistakeable.

-- 'Nana' Dallimore & Jackson, 304 (1923). The available description of 'Nana' is so inadequate that it does not merit independent recognition any longer.

-- **'Riet'** A very slow-growing plant raised from a witch's broom discovered by Mr. van Hoey-Smith on a tree near Bear Basin, Oregon, U.S.A. during a tour by the International Dendrological Society in 1971 and named by Sir Harold Hillier as a compliment to Mrs. van Hoey-Smith. From 'Compacta' and 'Intricata' it is quite distinct, the leaves and branches being very narrow.

Two other species, possibly regional variations of the single species, since they do not deviate therefrom in major respects, are possibly not in cultivation in Europe. They are:-

C. ***formosana*** Florin in Taxon **5** (8): 192 (1956); *Libocedrus formosana* Florin, Svenk. Bot. Tidskr. **24**: 126 (1930). *Heyderia formosana* (Florin) Li. *Habitat*: N. Taiwan (formerly known as Formosa.)

C. ***macrolepis*** Kurz in Journ. Bot. **2**: 196 (1873); *Libocedrus macrolepis* (Kurz) Bentham & Hooker f. in Gen. Plant. **3**: 426 (1880). *Habitat*: N.E. Burma, China: S. Yunnan; Hainan. Doubtfully hardy.

CATHAYA

Pinaceae

Cathaya Chundel & Kuang in Botanichesky Zhurnal **43**: 461 (1958). Monoecious trees with horizontal branches: branchlets short-tomentose the first year, later glabrous, with rather prominent leaf-cushions separated by grooves: winterbuds ovoid, without resin; leaves in the short, sterile spurs clustered, on shoots spirally arranged, linear, with 2 white stomatic bands beneath; margin narrowly revolute, ciliate the first year; cones on the shoots only, rather short, long persistent, with 13–16 scales; bracts short, acuminate: seeds 2 at each scale, with a rather long wing.

Two species in W. China, discovered in 1955; the genus was already known from palaeontological literature and it is to be considered as a relict of ancient times (like Metasequoia). It looks intermediate between Larix and Tsuga.

C. ***argyrophylla*** Chung & Kuang l.c. 265. Tree, 20 m; branches horizontal, with grey bark; branchlets brownish; winter-buds 6–8 mm long; leaves 4–5 cm long, 2.5–3 mm broad, on the spurs much shorter, dark green and slightly setulose above; white bands broader than the green ones, with 11–17 stomatic lines on each side, margin with acute cilia; cones 3–5 cm long, 1.5–2 cm broad, chestnut-brown, later dark brown; seeds 5–6 mm long, dark olive-green irregularly maculated; wings 10–15 mm long.

Kwangsi (W. China), discovered by the Kwangfu Lingchu Expedition in 1955; not yet in cultivation.

CEDRUS

Pinaceae

Cedrus Link in Linnea **15**: 537 (1841). The Cedars, as well as being amongst the most spectacular of trees are perhaps also the most well known, at least by name. A cedar can be readily recognised by the presence of both long shoots and of whorls of foliage on short spur-shoots, and can be readily distinguished from a larch by the evergreen foliage and the large, handsome cones. But the four species comprising the Genus are so much alike that some authorities have regarded them as a single species. It is therefore not surprising that distinguishing between them is often difficult.

Size and habit Tall, evergreen, monoecious trees, often of spectacular aspect in maturity and many fine examples exist in Britain. The shape of the crown on an old tree differs with the species and so is a good diagnostic feature.

Trunk Usually massive. *Bark*: Grey or brown, smooth or lightly furrowed on young trees; on old specimens deeply furrowed and scaly.

Branches Massive, wide-spreading or ascending, according to the species.

Figure 79. *Cedrus libani*. Cedar of Lebanon. The broad crown is characteristic. (10)

Branchlets Of two forms; long growth shoots and short lateral spurs.

Winter buds Minute, ovoid, with a few scales.

Leaves Needle-like, usually triangular in cross section, rigid, sharply pointed. On growth shoots the leaves are widely spaced and emerge spirally; on the spurs they are in dense fascicles.

Flowers Cedars flower in autumn, the male flowers occurring singly on the ends of the spurs, looking rather like miniature upright cones about 5 cm high; the female flowers ovoid, green or purplish.

Mature cones Large, upright, ovoid to ellipsoid-oblong, blunt or notched at apex, maturing the second or third year and breaking up when ripe, leaving a long, narrow spindle and a stout, woody base-plate. *Scales*: very broad, tomentose outside.

Seeds Irregularly triangular. *Wing*: Broad, membraneous.

Cultivation Cedars prefer well-drained, loamy soil, but will also grow in sandy clay, provided

Figure 80. *Cedrus atlantica*. Indistinguishable otherwise from the Cedar of Lebanon, the Mt. Atlas cedar maintains leader dominance throughout its life.

there is no stagnant moisture.

C. atlantica Manetti, Cat. Pl. Hort. Modic. Suppl. 8 (1844); *C. libani* var. *atlantica* Hook., Nat. Hist. Rev. 15 (1862). Atlantic or Atlas Cedar. *C. atlantica* is closely related to *C. libani*, of which species it is still by some authorities regarded as a *varietas*, and differentiation is often difficult. Mitchell (*Conifers in the British Isles* H.M.S.O., 60 (1972)) suggests that the distinctions between these two species are not absolute, but that in addition to the more conical habit and the upstanding branches or branch tips of *C. atlantica* they include the presence (in that species) of translucent yellow-white to clear spine on the leaf (visible only under a lens), a smaller cone with a broad cavity in the top and bark colour that is grey rather than the brownish colour more usual in *C. libani*. Mitchell's very detailed descriptions

Figure 81. *Cedrus brevifolia*. This species is quite distinctive in the smallness of its leaves. (10)

mention other points of difference, but he completely rejects the widely held belief that leaf colour has any reliable diagnostic value, because of there being so much variation between individual plants.

Habitat Atlas Mountains in Algeria and Morocco; Introduced by A. Sénéclauze who received the first cones in 1839. A large number of selections have received cultivar names. Zone 6. (See Illus. No. 80, 83, 84)

-- **'Albospica'** Sénéclauze. 47 (1868). Young shoots pure white.

-- **'Aurea'** Kent. 409 (1900). Golden Atlas Cedar. Pyramidal, 3–5 m; leaves shorter, distinctly golden yellow. Probably originated in Boskoop, Netherlands. Common in the trade, but liable to sun-scorch. There are now better

Figure 82. *Cedrus deodara.* The Deodar. The pendulous growth tips are characteristic. (10)

Figure 83. *Cedrus atlantica.* Foliage comparison with *Cedrus deodara.* (10)

Figure 84. *Cedrus atlantica.* Bark on trunk. (10)

Figure 85. *Cedrus libani.* Detail of cones and foliage. (10)

selections available.

-- **'Aurea Robusta'** den Ouden 89 (1949). Conical 10–15 m, more robust in growth than the preceding; leaves somewhat larger and paler yellow. Introduced into the trade by H. den Ouden & Boskoop, Netherlands in 1932.

-- **'Fastigiata'** Carrière in Rev. Hort. 32 (1890). Narrow pyramidal or nearly columnar, densely clothed branches ascending, erect: branchlets more or less upright, usually very short; leaves 2.5 cm long, the shorter ones light green above, bluish-green below. Raised by Lalande. Nantes, France.

-- **'Fez'** A form collected by Maurice Mason of King's Lynn, Norfolk, U.K. at Fez, in Morocco.

-- **'Glauca'** Carrière 374 (1867). *C. atlantica* f. *argentea* Murray Pin. Brit. **3**: 224 (1867). Blue Atlas Cedar. Leaves conspicuously bluish white. Frequently occurring in the wild and very common in cultivation.

-- **'Glauca Pendula'** Paillet ex Beissner in Gartenwelt, 221 (1900). Pendulous; Leaves conspicuously glaucous. Raised by Paillet at Chatenay, France. (See Illus. No. 86)

-- **'Morocco'** New cultivar. A selection with long, dark green needles and a conical habit. Distinctive. Raised from seed by Wm. Goddard, Floravista Gardens, Vancouver, B.C., Canada.

-- **'Mount St. Catherine'** New cultivar. An extremely slow-growing cultivar originating from a witch's broom found by Peter Taverna at Crafers, S. Australia. The colour of the foliage is normal, but the growth is extremely tight, with an annual growth of 5–7 cm.

-- **'Pendula'** Moreau ex Carrière in Rev. Hort. 160 (1875). Weeping Atlas Cedar. Branches grace-

Figure 86. *Cedrus atlantica* 'Glauca Pendula' at Les Barres Arboretum, near Paris.

fully pendulous. Raised by Moreau, nurseryman at Fontenay aux Roses, near Paris.

-- **'Rustic'**. As 'Rustica' in Monrovia Nurs. Cat. **3** 1962/3: A selected cultivar with an outstanding blue colour of the leaves.

-- **'Taverna'** New cultivar. Origin as 'Mount St. Catherine' above, but it makes a less satisfactory plant.

-- **'Wilkman'** Foliage rich green, otherwise normal to the species.

C. brevifolia (Hook. f.) Henry 467 (1908); *C. libani* var. *brevifolia* Hooker f. in Journ. Linn. Soc. Lond. **17**: 518 (1880). Cyprus Cedar. Shrubby tree, as broad as high; trunk of about 100 years old not more than 12 m and about 2 m in girth; crown on old trees branched umbelliformly; branches drooping; leaves 5–8 mm long, bluish-green; cones about 7 cm long, 4 cm wide, with a short umbo in the apical depression. This species is readily distinguishable from the other species by its small leaves which belie its capacity to form a medium sized tree. But grown in a container that restricts root development, it seems content to remain dwarf indefinitely, and because of its small leaves, which are nicely in scale, it makes a good subject for Bonsai.

Habitat Cyprus, in the mountains. Discovered by Samuel Baker in 1879. Zone 7 (See Illus. No. 81)

-- **'Bergman'** A form from seed that produces a compact, somewhat irregular tree, with short, dark blue-green leaves, and a rate of growth about 2/3 of normal.

-- **'Epsteinianum'** Welch, 149 (1979). A selection with leaves somewhat shorter than normal. It is very slow-growing, forming a plant with an irregular outline.

-- **'Hillier Compact'** den Ouden and Boom 58 (1965); *C. brevifolia* Compacta Hillier 12 (1964). A very dense bush with crowded branches and decidedly glaucous foliage; found in the garden of Mr. Archer, as a plant of 30 cm high and 70 cm across, being at least 20 years old. This, and the following cultivar were almost certainly cultivariants, since all such plants tend to throw up a leader and develop normally, sooner or later.

-- **'Horizon'** den Ouden and Boom 58 (1965); *C. brevifolia* Horizontalis Hillier 12 (1964). A small, very flat plant, probably the product of a side graft from a lateral growth. Found by Sir Harold Hillier in his garden.

C. deodara (Roxb.) Loudon, Hort. Brit. 388 (1830); *Pinus deodara* Roxburgh., Fl. Ind. Or. **3**: 651 (1832). Deodar or Indian Cedar (Gr. Brit.); Himalayan Cedar (Amer.). Tree, 50 m, gracefully pyramidal; branches horizontal; branchlets densely set, pale grey, pubescent; leaves ± 30 in a bundle, 25–50 mm long, straight, stiff, as thick as broad, light blue or greyish-green, sometimes silvery; cones solitary or 2 together on short branchlets, ovoid or oblong-ovoid, 7–10 cm long, 5–6 cm wide, apex rounded bluish, bloomed when young, reddish brown at maturity; scales numerous, almost woody, wedge-shaped, narrowed at the base, oblong to ovate, 5–6 cm wide, less tomentose than in other species; seeds obovate to ovate, narrowed at the base, 16–17 mm long, 6–7 mm wide, whitish; wing large, light brown.

Himalayas at an altitude of 1200 m and higher. Introduced into Britain by W. L. Melville in 1822. Zone 7. (See Illus. No. 82)

This has always been found to be a variable species and quite a number of variants were selected and given cultivar names during the 19th century. Hopefully, the clones now in cultivation under the names 'Albospica' and 'Aurea' can claim their birthright on the grounds that they have been in continuous cultivation since their introduction, but a number of other clones with names consisting of Latin adjectives are almost certainly not identifiable today. Following the general plan in this book, these are transferred to Appendix One in Volume Two. Of recent years a number of new selections have been made, and for these new cultivar names have quite properly been suggested. Re-use of obsolete names for new introductions is undesirable. (See Illus. Nos. 82, 83)

-- **'Albospica'** (Ottolander) Henry 477 (1908); *C. deodara spica* Ottol. in Sempervirens **3**: 260 (1874); *C. deodara albospicata* Annesley in Gard. Chron. **1**: 399 (1899). Tips of branchlets white.

-- **'Aurea'** Nelson. 56 (1866). Golden Deodar Cedar. Pyramidal, 3–5 m high, higher in a more favourable climate; leaves golden yellow in spring, becoming yellowish-green in autumn, bloomed greenish-yellow in winter.

-- **'Aurea Pendula'** Hillier, Manual 466 (1971). It forms a stately, broadly pyramidal tree with branches drooping, new growth golden, becoming yellow-green by winter.

-- 'Aurea Wells' An illegitimate name. Now known as 'Well's Golden'.

-- **'Bold Dwarf'** Dendroflora **20**: 67 (1983). A low-growing and somewhat globose form with grey-blue needles shorter than normal. Raised from seed on the G. Huizer Nurs, Boskoop, Holland. A better name should be found.

-- **'Clarke'** den Ouden and Boom 59 (1965); *C. deodara compacta* Clarke & Co, Cat. 1932, not Carrière (1867). Differs from the following cv. by the not decurved branches. Raised by W. B. Clarke, nurseryman, San José Calif. in 1928.

-- **'Compacta'** Carrière 369 (1867). Pyramidal, compact, a few metres high; branches crowded, decurving: branchlets drooping.

-- 'Contorta'. See 'Raywood's Contorted', below.

-- **'Cream Puff'** An upright-globose selection with dense, creamy-white foliage that does not sun-scorch. It grows slowly, but eventually makes a small tree. Raised from seed and introduced by Floravista Gardens, Victoria, B.C., Canada.

-- **'Deep Cove'** A pyramidal tree with young foliage stark white turning to blue-green, forming a dazzling contrast. Origination the same as 'Cream Puff'.

-- **'Descanso Dwarf'** A compact, rather than dwarf, upright plant with semi-pendulous branches and bright green foliage.

-- **'Eisregen'** A clone of the 'Paktia' group with dense and short leaves, a not fully-pendulous habit, and light blue-grey foliage.

-- **'Eiswinter'**. Another 'Paktia' selection with grey-green leaves, winter-hardy, otherwise not outstanding.

-- **'Emerald Spreader'**. A vigorous selection, descriptively named. It was at first distributed under the illegitimate name 'Viridis Prostrata'. It was introduced by the Harold W. Wells nursery, Mount Vernon, Washington, U.S.A.

-- **'Fastigiata'** Carrière 368 (1867). Columnar, similar to the Lombardy Poplar; branches ascending, less branched; branchlets equal, short; leaves widely set, very unequal in size, the longer ones spreading, the shorter ones directed forward, often concealing the large straight buds. Found by Dr. Turrel in a garden 12 miles from Toulouse, France. More than one clone is probably in cultivation.

-- 'Glauca' Schelle 17 (1909). Leaves bluish-green or silvery grey. Since the clone in question can no longer be identified this name is of no value, save as a collective term or group heading.

-- **'Gold Cone'**. Narrowly pyramidal, somewhat dull, with golden colour foliage. All terminal shoots are completely pendulous. Origin as 'Cream Puff', above. At 15 years the mother plant was over 3 m high and only 1 m wide, but its growth may have been affected by the removal of side shoots for propagation.

-- **'Gold Mound'**. New cultivar. This selection forms a broad pyramid with the young foliage at first a bright golden yellow.

-- **'Gold Strike'**. A broadly-conical form introduced by Floravista Gardens, Victoria, B.C. Canada.

-- **'Karl Fuchs'**. A clone of the 'Paktia' group of seedlings. The foliage is a nice blue colour and it is outstandingly hardy, allegedly even more so than the following. Hardy to –25° C.

-- **'Kashmir'**. A seedling raised on the J. Franklyn Styer Nursery at Concordville, Penn. in 1934 and selected for hardiness because it was "the only plant of 200 set out in nursery rows in 1930 which survived the winter of 1933-34, when temperatures dropped suddenly to 25° below zero. Standing alone, it also survived the similar winter of 1936-37". The foliage is a nice blue colour, but otherwise it is quite normal. The name was registered in 1968.

-- **'Klondyke'**. A clone that forms a broadly conical tree with both the leaves and the shoots, turning a very unusual chartreuse-gold in winter. Does not scorch. At 18 years, the mother plant was 4.5 m high and 3 m across. Origin as 'Cream Puff' above. It was noticed by Goddard in a seed-bed when visiting another nursery, whose owner would have thrown it out for dead. It turned out that he was colour-blind!

-- **'Lime Glow'**. A seedling selected by A. J. Teese of Yamina Rare Plants Nursery, Monbulk, Victoria, Australia. It is slow-growing, with yellow foliage soon changing to an unusual but attractive yellowish-green or 'lime' colour. The name was registered in 1980.

-- **'Limelight'**. This clone makes a pyramidal tree with drooping branches. The young leaves are lime-white, turning to pale green. The clone 'Klondyke' is preferable. Origination the same as 'Cream Puff'.

-- **'Mylor'**. Teese in Amer. Conif. Soc. Bull. **2** (4): 113 (1985). Originated from a witch's broom in Victoria. It is similar to 'Lime Glow' but is less bright.

-- **'Nana'**. Hornibrook 34 (1939). Dwarf Deodar Cedar. Dwarf bush, rather flat topped, compact, rounded. Found in the Botanic Gardens at Kew, then over 40 years old and still listed in the Kew catalogue, 1984. A number of clones answer to this description, since slow-growing seedlings occasionally turn up in seed-beds.

-- **'Paktia'**. This is a loose collective term used to cover a group of seedlings from seed collected in the Paktia Province in Afghanistan. All are outstandingly hardy.

-- **'Pendula'**. Beissn. in Mitt. d. d. d. **9**: 100 (1900); *C. deodara prostrata* Hort., not Nelson. Weeping Deodar Cedar. Pendulous; branches spreading over the ground, the ends upright. To obtain an ascending tree, a leader must be tied upright.

-- **'Polarwinter'**. A selected 'Paktia' seedling with a pendulous habit of growth, stout branches and green foliage. Other selections are 'Eisregen', with light blue-grey needles and a stiffer habit than normal, and 'Eiswinter', with pale grey-green needles.

-- **'Pygmy'**. den Ouden and Boom 61 (1965); *C. deodara 'Pygmaea'* Hillier 12 (1964); *Pygmy Deodar Cedar* Gotelli in Amer. Hort. Mag., Oct. 1960. A very small plant with conspicuously glaucous foliage; found in 1943 in a small nursery in California; after 17 years the plant had attained a height of 30 cm and a spread of 40 cm.

-- **'Raywood's Contorted'**. Raised form a seedling on the Raywood Nursery, Delamere, South Australia, and introduced to the trade by P. C. Nitschke. The annual growth is 10–15 cm; the foliage is not unlike that or *C. atlantica*. Each new year's growth develops at an angle of 10° to the old shoot, giving the plant a 'contorted' appearance – more curious than beautiful.

-- **'Repandens'** Clarke, Cat. 3 (1930). Habit strongly pendulous or almost prostrate; originated in the Clarke nursery, San José, Calif., U.S.A. It would appear to be doubtfully distinct from the older 'Pendula'.

-- **'Robusta'** (Lawson) Carr. (1855) 282; *Pinus deodara* var. *robusta* Laws. (1852) 23; *C. deodara gigantea* Hort., robust, 20 m tall; branches irregularly spreading, decurving; branchlets decurving, stout; leaves numerous, 5–6(–8) cm long, stout, dark glaucous blue.

-- **'Sander's Blue'** New cultivar. An erect, graceful tree with very pendulous branches and foliage that is a bright blue, a great improvement in this respect to the old cultivar 'Verticillata Glauca', which in other respects it resembles.

-- **'Scott'** Obtained from a witch's broom found, named and distributed by Peter Taverna of Upper Sturt, South Australia. The annual growth is from 15 to 30 cm, the habit is pendulous, and the foliage is blue green.

-- **'Shalimar'**. A very hardy selection raised from seed at the Arnold Arboretum, Jamaica Plain, Massachusetts, U.S.A. See "Arnoldia" **42** (4): 153–156 (1982.)

-- **'Silver Mist'** A semi-dwarf, forming a mound of neat, 'silvery' white foliage. Origin as 'Cream Puff' above. At 15 years the mother plant was 60 cm high 1 m across.

-- **'Snow Sprite'** Another of the Goddard introductions from Floravista. It is very slow-growing, forming a glistening white mound or small bush. Foliage is tinged with green in the interior of the plant. It is intermediate in vigour between 'Silver Mist' and 'White Imp'.

-- **'Verticillata'** Henry 477 (1908). Branchlets almost whorled; leaves green. A specimen is recorded in the Arboretum La Maulevrie, Angers, France and the name appears in the Kew Catalogue, 1984. From descriptions alone it is not possible to distinguish 'Verticillata Glauca' de Vos 227 (1887) described as a narrow ascending form, sparingly branched; branches horizontal with white interspaces which are filled with irregularly interplaced spurs; fully developed branchlets few, stout; leaves on long shoots in tiers, distinctly glaucous green.

-- **'Victoria'**. Forms a stately, conical tree with very long needles. The foliage is very dark green, at first with a whitish tinge.

-- **'Viridis'** (Lawson) Pepin in Rev. Hort. 84 (1853); As *PINUS* Lawson 23 (1852); *C. deodara tenuifolia* Hort. Branches less drooping; leaves somewhat straighter, vivid green, destitute of any glaucous hue.

-- 'Viridis Prostrata' Hort. Amer. See 'Emerald Spreader'.

-- **'Warrakilla'**. New cultivar. From a witch's broom found at Mylor, South Australia. It grows in a mushroom-like shape with tight blue foliage. Annual growth about 10–15 cm.

-- **'Waverly Ridge'**. Teese in Amer. Conif. Soc. Bull. **2** (4): 113 (1985). A bright yellow dwarf with a relatively upright and vigorous growth. Young growth is liable to sun-scorch. It arose as a witch's broom on an old tree on Waverly Ridge Road, Crafers, South Australia found by Peter Taverna in 1979 and was introduced by P. C. Nitsche, of Hahndorf, South Australia.

-- **'Well's Golden'**. Broadly pyramidal, with multiple leaders and irregular growth pattern but very dense growth. It was raised from seed on the Wells Nurs, Mount Vernon, Washington, U.S.A., and was at first distributed under the illegitimate name 'Aurea Wells'.

-- **'White Imp'**. A very diminutive plant with wholly white foliage. Raised by Wm. Goddard of Floravista Gardens, Vancouver Island, B.C., Canada.

-- **'Wiesemannii'**. P. Moll, Cat. 1936, front page with plate; in Möller's Gart. Zeit. **20**: 231 (1938). Pyramidal, compact, slow-growing; branches densely borne, bending over; leaves densely set, bluish-green. Originated ± 1930 in the nursery of P. Moll, Heisterbacherrott, Germany; named after Wiesemann, Superintendent of the Bonn Botanic Garden.

C. libani A. Richard in Dict. Class. Hist. Nat. **3**: 299 (1823); *C. libanensis* Juss. ex Mirb. in Mus. Hist. Nat. Paris **13**: 71 (1825); *C. libanotica* Link (1831) 480; *C. libanotica* Trew ex Pilger (1926) 329. Cedar of Lebanon. The Cedar of Lebanon, due no doubt in part to its Biblical associations and in part, perhaps, to the widespread distribution of majestic specimens throughout Britain, is one of our most well-known trees. One problem is to differentiate between *C. atlantica* and this species. The principal clues listed by Alan Mitchell, which he suggests must be considered *en bloc* rather than in isolation, are cited above under the former species. Even so, in young trees positive identification is difficult since both species start life as conical trees with a clean trunk, a single leader and an ascending branch system. Later, *C. libani* may or may not divide its leader, but in an open situation as it ages it will certainly develop stout, horizontal branches which, as leader dominance fails will give to the tree a wide-spread crown, either shallowly domed or completely flat-topped, the branches carrying flat dense masses of foliage. It is to this combination of massive long branches and these dense plates of dark green foliage that the tree owes its majestic appearance.

Tree, 25–40 m, growing slowly, pyramidal when young; the crown broad, flattened or umbelliform on old trees; trunk thick; bark blackish-grey, splitting or dividing; branches horizontal, stout, leading shoots upright or spreading; branchlets very numerous, densely arranged, more densely than those of *C. atlantica*, spreading in a horizontal plane, short, glabrous or irregularly pubescent; leaves and spurs 30–40 in a bundle, 20–35 mm long, stiff, quadrilaterally compressed, broader than high, pointed at apex, dark- or bright green; cones stalked, solitary, upright, barrel-shaped, 8–10 cm long, 4–6 cm across, often impressed at apex, sometimes resinous; scales numerous, densely imbricated, narrowed to base, about 5 cm across, outside slightly decurved, entire, dorsal side short downy pubescent; seeds oblong-ovoid, triangular, narrowed to base, compressed, 15–18 mm long; wing thinly membranous, half fan-shaped, straight truncate above, 25 mm long.

Asia Minor and Syria; The oldest specimen in cultivation is recorded from England ± 1638, but many specimens dating from the early 19th century exist. Zone 7. (See Illus. Nos. 79, 85)

-- **'Aurea'**. Sénéclauze 50 (1868); *C. libanitica* var. *aurea* Bail. 130 (1933). Pyramidal, moderate in height; big trees unknown; leaves golden yellow on a greenish ground, the colour appearing stronger in winter, forming a characteristic difference from golden forms of *C. atlantica*, in which the ground colour of leaves is greyish.

-- **'Comte de Dijon'**. Barbier, Cat. 1908. Pyramidal, compact, dense, slow-growing; branches and branchlets very short; leaves 2–2.5 cm long, finally straight, broadest in the middle, tapering to apex and base, sharp-pointed at apex. Raised by Barbier, nurseryman at Orléans, France.

-- 'Conica Nana'. Hort. Amer. See 'Nana Pyramidata', below.

-- **'Glauca'**. Carr. 284 (1855); *C. libani argentea* Loud. ex Gord. 44 (1858). Branches usually shorter; leaves silvery bloomy. Uncommon in cultivation.

-- **'Golden Dwarf'**. Den Ouden and Boom 64 (1965); *C. libani 'Prostrata'* Kruessm. 81 (1960); *C. libani 'Aurea Prostrata'* Hillier 13 (1964). Low growing with horizontal branches and golden-yellow foliage.

-- **'Gold Tip'**. Welch 150 (1979). A seedling found by Fred Bergman of Raraflora, Feasterville, Pennsylvania, and at first distributed by him as 'variegated form'. It forms a large, irregular bush with the foliage bright yellow early summer, fading to mixed yellow and green.

-- **'Green Knight'**. A very dwarf form of irregular growth, with leaves 25 mm long, very densely held, hiding theshoot. It has been unjustifiably distributed as 'Minuta'. **'Green Prince'** is similar, but somewhat faster growing. Both cultivars originated in America.

-- **'Heemstede'**. Van der Laar 209 (1986). Nom. A hardy selection of good form, with exceptionally deep-green leaves, made on the Draijer Nursery, Heemstede, Holland.

-- **'Nana'**. Loudon 2403 (1838); *C. libanitica* var. *nana* Bail. 130 (1933); *C. atlantica nana* den Ouden 91 (1949). Dwarf, pyramidal, compact,

Figure 87. *Cedrus libani* 'Nana'. Several slow-growing clones are in cultivation under this name.

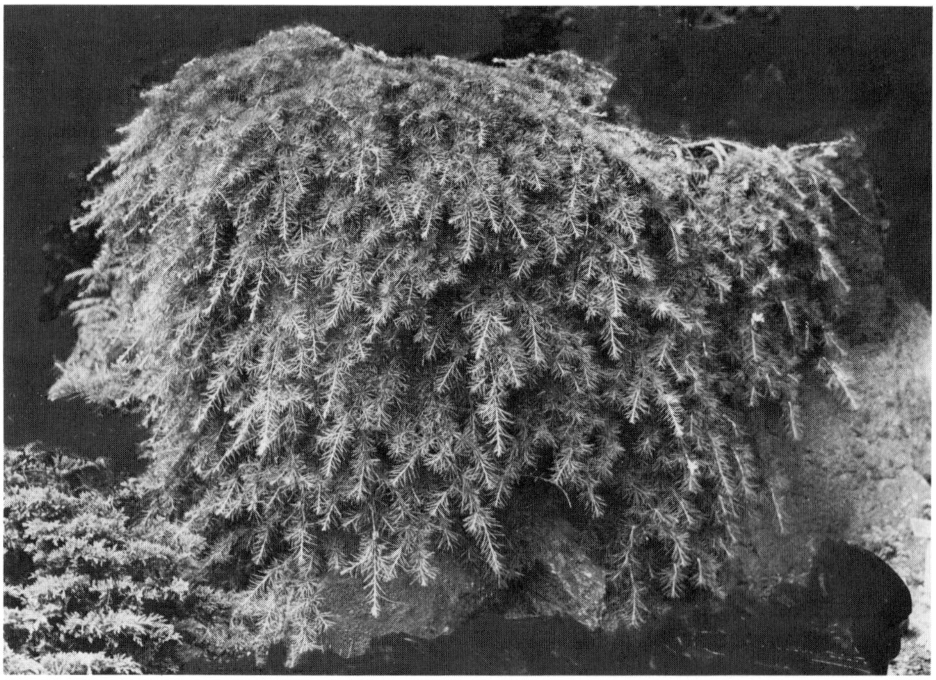

Figure 88. *Cedrus libani* 'Sargentii'. A slow-growing spreading clone stem-trained to form a low pendulous tree.

broad, in 1838 not more than 1.5 m; branches spreading, regularly disposed; leaves on the shoots 15–18 mm long, 1 mm broad, slightly sickle-shaped, apex pointed, slightly convex and with 5–6 irregular stomatic lines above, flat and light greyish-green beneath; on spurs 30–40 in a bundle and 10–25 mm long, thinner. (See Illus. No. 87)

-- **'Nana Pyramidata'**. Carr. 284 (1855). Dwarf, conical, slow-growing, 0.8–1 m; branches crowded, irregularly spreading; branchlets erect, slender; leaves short, green. Known in cultivation as '*Nana*'. Raised by A. Sénéclauze, nurseryman at Bourg-Argental, France, in 1827.

-- **'Pendula'**. Knight & Perry, Syn. Conif. 42 (1850), name only; Carr. 284 (1855). Tree with extremely pendulous branches and long leaves.

-- **'Sargentii'**. (Hornibrook) den Ouden 95 (1949); *C. libani* var. *pendula sargentii* Hornibr. 26 (1923); *C. libanitica* var. *sargentii* Hornibr. 33 (1939). Dwarf, shrubby, slow-growing, 1–1.5 m; trunk short; branches crowded, at first horizontal, afterwards decurving in different directions; branchlets crowded; leaves long, thick, with bluish bloom. Originated in the Arnold Arboretum. (See Illus. No. 88)

-- var. ***stenocoma***. *C. libani* var. *stenocoma* (Schwarz) Davis in J.R.H.S. **74**: 113 (1949); *C. libanitica* subsp. *stenocoma* Schwarz in Fedde, Repert. **54**: 26 (1944). Slender pyramidal, becoming columnar, flat-topped with age; branches horizontal, short; leaves silvery grey; leaves and cones intermediate in size between *C. libani* and *C. atlantica*. Found in 1938 in S.W. Anatolia, forming forests at elevations of 1000–1800 m.

CEPHALOTAXUS

Cephalotaxaceae

Cephalotaxus Siebold & Zuccarini ex Endl., Gen. Pl. Suppl. **2**: 27 (1842). Plumyew. A genus of 6 or 7 species, all from S.E. Asia, of which only 2 species are hardy in the temperate zone. They are all shrubs or small trees with branches opposite or in whorls; branchlets opposite, remaining bright green for three years; winter buds small, ovoid, blunt; bud-scales numerous, imbricate and persistent even when dry, at the base of the new year's growth. The leaves emerge spirally, but those on lateral shoots are twisted so as to lie more or less in ranks on each side of the shoot; they are linear, abruptly narrowing at the base to a short stalk and tapering gradually towards the tip to a fine point, not sharp to the touch; green above, with a prominent mid-rib and with two broad bands of stomata below. All species are dioecious, (or usually so) the male flowers hanging from the leaf axils along the young shoots and the female flowers carried on stalks arising from the scales at the base of the branchlets. The fruits, held on long stalks are olive-like, 2–3 cm long, consisting of a fleshy coating around an almond-like kernel (the seed); green, ripening plum-purple or brown. From Torreya it differs in having leaves that are soft rather than prickly pointed and in the male flowers hanging in small clusters. From Taxus it is distinct because of the much longer leaves and the different fruit.

C. drupacea. Now known as *C. harringtonia* var. *drupacea*.

C. fortunei Hooker in Bot. Mag. **76**: t. 4499 (1850); *C. fortunei pendula* Carr. 509 (1855); *C. fortunei mas* Carr. 509 (1855); *C. kaempferi* Anon. in Garten Nachr. 41 (1858) Chinese Plum. A tree to 12 m. In cultivation usually shrubby and seldom over 6 m; the stem near the ground mostly divided into 2–5 branches; bark reddish-brown, peeling off in irregular strips, leaving light brown scars; branches slender, spreading, tops often decurved; branchlets in whorls, long, thin, slightly decurving; winter buds small, scales pointed, glossy reddish-brown; leaves irregularly spreading in two ranks, widely set, lying parallel but frequently rather attractively curled or twisted; 5–9 cm long, 3–4 mm wide, (on young plants longer and broader and on female plants often shorter and narrower); a rich glossy green with a strongly raised midrib set in a groove above, and below with a less conspicuous mid-rib and 2 distinct white bands, each consisting of 18–22 stomatic lines, separated by a raised green mid-rib. The seed is 22–32 mm long, 10–15 mm broad, bluish-green, ripening to olive-brown.

Central and S.W. China; mountain forests of Hupeh, Yunnan, Kiangsi and Chekiang. Introduced into England (Standish)Nurseries, Bagshot) by R. Fortune in 1846, but still quite uncommon. Zone 7.

-- **'Grandis'** Hillier, Nurs. cat, 187 (1928). Several clones differing only in leaf-length have received descriptive cultivar names in the past but no longer seem identifiable. 'Grandis', a female clone with exceptionally long leaves was selected at the Hillier Nursery, Winchester, U.K. and is still listed by them.

-- **'Lion's Plume'**. A vigorous clone in which the leaves are frequently more than 15 cm long. Material was received at the Willowwood Arboretum of Rutgers University, Gladstone, NJ. in 1952 from Japan.

-- **'Prostrata'**. Den Ouden & Boom 66 (1965) suggested the cultivar name 'Prostrate Spreader' in the belief that the use of the epithet 'Prostrata' [Hillier, 15 (1964)] broke the rules, but in *Hillier's Manual* 467 (1971) we learn that this cultivar arose from a side-branch cutting on the Hillier Nursery, Winchester, U.K. before the 1914-19 war and this, of course, legitimatizes the earlier name. It is a low-growing shrub with wide-spreading branches and large, deep green leaves. Sir Harold Hillier's own comment that for a really bold foliage evergreen to give ground cover in a shaded position on either an acid or a chalk soil, this wide-spreading, large-leaved shrub would be ideal. Although a cultivariant, it would appear to be quite stable, since the mother plant at Winchester is seventy years old and covers an area of several metres.

C. harringtonia (Forb.) R. Smith, Plants Fir Tribe. ed. 2, 15 (1864); *Taxus harringtonia* Knight ex Forb. 217 (1839); *C. drupacea* var. *harringtonia* (Forb.) Pilg. 102 (1903); *C. pedunculata* S. and Z. 108 (1846); *C. drupacea* var. *pedunculata* (S. and Z.) Miq. The Earl of Harrington's Plum-Yew. A shrub or small tree, rarely higher than 10 m, usually with a broad, rounded crown; the bark flaking or peeling off in strips, branches spreading or arching in young plants, more upright in older specimens; branchlets crowded, winter-buds small, sharply pointed. The leaves are 3.5 cm long (shorter than in *C. fortunei*), are held rather stiffly in two-ranks; shiny bluish green above with the usual prominent mid-rib, and below with 2 greyish-green bands, each consisting of 13–15 stomata. The fruit (seed) is 2.5 cm long, without any projection at the tip; green, purplish when ripe. China and Japan. Sent by Dr von Siebold from Japan to Gent, Belgium, in 1829. Zone 6.

-- var. ***drupacea*** *C. harringtonia* var. *drupacea* (S. & Z.) Koidz. in Bot. Mag. Tokyo **44**: 98 (1930); *Taxus baccata* Thunberg, Fl. jap. 275 (1784), not **L.**; *C. drupacea* S. & Z. 108 (1846). The distinction between var. *drupacea* and the normal form (technically var. *harringtonia*) is difficult to establish from the foliage alone, since both shew minor variations between male and female plants and both can vary from shoot to shoot on the same plant. The carrying of the leaves stiffly upwards at a steep angle, so as to form a V-shaped parting along the shoot is a characteristic of var. *drupacea*, and such leaves tend to curl very slightly upwards and inwards. But this feature is not confined to this variety; some shoots on var. *harringtonia* may form this same parting, but in this case any tendency to curl is very slightly outwards. (See Illus. Nos. 89, 90)

-- **'Duke Gardens'** Name registered by R. F. Doren, USNA, Washington DC, U.S.A. in 1977 for a clone with a low, spreading habit with dark green foliage that originated at the Sarah P. Duke Gardens, Durham, NC, U.S.A. in 1958.

-- **'Fastigiata'** (Carr.) Schn. 162 (1913); *Podocarpus coraianus* S. & Z. in Ann. Soc. Hort. Pays-Bas 34, 1844: name only; *C. pedunculata* f. *fastigiata* Carr. 717 (1867); *C. harringtonia* var. *coraiana* Koidz. in Bot. Mag. Tokyo **44**: 98 (1930). Fastigiate Japanese Plumyew. Columnar shrub, 3–5 m; branches ascending, stiff, straight; branchlets few; leaves all radially set, 3–5 cm long, 3 mm broad, apex short, sharply acuminate, glossy dark green.

The only garden form which is hardy in cold areas. Introduced in 1830 by Dr. von Siebold to Gent, Belgium. Common in the trade. Propagated from cuttings or layers. (See Illus. No. 91)

-- **'Gnome'**. A dwarf form with shortly ascending stems and radially arranged leaves which forms a rather squat, dome-shaped plant. It originated as a sport on a plant of the cv 'Fastigiata' at the Hillier Nursery, Winchester, U.K. in 1970.

-- var. ***koreana*** *C. harringtonia* var. *koreana* (Nak.) Rehd. in J.A.A. **22**: 571 (1941); *C. koreana* Nak. in Bot. Mag. Tokyo **44**: 510 (1930). Korean Plumyew. Low, caespitose, slow-growing shrub, 1–1.5 m; it differs from '*Nana*' having somewhat narrower and more slender leaves, inclined to be more falcate. Native of Korea.

-- **'Korean Gold'** Yinger in Arnoldia **43** (4): 6 (1983). A strict, upright-growing cv. similar to **'Fastigiata Aurea'** (a clone in which the needles are margined by yellow) but with the young growth at first wholly yellow, turning chartreuse by mid-summer and becoming green by the winter. It was introduced by B. Yinger from the Shibamichi Nursery, Kawaguchi City, Japan in 1977. The retention of Japanese vernacular names often presents a problem. The *Cultivated Code* (ICNCP) in Art. 32 expresses a preference for this, but in this case the name *Ogon Chosen Maki* (meaning Golden Korean Podocarpus) under which the plant was being grown hardly qualifies as a cultivar name in the western sense, so the translation is justified.

-- var. **nana**. (Nakai) Rehder in J.A.A. **22**: 571 (1941); *C. nana* Nakai in Bot. Mag. Tokyo **33**: 193 (1919); *C. drupacea* var. *nana* (Nakai) Hornibrook (1939) 36; *C. pedunculata* var. *nana compacta* Froebel ex Beissn. in Mitt. D.D.G. **8**: 114 (1899). Dwarf shrub, 0.3–1.8 m; stem upright and spreading by suckers; seeds subglobose, 2 cm long, edible. A variety from

Figure 89. *Cephalotaxus harringtonia* var. *drupacea*. Detail of foliage and fruit.

Figure 90. *Cephalotaxus harringtonia* var. *drupacea*.

Figure 91. *Cephalotaxus harringtonia* var. *drupacea* 'Fastigiata'

North and Central Japan, sent to the Arnold Arboretum by E. H. Wilson in 1916. It forms a low shrub not exceeding 2–5 m tall with a creeping rootstock producing at intervals upright shoots which flower and fruit when about 1 m tall. It comes true from seed.

-- **'Prostrata'**. Hornibrook 27(1923); *C. drupacea* var. *prostrata* (Hornibrook) Fitsch. 57 (1930). Habit prostrate. Originated in the Hillier Nursery, Winchester, UK. before 1920. (See Illus. No. 92)

-- var. ***sinensis*** (Rehder & Wilson) Rehder in J.A.A. **22**: 571 (1941); *C. drupacea* var. *sinensis* Rehder & Wilson 3 (1914). Shrub; leaves linear, narrowing to the apex, dark green above, bluish-green beneath. A variety occurring in Central and W. China. Uncommon in cultivation.

C. pedunculata Sieber and Zuccarini (1846). Now known as *C. harringtonia.*

CHAMAECYPARIS

Cupressaceae

The false Cypresses are probably the most prominent star in the garden conifer firmament, so much

Figure 92. *Cephalotaxus harringtonia* var. *drupacea* 'Prostrata'

so that, flying low over any parkland or suburban residential area in any city within the hardiness range, you might well have difficulty in finding anywhere to land out of sight of at least one specimen of this much-planted Genus.

For many years the *Chamaecyparis* were included in the Genus *Cupressus* and even today some botanists question the need for the separation (and some conservative nurserymen ignore it). Dame Nature will not submit to too much regimentation, so often the limits of demarkation are unclear – hence the traditional conflict between taxonomists who are 'splitters' and their opponents the 'lumpers'.

Usually, the value of changes in classification becomes evident with time and further study. In the present case one species, *C. funebris*, has had to submit to change in one direction and it may be only time before another, *Ch. nootkatensis*, will have to do the same in the opposite direction. Apart from some morphological features that might suggest this latter change, practical nurserymen will tell you that this species will not 'take' on other *Chamaecyparis* understocks but must be grafted on *Cupressus* or *Platycladus*. Sap, like blood, is evidently thicker than water. The principal difference between the two genera as currently accepted is that *Cupressus* has much larger cones and has more seeds per scale (6–30).

Chamaecyparis Spach, Hist. Nat. Veg. Phan. **11**: 320 (1842); *Retinospora* S. & Zucc. 36 (1842); *Retinospora* Carrière 136 (1855). False Cypress.

Size and habit The Genus comprises 6–8 species of evergreen, monoecious trees varying from a small tree in Japan to very tall trees in W. North America and Taiwan.

Trunk Usually entire, but in some species often split near the ground.

Branches Horizontal or upswept, the final shoots sometimes nodding or pendulous.

Branchlets Soon turning brown. Foliage mostly frond-like, in flattened sprays (*Ch. thyoides* is an exception). The terminal shoots are nodding unlike those in *Cupressus* which are held erect.

Winter buds Hidden by foliage.

Leaves Juvenile foliage consists of needle-like or awl-shaped leaves, usually spreading. The leaves in adult foliage are small, overlapping in decussate pairs, scale-like and hard to the touch. In some species foliage with intermediate characteristics, (known as 'semi-juvenile'), is found.

Flowers Borne terminally on lateral branches, male flowers ovoid or oblong, yellow, rarely red, often conspicuous by their large number, stamens with 2–4 anther-cells; female flowers small, globular, inconspicuous.

Mature cones Globose, short-stalked, solitary, hard, ripening the first year. *Scales*: 6–8 (seldom 4, or 10–12), peltate, pointed or bossed in the middle. All species save *Ch. nootkatensis* ripen their seeds the same year.

Seeds 1–2, rarely up to 5 per scale, slightly compressed. *Wings*: Broad, thin. Cotyledons 2.

Habitat E. and W. coastal areas of N. America: Taiwan: Japan.

Cultivation Grows best in somewhat moist, well-drained sandy loam and in partly shaded positions, sheltered against dry winds. Propagation of the species is by seed. The cultivars are readily rooted from cuttings, save in a few cases of difficulty in which grafting must be employed.

Three of the species are very variable and have given rise to an almost overwhelming number of cultivars. In addition, occasionally a seedling will retain the juvenile type of foliage (with awl-like, spreading leaves) permanently, or at least indefinitely, and vegetative propagation of these has given rise to some very pretty garden forms, most but not all of which are also sufficiently slow-growing to be used as 'dwarf conifers'. This propensity induced the early botanists Siebold and Zuccarini to treat these forms as a separate genus *Retinospora* (Flora Japonica 1842). This served as a temporary home for all the persistent juvenile forms in this and some other Genera such as *Thuja*, but when their true allegiance was realised this Genus was abandoned and so should no longer be recognised. The allocation to the correct Genus has not been easy. In some cases it has been determined by the development on a plant of a tiny scrap of adult foliage, but in the difficult case of *Thuja orientalis* 'Sanderi' (long considered to be a form of *Ch. obtusa*) the use of gas chromatography eventually established the truth. [Gough and Welch in Bot. Journ. Linnean Soc. **77**: 217 (1978).]

KEY TO SPECIES OF CHAMAECYPARIS

All this variability much reduces the value of a key, which can in such circumstances only cover the characteristics normal to or typical of the species.

KEY TO SPECIES OF CHAMAECYPARIS

1	Leaves with white or glaucous markings beneath	2
1	Foliage without white markings	6
2	Leaves with well-defined X- or Y-shaped markings beneath; leaf-tips rounded. [Resinous]*	Ch. obtusa
2	Leaves with ill-defined or no markings beneath	3
3	Leaves green, tinged with bronze, sometimes with white markings beneath. [Rank, unpleasant]*	Ch. formosensis
3	Foliage green, no trace of bronzing	4
4	Leaves acute with distinct glands visible as translucent spots when the spray is held against the light. [Parsley]*	Ch. lawsoniana
4	Leaves acute, obscurely glandular, variable	5
5	Leaves yellow-green, black at tip; sprays like *Ch. lawsoniana* in miniature. [New-mown grass]*	Ch. funebris
5	Leaves rough to the touch; with bright white markings beneath. [Resinous]*	Ch. pisifera
6	Leaves dull green, without glands, Cones ripening the 2nd year. [Rank, unpleasant]*	Ch. nootkatensis
6	Branchlet system fan-shaped, not in flattened sprays. Leaves acute with a prominent gland on the back. Cones ripening the 1st year. [Ginger]*	Ch. thyoides

* Words in square brackets. [ ] refer to the scent of the crushed foliage.

Ch. boursieri Carrière 141 (1855). Now known as *Ch. lawsoniana*.

Ch. formosensis Matsum. in Bot. Mag. Tokyo **15**: 137 (1901); *C. formosensis* Henry 1149 (1910).

The chief claim to distinctiveness of the Formosan cypress is the great size (to 70 m) it attains in its native land. Habit and foliage are quite typical of the Genus, which means it shares characteristics with other species, especially *Ch. pisifera*, so when seen in cultivation, recognition of the Genus is simple, but of the species it is not. Its general air of well-being, however, the reddish-brown, peeling bark, the yellowish-green foliage overlaid with a coppery-bronze sheen above and with undefined white markings beneath and a rank odour of rotting seaweed given off by the foliage, if bruised, all help diagnosis, and if present the elipsoid (ie not globose) shape of the ripe cones is conclusive.

Mountainous areas of Taiwan at 1000–3000 m. Introduced by Admiral Sir Lewis Clinton-Baker, who forwarded it to Bayfordbury, U.K. in 1910. Zone 8. (See Illus. No. 95)

Ch. funebris (Endlicher) Franco. Agros. **24**: 93(1941); *C. funebris* Endlicher 58 (1847); *C. pendula* Lambert 97 (1803) not Thunberg.

Chinese Weeping Cypress. This species, long listed as *Cupressus* must integrate as best it may with the Genus to which it has been transferred because of its smaller cones. It makes a tree to 20 m with an upright trunk, shallowly fluted but with deeply fissured grey/brown bark; slender horizontally held or ascending main branches but with all final growth strongly pendulous; terminal shoots thin and slender, leaf-sprays flat; foliage hard and fine; leaves very small, 2–3 mm long, densely appressed but with apex free, sharply pointed, shining yellow/green with a black patch near the yellow

Figure 93. *Chamaecyparis lawsoniana.* Detail of foliage and cones.

Figure 94. *Chamaecyparis pisifera*. Detail of foliage and cones.

spine; cones often on a curved stalk, irregularly globose 8-20 mm broad, dark brown; scales 6–8 with a small mucro (point); seeds 3–5 in each scale, 3 mm long, glossy reddish-brown with scattered resin-knobs; wing narrow. Patches of soft juvenile foliage often occur on young trees.
Central China. It is cultivated in some parts of China but is rare in cultivation in Britain, being only marginally hardy. Zone 8. (See Illus. No. 171)

Ch. henryae Li (1962). Not now distinguished from *Ch. thyoides.*

Ch. lawsoniana (Murray) Parlatore in Ann. Mus. Stor. Nat. Furenze **1**: 181 (1864); *C. lawsoniana* Murray in Edinb. New Phil. Journ. **1**: 299 (1855). Port Orford Cedar, Lawson False Cypress.

Size and habit Tree to 50 m, narrowly conical if with a good leader, otherwise often round-topped with a group of sub-leaders.

Trunk Usually single and stout but in some trees soon dividing and then forking repeatedly. Since the propensity to grow in this way can be clearly seen in nursery stock (i.e. in plants in which the terminal growth spray characteristically lacks leader dominance) it is possible to plant only trees that will perform well in this respect. *Bark:* On young trees smooth, dark grey-green; on old specimens reddish brown, divided into thick scaly ridges or plates.

Branches Trees usually well furnished to the ground, branches ascending or horizontal, usually nodding at the ends.

Foliage Typical of the Genus, the facial pair of leaves much the smaller and carrying a central oval gland visible when held against the light; the underside with indefinite white lines.

Flowers Male flowers scarlet, sometimes conspicuous in Spring because of their abundance. Female flowers steel-blue, inconspicuous.

Mature cones Usually very freely borne, globose, 8 mm across; bluish green, ripening brown, retained on the tree even when dry and spent.

Seeds Oblong, appressed, glossy brown, broadly winged.

Habitat S.W. Oregon, N.W. California in USA.

Cultivation The Lawson cypresses are relatively hardy. All forms have, however, one drawback which is that they will never break into new growth from old wood. This in gardener's language means that they resent heavy pruning, so both formative and restrictive pruning (the former aimed at shaping the tree to our requirements and the latter being done in order to keep it down in size) must be done little and often, and always before it is necessary to cut into old wood. It also means that the species has little capacity to recover from severe damage, whatever its cause. So should an exceptionally severe winter or a strong gale (or the unwelcome attentions of a large dog) kill off all the green on one side of the tree it may seldom be killed outright but it will never 'break again' and recover its shape. This much reduces the value of this species for use as a shelter belt or for hedging. Zone 6. (See Illus. Nos. 93, 99)

The growth in popularity of the Lawson cypress as a garden tree is quite a horticultural success story. From a very restricted distribution 'in the wild' in the U.S.A. it is now probably the most widely planted for ornament of any conifer. It was first introduced to Britain in 1854 and from the very first sowings it must have demonstrated its capacity for variation, because by 1875 no fewer than 25 'named forms' were in the leading nurserymen's catalogues, covering all the variations in habit, colour and size that we know today. Unfortunately it has maintained this propensity ever since with the result that we now have a bewildering range of cultivars – far more than the widest needs of our gardens. And still they come!

As all these cultivars are seedling mutants they are reasonably stable, i.e. they do not 'revert', but

Figure 95. *Chamaecyparis formosensis*. Detail of foliage.

Figure 96. *Chamaecyparis thyoides*. Detail of foliage, with cones.

the invariable tendency of peculiarities noticeable in a young plant to become less and less pronounced with age means that many of the older 'named forms' are now indistinguishable and not worth retention. The names of a number of such cultivars described by den Ouden and Boom are listed in Appendix One in Volume Two.

-- 'Alba Keessen'. See 'Erecta Alba'.

-- 'Alba-spica'. *Ch. lawsoniana alba-spica* R. Smith. Plants Fir Tribe ed. 3, 14 (1867); *Ch. lawsoniana alba spica* 'Young's Variety' Gordon 424 (1875). Bright green foliage densely and regularly mottled with silvery-white specks. The name was used for a clone of continental origin and also for an alleged improved selection raised on the M. Young Nursery, Godalming, Surrey, U.K.

-- **'Albo-variegata'** (Veitch) Nicholson 303 (1884) and (under *Cupressus*) Veitch 232 (1881); *Ch. lawsoniana* f. *albopicta* Schneider 168 (1913). Dwarf, forming a compact conical bush to 3 m; foliage deep green, profusely speckled with white or creamy-white. With age the plant tends to become round-topped. Here and there about the plant are small sprays of pure white. These give quite a distinctive appearance but are very prone to damage by frost. It is an attractive cultivar deserving to be more widely planted.

-- **'Alumigold'** Maks ex de Vogel in Dendroflora **4:** 61 (1967). A broadly conical form similar to 'Alumii' (on which it arose as a sport) but denser, and with the young surface growth clear yellow, overlaying the blue-green older foliage in the interior of the tree, giving a two-tone effect.

-- **'Alumi'** Beissner 73 (1891). One of the earliest selections, of unrecorded origin but one of the best and still widely planted. As a young tree it is narrowly conical, with a fastigiate branch-system and flat, narrow sprays held stiffly erect on pink or brownish shoots. The wholly adult (scale-like) foliage is mid-green, but a heavy dusting of white gives the tree a distinctly blue-grey appearance. With age a tree broadens because basal branches form sub-leaders. Although the earliest recorded spelling was 'Alumi', the spelling 'Allumii' is now in general use.

-- 'Alumii Magnifica'. Name illegitimate. A hardy, vigorous clone originating in France. Of slim, upright habit and neat outline similar to 'Alumii' but having the colour of 'Columnaris'. A new name is needed.

-- 'Alumii Nana Compacta'. Krüssmann, 70 (1979). An illegitimate name for a spreading clone of similar colour to 'Alumi', distributed by the Minier Nursery of Angers. France. An acceptable name must be found.

-- **'Argentea'**. A name applied to several clones no longer in circulation and therefore of uncertain application today.

-- **'Argenteo-variegata'** (Fraser) Beissner 75 (1891) and (under *Cupressus*) Fraser, Handy Book of Ornamental Conifers 54 (1875); *C. lawsoniana argentea* Gordon 24 (1862). A rather robust form; branchlets and leaves interspersed with a silvery whiteness. It originated in the Lawson nurseries, Edinburgh. (See Illus. No. 98)

-- 'Armstrong' Hort. New Zealand. Harrison says [36 (1975)] that this appears to be identical with 'Aureovariegata'.

-- **'Ashton Gold'** Krüssmann, 70 (1979). A clone similar to 'Lutea' but said to produce a better crown. The habit is upright, with ascending main branches; the foliage is an attractive chrome yellow. It was received from an unrecorded source in Ireland and introduced by the Jan Spek Nursery, Boskoop, Holland.

-- **'Aurea'**. (Gordon) Beissner 452 (1884) and (under *Cupressus*) Waterer ex Gordon 24 (1862). A somewhat misleadingly named cultivar described as "A pretty variegated variety with some of the lesser sprays and leaves, scattered all over the plant, of a golden yellow. A very desirable variety when well variegated which originated in the nursery of Mr John Waterer, at Bagshot." Not listed today, it was almost certainly the plant in circulation under

Figure 97. Some *Chamaecyparis obtusa* cultivars. A – 'Kosteri'; B – 'Nana'; C – 'Coralliformis'; D – 'Nana Gracilis'; E – 'Filicoides'. All cultivars have the characteristic bluntly-rounded tips to the shoots.

Figure 98. *Chamaecyparis lawsoniana* 'Argentovariegata'.

the much more descriptive name **'Aureo-variegata'** (*Cuoressus lawsoniana aureovariegata*) Waterer ex Sénéclauze 69 (1868).

-- **'Aurea Densa'**. [See Welch 154 (1979)] and **'Minima Aurea'** Hornibrook 43 (1939) are popular and reliably dwarf forms with neat golden-yellow foliage. 'Aurea Densa' forms a cone with a blunt top which is firm to the touch. 'Minima Aurea' makes a broader plant and the foliage is quite soft, otherwise they are very similar. **'Lutea Nana'** Den Ouden and Boom (1965), less well known, is somewhat stronger-growing, is paler yellow and reputedly is more winter-hardy. (See Illus. Nos. 100, 101)

-- **'Azurea'**. A sport from 'Allumii' with extremely blue foliage in large flattened sprays. Listed Mitsch Nurs. Aurora, OR, U.S.A. in 1977.

-- 'Backhouse Silver'. See 'Pygmaea Argentea'.

-- **'Barabit's Globe'**. New name. *Ch. lawsoniana* var. *glauca globus*. Barabits in Magyar Fenyoujdonsagok (New Hungarian Trees) 55 (1965); 'Glauca Globus' Barabits ex Krüssmann 75

Figure 99. *Chamaecyparis lawsoniana*. Bark on trunk.

Figure 100. *Chamaecyparis lawsoniana* 'Aurea Densa' A 1 m plant with the characteristic blunt crown.

(1979). A very attractive globose form that arose as a seedling in the Botanic Garden at Sopron, in Hungary. The folliage is very dainty and of an attractive blue green. The plant retains its globose shape, to 1.5 m. It is an attractive plant that would make a fine specimen on even a small lawn. The names currently in use are illegitimate in form.

-- **'Barry's Silver'**, introduced by South Taranaki Nurseries, Hawera, New Zealand, was raised from a sport found on an old plant of 'Silver Queen' growing near the nursery. It forms a bushy to upright plant to about 1 m, bearing only plumose, semi-juvenile foliage. Mr Roland Barry writes that grown in partial shade it is very silvery, especially during the winter; but on trees in full sun the foliage has a cream tinge.

-- 'Bleu de Nantes', 'Bleu Nantais'. These names are usually anglicised to 'Blue Nantais', which see.

-- **'Blom'**. Den Ouden 104 (1949) (as 'Erecta

Figure 101. *Chamaecyparis lawsoniana* 'Minima Aurea'. A plant 1 m high at 15 years. Softer to the touch than 'Aurea Densa'.

Blom'). Arose as a sport of 'Allumii' about 1930 and was at first distributed under the name 'Erecta Blom'.

-- **'Blue Gem'**. A very blue sport from 'Ellwoodii' found on the South Taranaki Nursery, Hawera, New Zealand in about 1960. It makes a slow-growing conical plant with foliage softer and more plumose than 'Ellwoodii'. Although of independent origin, it is virtually indistinguishable from 'Chilworth Silver'.

-- **'Blue Gown'**. Hogger, Hogger's Nurseries Ltd. Cat. 1960 (or earlier); *Ch. lawsoniana 'Hogger'* den Ouden 107 (1949). Conical; branches and branchlets spreading and regularly borne; leaves glaucous; almost as blue as 'Pembury

Blue', but it has a neater habit. This makes it very suitable for hedges.

-- **'Blue Jacket'**. Young, Nurs. Cat. 1875; 'Milford Blue Jacket' Dall. & Jacks. 230 (1948); *Cupressus lawsoniana* 'Blue Jacket' Dallimore in Chittenden. 16 (1932) Nom. A broadly conical tree with crowded branchlets and foliage in fine sprays with leaves small, bluish-green above, the underside rendered bluish-white by the markings. Raised by Maurice Young Nurs, Milford, Surrey, U.K. Not common, but still available in the trade.

-- **'Blue Nantais'**. Name registered by Renault Nurs. Gorron, France with the Museum Nationale Histoire Naturelle in 1971. A mutation from 'Ellwoodii' with semi-juvenile foliage and a silvery-blue colour. It is slower-growing and more bushy than the rather similar 'Chilworth Silver'.

-- 'Blue Plume'. See 'Plumosa Glauca'.

-- 'Blue Ribbon'. See 'Dart's Blue Ribbon'.

-- **'Blue Surprise'**. De Beere ex de Vogel in Dendroflora **5**: 69 (1968). A fixed juvenile form raised from seed by A. P. J. de Beer, Tilburg, Holland that forms a cone-shaped plant with short vertically held branches and blue-grey leaves 3–9 mm long, 075–1.5 mm wide, thin and sharply pointed.

-- **'Broomhill Gold'**. Raised and introduced by Stewart's Nursery, Ferndown, Wimborne, UK. A vigorous clone with an upright branching system, forming a nice, narrow column with a good golden colour in summer, fading to a greenish yellow. Doubtfully an improvement over some existing clones.

-- **'Caudata'** Hornibrook 41 (1939) and under *Cupressus* in Gard. Chron. 273 (1934). Dwarf, monstrous shrub, conical, irregularly dense, stout-growing: branchlets crowded, bearing apically dense tufts of foliage like 'Witches' brooms; these show 2 types of growths, one has scale-like leaves, tightly appressed to the branchlets except their tips and these tufts throw out occasionally elongated branches tail-like in appearance varying in length from 10–22 cm; the appearance of the other type of tuft is tighter and more compact, the leaves being appressed throughout their whole length and in this type the tail-like branches are absent. Raised by Goudkade Bros, nurserymen, Boskoop, Netherlands. It is merely a curiosity.

-- **'Chilworth Silver'**. Hillier, Manual Trees and Shrubs 468 (1971); 'Nyewoods' Corley in Gar. Chron. **163, 2:** 29 (1968) and Welch in Napier, Conif. Brit. Is., 49 (1972). This is one of a number of grey or blue sports from the popular 'Ellwoodii' that have appeared. It forms an attractive, slowgrowing, upright bush with densely packed, silvery-blue foliage. It originated as a sporting branch found on one of the original trees of 'Ellwoodii' in the village of Chilworth near Southampton mentioned by Hornibrook, and young plants were given by the late A. H. Nisbet of Gosport to H. G. Hillier, who introduced it to the trade. Nisbet would not agree to the name 'Nisbets' Silver', preferring it should commemorate the place of its origin. Although the name suggested by R. S. Corley may have priority the plant has been widely distributed as 'Chilworth Silver' and in view of its origin given here I suggest the retention of that name.

-- **'Chingii'**. A dense, narrowly conical selection that originated on the Millichamp Nursery, Ashburton, New Zealand in about 1949 and named for J. Ching, a member of the staff at that nursery. The ultimate branchlets are gracefully pendulous; the colour is a soft lemon yellow, somewhat lighter than 'Stewartii' or 'Golden King'. It does not scorch in full sun.

-- **'Columnaris'**. Spek ex Den Ouden, 103 (1949); *Ch. lawsoniana* 'Columnaris Glauca' Hort. A narrowly conical, at first columnar tree with thin, steeply ascending branches and branchlets. Foliage dense and regular, the leaves dark green above, somewhat glaucous towards the tip and markedly so beneath, giving the tree a pale grey appearance. As often in fastigate selections, the tree tends to open out as it matures. Raised by J. Spek, nurseryman at Boskoop, Holland ± 1940.

-- 'Columnaris Aurea'. See 'Golden Spire'.

-- '**Compacta**' R. Smith, Pl. Fir Tribe, ed. 3 , 15 (1867); var. *compacta* (de Vos) Beissn. Zierg. Ed. 2, 452 (1884). This is a slow-growing dense form with fan-shaped leaf sprays which when young could very well be taken for a strongly growing 'Minima'. It forms a neat, broadly conical plant. The specimen at the foot of the rockery at Wisley described by Hornibrook as measuring 1.5 by 1.2 m in 1927 has now reached 4 by 3 m and has retained its shape and denseness well.

-- '**Croftway**' Krüssmann 72 (1979). A selection which develops into a medium-sized tree of conical habit, with foliage that is dark green but with the tips of the young growth at first a creamy-white.

-- 'Croftway Green'. See 'Croftway'.

-- '**Dart's Blue Ribbon**'. A recent selection on the Darthuiser Nursery, Holland in which the growth is more vigorous and the blue colour of the foliage more intense than in the cultivar 'Columnaris'. It is difficult to distinguish from 'Grayswood Pillar'.

-- '**Delorme**'. Delorme ex Grootendorst in Dendroflora **6:** 77 (1977). A broad, conical tree with horizontal main branches and foliage uniformly green throughout. Introduced by the Minier Nursery, Angers, France.

-- '**Doone Valley**'. Name registered by J. W. Archer in 1981. Of conical habit, moderate growth and foliage that is pale green. An alleged cross between 'Fletcheri × 'Wisselii'.

-- '**Dow's Gem**'. Welch in Wansdyke Nurs. Cat. 4 (1964); 'Dow's Variety' Hort. Amer. and Dow's Nurs. Cat.?; 'Noble's Variety' Hort. Eng. This attractive plant makes a spreading bush, (eventually becoming quite large), with dark green foliage with a heavy white bloom on the

Figure 102. *Chamaecyparis lawsoniana* 'Dow's Gem'.

Figure 103. *Chamaecyparis lawsoniana* 'Duncanii'. Forms a spreading plant several metres across.

underside borne in rather coarse, heavy, open sprays. It makes a fine specimen on a large lawn. Raised from seed at Dow's Nursery, Oakdean, California, U.S.A. (See Illus No. 102)

-- **'Duncanii'**. Duncan and Davies Nurs. Cat. 108 (1953/54). A large, flat-topped, spreading bush with fine, threadlike, glaucous foliage. It was found, between 1912 and 1922, by the late Mr. V. C. Davies in a plantation of seedlings intended for hedging and was named by him in honour of Mr. J. R. Duncan, a founding partner of the firm. It makes a fine specimen of no great height on a large lawn where it has room to develop into a plant several metres across. (See Illus. No. 103)

-- **'Dutch Gold'**. Makes a broadly conical tree with soft, elegant, golden foliage that keeps its colour troughout the year. It was raised by the propagation by Mr. Fred Barcock of Drinkstone, Bury St. Edmunds, Suffolk of a single plant left with him by an unrecorded Dutch pursery representative.

-- **'Elegantissima'**. (Gordon) Schelle, 294 (1909) not Hillier; *Cupressus lawsoniana elegantissima* Barron ex Gordon, 424 (1875). A selection made by Wm. Barron and Son of Barrowash, U.K., which carried branches and foliage of a canary-yellow throughout the year. This clone is no longer identifiable, but the name is now in use in Britain for a beautiful small tree of broadly conical habit, with pale yellow shoots and broad, flattened, dropping sprays of silver-grey to greyish-green foliage, raised on the Hillier Nursery, Winchester, U.K. before 1920.

-- **'Ellwoodii'**. Hornibrook, 38 (1939) and (under *Cupressus*) in Gar. Chron. 50 (1929). This is the popular juvenile form that arose from a chance self-sown seedling in Swanmore Park, Bishop' Waltham, U.K. and named in honour of Mr Ellwood, the gardener there.

This ubiquitous clone makes a narrowly-conical or broadly-columnar plant with a strongly fastigiate branch system and attractive dense, semi-juvenile foliage that is glaucous blue-green, at its brightest in the autumn. It

Figure 104. *Chamaecyparis lawsoniana* 'Ellwoodii'. The most popular 'dwarf conifer'. (But see Illus. 105)

Figure 105. *Chamaecyparis lawsoniana* 'Elwoodii'. The original plant at 'Nyewoods' near Southampton, U.K., now 12 m high.

is very easily propagated and for several years from planting it is very slow-growing and has no real competitor as an eye-catching little plant, but it has probably done more to discredit dwarf conifers as a class in the eyes of the gardening public than any other plant because of its propensity for sooner or later throwing off all restraint and growing at a rate that at once puts it out of that class. Not only so, the resulting vigorous foliage is quite out of character, so should on no account be used as propagating material. From 'Fletcheri', 'Ellwoodii' differs in its closer and more stiffly erect habit and in the much more glaucous foliage, much paler in the interior of the plant. A number of coloured sports and variegated selections have been named, **'Chilworth Silver'** being one of the best.

-- **'Ellwood's Gold'**. Corley, Gar. Chron. **163** (2): 20 (1968); 'Ellwoodii Aurea' Hort. This is a sport with habit and growth similar to 'Ellwoodii' but the foliage is bright yellow in early summer, becoming less marked by autumn. It has become very popular.

-- **'Ellwood's Pillar'**. Grootendorst in Dendroflora **13–14:** 68 (1967). A presumed mutation from 'Ellwoodii' that is denser, slower-growing and with a more columnar habit of growth. At the annual trade exhibition at Boskoop in 1976 it received a silver medal. I cannot say how stable it will prove to be.

-- **'Ellwood's Pygmy'**. Welch, Wansdyke Nurs. Cat. 4 (1964); 'Ellwoodii Nana' Hort. This is to 'Ellwoodii' as 'Fletcher Nana' is to 'Fletcheri', and doubtless arose in cuiltivation in the same way, as described on p. 25. If care be taken not to let strong leader-growth take over, a young plant will settle down and make a very slow-growing little shrub less than 1 m high, by as much across.

-- 'Ellwoodii Variegata' Hort. An illegitimate name used loosely for any variegated sport from 'Ellwoodii'.

-- **'Ellwoods White'**. Den Ouden and Boom, 18 (1965). This is a slow-growing mutation from 'Ellwoodii' with a clear creamy-white variegation. A similar mutation has occurred more than once and several forms are in cultivation. One such, **'Fleckellwood'** Barry Nurs. Cat. (1970) – spelt 'Fleckel wood' originated at the South Taranaki Nurseries, Hawera, New Zealand. It is grey-green with a small, evenly distributed creamy-white fleck throughout and has a rate of growth that is about two thirds that of 'Ellwoodii'. I have seen other variegated forms that are not worth distinguishing, even if they are distinguishable.

-- **'Erecta'** Knaphill Nurs, Cat.; under *Cupressus lawsoniana* R. Smith, Plants Fir Tribe ed. 3, 15 (1867). This was one of the earliest selections to have been recorded. Since Richard Smith clearly distinguished between 'Erecta' and 'Erecta Viridis' there should be no confusion. But better forms have been introduced, and authentic surviving trees of 'Erecta' are very uncommon.

-- **'Erecta Alba'** Keessen ex Beissner 73 (1891). Pyramidal, tall; branches spreading, stout; tips of the branchlets conspicuously white; leaves near the base of branchlets passing to green. Raised and introduced by W. Keessen Jz. & Sons, Terra Nova Nursery, Aalsmeer, Netherlands.

-- **'Erecta Argenteovariegata'**. (de Vos) Beissner in Mitt. d.d.d. Ges. **13:** 87 (1907) and (under *Cupressus*) de Vos in Fl. and Pom. **1:** 18 (1876). A sport of '*Erecta Viridis*'; leaves green, variegated with white. Found by A. Koster Mz., nurseryman at Boskoop, Netherlands in ± 1874. Although seldom offered, this is still in cultivation. The creamy-white variegation in bold patches is very effective, but the plant shares with its parent the habit of becoming bare at the base.

-- **'Erecta Aurea'**. (de Vos) Hornibrook 50 (1939) and (under *Cupressus*) de Vos in Fl. and Pom. **1:** 18 (1876). Dwarf, conical, elegant, slow-growing; branches and branchlets erect; leaves golden yellow. Raised by a florist in the Netherlands in 1874; still in cultivation. This is a descriptively named form that seems to have turned up more than once. As now known it is slow-growing and attractive when young on account of its habit and bright colour which is perhaps the brightest gold of any. It is not very hardy and with age becomes a large bush to 3 m, frequently seen much the worse for damage by snow or heavy winds.

-- **'Erecta Filiformis'**. The plant known under this name is descriptively named, although possibly not the original clone raised by R. Neumann, Waldorf in Saxony, in 1896. It differs from 'Filiformis Compacta' because of its fastigiate habit, and from 'Masonii' by its thread-like foliage.

-- **'Erecta Viridis'**. R. Smith, Plants Fir Tribe ed. 3, 15 (1867). From the date of its introduction this would have been one of the earliest selections – possibly from the first importation of seed in 1854. It is still one of the most widely planted, although whether any of the (probable several) clones in production today is the rightful claimant to the name is doubtful. All start life as most attractive little plants with a very fastigiate branch-system and rich, deep-green foliage. But, like all fastigiate trees it is liable to damage from heavy snow and prone to become very bare at the base. The wretched state into which some old trees get would suggest the need for some authoritative studies aimed at eliminating all but the best available form. It was raised and introduced on the Anthony Waterer nursery, Knaphill, Surrey, U.K.

-- **'Felix'**. Felix and Dijkhuis Nurs. Cat. 35 (1941) and den Ouden 107 (1949). Pyramidal and robust; branches and branchlets spreading; branchlets densely set, directed forward, fine; leaves glaucous-blue, tips loosely appressed. In growth similar to *'Triomf van Boskoop'*, but branchlets and sprays much finer and rather soft to the touch. Raised by Felix & Dijkhuis, Nurserymen at Boskoop, Netherlands, in 1937.

-- **'Filifera'**. de Vos, 233 (1887). This and the following cultivar are both spectacular trees with a more or less pendulous habit and because of this similarity in description they are sometimes confused, but unnecessarily since they are quite distinct. Alan Mitchell [74 (1972] describes 'Filifera' as "a narrow, conic tree of somewhat pendulous habit and rich pale-green foliage with long, slender curved, distant branchlets, in lace-like open sprays." He adds that the ultimate shoots are sparse, to 3 cm long in somewhat flattened sprays and that each scale-leaf has a narrow white margin. (See Illus. No. 107)

-- **'Filiformis'**. (Fillot) Nicholson 303 (1884); under *Ch. boursieri* Fillot in Rev. Hort. Belge **3**: 219 (1877). Mitchell's description of this cultivar is "A tree usually of several stems, broadly columnar or squat, with arching branches from which hang long, threadlike branchlets bearing occasional bunches of grey-green or yellowish-green dense branchlets." Since he also gives very clear line drawings, there need be no difficulty in distinguishing the two forms. (See Illus. No. 108)

-- **'Filiformis Compacta'**. Beissner 77 (1891); *Ch. lawsoniana globosa filiformis* Hort. Dwarf, globose, very broad; branches crowded, spreading; branchlets drooping, not unlike those of *Ch. pisifera 'Filifera'* but much finer and smaller; leaves, small, tips free, dark bluish-green. There are two cultivars one meets under this name, both being, as one would expect, plants with thread-like foliage. The correct name for an upright, tall-growing form with multiple stiff

Figure 106. *Chamaecyparis lawsoniana* 'Filiformis Compacta'. A plant 1.5 m at Wisley Gardens, Woking, U.K.

Figure 107. *Chamaecyparis lawsoniana* 'Filifera'. Detail of foliage. (cf. Illus No. 108)

and erect branches is **'Erecta Filiformis'** (Neumann) Beissn. Handb. Ed. 2, 544. (1909), whereas the true 'Filiformis Compacta' is a globose bush with very fine cordlike branches and tiny dark blue-green leaves, in which the branches seem barely strong enough to bear their weight and flop about in all directions, giving the bush the appearance of a wet mop.

This cultivar need never be confused with *Chamaecyparis pisifera* 'Filifera' since the foliage is very much thinner and is a blue-green, nor with any of the thread-leaf forms of *Thuya occidentalis* which are all much coarser and stiffer and have the characterictic rank odour if the foliage is bruised. The foliage is similar to but much finer than that of 'Duncanii'. (See Illus. No. 106)

-- **'Fletcheri'**. Hornibrook 31 (1923). A very popular, conical tree with reddish ascending main branches and dark blue branchlets; foliage greyish-green, often with a mauve tinge in winter, the leaves decussate, intermediate in character between the completely juvenile and the adult, scale-like leaves; spreading upwards, the top third of their length slightly incurved. It originated as a mutation on the Ottershaw nursery near Chertsey, Surrey, U.K. (and was introduced to the trade by Fletcher Bros). It is an example of semi-juvenile or intermediate foliage that has become 'fixed'. Although not one of the most exciting forms it propagates easily and is a good stand-by in the trade. From 'Ellwoodii' it differs in the intermediate character of the foliage, in the lighter, greyish (rather than blue-green) colour of the leaves and in its greater rate of growth. 'Fletcheri' soon makes up into a small tree; it therefore does not have even the questionable right of 'Ellwoodii' to be sold as a 'dwarf conifer'.

-- **'Fletcheri Nana'**. Hornibrook 40 (1939). Murray Hornibrook records the production of more or less stable low, bun-shaped or globose plants resulting from the selection of "the lowest, underside shoots" (of Fletcheri) and he raised

Figure 108. *Chamaecyparis lawsoniana* 'Filiformis'.

this name for such plants, but with the remark that he believed them to be merely cultivated states. He was therefore possibly the first to draw attention to this phenomenon. The selection of propagating material in many groups of plants, conifers among them, can have a profound effect on the early (and even the eventual) behaviour of the resulting plants. Not only, as in the case noted by Hornibrook, do weakly-growing side shoots emerging directly from older wood produce plants correspondingly lacking in vigour and leader dominance, the selection of over-vigorous growth from a dwarf plant has the opposite effect and can, if persisted in, destroy its dwarf character in just a few 'cuttings generations'.

Another aspect of this phenomenon arises with genera such as *Picea* in which leader growth and side branches differ structurally and which have varying degrees of difficulty in initiating leader growth. For situations where this aspect of plant life is deliberately made use of to produce garden plants with an induced habit of growth, the name cultivariant is used, and sometimes it is convenient to use a cultivar name. But it must be borne in mind that such plants are in no sense a separate clone – they remain individuals, just as much as does a 'fan-trained' peach tree or a 'standard' rose-bush.

-- **'Fletcher's Compact'** Den Ouden and Boom, 80 (1965) is a name raised by Den Ouden and Boom for another cultivariant – much slower in growth than the normal 'Fletcheri' but retaining its upright shape. Another clone sometimes met with, and difficult to separate is **'Kestonensis'**. *C. lawsoniana fletcheri Keston variety* Reuthe, Nurs. Cat. (1935). This was discovered on the Reuthe Nursery at Keston in Kent in the 1920s in a bed of Lawson Cypress seedlings. It is more squat and bushy than 'Fletcheri' and has a similar type of foliage, but is much more-slow-growing.

-- **'Fletcher's White'** Den Ouden and Boom, 81 (1965) is a variegated form of 'Fletcheri'. It carries a nice, bold variegation of creamy-white. Plants grow rapidly and this form will probably turn out as big a plant as the 'Fletcheri' form which it was derived as a sport. It is not altogether hardy, so requires that elusive site in the garden – an open but sheltered spot.

-- **'Forsteckensis'**. *Ch. lawsoniana forstekiana* Beissner 79 (1891) spelling corrected to *forsteckensis* – Mitt. d.d.d. G. **13:** 146 (1904). This popular and well tried variety is indispensable in any collection. When well grown this cultivar produces a dense globose or broadly conical bush not over 60 cm high with main branches very densely set, main branches short, and all its growth very congested. It increases in size by sending out longer tips of normal foliage. These will eventually thicken up but the bush can be kept to any desired size and with foliage almost unbelievably congested and moss-like by the systematic removal of these longer growths. It originated at Forsteck near Kiel, Germany.

There is, however, a much looser form in cultivation which forms a much larger plant. Hornibrook expresses the opinion that this is the result of propagation from the stronger branches referred to above, but I incline to the view that it is a separate clone. Of this variety there is a wonderful old specimen in Goatcher's Nursery at Washington in Sussex which is (I write from memory) over 2 m high and wider than it is high. This form is just as attractive as the slower-growing one but not, of course, suitable for use in the same position in the garden. A form occasionally met with, **'Tilgate'** Ingwersen, Nurs. Cat, (?) is very similar.

From time to time variegated branches appear on plants of 'Forsteckensis', but I have never yet come across a case where the variegation has been stable when used for propagation.

-- **'Fraseri'**. Beissner, 12 (April 1887); *Ch. lawsoniana fraseri glauca* Lombarts Nurs. Cat. 8 (1893–94). This clone forms a rather slender columnar tree, narrow at the base and with ascending branches and branchlets, so that the foliage sprays are held in flattened, vertical planes. From 'Allumi' it differs in the dull grey-green colour of the foliage. It is a hardy clone.

-- **'Gimbornii'**. Groodendorst ex Hornibrook, 50 (1939). This forms a dense, compact, oval little bush seldom seen more than 1 m high. The main branches are thick, stiff and erect and the branchlets short. The foliage is a glaucous blue-green with the tips of the growing shoots purplish-blue during most of the year. It is similar to 'Nana Glauca' but it has finer foliage in narrower sprays. This cultivar originated as a seedling on the Von Gimborn Estate at Doorn, Holland. The mother plant reached a height of nearly 2 m but was killed outright by the very severe winter 1962–63. (See Illus. No. 109)

-- 'Glauca'. (Lawson) Hochstetter, 79 (1882); *Cupressus lawsoniana glauca* Lawson, Nurs.

Figure 109. *Chamaecyparis lawsoniana* 'Gimbornii'. Will reach 2 m in time.

8 (1874–75). This name was applied to several early selections with glaucous blue foliage that can no longer be identified with certainty. The word is sometimes included, quite unnecessarily, in the names used for later selections, (e.g. 'Felix', 'Kooy', 'Spek'.) It is therefore now only of value as a group or collective name.

-- **'Glauca Elegans'**. Schelle 292 (1909); *Ch. lawsoniana robusta glauca* Hort., not Beissner; *Ch. lawsoniana robusta elegans* Hort. Columnar, slender; branches ascending; branchlets erect, tips conspicuously white in summer, becoming glaucous in winter; leaves rather large, rough (giving a robust appearance to the tree, whence the name robusta in the synonyms), glaucous bloomy. A well-known form, often wrongly named in the trade. Origin unknown.

-- 'Glauca Globus'. See 'Barabit's Globe'.

-- **'Glauca Lombartsii'**. Lombarts, Nurs. cat. 46 (1925/26). Pyramidal, elegant, sparingly branched, branches spreading: sprays short, thin; leaves very small, thin, bluish-bloomed, paler on the under side of branchlets, and with the markings inconspicuous. Raised by Pierre Lombarts, nurseryman at Zundert. Netherlands, in 1910, and distributed by him in 1926.

-- **'Glauca Veitch'**. Veitch ex den Ouden 15 (1937), name only; 108 (1949). Pyramidal; branches and branchlets spreading, tips decurving; leaves small, green, soft blue bloomed, on under side of branchlets paler, markings inconspicuous. Probably raised in the Veitch nurseries and distributed as *Cupressus lawsoniana glauca.*

-- **'Globosa'**. Den Ouden, 108 (1949) and (under *Cupressus*) de Vos in Fl. and Pom. **1:** 17 (1876). Several globose clones are probably in cultivation besides the well known 'Minima Glauca' and, understandably, several of these may have attracted a reference to their shape at their christening. This cultivar is apparently a very old variety introduced from continental cultivation into the United States in 1963. Bergman (Pl. and Gar. **21** (1): 30 (1965) described it as a "Broadly globose shrub with ascending branches and rather thick, short branchlets, the tips of which recurve slightly. Leaves bright green. At 8 years, 6 inches high and 10 inches wide."

-- 'Globus'. See 'Barabit's Globe'.

-- **'Gnome'** Corley, Gar. Chron. **167, 23:** 31 (1970). This is a very diminutive bun-form raised from seed at Warnham Court, Horsham, Surrey, by the head gardener, W. Hart, and introduced to the trade by Heath End Nurseries, Farnham, U.K. It is similar to but much more slow-growing than the popular 'Forsteckensis' but is irregular in growth and apt to put out coarse growth and lose its attractiveness. If any such growth is cut away it will settle down as an attractive and congested little globose to broadly conical plant. A light annual clipping in the Spring will result in an extremely dense appearance.

-- **'Golden King'**. K. Wezelenburg & Son, Nurs. Cat. 1931. Conical, robust; branches spreading, tops decurving; branchlets broad, arranged in a frondlike manner; foliage golden yellow, bronzy-yellow in winter, yellowish-green in the underside of the branchlets. Raised as a seedling of *'Triomf van Boskoop'* by the firm K. Wezelenburg & Son, Hazerswoude near Boskoop, Netherlands.

-- **'Golden Light'**. Name registered in 1974 by R. J. Belding for a yellow form raised from seed sown in 1969. The habit is stated to be conical; the foliage is a pale lemon-yellow in summer, darkening to a bronzed dark yellow. It is reported to be hardy.

A feature of seedlings exhibiting some peculiarity is the tendency for that peculiarity to lessen or completely disappear with age. This is particularly so with the colour forms of this species. Several of the older selections have become indistinguishable from one another with the passage of time, and it is certain that many of the present ridiculous flow of golden seedlings will suffer the same fate. It is highly desirable that such names be accepted and/or registered on some provisional basis only until the distinctiveness of the plant in maturity has been established. 5 years from seed is much too soon for an assessment of any plant's potential.

-- **'Golden Pot'**. Introduced by D. W. Hatch Nurs. Honiton, U.K. and registered by him in 1983. A sport from 'Pottenii', similar in the softness of the foliage, but slower-growing and forming a more rounded bush. The colour is bright golden-yellow, persisting into the winter. There has apparently been some misunderstanding regarding this clone, since plant variety rights were later issued for it under the name 'Pot of Gold'.

-- **'Golden Prince'**. Hooftman ex Krüssmann, 77 (1983). A vigorous clone with yellow-brown branchlets and somewhat pendulous golden-yellow foliage; claimed to be an improvement on 'Golden King'. Introduced by W. J. Hooftman of Boskoop, Holland in 1978.

-- **'Golden Spire'**. Bolwyn ex Krüssmann, 75 (1979). A clone that is indistinguishable from, if not identical with 'Kelleriis Gold'. It was for a short time distributed under the illegimate name 'Columnaris Aurea'. It should not be confused with *Chamaecyparis obtusa* 'Gold Spire'.

-- **'Golden Triumph'**. Mauritz ex Krüssmann, 75 (1979). An upright, almost columnar plant with numerous upright branches and a good yellow colour in summer turning to tawny bronze by winter. It was introduced by Mauritz Nurs. Bussum, Holland in 1972 and is widely grown on the continent.

-- **'Golden Wonder'**. Spek in Journ. Roy. Hort. Soc. **88:** 22 (1963). Raised on the Spek Nurs., Boskoop, Holland, in 1963. Similar to 'Stardust' but slower-growing. The yellow colour is well maintained throughout the year.

-- **'Goldfinger'**. Name registered by Goatcher and Sons Nurs. Pulborough, U.K. in 1978. A seedling from 'Stewartii' and of similar colouring, but dense-growing, with a fastigiate branch system.

-- **'Gold Splash'** Harrison (N.Z.) 4 (1975). A variegated sport of 'Fletcheri' that originated in New Zealand. It has irregular splashes of yellow foliage, most noticeable in contrast to the normal green in Spring and early Summer.

-- **'Gracilis'**. Nelson, Pinacese, 18 (1866). The word means 'slender', so this, together with 'Gracilis Aurea', 'Gracilis Aurea Pygmaea', 'Gracilis Glauca', 'Gracilis Nana', 'Gracilis Nova', 'Gracilis Pendula' and 'Gracillima' were early selections presumably considered to have this characteristic, but none seem to have survived in cultivation. It is not unusual for a promising seedling to develop into a commonplace plant, in time.

-- **'Grandi'**. Clarke & Co., Nurs. Cat. 2 (1950). Dwarf, flattened, globular shrub; branches, more or less horizontal; branchlets spreading, some sprays twisted showing the glaucous colour of the underside of the leaves; leaves dark glaucous-green, bluish on the underside of the branchlets. A dark glaucous dwarf form widely distributed on the Pacific Coast of America. Introduced by W. B. Clarke & Co. San José, California. It is very similar to 'Tamariscifolia'.

-- **'Grayswood Feather'**. An attractive, slow-growing form that should become popular. The foliage is dark green on acid soils but can be quite yellow on alkaline soil. It was introduced by D. W. Hatch, Chantrey nurseries, Honiton, Devon, U.K. Each of the 'Grayswood' series arose of the garden of Mr. G. L. Pilkington, in Haslemere, Surrey.

-- **'Grayswood Gold'**. A semi-dwarf columnar selection with foliage colour deep chrome yellow, becoming almost orange at the growing tips.

-- **'Grayswood Pillar'**. Hillier's Manual. 470 (1971). Similar to 'Columnaris Glauca' in colour, but very much more narrowly fastigiate; a plant 9 m high may remain less than 50 cm wide. It originated as a mutation on the cultivar 'Blue Jacket'.

-- **'Green Globe'**. Harrison, 42, 1975. This arose as a seedling selected at Palmers Nursery, Glen Eden, Auckland, N.Z. about 1950. It is extremely slow-growing, forming a rounded ball of tiny foliage, taking many years to reach to football size. It shews little tendency to put out coarse, uncharacteristic growth, but all such diminutive plants must be watched, since any development of this kind soon takes over and destroys the plant's appeal.

-- **'Green Hedger'**. Jackman, Planters' Handbook No. 4: 154 (1939). A regular growing, dense and pyramidal form of a bright green colour. Originated as a seedling in the Jackman's nursery, Knaphill near Woking, England and is recommended by that firm for hedges.

-- **'Green Pillar'**. Hogger's Nurs. Cat., Felbridge, Sussex, U.K. (1960).

A variety raised by J. Hogger about 1938. From 1947 until 1964 it (or an indistinguishable mutation) was distributed by Geo. Jackman Nursery, Woking, Surrey, as *C. lawsoniana erecta* (*Jackman's Variety*).

Habit upright, forming an excellent medium-sized pyramid; branches and branchlets strictly ascending; leaves bright green with a golden tint most noticeable in early spring. According to the firm Hogger the plant does not appear to be damaged by a heavy fall of snow.

-- **'Green Spire'**. Either another clone introduced by the same nursery and doubtfully distinguishable from 'Green Pillar' or (more probably) merely a redundant name for that cultivar.-

-- **'Green Wall'**. Sequi ex Grootendorst in Dendroflora **15–16:** 53 (1979). A small, columnar form with good green foliage that originated in France. It is rather similar to 'Green Pillar'.

-- 'Greycone'. A recently selected clone with juvenile foliage said to be resistant to wind and salt spray. The colour is grey-green with tints of mauve and the habit is to form a neat conical shape. A better name should be found.

-- **'Headfort'**. Hunt, Conif. Brit. Is. 112 (1972); *Cupressus lawsoniana var. Headfortii* Dallimore 17 (1932). A medium-sized tree of graceful habit with spreading branches and large, loosely borne sprays of blue-green foliage, silvery-white beneath. Named in honour of Lord Headfort of "Kells", County Meath, Eire.

-- **'Henri Smits'**. Smits ex van de Laar in Dendroflora **19:** 88 (1982). A distinctive form with the usual golden-yellow colour of many other varieties, held well throughout the year, but with an unusual habit. It forms an open, broadly conical tree with horizontal main branches arching over and pendulous at the tips, the effect being similar to *C. obtusa* 'Crippsii'.

-- **'Hillieri'**. Hillier Nurs. Cat. 104 (1950): *Cupressus hillieri* Hillier, Nurs. Cat. 188 (1928); under *Cupressus lawsoniana* Bean **3:** 123 (1933). A medium-size tree with a dense, conical habit, horizontal branches and foliage in large, floppy, feathery sprays; a bright pale golden yellow at the outer surface, greenish within. It was selected by Edwin Hillier before 1920 and is said to be the most beautiful golden cypress, but it is less hardy than some clones.

-- 'Hogan'. See 'Oregon Blue'.

-- **'Hollandia'**. Bailey, Cultivated Conifers. 167 (1933). An early selection, having been introduced by Koster and Co. Nurs., Boskoop in or around 1895. It forms a strong-growing, conical tree of unusual habit. The main branches are thick, horizontal and arching over and carry fan-shaped sprays of foliage held upwards and forwards above the shoot. The foliage is dark green. This clone is hardy, but to maintain its individual character it is essential not to prune trees after they have developed their main framework

-- **'Howarth's Gold'**. Grootendorst in Dendroflora **6:** 77 (1969). This clone makes an attractive, narrowly conical bush even when quite small, with dense but soft foliage. The golden-yellow colour is at its best in early summer.

-- **'Hughesii'**. A cream coloured seedling raised by W. Stan. Hughes of Blue Mountain Nurseries, Tapanui, West Otago, N.Z. in the early 1930's. It has had a steady but limited sale in New Zealand since its first distribution in about 1939.

-- **'Imbricata Pendula'**. R. E. Harrison, Trees and Shrubs. 88 (1959). A clone raised from seed in Palmerston North, N.Z. about 1930. It makes a medium tall tree with a strikingly graceful habit. The foliage is wholly filiform and completely pendulous, exposing the upper sides of the pendent main branches. Harrison gives a good illustration. No doubt in years to come it will vie with the older 'Filiformis', but it never develops the long, cordlike shoots of that variety.

-- **'Intertexta'**. (R. Smith) Nicholson 304 (1884) and (under *Cupressus*) R. Smith 15 (1872). An ascending form; branches loosely borne and regularly disposed. Tops pendent, branchlets widely set, interlaced, drooping, grooved; leaves on older branchlets loosely appressed, axial-leaves sometimes 1.5–2 cm long, thick, sharply long-pointed, bluish-green, slightly glaucous, bloomy. Young plants rather tender. Originated in the Lawson's nursery, Edinburgh, about 1869. The coarse, open structure and the divaricate, glaucous grey foliage makes this a striking tree in maturity. (See Illus. No. 111)

-- **'Ivonne'**. Van de Laar in Dendroflora **18:** 69 (1981). Another small-growing, narrowly conical tree with yellow to yellowish-green foliage. It is said to be winter-hardy and to hold its colour well throughout the year.

-- **'Juvinalis Stricta'**. Harrison, Ornam. Conif. 44 (1975). The plant thus named, and illustrated by Harrison, must be what was sent to me at Wansdyke Nursery, Devizes, U.K. some years ago as *C. thyoides leptoclada* which turned out to be a Lawson with juvenile foliage very similar to 'Fletcheri', except for its green colour. Harrison records that it originated as a seedling in a row of *Chamaecyparis lawsoniana* plants many years ago on a nursery in Rotorua, New Zealand.

Figure 110. *Chamaecyparis lawsoniana* 'Kilmacurragh'. A columnar clone much superior to 'Erecta Viridis'.

-- 'Keessen' or 'Alba Keessen'. See 'Erecta Alba'.

-- 'Kelleriis Aurea' (syn. 'Aurea Kelleriis'). See 'Kelleriis Gold'.

-- **'Kelleriis Gold'** Poulsen ex Grootendorst in Dendroflora **6:** 77 (1969). A seedling raised by D. T. Poulsen, Thomsens Planteskole, Denmark in about 1938. It is a very hardy, yellow-green, upright-growing clone. It was at first distributed as 'Aurea Kelleriis' or 'Kelleriis Aurea'.

-- **'Killiney Gold'**. Columnar with terminal shoots pendulous. Foliage is an attractive golden-

Figure 111. *Chamaecyparis lawsoniana* 'Intertexta'.

Figure 112. *Chamaecyparis lawsoniana* 'Minima Glauca', shewing the characteristic rounded crown.

Figure 113. *Chamaecyparis lawsoniana* 'Nana', shewing the characteristic bluntly conical crown.

yellow. Introduced by Watson's nursery, Killiney, Co. Down, Ireland.

-- **'Kilmacurragh'**. Hillier, Suppl. List 7 (1951). Columnar, tapering gradually to apex, dense, moderately strongly growing; branches ascending at angles of 30°–40°, very short; branchlets densely set, short; sprays somewhat irregularly disposed, with their outer margins flat and rounded; leaves dark greyish-green. Raised at Kilmacurragh, Co. Wicklow on the estate of Thomas Acton. (See Illus. No. 110)

-- **'Knowefieldensis'**. Beissner in Mitt. d.d.d. Ges. **20:** 172 (1911); Hornibrook 34 (1923) and ed. 2, 53 (1939): 'Knowefieldensis Glauca' Hort. As grown in the Pygmy Pinetum at Devizes, Wilts, U.K., it is a wide-spreading, flat-topped, untidy looking plant. The foliage is very thin and fine - a deep, rich green above and only slightly glaucous beneath. The leaves are very small, densely set and appressed for most of their length, only the tips being free (and that not noticeably so as they are in 'Nidiformis' and 'Tamariscifolia'). They are borne in flat, parallel-sided sprays about 2.5 cm wide, on side shoots regularly set about 2 cm part and all emerging at a constant angle of about 45° and regularly decreasing in length towards the dominant terminal shoot. The resulting triangular outline at once distinguishes this form from 'Nidiformis', and from 'Tamariscifolia' it is distinct by the mainly appressed leaf tips and the crowded, overlapping growth. Raised in Little & Ballantine's Nursery, Carlisle, England.

-- **'Kooy'**. Valk. Suring. in Jaarb. Ned. Dendr. Ver. 143 (1927). conical, dense; branches spreading, regularly borne; branchlets thin, tips not much decurving; sprays fine; leaves on both sides of branchlets with glaucous-blue markings. Raised by G. Kooy & Son, nurserymen at Boskoop, Netherlands, ± 1925.

-- **'Krameri'**. Beissner, 33 (1887). This is a globose or shrubby variety of no great beauty, reaching to several metres high, with its main branches set horizontally, partly twisted and pendulous. The branch system is very irregular and contorted, the final growth cord-like, glossy dark green.

-- **'Lane'**. den Ouden 110 (1949); *Ch. lawsoniana Lane's aurea* Hort.; *Ch. lawsoniana lanei* Hort. Narrowly columnar, rather slow-growing: branches spreading; branchlets crowded, thin; sprays feathery. thin; leaves golden yellow, yellow-green on underside of branchlets. Still one of the most popular of the golden-yellow Lawsons. Introduced by Lane's nurseries, Great Beckhamsted, England, about 1938.

-- **'Lemon Queen'**. A rather slow-growing tree with the foliage a rather unusual pale sulphur or lemon colour. Not to everyone's taste.

-- **'Little Spire'**. A seedling raised by J. W. Archer from 'Fletcheri', with 'Wissellii' as the pollen parent. Perhaps it would best be described as a compact, rather than a dwarf form of 'Wisselli' since its ultimate height is unknown. Introduced by Don Hatch, Chantrey Nurs, Honiton, Devon, U.K. in 1972.

-- **'Lombartsii'** Lombarts, Nurs. Cat. 30 (1915). A vigorous clone with a dense habit of growth; branches spreading, stout, tops decurving; branchlets stout, broadly frond-like; leaves greenish-yellow, slightly bluish-bloomed, dark yellow in winter, yellowish-green and blue-bloomed on under side of branchlets. Raised by Pierre Lombarts, nurseryman at Zundert, Netherlands in 1904, and introduced in 1915. Still very rare in Britain.

-- **'Luna'**. Van de Laar in Dendroflora **20:** 77 (1983). Arose as a mutation on 'Silver Queen' found by A. Blanken of Boskoop, Holland. It forms a fairly dense, broadly conical tree with adult foliage that is yellow with white tips during the growing season, fading to a grey-green by the winter. Claimed to be an improvement on the older cultivar because of its more compact habit and the tolerance of the bright young growth to full sunlight.

-- **'Lutea'** (Rollisson) Nicholson, 304 (1884); under *Cupressus lawsoniana* Rollisson in Gard. Chron. 6 (1873). The earliest recorded golden

Figure 114. *Chamaecyparis lawsoniana* 'Lycopodioides'. The twisted foliage is unique.

cultivar and still one of the best. 'Lutea' forms a broadly columnar tree tapering to a narrow crown. The main branches are spreading but short; the foliage is dense, the leaf sprays being broad, pendulous and curled, a bright golden yellow where fully exposed to the light, elsewhere whitish, and green within the interior of the tree. Raised in the G. & W. Rollisson Nurs, Tooting, London before 1873. Young plants, to about 1.5 m, appear loose and open, but growth soon becomes dense and some magnificent specimens can be seen in several collections in U.K.

-- 'Lutea Nana'. See p. 203, under 'Aurea Densa'.

-- **'Lycopodioides'**. Van der Elst in Mitt. d.d.d. Ges. 54 (1896). This is not a dwarf form, but its curiously twisted foliage makes it an interesting plant for several years. It might perhaps best be described as a thread-leafed form in which the threads are curled and twisted in all directions, somewhat in the manner of *Ch. obtusa* 'Coralliformis', from which it can at once be distinguished by the pointed leaves with their free standing tips. Raised from seed by van der Elst, Tottenham Nurs, Dedemsvaart,

Netherlands ± 1890. (See Illus. No. 114)

-- **'Maas'**. Lombarts. Nurs. Cat. 22 (1957/58). Conical; branches spreading and ascending. Resembling *'Triomf van Lombarts'*, but the foliage much finer and yellowish-green. Raised and distributed by Pierre Lombarts, nurseryman at Zundert, Netherlands.

-- **'Melfard'**. A Danish selection with a curious lop-sided habit of growth which Krüssmann considers to be unattractive.

-- **'Merrist Wood'**. Yates in Roy. Hort. Soc. Journal **97:** 272 (1972). A selection made at the Surrey Agricultural College, Merrist Wood, U.K. shewing good resistance to *Phytophthera*. It is in other respects typical of the species.

-- **'Mini Globus'**. Van de Laar, Naamlijst. 211 (1986). Name illegitimate. Probably a synonym of 'Barabit's Globe'.

-- **'Minima'**. (R. Smith) Sudworth 90 (1897) and (under *Cupressus*) *lawsoniana* R. Smith ed. 3, 15 (1867). Dwarf, flat-globular, compact; branches erect; branchlets slightly curled and perpendicular or sideways disposed; leaves densely set together, small, somewhat yellowish-green, distinctly marked white on underside of branchlets. Welch [159(1979)] sets out the distinction between the members of this and the 'Nana' group of dwarf garden forms.

-- **'Minima Argenteovariegata'** Beissner in Mitt. d.d.d. Ges. **9:** 64 (1900). Variegated mutations occur but to my knowledge none has yet proved stable when propagated.

-- **'Minima Aurea'**. Hornibrook in Gar. Chron. 49 (1929) and Dw. Con. ed. 2, 43 (1939); *minima aurea Rogersi* Hort. See p. 203 under 'Aurea Densa'. It was raised from seed by H. W. Rogers & Son, Southampton, U.K. and has no relationship with 'Minima'. It is one of the best truly dwarf conifers and deservedly popular. (See Illus. No. 101)

-- **'Minima Glauca'** (R. Smith) Beissner 452 (1884) and (under *Cupressus*) R. Smith ed. 3,15 (1867). Similar in every way to 'Minima' except that a heavy glaucescence gives a distinctly blue-grey tone to the bush. This glaucous form is the one more frequently seen. (See Illus. No. 112)

-- **'Moerheimi'**. Ruys, Cat. Moerheim nurseries **46:** 50 (1934). Pyramidal; branches and branchlets spreading; leaves yellow, becoming yellowish-green within the plant. Raised by B. Ruys, Moerheim nurseries, Dedemsvaart, Netherlands.

-- **'Moonlight'**. Watson ex Krüssmann, 82 (1971). A form originating from Eire. With an open habit resembling some forms of *Ch. obtusa*, with yellowish-grey foliage.

-- **'Naberi'**. Naber ex den Ouden 112 (1949). A conical tree of rather open habit, with foliage in vigorous sprays held at varying angles, often edgeways.

-- **'Nana'**. (Pepin) Hochst., Conif. 80 (1882) and (under *Cupressus*) Dauvesse ex Pepin in Rev. Hort. 159 (1864): *Ch. lawsoniana densa* Hort. Dwarf, squatly-conical shrub with a leader, at least when older; branchlets more horizontally spreading; foliage slightly bluish-green. Raised in 1861 by Dauvesse, nurseryman at Orléans, France. It differs form 'Minima' by the acute top of the plant ('Minima' is round or flat-topped) and by the horizontal rather than vertical disposition of the branchlets. (See Illus. No. 113)

-- **'Nana Albospica'**. [*C. lawsoniana nana alba* (R. Smith) Nicholson. Dict. Gar. **1:** 304. 1884; *nana alba* Veitch, Man. 233. 1881; *albo-spica nana* (Gordon. Pin. Ed. 2, 424. 1875; *nana albo-spicata* Beissner. Handb. 78. 1891; *alba spica nana* Hornibrook. Dw. Con. Ed. 2, 41. 1939; *nana alba* Hornibrook. Dw. Con. Ed. 2, 44. 1939.] Seedlings (sometimes slow-growing and therefore dwarf) with milky-white young growth turn up occasionally in the seed beds. At first they are attractive and numerous attempts have been made by early writers to find descriptive names for novelties of this kind, but with age they tend to have less appeal and

I can only trace one form in cultivation today. Presumably all the other forms (if in fact, they were different) are now lost. The form we have is a slow-growing dwarf, conical plant, dense in growth. The main branches are short; the branchlets short and thin. The leaves are small, white at the growing points, elsewhere pale green. This variety pays for planting in rich soil as it is only when it is growing strongly that it has the 'white appearance to the whole plant' described by Hornibrook, and that only during the growing season. This effect is enhanced by rich soil and an annual clipping during the Spring to encourage vigorous new growth.

The orthography of the present name is obscure but I have selected it as the one that best preserves established usage.

-- **'Nana Glauca'**. (Veitch) Nicholson, 304 (1884); under *Cupressus* Veitch, 233 (1881). Similar to 'Nana', but foliage coarser and glaucous.

-- 'Nana Glauca Rogersii'. See 'Rogersii'.

-- **'Nestoides'**. An American selection, doubtfully distinguishable from 'Tamariscifolia'. It forms a flat-topped bush with very dense, soft-looking foliage of a rich mid-green, held in broad sprays that exhibit little or no leader dominance, but maintain a branch divergence of 45°–50° noticeably uniform throughout the plant. It was introduced by Jack Manten of Vancouver, BC Canada.

-- **'New Golden'**. Lombarts, Nurs. Cat. 22 (1957). An upright-conical, rather open-branched golden form of no great distinction, possibly no longer in cultivation.

-- **'New Silver'**. An uncommon cultivar similar in growth to 'Pottenii' but it better retains its tight growth and narrow crown, while the down-turned spray-tips are silvery grey-blue. It deserves to be better known and to be given a more acceptable name.

-- **'Nidiformis'**. Beissner in Mitt. d.d.d. Ges. **10:** 74 (1901) and Handb. ed. 2, 552 (1909) (*plumosa nidifera nana* Hort, pro syn.); *Chamaecyparis* × *nidifera* Hornibrook Rep. Conif. Conf. 72 (1932). Later writers, following Hornibrook, appear to have confused this form with *Chamaecyparis nootkatensis* 'Nidifera'. Welch 163–5 (1979), gives a detailed historical account of how this confusion arose and was cleared up. The true 'Nidiformis' is a widespreading shrub, eventually covering several metres. When young, a low spreading plant very similar to 'Tamariscifolia' (for which cultivar it is often mistaken) vaguely suggesting a large bird's nest (hence the name), with radially spreading and horizontal or arching main branches and decurving branchlets. The foliage looks at a glance to be coarser than that of 'Tamariscifolia' and bears blue or grey-green foliage a little glaucous above and very glaucous beneath, much flattened (1.5 mm wide); the leaves very crowded, the lateral pair standing off at the free tips at 45°, the facing pair smaller, appressed, in shape a 45° triangle (variable), seldom glandular. The growth is in wide sprays in which many of the side branches twist out of the plane of the spray and almost all the ultimate shoots curve, some up, some down, as though the spray had been gone over with a flatiron when wet, and in drying each little shoot had curled as it felt inclined.

-- **'Olbrichii'**. Froebel ex Beissner in Mitt. d.d.d. Ges. **13:** 86 (1904). Columnar, densely branched, and with a narrow top and drooping leading shoot, branchlets feathery: leaves loosely appressed, apex pointed, bluish-green. Not quite hardy. Raised by Froebel, nurseryman at Zurich, Switzerland. Presumably 'Albrechii' and 'Albechii' are misspellings.

-- **'Oregon Blue'**. A very vigorous. fast-growing clone that makes a broadly-conical tree of an outstanding silver-blue colour. It was raised from seed on Arneson's Nurs, Canby, Oregon, U.S.A. raised and introduced by the Richard Bush Nurs, Canby, Oregon, U.S.A.

-- **'Parsons'**. Hillier ex Den Ouden and Boom 92 (1965); Hillier 18 (1964). A beautiful dense, dome-shaped shrub, built up from many layers of arching branches superimposed one on another.

-- **'Patula'**. Beissner in Mitt. d.d.d. Ges. **12:** 57 (1903). A graceful, bushy, broadly conical tree of medium size with distinctive foliage, having very long slender ultimate shoots in narrow fanlike sprays curving outwards. Leaves are appressed, dark grey-green or sea-green, yellowish beneath with white bases. Very rare in cultivation.

-- **'Pelt's Blue'**. A recent introduction in Germany. Similar to 'Columnaris' but with foliage of a more intense blue. Listed Kordes Nurs. Cat. 156 (1984–85).

-- **'Pembury Blue'**. Jackman, Planters' Handb. **25:** 75 (1963–64). One of the most beautiful of the glaucous cultivars and quite unique in the foliage colour, which in the first year is a silvery dove-grey, darker the second year. At first rather open, it soon builds up into a compact, shapely plant. It was raised on the Baggesen Nurs, Pembury, Kent. U.K. and was introduced to the trade by Geo. Jackman and Son, Woking, Surrey, U.K.

-- 'Pendula'. This descriptive adjective, either alone or in combination with others such as *alba, aurea, nova* and *vera* was early used for clones which, although living specimens may survive (and possibly even justify reintroduction) are not in current production, so the name should only be used in a group or collective sense.

-- **'Pick's Dwarf Blue'**. Forms an upright-ovoid bush similar in shape and its neat foliage to 'Gimbornii', but the white markings on both sides of each leaf give a grey- or blue-green tone.

-- **'Pixie'**. Streng in Dendroflora **11–12:** 57 (1977). Another diminutive, globose seedling with dense, fine growth. The colour is bluish-green, which may serve to distinguish this from other, similar, clones.

-- **'Plumosa Glauca'**. Krüssmann, 95 (1960); 'Blue Plume' den Ouden & Boom 74 (1965). The latter name was suggested because the date of appearance of the former name raised a presumption of its illegitimacy. Krüssmann now, [80 (1985)] traces its history back to 1914, thus reversing the presumption. It forms a broadly columnar tree with attractive intensely blue-green foliage, but (perhaps because it was not reliably hardy) no longer seems to be in current production.

-- 'Pot of Gold'. See 'Golden Pot'.

-- **'Pottenii'**. Hillier ex Dall. & Jacks. 233 (1948); *Cupressus laws. var. pottensoi* Dall. and Jacks. 203 (1923); *Cupressus laws. pottenii* Hillier Nurs. Cat. 10 (1924). A narrowly columnar small tree with long, thin, vertical branches and branchlets with the tips nodding. The foliage is very uniform, the small, partially juvenile leaves, pointing forward, being yellowish green and very soft to the touch. Originated on the Potten Nurs. at Cranbrook, Middx. U.K. This cultivar is popular because it propagates easily and (in nurserymen's parlance) 'furnishes well' when young and builds up into a spectacular small tree, ideal for a Flower Show. But it is not reliably hardy, and after reaching, ten feet or thereabouts it tends to fall apart, the long, slender branches being unable to recover from damage resulting from snow or gales.

-- **'President Roosevelt'** and **'Winston Churchill'** were golden-yellow selections made on the then Hogger Nurs. East Grinstead, Sussex, U.K. and introduced in about 1945. Although presumably differing as young plants, they can with age no longer be distinguished from one another or, convincingly, from some other golden selections. It is probable that some of today's novelties will suffer the same loss of distinctiveness in time and also become not worthy of retention.

-- **'Pygmaea Argentea'**. Beissner 13 (april 1887; *pygmea* (sic) *argentea* Backhouse, Nurs. Cat. 37 1907/8; 'Backhouse Silver' Hort. This popular cultivar is a dwarf form of very slow growth, forming a small globose or squatly conical bush with its main branches upright. The foliage is green, pale creamy-white at the tips and very densely set. When the plant is growing strongly the foliage is almost white in early summer and the whole bush then has the appearance of having been turned upside down when wet into a barrel of flour, but this foliage will scorch in full sun.

-- **'Pygmy'**. Jackman, Planters Handb. **30:** 76 (1968). Best described as a miniature version of the well-known 'Minima Glauca', with the same rounded shape, colour and neat dense habit, but with much less than half its rate of growth. A plant to be safely recommended for even the smallest garden or rockery.

-- 'Pyramidalis'. Apart from the absurdity that no conifer has ever grown to a pyramidal shape, the limitations on the value of this name (and others made by adding descriptive adjectives such as *alba*, *alba nana*, *glauca*, *lutea* and *lutea gracilis*) are the same as stated above under 'Pendula'. (See p. 225)

-- **'Rijnhoff'**. Krüssmann, 84 (1983). A new selection similar to 'Nidiformis'. It forms a flat-topped, spreading bush with bright-green, lace-like foliage in overlapping sprays lying in all directions. Originated at the K. van Rijn Nurs., Hazerswoude, Holland.

-- 'Robusta' and its derivatives. As 'Pendula' above. See p. 225.

-- 'Robusta Glauca' Hort. A synonym for 'Glauca Elegans'.

-- **'Rogersii'**. Den Ouden and Boom, 94 (1965); *C. lawsoniana nana Rogersii.* Den Ouden, Naaml. 15 1937 (Nom.); Rogers and Son. Nurs. Cat. c. 1930 is a glaucous blue, globose variety with thread-life foliage raised from seed by Rogers and Sons of Red Lodge Nursery, formerly at Southampton but now at Chandlers Ford, Hampshire. It will grow to 2 m in time. The colour is similar to the well known tall variety 'Allumii' but the blue is not quite so intense and it seems to become less marked as the plant ages.

-- **'Royal Gold'**. Name registered by J. Allen of Guernsey in 1975. It forms an upright-ovoid plant with dense golden-yellow foliage to which the very small leaves give a dainty appearance; even young plants being well furnished to the ground. It was discovered in a garden at Taunton, U.K. by Humphrey Welch. The colour is well maintained throughout the year. It makes a very neat hedge, where the colour is acceptable.

-- 'Silver Gem'. See 'Silver Queen'.

-- **'Silver Threads'**. Name registered in 1983 by D. W. Hatch, Chantrey Nurs, Honiton, U.K. for a variegated sport from 'Ellwood's Gold' that was found and introduced by him. It is slower growing than 'Ellwood's Gold' and has creamy-white flecks among the golden-green foliage.

-- **'Silver Queen'**. Hort ex Beissner, 75 (1891); under *Cupressus boursieri* van Geert in Siebold-ia **9:** 155 (1883). Mitchell [77 (1972)] tells us that this name is misleading, since although the young leaves of this narrowly conic and (in old specimens) rather open-crowned tree are silvery-grey for a brief period, it "spends most of the year with its leaves a rather sad yellowish colour above and not very white bases to leaves below. Krüssmann [85 (1983)]states that **'Silver Gem'** is similar, with a better white colour in Spring, but less hardy.

-- **'Smithii'**. (J. Smith) Dallimore & Jackson 230 (1948); *Cupressus lawsoniana lutea smithii* J. Smith & Son, Cat. (1898); *Ch. lawsoniana lutea 'Smith'* den Ouden, 110 (1949); *Ch. lawsoniana darleyensis* Beissner, 47 (1909), not J. Smith. Conical. open, rather slow-growing; branches spreading; branchlets arranged in an irregular manner; leaves canary-yellow. Rather tender. Originated in the nursery of James Smith & Son, Darley Dale, England.

-- **'Snow Flurry'**. Harrison, Ornam. Conif. 50 (1975). A creamy-white sport that arose on a plant of *Ch. lawsoniana* that carried juvenile foliage and which was introduced by the South Taranaki Nurs, Hawera, N.Z. in the mid 1960s. It grows slowly at first but will become large in time. According to the picture he has given to us, the variegation is of a good colour, but

Harrison warns us that it scorches in wind and sun, so needs a sheltered location.

-- **'Snow White'**. This arose as a mutation on a 6-yr old plant of 'Ellwoodii'. All the young growth is pure white, and on trials has not suffered from sun-scorch. It was introduced to the trade by R. Murray, Allandale Nursery, Willerby, E. Yorkshire, in 1985.

-- **'Somerset'** Scott, Nurs. Cat. 94 (1967). A small, upright-conical to columnar small tree with mostly blue-grey juvenile foliage yellow tinged during the summer. It originated from a tree of 'Westermanni' on Scott's Nurs. Merriott, Somerset, U.K. that suddenly began putting out this type of foliage, which it has since continued to do.

-- **'Southern Gold'**, was raised by Ken Burns, Timaru, N.Z. and introduced to the trade by Duncan and Davies Nurs, New Plymouth. It is slow-growing and somewhat pendulous. The lemon-colour foliage needs shelter from wind and looks at its best in semi-shade.

-- **'Spek'**. Spek ex den Ouden 107 (1949). Pyramidal, strong-growing; branches stout, stiff, tops decurving; branchlets at the tip light brown with a ruby hue; foliage glaucous-grey, rough to the touch; probably the most glaucous form of the species. Raised in the nurseries of J. Spek, Boskoop, Netherlands, introduced into the trade in 1942.

-- **'Stardust'**. Langenburg ex de Vogel in Dendroflora **3:** 74 (1966). A selection from seed, raised by L. C. Langenberg of Boskoop, Holland, that is becoming very popular. It has the usual golden-yellow colour, even within the tree, but has a strongly erect branch system and many upright branchlets, so makes a nice compact upright plant.

-- **'Stewartii'**. Schelle in Mitt. d.d.d. Ges. **29:** 39 (1920). An old and hardy cultivar still widely planted even although surpassed for colour by several newer selections. It forms a regularly conical tree with ascending branches and branchlets with every growing tip curling outwards. The young leaves at the surface are bright yellow, but the interior of the tree is green and the whole tree becomes yellowish-green by winter. The leaves have an almost glossy sheen, as though dipped in oil or varnish. Raised by D. Stewart & Son, Bournemouth, U.K.

-- **'Summer Snow'**. Raised as a chance seedling at the T. H. Harraway Nursery at Warminster, Wiltshire, U.K. It is a slow-growing plant with its young foliage creamy-white at first but this soon fades. A regular light clipping in April produces (as it does in all clones with this colouration) vigorous new growth of almost dazzling brightness.

-- **'Tamariscifolia'**. Hornibrook 31 (1923). In this cultivar the trunk is always entirely absent, and its place is taken by a number of main branches which grow upwards and outwards in such a way that although the plant will reach to 2 m high in time the top is always more or less flat and the width is about twice the height. The branching is very irregular and the branches appear to tumble about in all directions. The foliage is a light midgreen, slightly glaucous above, noticeably so below, much flattened as in 'Nidiformis' but wider (2 mm). The leaves are longer and narrower, the lateral pair standing off at 30° and the facing pair in the shape of a 30° triangle, nearly always noticeably glandular. The foliage is in wide sprays in which the length of side branchlets reduces regularly towards the dominant leader, resulting in a triangular outline quite different to the bluntly rounded spray of 'Nidiformis' and although there is much overlapping the spray lies in a flat plane. (See Illus. No. 115)

-- **'Tharandtensis Caesia'**. Buettner in Mitt. d.d.d. Ges. **43:** 431 (1931) and Hornibrook 48 (1939); 'Forsteckensis Glauca' Hort. This could broadly be likened to a coarse, upright-growing form of 'Forsteckensis' with grey-green to blue-green leaves. This cultivar is much the better for having its long shoots pruned well back occasionally. This keeps the bush dense and postpones the degeneration into more or less typical *C. lawsoniana* foliage which is usually

Figure 115. *Chamaecyparis lawsoniana* 'Tamariscifola'. There may be several clones in cultivation under this name.

the ultimate fate of all these seedling forms. I have found it unsatisfactory on soil that had been dressed with lime.

-- **'Tilgate'**. Reuthe, Cat. Dwarf, flat-globular; foliage greyish-green. Probably originated at the Tilgate Plant Research Station, England. Distributed by R. G. Reuthe, Keston, Kent, England. It looks identical with the loose form of 'Forsteckensis'.

A plant entered in the Exhibition at Boskoop, Holland, in 1969 was misidentified as 'Tilgate'. It was either 'Tamariscifolia' or a similar clone. Unless it is the former, a new name must be found, since the continued misuse of the name 'Tilgate' is not permissible.

-- **'Triomf van Boskoop'**. Beissner in Mitt. d.d.d. Ges. **7:** 18 (1898). (= **'Triomphe de Boskoop'** Hort. Angl.) One of the best selections for a specimen tree in a dominant situation. Mitchell's description [77 (1972)] cannot be bettered. "Broad-columnar, open crowned, level-branched and usually with a single bole to the tip, this cultivar is more vigorous and of much better form than the type tree. Bark smooth until bole is quite large. Large open sprays, outer leaves pale blue-grey or whitish-grey; inner, dull blue-grey to grey-green. The open crown allows the bole to be visible nearly to the top. Widely planted; an excellent tree." Raised by D. Grootendorst ± 1890.

-- **'Triomf van Lombarts'**. Lombarts, Cat. 31 (1940–41); Lombarts ex den Ouden 116 (1949). Pyramidal and fast-growing; branches and branchlets spreading; foliage golden yellow. Raised by Pierre Lombarts, nurseryman, Zundert, Netherlands.

-- **'Van Eck'**. den Ouden and Boom, 97 (1965). Conical, moderate in growth, dense; branches spreading: branchlet-tips yellow, fading from slightly glaucous to yellowish-green inwards to the centre of the plant. Selected from a seedbed in 1934, and distributed by G. van Eck, Boskoop, Netherlands.

-- 'Van Pelt's Blue'. See 'Pelt's Blue'.

-- **'Van Tol'**. Den Ouden and Boom, 97 (1965). In growth similar to *'Alumii'*, but leaves golden yellow, yellowish-green in winter. Raised from seed of *'Alumii'* in the nursery of P. van Tol, Boskoop, Netherlands in 1946. Undoubtedly spontaneously fertilized by *'Triomf van Lombarts'* standing in the neighbourhood.

-- **'Versicolor'** (J. Coninck) Beissner 75 (1891) and (under *Cupressus*) J. Coninck, Nurs. Cat. Tottenham 5 (1888). Pyramidal and broad; branches spreading; branchlets spreading, decurving; sprays and leaves spotted with creamy white or bright yellow. Unusual, but of no particular beauty. Raised by C. Jongkind Coninck, Tottenham Nurseries, Dedemsvaart, Netherlands, ± 1882.

-- **'Viner's Gold'**. Krüssmann, 79 (1979). Yet another golden-yellow selection making an attractive columnar tree with lateral branches outspread and rather dense.

-- **'Wallis Gold'**. A very good golden-yellow selected on the Wallis Nursery, from a consignment of 'Stewartii' plants imported from Holland in 1958.

-- **'Wansdyke Miniature'**. Welch, 162 (1979); *Picea abies* 'Wansdyke Miniature' Den Ouden and Boom, Man. 251. 1965. This plant, although mistakenly reported as a *Picea abies* form by Dr. Boom, was well described by him as 'A compact and regular form, just a miniature of a normal *Chamaecyparis lawsoniana*; stem thick and strong for the size of the plant; foliage bluish-green. Found by H. J. Welch, Wansdyke Nursery, Devizes, England'. It is unremarkable as a small plant but in time, because of its thick trunk and columnar shape it develops into a veritable scale model of a mature Lawson cypress and so would be an excellent plant for a scenic railway or model village.

-- **'Westermannii'**. (Van Hulle) Beissner; 75 (1891); *Ch. boursieri westermannii* van Hulle in Tijdschr. Boomteeltkunde 113 (1890). Pyramidal, moderately dense; branches spreading; branchlets spreading or somewhat pendent; sprays golden- or light-yellow when young; leaves on the tips of branchlets light yellow, becoming yellowish-green towards the base. Raised by Jac. Jurrissen & Son, Naarden, Netherlands, ± 1880. For the first time exhibited at Gent, Belgium in 1888. Still listed but not always supplied true.

-- **'White Spot'**. A mutation from a very old cultivar 'Monumentalis Nova'. A tall-growing tree with the new foliage white or white-speckled, but this colouration soon disappears. It was introduced by the D. T. Poulsen Nursery, Denmark (1943).

-- 'Winston Churchill'. See 'President Roosevelt'.

-- **'Wisselii'**, van der Wissel ex Witte in Sempervirens **22:** 560 (1893); *Ch. lawsoniana* f. *fragrans conica* Beissner. (1891) Mitchell 77 (1972) has given us a description that would be difficult to improve upon. "Columnar, very narrow to fairly broad, with pointed long spire or spires and level branches turning up to small turret-like spires all round the crown. Young trees often rather gaunt. From distance, deep blackish blue, turning crimson with dense masses of male flowers in April, and brown as these shed pollen but remain on the tree. Small branches wiry; shoots soon copperbrown becoming dark greenish-brown. Branchlets dense, spread at all angles round vertical shoots; congested, partly fasciated, with ultimate branchlets slender, ± alternately long and short, in flattened sprays; light grey-green dusted bluish-white outside, dark yellow-green inside. Fairly common. A deceptive tree, looking as if it would always be very slow-growing, as it is for some years, but then it becomes enormously vigorous, in height, and especially in girth."

Raised by van der Wissel, nurseryman at Epe, Netherlands in 1888, it is indistinguishable from *C. lawsoniana fragrans conica* of Beissner which was raised by Buetner at the forestry station at Tharandt, Germany, ± 1885.

-- **'Wisselii Nana'**. Groot. in Hornb. Dw. Con. Ed. 2, 52 (1939) is a charming little extremely slow-growing cultivariant from the well known tall cultivar 'Wisselii'. Cuttings of 'Wisselii' taken low down on the tree from weak shoots should give rise to plants of 'Wisselii' Nana. Such are, however, unstable, and tend to shoot up in time.

-- **'Witzeliana'**. Spaeth, Cat. (1934); *Ch. lawsoniana erecta* Witzel ex den Ouden, 105 (1949). Columnar in habit like *Juniperus communis* 'Hibernica', 3–5 m tall; branches crowded, ascending, covered by their foliage; branches erect; leaves small, dark green. According to Spaeth very hardy, but not in Boskoop.

-- **'Yellow Transparent'**. Den Ouden & Boom, 98 (1965). A sport of 'Fletcher' differing in the colour of the foliage which is transparently yellow in summer, brownish in winter; also the size is somewhat smaller. Originated in the nursery of H. van 't Hof at Boskoop, Netherlands, ± 1955.

-- *Ch.* × *nidifera*. See *Ch. nootkatensis* 'Nidifera'.

Ch. nootkatensis (D. Don) Spach, 333 (1842); *Cupressus nootkatensis* D. Don in Lamb. 18 (1824); *Ch. nutkaensis* Lindl. & Gord. in Journ. Hort. Soc. Lond. **5:** 207 (1850); *Thujopsis borealis* Fisch. ex Carr. 113 (1855). Alaska Cedar; Nootka False Cypress.

Size and habit Narrowly conical tree, 15–30 m, usually open but with a good, usually narrow, crown.

Trunk Usually single, stout. *Bark*: Orange- to grey-brown, irregularly longitudinally or spirally fissured, separating into large, thin scales.

Branches Ascending or spreading horizontally.

Branchlets Flattened or quadrangular, arranged in vertical planes on the upper part of the tree, usually more or less pendulous elsewhere.

Winter buds Hidden.

Leaves Scale-like, in 4 ranks, closely appressed, spreading on vigorous shoots, dull, dark grey-green with pale margins, abruptly pointed, keeled or rounded on the back and usually not glandular. They emit a heavy odour when bruised, which Mitchell says is like ox-eye daisies.

Mature cones Stalked, sub-globose, maturing the 2nd year, about 1 cm diam, yellowish-green, ripening to reddish-brown, bloomy. *Scales* 4–6 with a erect pointed boss (umbo) and often with resinous glands.

Seeds 2–4 under each scale, broadly winged on both sides; the wings notched above.

Habitat Pacific Coast region of N. Amerika, occurring in the Cascade Mountains of N. Oregon, Washington, British Columbia and Alaska to Prince William Sound. Discovered at Nootka Sound by Menzies in 1793. Introduced into Europe through the Botanic Garden, St. Petersburg, first into France in 1851. Zone 4. (See Illus. Nos. 117)

Cultivation It is not free-fruiting so is not usually raised from seed but from cuttings or by grafting, using *Cupressus* or *Platycladus orientalis* as understock, the latter being recommended for cold areas, since it provides a hardier root system. It does not 'take' on *Chamaecyparis*. It is very tolerant of soil and growing conditions, but it thrives best in the moist parts of low-lying land and along streams.

This species is more remarkable for its hardiness than its beauty, the very dull colour of the foliage putting it at a great disadvantage compared with the Lawson cypress. But as one of the parents, it has contributed the former valuable quality to the bi-generic hybrid × *Cupressecyparis leylandii*.

The facility with which this cross seems to occur suggests an affinity of this species with the genus *Cupressus*. (See Illus. Nos. 117, 119)

-- **'Aurea'**. (Beissn.) Beissn. 538 (1909); *Ch. nut-*

Figure 116. *Chamaecyparis nootkatensis* 'Pendula'. Similar forms turn up in seedbeds, so several clones are in cultivation. (10)

Figure 117. *Chamaecyparis nootkatensis*. Nootka cypress. (10)

Figure 118. *Chamaecyparis nootkatensis* 'Nidifera'. Note the foliage difference from *Chamaecyparis lawsoniana* 'Nidiformis'.

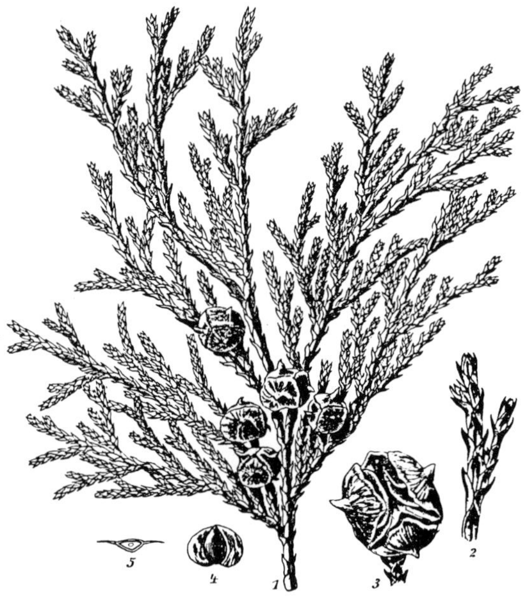

Figure 119. *Chamaecyparis nootkatensis*. Detail of foliage and cones. (2)

kaensis f. *aurea* Beissn. 82 (1891); *Ch. nootkatensis* var. *lutea* (Webst.) Rehd. 730 (1914); *Cupressus nootkatensis* var. *lutea* Webster 44 (1896). Pyramidal, moderately growing; sprays and leaves yellow when young, becoming yellowish-green with age.

-- '**Aureovariegata**'. (Young) Dall. & Jacks. 239 (1948); *Thujopsis borealis aureovariegata* Young, Cat. **1875:** 16; *Cupressus nutkaensis aurea variegata* Young ex Gord. 425 (1875); Variegated with a good portion of the lesser branchlets of a light bronzy-yellow colour, much brighter in summer. Raised by Maurice Young. Milford, England.

--'**Compacta**'. K. Koch, Dendrol. **2,2:** 166 (1873); f. *compacta* Rehd. Bibl. 51. (1949); *C. nutkaensis compacta viridis* Schelle, Wint. Nadel. 306 (1909). This forms a dense rounded bush to about 2 m high with typical foliage. It is very hardy, but not particularly outstanding as a garden plant.

-- **'Compacta Glauca'** Welch, 120 (1966); 'Glauca Compacta' Bergman in Pl. and Gar. **21** (1): 312 (1965); *C. nutkaensis compacta glauca* Schelle, Wint. Nadel. 306 (1909). Similar to 'Compacta' except that the leaves are glaucous. Mrs. Bergman remarks that semi-juvenile foliage is frequently present on the inner growths.

-- **'Glauca'**. (Mueller) Schn. 167 (1913); *Ch. nutkaensis* var. *glauca* Mueller in Walp., Ann. Bot. Syst. **5:** 796 (1858). Pyramidal; leaves very glaucous. Common in cultivation.

-- **'Jubilee'**. Van de Laar in Dendroflora **20:** 68 (1983). A narrowly upright selection of 'Pendula' originating at the Bock Nurseries, Canada.

-- 'Lutea'. See 'Aurea'.

-- 'Nana'. A collective name loosely used by Silva Tarouca [167 (1913)] for the dwarf cultivars of this species.

-- **'Nidifera'**. *Thujopsis nidifera* Nich. Gar. and For. **2:** 568 (1889); *Thuja borealis* var. *nidifera* Mast. Gar. Chron. **7:** 108 (1890); *Cup. nutkaensis nidiformis* Hort. in Beissn. Handb. 83 (1891); NON *Chamaecyparis lawsoniana nidiformis* Beissn. Handb. ed 2, 552. 1909; NON *Cupressus nidifera* Hornb. Rep. Conif. Conf. 72 (1932).

The story of how the correct identity of this plant has been established after nearly a century of confusion and of how it has again been brought into cultivation will be found in Welch [163 (1979)]. It makes a large, spreading bush or tree several yards across, with foliage typical of the species and quite distinct from any form of *Ch. lawsoniana*. (See Illus. No. 118)

-- **'Pendula'**. (Nich.) Schn. 167 (1913); *Ch. nutkaensis* var. *pendula* Nich. 304 (1884), Pendulous; branches widely borne, horizontal, curved upward near the ends; branchlets pendulous, long, held almost perpendicularly in two rows; leaves green. Specimens from seed with varying degrees of pendulous final growth are not infrequuent, and certainly more than one clone is in cultivation under this name. They vary considerably in landscape value in maturity, so only proven clones should be propagated. (See Illus. No. 116)

-- **'Tatra'**. Machala in Dendroflora **10:** 41 (1973). A form raised by Fr. Machala, Zehusice in Czechoslovakia that makes a robust, dense, broadly conical tree with numerous upright-growing main branches. The foliage is bluish-grey, especially during the summer: a better colour than 'Glauca' and finer and softer to the touch.

-- **'Variegata'**. K. Koch, 166 (1873); *Ch. nootkatensis* var. *argenteovariegata* (Veitch) Dallimore & Jackson, 239 (1948); *Cupressus nootkatensis argenteovariegata* Veitch, 23 (1881); *Ch. nutkaensis albovariegata* Nicholson, 221 (1901). Foliage variegated with white. Not very stable.

Ch. obtusa (S. & Z.) Endlicher, 63 (1847); *Retinospora obtusa* S. & Z. 38 (1844). Hinoki False

Figure 120. *Chamaecyparis obtusa.* Detail of foliage and cones. (2)

Cypress. This species in its typical form is usually only found in botanical collections, garden planting being entirely limited to the many cultivars that are available. These vary in colour and habit, and in size they cover the whole range down to plants that are so diminutive that a mature specimen will be no larger than a football. This variability limits the value of the following species description, but the (usually) bright green foliage and the bluntly-rounded tips of the leaves (to which the species owes its name) is recognisable in all cultivars.

Size and habit A broadly conical tree with a rounded crown; in Japan to 50 m, in cultivation to 25–35 m.

Trunk Stout, cylindrical, 5–6 m (average 3 m) in girth. *Bark*: Bright reddish brown, rather smooth but slightly furrowed longitudinally and peeling off in thin strips.

Branches Horizontal, often up-swept.

Branchlets Shoots thin, foliage densely arranged in flat planes. New shoots sometimes project noticeably.

Winter buds Hidden.

Leaves Shiny, bright green; scale-like, obtuse (i.e. blunt-ended, not pointed); decussate, in dissimilar pairs: the lateral pair 2 mm long with thick, incurved tips and (below) X-shaped white, waxy markings along the edge of the leaves; the facial pair smaller 1.5 mm long, obtuse or bluntly rounded at the top; glandular; scented.

Mature cones Short-stalked, globose, 8–12 mm diam. green, ripening to orange-brown. *Scales*: 8 (rarely 10), depressed on the back and with a small mucro on each scale.

Seeds 2 (occasionally up to 5) on each scale, convex or nearly triangular on both margins, often with two narrow wings.

Habitat Japan and Taiwan. Introduced by Dr. von Siebold into the Netherlands (Leiden) and in 1861 by J. G. Veitch into Britain.

Cultivation Growth is very slow, especially so in the dwarf forms. It is sufficiently tolerant of environment to give us good garden plants, but it dislikes a dry climate and dryness at the roots. Zone 4. (Illus. No. 120)

The species is variable from seed and numerous selections with more or less monstrous growth or much reduced growth-rate have been reported from Japan, although few of these are now in western culture. Such aberrants seldom set seed, but before the second world war a number of seedlings were raised in the U.K. from seed collected from a plant of 'Nana Gracilis', itself already a very slow-growing form, and from these were selected some extremely diminutive forms (often referred to as the 'tennis-ball' cultivars). In 1966, from another such freak seed-crop, Joel W. Spingarn of Baldwin, Long Island, New York, raised a batch of very slow-growing plants from amongst which he selected and named nine cultivars, including several golden forms, all of which he claims to be hardier than the original forms. Since then, other crops of seeds have resulted in seedlings galore. Although each seedling may be interesting in itself, no justification exists for introducing additional cultivars and proliferating names in this group. We already have more than enough!

-- **'Albospica'**. Nich. 304 (1884); as *albo-spicata* Beissner 555 (1909); *Retinospora obtusa albo spica*. R. Sm. ed. 3, 37 (1867). A compact form with the young foliage during the Sring and Summer months pure white, turning in Autumn to bright green.

-- **'Argentea'**. (Fortune ex Gordon) Henk. & Hochst. 253 (1865); *Retinospora obtusa argentea* Fortune ex Gordon 94 (1862); *Ch. obtusa argenteovariegata* Schelle 302 (1909). Branchlets and leaves partly silvery white and mixed evenly with green throughout the plant. Introduced by R. Fortune from Japan where it is commonly cultivated, to the Standish Nurseries, Bagshot, England, ca. 1860.

-- **'Aurea'**. (Fortune ex Gord.) Henk. & Hochst. 253 (1865); *Retinospora obtusa aurea*. Fort. ex Gord. 93 (1862). Pyramidal; sprays and leaves rich golden regularly interspersed with green throughout the plant. Introduced ± 1860 by R. Fortune to the Standish Nurseries, Bagshot, England from Japan as a cultivated plant. It is seldom planted today, since the all-golden 'Crippsii' is a great improvement.

-- **'Aurora'**. Koster ex Van de Laar, Dendroflora **21:** 81 (1984). A sport from 'Nana Gracilis', similar in habit and eventual size but with yellow (fading to yellow-green) foliage, intermediate in colour between the mother plant and 'Nana Aurea'. It was discovered on the Koster and Sons Nursery, Boskoop, Holland, before 1940 and was introduced to the trade by J. C. van Klaveren, also of Boskoop.

-- **'Barkenny'**. A seedling raised by Joe Reis of Long Island, New York. It forms a very slow-growing, squatly conical bun, with very tight foliage.

-- **'Bassett'**, Welch, 131 (1966). This is one of the Red Lodge seedlings, less well known than some. It is not unlike 'Juniperoides' but forms a much taller plant with ascending branches and the foliage is a dark green.

-- **'Bess'**. Another of Joe Reis's seedlings, named for his wife Elizabeth, very slowly forms a narrow column of dense, dark green foliage reaching about 50–60 cm in 12 years. Quite a distinctive plant.

-- var. ***breviramea***. (Maximowicz) Regel, Russk. Dendr. ed. 2: 26 (1883); *Ch. breviramea* Maxim. in Mél. biolog. **6:** 25 (1866). A rather tall tree; crown narrow: branches horizontal, short; sprays and side-branchlets compacted together; leaves thick, glossy green en both sides, without the white markings on the under side; cones rather small. According to Regel this variety is intermediate between '*Filicoides*' and the typical form of the species. It grows together with *C. obtusa* in Northern Kiushu and is commonly cultivated in Japan.

Figure 121. *Chamaecyparis obtusa* 'Minima'. Two mature plants in the open ground at the Royal Botanic Garden, Edinburgh.

-- '**Caespitosa**'. Rogers ex Hornb. Dw. Con. 44. (1923). This is another of the Red Lodge introductions, one of the smallest. It forms a tiny, dense bun of very blue-green, cup-like sprays with tiny, tightly appressed, scale-like leaves. It is one of the smallest conifers in existence. Not unlike it for size is '**Minima**' *Ch. obtusa* var. *tetragona minima* Hornb. Dw. Con. 43 (1923), corrected to var. *minima* in ed. 2, 69 (1939). In this variety the foliage is of a very light green, the cup-like cupressoid sprays of 'Caespitosa' are absent and in their place there is a close mass of ascending branches and branchlets that are distinctly, tetragonal (i.e., four-sided) in cross-section. It is, if anything, even slower growing and smaller than 'Caespitosa', and usually forms a more dense, hemispherical cushion, a twenty-year-old plant being still but a few inches high. (See Illus. No. 121)

-- '**Chabo-yadori**'. Hillier, Man. 473 (1971). When young this is an interesting little bush because it carries a mixture of juvenile and adult foliage, but it tends to a loose, untidy habit and with age loses its appeal. It is sometimes distributed as 'Ericoides'. (See Illus. No. 124)

-- '**Chilworth**'. Welch, 131 (1966). This is another of the Red Lodge Nurs seedlings. It has not become so well known as some of the other forms but is sufficiently distinctive to justify its becoming better known than it is. It forms an upright and quite a tree-like little plant with light green, tiny leaf sprays which are recurved. It is a form which should appeal to Bonsai addicts. It takes on a dull brownish tinge in winter.

-- '**Chimaanihiba**'. Welch, 172 (1979): 'Pygmaea Densa' Hillier, Man. 474 (1971). This might be described as a slower-growing form of 'Pygmaea'. Other spellings are found.

-- 'Columnaris' Hort. Amer. (in part.) See 'Green Diamond'.

-- '**Compacta**'. (Gordon) Hartweg & Ruempler, 660 (1875); *Retinospora obtusa compacta* Gordon, 368 (1875). Conical, compact, rather broad; branches spreading; branchlets rather elongated; leaves green. An old variety, descriptively named, but little planted today.

-- '**Confucious**'. Originated as a mutation in the Duncan and Davies Nurs, New Plymouth, N.Z. on a plant of 'Nana Aurea'. Differing in a slightly larger, looser and faster growth; with a very bright variegation.

-- '**Contorta**'. Den Ouden. Conif. 120 (1949). This forms a slow-growing, conical plant, eventually to 1.5 m, with dense foliage which is twisted and contorted in what, to my eye, is not a particularly attractive manner. The colour is deep green, bronzing in winter. Raised from seed of '*Nana Gracilis*' by H. den Ouden & Son, Boskoop, Netherlands and introduced in 1945. Any tendency to throw out coarse, uncharacteristic growth should at once be checked.

-- '**Coralliformis**'. Beissner Ed. 2, 562 (1909); Hornibrook Ed. 2, 64 (1939); 'Tsatsumi' Slavin, Rep. Conif. Conf. 98 (1932). This is usually seen as a low, rounded bush in which the normal sprays are entirely absent and instead the foliage is thread-like. The branches and branchlets are curled and twisted in a way that is said to be suggestive of coral and this constitutes the main attraction of the plant. It should not be confused with 'Lycopodiodes' as its foliage has no trace of the coarse, bloated appearance of that variety, nor with *C. lawsoniana* 'Lycopodioides', as it lacks the pointed, free standing leaf tips of that variety, although the twisted growth is very similar. From 'Torulosa' it differs in not developing the fasciations that characterise that variety. As a young plant it is rather weedy looking and insignificant but it fills out in time, especially if any lanky growth is shortened back regularly. It makes a better plant if tied up when young until it has formed a leader and it will in time reach 2 m or more in height. (See Illus. Nos. 97, 127)

-- '**Crippsii**'. Rehder in Möller's Deutsche Gärtnerz. **16:** 300 (1901); *Cupressus obtusa aurea crippsii* Hort. Pyramidal, rather broad; branches spreading; branchlets broadly frond-like, tops decurving; leaves rich golden yellow, changing into green within the plant, yellowish-green on underside of branchlets. Raised and distributed by Thomas Cripps & Son. Tunbridge Wells, England. Possibly the most beautiful of all the golden conifers.

-- **'Dainty Doll'**. Spingarn ex Welch, 167 (1979). A conical but round-topped plant with an irregular, upright habit, with sprays of foliage arching outwards. Compact, but it may in time develop to the size of 'Nana Gracilis'. It has good Bonsai potential. An account of the origination of this and the other seedlings raised by Joel W. Spingarn will be found in my earlier book (Manual of Dwarf Conifers. 166 (1979). (See Illus No. 122)

-- 'Densa' Hornb. 40 (1923), as var. *nana densa*. Since Hornibrook changed his mind about this between his two editions, Hillier [21 (1964)] disagreed, and neither plant can now be positively identified, this name can well be discarded.

-- **'Drath'**. Draht ex Grootendorst, Dendroflora **15–16:** 53 (1979). A conical plant with dark green foliage somewhat monstrous, less so than 'Lycopodioides.' This may well be the 'Troubetzkoyana' of Hornibrook [ed. 2, 65 (1939)]. The spelling 'Draht' is phonetic..

-- **'Elf'**. Spingarn ex Welch, 167 (1979). A dense, hemi-spherical little bush with ascending main branches and very dense, dark-green foliage in forward-pointing sprays often held in a vertical plane. Leaves are slightly glaucous on the underside. Annual growth: 10 cm.

-- **'Erika'**. This very slow-growing little bush with sage-green juvenile foliage and a twisted, 'wiry' look, arose as a mutation on an old plant of 'Nana Gracilis'. It was raised by Don Hatch, Chantrey Nursery, Honiton, Devon. U.K. and the name was registered by him in 1983.

The mother plant after ten years was 60 cm high. It has never developed any adult foliage. A form with fixed juvenile foliage was given the name 'Ericoides' in 1862 but is long since

Figure 122. *Chamaecyparis obtusa* 'Dainty Doll'. 25 × 20 cm. One of the J. W. Spingarn introductions.

Figure 123. *Chamaecyparis obtusa* 'Nana Lutea'. The young growth is a light creamy-yellow.

Figure 124. *Chamaecyparis obtusa* 'Chabo-yadori'. Plants develop adult foliage increasingly with age.

Figure 125. *Chamaecyparis obtusa* 'Kosteri'. A plant that would have been stem-trained when young, to form a central stem.

Figure 126. *Chamaecyparis obtusa* 'Mariesti'. Seldom exceeds 1 m. Several mutations have been propagated but are doubtfully stable.

Figure 127. *Chamaecyparis obtusa* 'Coralliformis'. The threadlike foliage is distinctive.

Figure 128. *Chamaecyparis obtusa* 'Nana Gracilis'. The well-known favourite that will reach 3 m. 'Nana' is an altogether smaller plant.

Figure 129. *Chamaecyparis obtusa* 'Lycopodioides'. The clone 'Lycopodioides Aurea' differs only in colour.

lost to cultivation. Mr. Hatch's action in raising a new cultivar name for the fresh mutation should be followed whenever such a situation arises. 'Re-typification' is in every way undesirable.

-- **'Fernspray Gold'**. Harrison, Orn. Conif. 55 (1975); *Cupressus obtusa* var. *filifera aurea* Kent in Veitch, 221 (1900). Although the identification of this attractive cultivar with the older name cannot be established, the description tallies. It is probably Kent's plant, long lost to cultivation in Europe but preserved in New Zealand, where for many years it was distributed by the Duncan and Davies Nursery, New Plymouth, N.Z. mistakenly under the name 'Tetragona Aurea' until the present name was established in association with the Nomenclature Committee of the New Zealand Nurserymen's Association.

-- **'Filicoides'**. Hort. in Veitch 241 (1875); *Retinospora filicoides* R. Sm. ed. 3, 36 (1867); *R. Nobleanum* Sieb. Nurs. Cat. 1868/69. Sent from Japan by Dr. von Siebold to Leiden, ± 1860 and by J. G. Veitch to Britain in 1861. This is a fairly strong-growing cultivar usually seen as a small and rather open tree, but it can (by systematic shortening of all strong shoots) be grown as an attractive bush. It is characterised by the development of side shoots from the axial buds of the lateral leaves only, so has flat, fernlike sprays. This peculiarity of foliage is quite distinctive, so 'Filicoides' is unlikely to be confused, except with very similar foliage on *Thuja occidentalis* 'Spiralis'. This latter can at once be distinguished by the acrid smell of the bruised foliage characteristic of that species. **'Compact Fernspray'** Corley, Gar. Chron. **156** (10). 247 (1964); 'Filicoides Compacta' Hillier, Dw. Con. 20. (1964); 'Pygmy Fernspray' Welch, Dw. Con. 133 (1966); 'Filicoid Dwarf' Den Ouden and Boom, Man. 104. (1965). This is a diminutive form of the above. It is a very attractive form with foliage similar but smaller in all its parts. It is difficult to propagate from cuttings. (See Illus. No. 97)

-- 'Filicoides Compacta'. See 'Compact Fernspray'.

-- **'Fontana'**. Verboom in Dendroflora **8:** 67

(1981). A mutation from 'Nana Gracilis' with a much looser habit and bright-green in colour.

-- var. ***formosana***. (Hayata) Rehder ex Bailey, Stand. Cycl. Hort. **2:** 731 (1914); *C. taiwanensis* Matsum, et Suzuki, Sylva **4:** 57 (1933). A tree 30–40 m, with branchlets more slender and thinner than in the type: the seeds also are smaller. The leaves are lustrous pale green on both sides, somewhat pointed and tightly appressed. Not to be confused with *C. formosensis*, although often associated with that species 'in the wild'.

-- 'Gerda von Gimborn-Dietz'. Changed to 'Gimborn's Beauty'.

-- **'Gimborn's Beauty'**. A strong-growing, fairly open form carrying a mixture of very glaucous juvenile foliage and broad sprays of adult foliage. Any long, stringly main shoots that develop should be removed.

-- **'Gnome'**. Springarn ex Welch, 167 (1979). Another very diminutive, bun-forming selection raised from seed by Joel Spingarn.

-- **'Gold Drop'**. A yellowish-green form, somewhat globular or mounding when young. A semi-dwarf, listed in U.S.A.

-- **'Golden Fairy'**. Spingarn ex Welch, 167 (1979). Forms a dense conical plant, broader than 'Leprechaun' and with the bright golden-yellow sprays rather more open. 40 cm high × 35 cm wide after 15 years. A very handsome little plant.

-- **'Golden Filament'**. Spingarn ex Welch, 167 (1979). Makes a loose-growing bush with long main branches and filiform foliage in open, lace-like sprays – a mixture of green and gold. Quite distinct from any other *C. obtusa*.

-- **'Golden Nymph'**. Spingarn ex Welch, 167 (1979). A compact, squatly conical form, with golden foliage and a green cast. 'Golden Fairy' is much more showy for colour. 1 m high, 55 cm across in 15 years.

-- **'Golden Sprite'**. Spingarn ex Welch, 167 (1979). This, the most distinctive of Joel Spingarn's selections forms at first a dense globose to hemispherical bush to 12 cm, eventually a plant of somewhat irregular outline with foliage in dense, flattened sprays arching outwards from the centre, often down-cupped. The foliage is golden yellow at the surface and greenish inside the plant. 30 mm high, 45 mm wide at 15 years. The only golden 'tennis-ball' conifer.

-- **'Goldilocks'**. Grootendorst in Dendroflora **15–16:** 53 (1979). A sport from the cultivar 'Nana Lutea' that is at first very slow-growing but soon produces a leading shoot and forms a conical plant, smaller than 'Crippsii' but with the same clear golden-yellow colour. It was introduced to the trade by the Grootendorst and Sons Nursery, Boskoop, Holland.

-- **'Gold Spire'**. Konijn ex den Ouden & Boom, 104 (1965). Narrow-conical and dense, to 5 m; sprays irregularly arranged, with lemon-yellow tips both in summer and in winter. A very conspicuous form found in the Blijdenstein Pinetum, Hilversum, Netherlands and introduced into cultivation by the firm L. Konijn & Co, Reeuwijk near Boskoop, 1963.

-- **'Gracilis'** Rehd. 216 (1923); under *Retinospora* Sieb., Cat. (1868), name only. Pyramidal, 3–5 m, narrow, compact; branches spreading; branchlets irregularly disposed; sprays shell-shaped; leaves short, densely set, dark green, somewhat glossy light on under side of branchlets. Introduced from Japan by Dr. von Siebold (Leiden), in ± 1862. Common in cultivation.

-- **'Gracilis Aurea'**. (Gordon) Nich. 304 (1884); under *Retinospora* Veitch ex Gordon 428 (1875) Pyramidal, 2–3 m, branches spreading; branchlets elongated, broadly fan-shaped, tops nodding, lowest leaves golden yellow at the base, gradually changing into green, those of the under side yellowish white. Raised by James Veitch & Son, Coombewood, Surrey, England. Uncommon in cultivation.

-- **'Graciosa'**. Konijn ex den Ouden & Boom, 105 (1965). *Ch. obtusa 'Loenik'* Hort. A sport of

'*Nana Gracilis*', differing in the more robust and loose habit (to 3 m); foliage brighter green. Originated in the nursery of L. Konijn & Co, Reeuwijk near Boskoop, Netherlands, ± 1935.

-- **'Green Diamond'**. Konijn ex van de Laar in Dendroflora **13–14:** 75 (1977). An elegant, upright-growing tree with wide-spreading branches and pendent growth tips. Dark green foliage with bluish-white markings beneath, conspicuously glossy during the winter months and with red-brown shoots. It was received from the USA by the Konijn Nurs, Ederveen, Holland, about 1965 under the illegitimate name 'Columnaris'.

-- **'Hage'**. Den Ouden 123 (1949). Dwarf, pyramidal, dense, compact, slow-growing to 1 m; branches spreading: branchlets crowded, twisted, tips slightly fimbriate; leaves very small, bright green. Introduced by Wm. Hage and Co. Nursery, Boskoop, Holland about 1928. Not always supplied true, being difficult to distinguish from 'Nana' and other cultivars as a young plant.

-- **'Intermedia'**. Den Ouden, Naaml. 17 (1937). This is one of the Red Lodge forms, which as its name indicates is a pygmy form intermediate between 'Nana' and the very minute 'tennis-ball' cypresses. It is, in my opinion, the smallest form that can be considered practicable for the open garden, except in sheltered districts. The foliage is a very dark green.

-- **'Junior'**. A seedling raised by Joe Reis, Long Island, NY (U.S.A.) which makes an extremely tight, dark-green cushion, only reaching 60 cm across in 20 years. It is sometimes found listed as 'J. R.' or as 'Hinoki Junior'.

-- **'Juniperoides'**. Hornibrook 44 (1923). This is another very dwarf globose form from Red Lodge with dark green foliage in open and rather wiry looking, fan-shaped, decurving sprays.

Not to be confused therewith is another form. **'Juniperoides Compacta'** Hornb. Dw. Con. ed. 2, 68 (1939) as the name would suggest, is generally similar in appearance but of more compact growth. The foliage is a yellow-green and the tiny, confused leaf sprays are down-cupped, with the tips drooping.

-- **'Kojolcohiba'**. Welch [174 (1979)] expresses the opinion that this attractive and vigorous clone received from New Zealand is, in fact, stock of the well-known 'Tetragona Aurea' that has escaped some virus-like infection that has enervated that old cultivar in Europe, and he suggests that the two names should be retained so as to maintain this distinction, since the slow-growing form still has value as a golden 'Dwarf conifer'.

-- **'Kosteri'**. Hornibrook 62 (1939), as var. *Nana Kosteri*, is one of the most attractive forms. In habit and foliage it might be regarded as intermediate between 'Nana Gracilis' and 'Pygmaea', but it is quite distinctive because every spray of the plant is twisted, one lateral turning down and the lateral opposite turning up. This gives the plant a most distinctive appearance and it is quite characteristic of this variety, which with its almost glossy, mid-green foliage is therefore one of the easiest to recognise. Left to itself it forms a rather sprawling bush not unlike 'Pygmaea' but I consider that it is best grown with a central stem trained up as a leader. If this be done the horizontal branches spread out in layers like a wedding cake and the attractiveness of the foliage is then seen at its best. The layer effect referred to can be accentuated by a little discreet thinning out of the foliage. **'Oregon Crested'** is a little-known sport of 'Kosteri' stronger in growth but with similar foliage. (See Illus. Nos. 97, 125)

-- **'Leprechaun'**. Spingarn ex Welch, 167 (1979). A selected seedling that forms a squat, conical bush to 50 cm, rather wider than high, with foliage of somewhat tufted appearance, due to areas of denser growth, 45 cm high and 35 cm wide after 15 years.

-- **'Little Ann'** is another very slow-growing seedling raised by Joe Reis. It makes a dwarf, narrowly conical bush with fine, dark-green foliage.

-- **'Little Markey'**. This is yet another of Joe Reis's seedlings. It makes an almost fastigiate plant 1 m high, with golden-yellow foliage in tiny, cupped sprays. This combination of habit and colour is very unusual, making this new form a great acquisition. 'Meroke' is probably a synonym.

-- 'Loenik'. An incorrect name for 'Graciosa'.

-- **'Lycopodioides'**. (Gordon) Carr. 132 (1867); *Retinospora lycopodioides* Standish ex Gordon, 92 (1862). This plant, more curious than beautiful, forms a globose shrub when young and eventually becomes a tall, open shrub to 2 m. The main branches ascend and are rather few and irregular. The branchlets are thick and often cockscomb-like at the tips; the leaves are irregular, dense and set all round the branches. The foliage is dark green and the monstrous growth gives the whole plant a curiously bloated appearance. The variety tends to form a rather open and gaunt tree but this can be checked by pruning back strong growth when young. It was introduced in 1861 from Japan by Dr. von Siebold (Leiden) and sent by R. Fortune to the Standish-nurseries, Bagshot, England. (See Illus. No. 129)

-- **'Lycopodioides Aurea'**. Beissn. in Mitt. d.d.d. Ges. **19:** 99 (1906). This is a much better garden plant than the preceding. It is similar but slower-growing and with butter-yellow growths. Introduced from Yokohama by J. Boehmer into Germany in ± 1890.

Several forms originating in Japan and with more or less monstrous foliage were introduced into Western cultivation during the last century and their vernacular names were recorded. [Beissn. Handb. Ed. 2, 561–2. 1909.] but are not traceable in Europe today. Unless such can now be positively identified in Japanese cultivation, new names should be found for any such forms now introduced.

-- **'Lynn's Golden'**. Dilatush, A.C.S. Bulletin **3** (4): 119 (1986). A golden-yellow plant most of the year but turning orange during the winter months, raised from seed by H. Weissenburg, Baldwin, NY.

-- **'Mariesii'**. Kent in Veitch, Man. ed. 2, 222 (1900); 'Mariesii' Hort. Angl. ex Beissner ed. 2, 561 (1909); 'Nana Albovariegata' Beissner 96 (1891) and Hornibrook 41 (1923); 'Nana Argentea' Hornibrook 61 ed. 2, (1939). The rightful claimant to this name is doubtful since the earliest descriptions make no mention of its being dwarf. But, following Hornibrook, it is now universally used for an attractive, slow-growing clone that makes a rounded or conical bush to about 1 m high, characterised by the leaves being yellowish-white to milky-white in summer, rather more yellowish-green in winter. The foliage is very thin and the bush tends to be open and sparsely covered unless checked by the shortening of the growing tips. On its own roots 'Mariesii' grows very slowly, but when grafted the vigour, colouring and even habit can change, according to the understock used. This may be the origin of the conflicting descriptions, but unless a slower-growing clone now being grown in America as 'White Dwarf' or 'Dwarf White Tip' is a new selection, the name **'Nana Argentea'** suggested by Hornibrook should be used. (See Illus No. 126)

-- **'Maries Gold'**. A mutation from 'Mariesii' with a yellow instead of white variegation.

-- 'Meroke'. See 'Little Markey''.

-- **'Minima'**. See above under 'Caespitosa'. (See Illus. No. 121)

-- **'Nana'** 'Carrière, Traité'. ed. 2, 131 (1867). This is a very old cultivar of extremely slow growth, as a result of which stronger-growing forms, particularly 'Nana Compacta' or 'Nana Gracilis' are sometimes offered for sale in its place. 'Nana' forms a very dense, flat-topped bush with a mainly horizontal branch system and foliage in densely packed tiers in neat, rounded, often cupped sprays of dark, almost black-green. A very old specimen will be less than 1 m high by as much or less across. Introduced from Japan by Dr, von Siebold (Leiden) between 1860 and 1862. (See Illus. No. 97)

-- **'Nana Aurea'**. Carrière 131 (1867); under *Retinospora* Veitch ex Gordon 428 (1875). This is a more vigorous plant than 'Nana', reaching to 2 m with age. The fan-shaped foliage sprays are golden-yellow on the outside of the bush and yellowish-green away from the light. Introduced from Japan by J. G. Veitch into Britain. Common in cultivation.

-- **'Nana Compacta'**. Krüssmann, 101 (1960); under *Retinospora* Barron, Nurs. Cat. 7 (1875). This form is sometimes distributed in Britain as 'Nana', from which it is quite different. It is a bright green form with the leaves glaucous beneath and it forms a dense, rounded plant.

Although in colouring it is similar to 'Nana Gracilis' it is much slower and more open in growth and is without the tendency to form an upright plant which characterises that cultivar.

-- **'Nana Gracilis'**. (R. Smith) Beissn. 96 (1891); under *Retinospora* R. Smith, Plants Fir Tribe. ed. 3, 37 (1867). Despite its misleading name this is a very attractive, dense growing form reaching 3 m high in time, making a pyramidal bush with rich lustrous green, healthy looking foliage and a somewhat rugged and picturesque outline. It is too well known to need much description and should be in every garden where there is room for it. It is worth a little trouble to get a plant on its own roots. (See Illus. No. 97, 98)

-- **'Nana Lutea'**. Welch, 138 (1966). This is a very nice yellow form which arose as a sport on a plant of 'Nana Gracilis' and was introduced by Jan Spek of Boskoop, Holland. It forms a neat, compact little bush with leaf sprays about equivalent in vigour to 'Nana' but of a clear pale, golden yellow, uniform throughout the bush and well maintained throughout the year. (See Illus No. 123)

-- **'Opaal'**. Konijn ex Krüssmann 83 (1960). A selection similar in growth to 'Nana gracilis', but in colour it is light green with irregular patched of creamy-yellow foliage. It was for some years distributed by the raiser as *Ch. obtusa aureovariegata.*

-- **'Pixie'**. Spingarn ex Welch, 167 (1979). An extreme miniature, forming a flat globe with very tight foliage (even when grafted) and an annual growth of 6 mm. 75 mm high and 100 mm wide at 15 years.

-- **'Pygmaea'**. (Gordon) Henk & Hochst. 253 (1865); under *Retinospora* Gordon 94 (1862). This is one of the most desirable forms and a well grown plant always draws attention to itself. It very slowly forms a low, spreading bush always much wider than high, with the flat, fan-shaped sprays lying closely one above the other. Introduced form Japan into Britain by R. Fortune who sent it to the Royal Nurseries at bagshot in 1861. Common in cultivation, but grafted plants are apt to lose character and become globose or upright. (See Illus No. 130)

-- **'Pygmaea Aurescens'**. Hornb. ed. 2, 63 (1939). The difference between this and the last-described cultivar is that it is very much stronger-growing and will throw out strong, horizontal shoots quite unlike the fan-shaped leaf sprays of 'Pygmaea'. Also, during the greater part of the year 'Pygmaea' is quite green and 'Pygmaea Aurescens' is brownish-green; in the autumn and winter 'Pygmaea' takes on a slight bronze colour whereas 'Pygmaea Aurescens' turns a rich copper-bronze. There should be no difficulty in telling them apart. It originated as a sport of 'Pygmaea' in the nursery of the firm of K. Wezelenburg, Hazerswoude near Boskoop, Holland.

-- **'Reis' dwarf'** Welch, 172 (1979). This is a curious form. The foliage is similar to that of 'Kosteri' and for a time it forms a neat bun-shaped plant, but it will eventually throw up long, straggling shoots. Unless these are cut away it loses its original habit and becomes an upright, rather open shrub. The congested type of foliage reappears on these longer shoots, so the plant lends itself to being trained into quaint and curious shapes. It was raised by Joseph Reis of Long Is. New York.

-- **'Repens'**. Den Ouden, Conif. 123. (1949). This is another prostrate form much the same in habit as 'Pygmaea' but very much stronger and

Figure 130. *Chamaecyparis obtusa* 'Pygmaea' in a hollow tree-trunk at the Pygmy Pinetum. Devizes, U.K.

Figure 131. *Chamaecyparis obtusa* 'Repens'.

coarser in growth and always green in colour, with no trace of bronze. It originated in the nursery of Wm. Hage & Co., Boskoop, Netherlands, as a sporting branch of 'Nana Gracilis', to which it is apt to revert, and was introduced into the trade in 1929. (See Illus. No. 131)

-- **'Rigid Dwarf'**. Den Ouden and Boom 109 (1965); 'Nana Rigida' Hillier, Dw. Con. 23 (1964) and Man. 474 (1971). This is a distinct form, of dense upright growth and of a stiff and rigid appearance. The main branches sweep up to a nearly vertical position, the foliage is dark green with glaucous markings beneath.

-- 'Sanderi'. See *Thuja orientalis* 'Sanderi'.

-- **'Spiralis'**. Hornb. Dw. Con. ed. 2, 69 (1939). This is an upright-growing form with a stiff habit; the sprays are cup-shaped and curiously twisted around the smaller branchlets; a very distinct and oustanding form, but not a particularly attractive one, at least when young.

-- **'Split Rock'**. A sport that arose in the Split Rock Nurs, NJ. U.S.A. on a plant of 'Nana Gracilis'. It is very slow-growing and at first carries juvenile foliage of an intense blue, but as it ages it tends to develop adult foliage and loses character completely in time, unless this is regularly cut away.

-- **'Stoneham'**. Hillier, Dwarf Conifers. 23 (1964) (as 'Nana Stoneham'). A diminutive selection from the Red Lodge seedlings, making a very tight little plant with its branches in tiers.

-- **'Suiryuhiba'**. Hillier, Man. 474 (1971). A plant received by R. S. Corley from Japan which is similar to, or identical with, 'Coralliformis'. As is often the case with Japanese vernacular names, other spellings are found.

-- **'Tempelhof'**. Den Ouden and Boom Man. 110 (1965). This makes a compact, broadly ovoid to pyramidal bush eventually reaching 2.5 m. The leaf sprays are fan-shaped, green to yellowish-green, often with a brownish tint (similar to 'Pygmaea Aurescens'), especially during the winter.

-- 'Tetragona'. Hornibrook 42 (1923); NON Rehder in Journ. Arnold Arb. **1:** 52 (1919); *C. obtusa* f. *Barronii* Rehder in Journ. Arnold Arb. **26:** 68 (1945); *Retinospora tetragona*, (*Chamaecyparis thujaformis* pro syn) R. Smith, Pl. Fir Tribe. ed. 3, 38 (1867). Since this (green) form "neither does, nor did, exist" all names are invalid under Art. 42 of ICNCP. See 'Tetragona Aurea'.

-- **'Tetragona Aurea'**. Nich. Dict. Gard **1:** 304 (1884): *Retinospora tetragona aurea* Barron Nurs. Cat. (facing) 12 (1875). A conical, compact, slow-growing plant with branches erect; branchlets crowded, short, irregular, tips congested, sometimes cockscomb-like, secondary branchlets short, four-sided; leaves oval, pointed, closely and regularly imbricated in four rows, foliage bronzy-yellow. Introduced from Japan by Wm. Barron, Elvaston nurseries, Barrowash and by R. Smith, Worcester, England, in ± 1870. It is usually seen in gardens as a small or medium-size bush, but it would seem to respond considerably to the selection of propagating material (or perhaps, choice of rootstock in the case of grafted plants) because whilst there is, for example in the late G. L. Pilkington's garden at Grayswood Hill, Haslemere, Surrey, a plant of this variety about 6 m high, I have seen minute and obviously old and slow-growing bushlets only a few centimetres high. It is probable that virus infection plays a part in this variability. See 'Fernspray Gold' and 'Kojolcohiba'. The colour in full sunlight is a rich bronze-gold but it only needs a relatively small amount of overshading from nearby trees or buildings for this to be reduced to a poor yellowish-green. On the other hand, in *full* shade it develops a dark, rich, blue-green which is quite attractive. Doubtless it was this characteristic that gave rise to the long-held, mistaken idea that there were two colour forms, but proper attention given to the description given in the very first catalogue 'blurb' that appeared would have avoided this. In the Wm. Barron catalogue for 1875, plants are offered at £ 2–2s and 31s 6d, according to size. These must have been quite substantial sums in those days, but nevertheless the plant is recommended to customers for 'bedding purposes'. Perhaps nurserymen needed to be optimists, even in those days!

-- **'Tonia'**. Den Ouden, Conif, 123 (1949). This is a variegated sport from a plant of 'Nana Gracilis' which originated in the nursery of William Hage & Co. of Boskoop, Holland, in about 1928. Plants I have seen are much weaker in growth and with a smaller leaf spray than 'Nana Gracilis'. The variegation is uncertain and spasmodic, and apparently young plants may make several years' growth before any white patches appear. These are a clear white contrasting well with the dark green normal foliage, but I consider the plant more a horticultural curiosity than a thing of beauty. **'Snowkist'** is a sport from 'Tonia' with the same variegation but very much slower growth. It arose on the Floravista Nurs, Vancouver Island, BC, Canada.

-- **'Torulosa'**. Bergm. Pl. and Gar. **21**, 33 (1965). An upright, tangled shrub with erect leading shoots and with branches contorted and twisted, with filiform foliage not unlike 'Coralliformis' but having small fasciated tips here and there. Doubtless Mrs. Bergman accepted a name already in general use, but since the word torulose has quite a different meaning I rather fancy the original coiner of the name meant to choose 'Tortuosa'. The Code permits the correction of spelling mistakes, but scarcely a howler of this kind. Plants sometimes listed as '*torulosa nana*' seem to be merely young plants of 'Torulosa' - probably propagations from weak side-cuttings.

-- **'Tsatsumi'** Slavin, 98 (1932). A dwarf, compact tufted, broad and slow-growing form; branches spreading; branchlets tufted, stout, somewhat twisted, thread-like; leaves dark green. The Rochester Park, NY. U.S.A. received it from the Elm City Nurseries as *Cupressus tsatsumi* in 1919; it is very similar to '*Coralliformis*' and is perhaps the same.

-- **'Verdonii'**. Another of the seedlings raised on the Red Lodge Nurs., Chandlers Ford, U.K. It has a compact, upright habit. The yellow foliage tends to bronze in winter.

-- **'Wells Special'**. Mitsch, Nurs. Cat. 2 (1977). A strong growing form similar to 'Nana Gracilis', and of an attractive green colour.

Ch. pisifera (S. & Z.) Endklicher 64 (1847); as *Retinospora pisifera* S, & Z. 39 (1844) Sawara False Cypress. This is another Japanese species that is variable and has given rise to a very large number of cultivars. These, being very popular in gardens, are much more widely met with than is the species itself in its typical form. These garden forms include the usual colour and habit variations and also forms in which the juvenile type of foliage first developed by seedlings has been retained and other forms with an intermediate (Semi-juvenile) type of foliage. They do not all remain dwarf, some of them being prepared to grow to a height of 9 m or more in this country. As a tree of this size with juvenile foliage is quite attractive they have their garden value but they cannot be regarded as dwarf forms. They do, however, stand heavy pruning, so can be kept within bounds to some extent.

The following description relates to the species in its characteristic form, bearing wholly adult foliage.

Size and habit Narrow to broadly conical tree, to 50 m, with an open crown.

Trunk Often straight and cylindrical, to 5 m in girth, but in some specimens dividing quite near the ground, probably as the result of a damaged leader.

Bark: Rather smooth, dark red-brown, with fine longitudinal fissures eventually peeling off in thin strips or bundles.

Branches Horizontal, often sweeping up to form sub-leaders.

Branchlets Slender, flattened, the foliage 2-ranked and arranged in separate, dense, fern-like sprays, often somewhat pendent.

Figure 132. *Chamaecyparis pisifera.* Detail of foliage and fruit. (2)

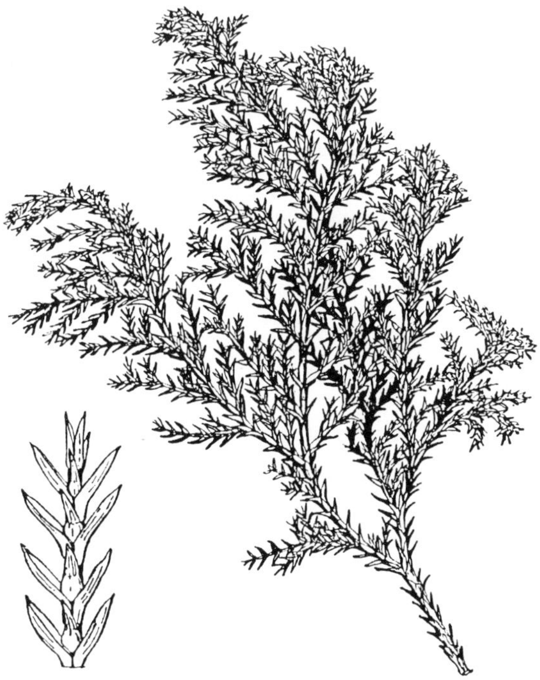

Figure 133. *Chamaecyparis pisifera* 'Plumosa'. Detail of foliage. (5)

Figure 134. *Chamaecyparis pisifera* 'Squarrosa'. Detail of foliage. (5)

Winter buds Hidden.

Leaves Scale-like, to 3 mm long, acutely pointed, decussate, in unequal pairs, the facing pair flattened, the lateral pair folded or keeled, shiny green above, white beneath, slightly spreading, obscurely glandular, heavily resin-scented. Note. Juvenile leaves are awl-like 6 mm long, decussate, blue-green above, white beneath.

Mature cones Ripening the 1st year; crowded, in clusters on short stalks, globose, 6-8 mm diam, pale green, ripening to dark brown. *Scales*: 8-12, soft-woody, upper side wrinkled, the middle depressed, with a tiny mucro at the depression.

Seeds 1-2 per scale, ovoid, bulbous on both sides, glandular; wings broad, translucent.

Habitat Japan, Hondo Island, in the prov. Mino and Shinano. Introduced form Japan by Dr. von Siebold (Leiden) between 1859 and 1861 and by J.G. Veitch into Britain in 1861.

Cultivation The species does not bear fruit freely in cultivation, but all forms propagate readily from cuttings, the cultivars bearing juvenile foliage being amongst the most willing to root of any of the conifers. Zone 4. (See Illus. Nos. 94, 132–4)

The invention of the supposed genus *Retinospora* [Sieb. and Zucc. Flora Japan **2:** 36. (1842)] to embrace a number of forms of conifer in Japanese cultivation, coupled with the habit that seemed prevalent at the time of applying the epithet *ericoides* (i.e. heath-like) to everything in sight bearing persistent juvenile foliage resulted in nomenclatural problems from which we are still not altogether free. For when, in European cultivation, the different forms went their individual ways, some developing patches of adult foliage and some even setting infertile cones, it became evident that *Retinospora* had been a mistake and the subsequent allocation of each form to its proper species (and also genus, for *Thuja* was involved as well) has been a matter of considerable difficulty. This difficulty has been largely overcome with time and further knowledge of the plants, but the sorting out of the nomenclature along conventional lines is a virtual impossibility owing to the undiscriminating use by the nineteenth century writers of the many names now to be found in the literature. If botanical writers are under a basic misunderstanding regarding both the generic and the specific allegiances of the plants they describe, plants which they only know of in the writings of others or at most as young plants, then it is too much to expect that they will leave behind them much of any real value to posterity. The treatment, here, therefore, particularly as regards the synonyms is not beyond argument, but I think it carries with it at least an air of probability and (being based on a knowledge of the plants now in cultivation rather than upon a probing into the probable intentions of writers of long ago) it will, I hope, provide a nomenclatural basis for present day use.

-- '**Argenteovariegata**'. Nich. 304 (1884); as *Ch. pisifera argentea* (Gord.) Carr. 132 (1867), (as to the name, description belongs to '*Plumosa Argentea*'); as *Retinospora pisifera argentea* Fort. ex Gord. 95 (1862). Some of the branchlets silvery-white, regularly intermixed all over the plant. Sent by R. Fortune form Japanese gardens tot he Royal Nursery of Bagshot, England, in 1861.

-- '**Aurea**'. (Fort. ex Gord.) Henk. & Hochst. 255 (1865); under *Retinospora*. Fort. ex Gord. 94 (1862), Pyramidal; leaves golden yellow, green within the plant. Introduced form Japan by Dr. von Siebold (Leiden) between 1859 and 1861, and sent by R. Fortune to the Royal Nurseries, Bagshot, England in 1861. Very common in cultivation.

-- 'Aurea Nana' Beissn. Handb. 91. (1891); 'Aurea Nana' ('*Aurea Nana* Fretsii pro syn.); Hornibrook, Dw. Conif. ed 2, 76 (1939). Now known as 'Strathmore'.

-- '**Aureovariegata**'. (Ottol.) Beissn. 91 (1891); under *Retinospora* Ottol. in Sempervirens **3:** 261 (1874). Very slow-growing; leaves golden variegated. Uncommon in cultivation.

-- **'Boulevard'**. Boulevard Nurs. Cat. 3. (1960); *Retinospora* 'K and C'. Kempenger and Christiansen, Nurs. Cat. (1934). 'Cyano-viridis' Hort. This is quite a newcomer, and seldom does a fresh introduction amongst garden conifers make such a rapid advance in popularity. It arose as a sport on *Chamaecyparis pisifera* 'Squarrosa' many years ago, and although it is not a dwarf form it is deservedly popular for its attractive foliage and wonderful colour which, when well grown, is a beautiful and outstanding silvery blue-grey, and as it seems to stand heavy pruning it should be possible to keep it within bounds in a given situation for a good many years. On one American nursery I visited, I noticed that all the stock plants had developed adult foliage. I have not seen this anywhere in Europe, but whenever it appears it should at once be cut away.

In dry soils and in soils containing lime the true colour fails to develop and instead the plant assumes an unattractive dirty brown, but whatever the soil the colour always seems to be better in shade and although it seems thoroughly hardy the plant will give of its best colour when grown in a cold house, especially during the winter months. Sad confirmation of the depravity of man comes in a report that plants grown 'soft' in this way are sold to the holiday makers in Cornwall as house plants, under the name 'Cornish Blue'.

-- **'Clouded Sky'**. Den Ouden & Boom, 112 (1965). Pyramidal, loosely branched, strong-growing; branches spreading; foliage in the first year rather similar to that of the typical form of the species, but with acicular leaves on the lower part of branchlets; leaves afterwards all acicular, distinctly blue on both surfaces. Originated as a sport of 'Squarrosa' in the nursery of L. Konijn & Co., Tempelhof, Reeuwijk near Boskoop, Netherlands.

-- **'Compacta'**, Hartw. u. Ruempl. Baume u. Str. 661 (1875). This distinct cultivar has been much neglected by authors and even Hornibrook does not mention it. It forms a fairly compact, low-growing, bun-shaped bush bearing wholly adult foliage, mostly in compact, tightly packed sprays which are somewhat recurved, but occasionally it sends out much stronger growth which does not develop the normal congested foliage until the next growing season, and by this means the plant increases in size. As well as the following, it has developed mutations marked by reduced vigour. See under 'Nana' and 'Minima'.

-- **'Compacta Variegata'** Hornb. Dw. Con. ed. 2, 75 (1939). This name was published for a sport from 'Compacta' with a fairly bold variegation in yellow to pale yellow flecks or splashes. In habit the plant is looser than the green form and in time will form a bush 1.5 m high by 2 m or more across. There is no general golden sheen over the whole plant, but the foliage on a large plant can very considerably. (See Illus. No. 138)

-- **'Cream Ball'**. Similar to 'Snow', coarser, less white and hardier. It is possibly a partially reverted form.

-- 'Cyano-viridis'. Now known as 'Boulevard'.

-- 'Dwarf Blue'. Superfluous name for 'Squarrosa Intermedia.

-- **'Ericoides'** Reg. [Russ. Dendrol. ed. 2, 27. (1883); Hornb. Dw. Con. ed. 2, 70. (1939); *Retinospora squarrosa dubia* Veitch, Man. 250. (1881). This is described as a plant not unlike *Ch. thyoides* 'Ericoides'. The only specimen known to Hornibrook is no longer in existence and so this form has been long regarded as lost to cultivation. Should it ever turn up, the very full description given by Hornibrook should provide adequate means of identification.

-- f. ***filifera***. (Sén.) Voss. Vilmor. Blumengärt. **1**: 1228 (1896). This is a botanical designation embracing all the *forms characterised by thread-leaf (filiform) foiliage in place of the usual leaf spray*. The cultivar that gave its name to this *forma* is not a dwarf.

-- **'Filifera'**. (Sénécl.) Hartw. & Ruempl. (661 (1875); *Ch. filifera* Veitch ex Sénécl. 54 (1868);

Retinospora R. Smith, 36 (1867) Pyramidal or conical, to 5 m broad; branches spreading; branchlets decurving; sprays pendulous, thread-like; leaves awl-shaped, distant, green above, glaucous beneath. Sent from Japan by R. Fortune to the Royal Nurseries, Bagshot, England in 1861 and by Dr. von Siebold to Leiden in ± 1865. Any use of the word in the name of a cultivar introduced since 1st January 1959 is illegitimate.

-- **'Filifera Argenteovariegata'**. Beissn. 91 (1891). Dwarf; pyramidal or globular (varying in cultivation); branchlets white-variegated. The white flecks often brownish if affected by the sun or by frost.

-- **'Filifera Aurea'**. (Veitch) Beissn. 91 (1891). This cultivar, like 'Filifera', is not a dwarf and will form an erect plant 4–5 m high in time.

-- 'Filifera Aurea Nana' Hort. See 'Golden Mop' and 'Sungold'.

-- **'Filifera Aureovariegata'** Beissn. 91 (1891). This is a form with a creamy-yellow variegation in bold splashes all over the plant which is fairly strong-growing, forming in time a large bush 1–1.5 m high. It is then an attractive form and should be more widely planted than it is, but like all the thread-leaf forms it is rather insignificant at first and calls for a little patience.

-- **'Filifera Gracilis'** (Gord.) Beissn, 91 (1891). This is a very attractive and uncommon form with foliage an unusual yellowish-green. Its foliage is slightly more slender and noticeably more regular than in the other 'Filifera' forms and it remains a dense little bush that should be seen more frequently than it is. It is occasionally offered as '*Filifera Flava*' Schelle, Winterh. Nadel. 301 (1909) which was described as being of a sulphur yellow colour and appears to be lost to cultivation.

-- **'Filifera Nana'** Beissn. Mitt. d.d.d.G. **6:** 58 (1897). This is only a slower-growing bush form of 'Filifera', from which it probably arose by bud mutation. This has evidently occurred more than once in different places, since plants in cultivation vary noticeably, but insufficiently to justify separation. (See Illus. No. 139)

-- **'Golden Mop'** Welch, Dw. Con. 146 (1966). 'Filifera Aurea Nana' Hort. I do not agree with Den Ouden and Boom [Man. 113. 1965] that this is the same as 'Filifera Aurea', since the habit is lower and the foliage is very much finer. Although slow to establish, this form develops into a dense, globose bush never over 1 m high and of the most wonderful gold colour, but unfortunately it is prone to sunburn. Although considered by the purists to be out of place in an alpine garden, it is very effective in a group of dwarf conifers or when grown as a specimen in a sunny spot.

-- **'Gold Spangle'** den Ouden 18 (1937). Pyramidal, dense, 5 m; branches spreading; branchlets either short, curved or even slightly decurved, with golden yellow foliage, part normal part thread-like. Originated as a sport of 'Filifera Aurea' in the nursery of Koster Bros, Boskoop before 1900 and introduced by A. Mesman, Boskoop, Netherlands.

-- 'Green Velvet' Welch, 178 (1979). Although I was responsible for unloosing this name in 1979, I now consider it insufficiently distinctive to warrant retention. My apologies!

-- **'Lemon Thread'**. A fast-growing, broadly upright plant with a rich golden colour. It originated from the development of filiform foliage (as in C. pisifera 'Filifera Aurea') on a tree of *C. pisifera* 'Lutescens' on the Mitsch Nurs., Aurora, OR. U.S.A. Both types of foliage are carried.

-- f. ***leptoclada*** (Endlicher) Rehder, Bibl. 56 (1949) in part; *Retinospora leptoclada* Sieb. Cat. Prodr. 6 (1861) (in part); *Retinospora squarrosa leptoclada* (Sieb.) Gord. Pin. 297. (1858) and *Retinospora leptoclada* (Zucc.) Gord. Pin. Suppl. 91 (1862); *Chamaecyparis pisifera* var. *squarrosa intermedia* Hornb. Dw. Conf. 49 (1923) and var. *squarrosa minima* l.c.; NON *Retinospora ericoides leptoclada* mast. Journ. Linn. Soc. 352. 1896). This early and rather phoney category covers a group of *forms in* *cultivation characterised by the persistence of at least some glaucous juvenile foliage consisting of very small leaves, often only at the base of an old tree.*

The epithet *leptoclada* (Latin = with slender shoots) appears very early in the literature and has been the cause of a great deal of confusion. It first turns up in a catalogue of 1861 which I have not seen and the plant was evidently an importation from Japan, said to be one of several there called 'Nazu' (= a dwarf). Gordon confused it with what we now know as *Chamaecyparis thyoides* 'Ericoides', thinking the latter was (his words) 'only the primordial form of *Retinospora leptoclada*' and Masters confused it with our *Chamaecyparis thyoides* 'Andelyensis'. Since Hornibrook (who also knew only small plants) did a very similar thing, Gordon's mistake was understandable; Master's was not and the perpetuation by Kent in Veitch [Man. Ed. 2, 232. 1900] of the same mistake still bothers us. Hornibrook's mistake is capable of a very simple explanation, as we shall see.

I am unable to make up my mind whether or not recognition of the group as a botanical form is justifiable or whether we are in fact only dealing with a single clone and a number of rather unstable mutations. I much incline to the latter conclusion, in which case we are dealing with a clone (or, as the case may be, a group of clones) which in maturity carries for the most part adult foliage of a most uninteresting kind which no one would want to propagate and which is (or are) only kept in cultivation at all by propagation by means of cuttings consistently selected from the attractive, glaucous, juvenile type of foliage. This type of foliage is usually so little in evidence in a predominately adult tree that the idenity of the two types of growth is often overlooked. So although Gordon gives a good description of an older plant, Hornibrook, thinking only of young specimens, describes his newly named 'Squarrosa Intermedia' as 'A most curious form of unknown origin with mainly or entirely juvenile foliage'.

So there the matter rests. I have seen old plants in several countries with hardly any juvenile foliage to be seen. So now those Superintendents and Curators whom I have doubtless outraged by saying that their plants were 'Squarrosa Intermedia' gone to the devil will be able to label them *Chamaecyparis pisifera* f. *leptoclada* with my blessing and – no doubt – much amusement.

The ability of young plants propagated from this juvenile growth to retain this horticulturally attractive quality is not at all complete, and all the cultivars in the group are apt to throw out adult or semi-adult growth and this must be rigorously cut out if the plant is to retain its early beauty. They are impossible to distinguish from one another by a study of the foliage alone, although (perhaps) differing in later behaviour. This may be due to clonal differences or merely be the result of several 'cutting generations' of carefully selected propagating material, possibly with cultural and climatic differences also playing a part.

-- 'Leptoclada'. In view of what is said in the previous paragraph, this name should not be used in any clonal sense. Unless identity with later selections such as 'Dwarf Blue', 'Ruck's Globe', 'Squarrosa Minima' can be established, the name '**Squarrosa Intermedia**' should be used for any such plants.

-- 'Miko' ('Mikko'). See 'Snow'.

-- '**Minima**' Dallimore & Jackson, 248 (1948), NON 'Squarrosa Minima' Hornb. This name, although accompanied by a not very explicit description, is now in use for a mutation in the 'Compacta'/'Nana' complex that is still more diminutive and tighter in foliage than 'Nana'. Of this attractive little plant, coloured and variegated sports have been introduced, some of them under Latin-form names the legitimacy of which depends upon their date of introduction.

-- '**Minima Aurea**' Hort. Amer., makes a plant about 20 cm high, 50 cm across after 15 years with a nice golden gleam across the whole plant. From it Wm. Goddard of Floravista on Vancouver Island, Canada, selected and introduced two further mutations, '**Silver Lode**' and '**Gold Dust**' with white and yellow variegations, respectively. In Britain **'Minima Variegata'** has a clear creamy-yellow variegation. Here also should be mentioned two names recorded by R. S. Corley of High Wycombe, UK [Bull. A. G. S. **38** (2): 155 (1970)] for plants he received from Japan. **'Pygmy'** seems very close to 'Minima' and was rather an unfortunate name, but **'White Pygmy'** is described as having light green foliage with pronounced white young growth.

-- f. ***nana*** Beissn. Handb. 91 (1891). A group, recognisable both botanically and horticulturally, of *very diminutive bun-like little plants, carrying adult foliage.*

-- **'Nana'** Beissn. Handb. 91 (1891); Hornb. Dw. Con. ed. 2, 75 (1939). This is one of the choicest of dwarf conifers. It makes a low, bun-shaped bush with dark bluish-green foliage, very glaucous beneath, in little fan-shaped sprays. **'Parslori'** Hillier, Man. 475 (1971) is a form that came to me from an Australian source. So far in the Pygmy Pinetum it has made a very neat, flattened, bun-shaped plant lower than 'Nana' and with lighter green foliage.

-- **'Nana Albovariegata'** Welch, Dw. Con, 147 (1966) is a form not mentioned by Hornibrook, with a small, white and not very effective variegation. It sometimes carries a wholly white spray of foliage in the interior of the bush. The name, despite its Latin form, might scrape acceptance under Recommendation 12A of the *Cultivated Code*; the relationship of this Recommendation to Article 27 is obscure, but to invent a new 'fancy' name for an unimportant mutation of the present kind would seem to me to be absurd, rules or no rules.

-- **'Nana Aureovariegata'** (R. Smith) Beissn. 454 (1884); under *Retinospora* R. Smith, Pl. Fir Tribe ed. 3, 38 (1867). This, as quite correctly described and named by Hornibrook, is a form of 'Nana' making the same very dense bun-shaped cushion but with a wonderful golden sheen all over the plant. It looks its best in sunshine and is a most desirable variety, reliably dwarf and slow-growing, but is apt to revert to the green form in patches. These should be cut away at once.

-- **'Nana Variegata'**. Carrière, 134 (1867). This is a variegated form of 'Nana' similar to 'Compacta Variegata' but smaller in every way. Details of a third group, even more diminutive than 'Nana' will be found under 'Minima'.

It is possible that all these forms (perhaps 'states' is a better word) have a common origin and that the forms of 'Compacta' were derived from the 'Nana' forms (or *vice versa*) by the selection of cutting material. This theory is perhaps borne out by the fact that the variegated forms are variable and that intermediate forms can often be found in different parts of the same bush. For this reason, care should be taken in pruning and the selection of propagating material to preserve the types I have described. Even 'Nana', which Hillier [Dw. Con. 27. (1964) tells us is a 'very consistent variety', does not always behave itself as well as it might, so should be watched for growth out of character.

-- **'Pici'**. Barabits, Deutsche Baumsch. **5:** 139 (1965). This is a form raised from seed in 1953 by Elemer Barabits of Sopron in Hungary. It is slower growing than 'Squarrosa Dumosa' but is very similar to that variety in summer. In winter it turns a warm bronze but is quite hardy. The mother plant is in the Botanic Garden at Sopron in Hungary.

-- f. ***plumosa*** (carr.) Beissner, Syst. Eintheil Conif. 13 (1887). This botanical epithet covers *all forms with persistent semi-juvenile foliage. The free standing part of the leaf is shorter and closer to the stem. Leaves are folded along the midrib and there may be some slight difference between the facing and lateral pairs. A definite branching system is more or less in evidence.*

The word also serves as the logical group name for an important group of cultivars. 'Plumosa' is used in this connection to indicate a semi-juvenile type of foliage that is intermediate between the juvenile 'squarrosa' type of foliage with its long, free standing leaves and the small-leaved adult type, and it has no connection with the botanical term 'plumose (which means 'spreading on either side of a common axis, like a feather.')

In most of the 'Plumosa' forms the leaf is much shorter and is pressed close to or attached to the stem for part of its length with only the tip standing free, and the branching system shows a definite pattern, with the tendency for the laterals to lie in the same plane, giving just the suggestion of being plumose. Unfortunately, consistency has not always been achieved in the naming of these forms and there are several cases where the epithet used belies the group to which the cultivar belongs. The use of the word in the name of a cultivar introduced since 1 January 1959 is illegitimate.

-- **'Plumosa'**. (Carr.) Otto in Hamb. Gart. u. Bl. Zeit **24:** 141 (1868); under *Ch. obtusa* Carr. 791 (1867). Conical, to 10 m; branches ascending, ultimate branchlets frond-like, feathery; leaves slightly spreading, awl-shaped, 3–4 mm long, bright green outside, slightly brown in winter, whitened beneath. (See Illus. No. 133)

-- **'Plumosa Albopicta'**. (Veitch) Nich. 304 (1884); *Retinospora plumosa albopicta* Veitch 248 (1881). This is a slow-growing form which makes an upright bush eventually to about 2 m, with exceptionally small leaves. These are dark green with numerous small areas of white which give the whole plant a speckled appearance.

It is attractive, especially as a young plant, but as the variegation is quite ineffective seen at a distance it should be planted close to a path or in a spot in which it would be observed form close at hand. The variegation is retained throughout the year, but is brightest in the spring and summer.

-- **'Plumosa Argentea'** (Sénécl.) Otto in Hamb. Gart. u. Bl. Zeit. **24:** 202 (1873); *Ch. plumosa argentea* Sénécl. 57 (1868). In growth similar to 'Plumosa'; but with the young growth creamy-white, turning green the second year.

Sent from Japan by R. Fortune to the Royal Nursery at Bagshot, England in 1861.

-- **'Plumosa Aurea'**. (Sénécl.) Otto in Hamb. Gart. und Bl. Zeit. **29:** 202 (1873); *Ch. plumosa aurea* Sénécl. 57 (1868). Similar to the preceding, but foliage golden yellow. Same origin and very common in cultivation. Many selections have been named (See below), one under test at North Carolina State Arboretum being **'Solar Flare'**, reported to retain its bright golden colour well.

-- **'Plumosa Aurea Compacta'**. *C. pisifera plumosa aurea compacta* Hort. ex Beissn. Handb. ed. 2, 570 (1909). The early descriptions of this and the following form are too brief to be of much help. Since the descriptions in my earlier book at least related to plants in current production probably the best thing is to leave well alone. Nomenclature must be our servant, not our master. 'Plumosa Aurea Compacta', then, forms a slow-growing, globose (later sometimes upright) bush to ultimately 1.5–2 m. The foliage is a golden-yellow and is much nearer to 'Squarrosa' than to 'Plumosa'. Each leaf has two strongly-marked bands of white on the underside.

It is extremely difficult to distinguish between young propagations of this cultivar and 'Plumosa Rogersii' described later, but the white marking on the underside of the leaves of the latter is slightly less prominent and also it will soon throw up a leader and begin to assume its characteristic conical shape, whereas 'Plumosa Aurea Compacta' soon develops into a rounded bush.

-- **'Plumosa Aurea Nana'** Hort. ex Beissn. Handb. 88 (1891). This is just a slow-growing form of 'Plumosa Aurea', presumably derived from it by selection of weak-growing cutting material. It bears rather small and typically 'Plumosa' foliage, which in the early summer is a beautiful golden-yellow. It is variable in the trade, and there are probably several clones in cultivations. Some forms later turn to a dull green but the best form retains its golden colour throughout the year. (See Illus. No. 136)

-- **'Plumosa Compacta'** Den Ouden, Conif. 128 (1949). This is a rather strong-growing form mentioned by Den Ouden which does not seem to be in cultivation. I have not seen it. The name should not be used for a dwarf form of any kind.

-- **'Plumosa Compressa'**. Hornb. Dw. Con. ed. 2, 74 (1939); *Thuja occidentalis* 'Silveriana Nana' (Spaan) Hort. Amer. This is probably the smallest form of *C. pisifera*. It was raised from a sporting branch of 'Squarrosa' by Van Nes of Boskoop. It is a most popular and attractive slow-growing little plant, bearing both 'Squarrosa' and 'Plumosa' type foliage of diminutive size which seems to vary in colour in different situations and seasons.

In its young state it looks like parsley or a tight, glaucous green moss, and by a little systematic clipping in early summer it can be induced to form a small bun of an almost unbelievably solid appearance and tightness of foliage but, normally, older plants show variation of foliage and colour. Because of this variability, propagations have been made from time to time of interesting branches, and to these have been given a wonderful variety of names. It would appear that as these names become more complex they became more difficult to memorize and more easy to quote differently next time, for one meets every possible combination of the words 'Nana', 'Compacta', 'Aurea', 'Glauca' and others. I once received a nice little plant labelled *Chamaecyparis pisifera* 'Plumosa Flavescens Aurea Compacta Nana' which seemed to me to be carrying the joke a little too far. The variations certainly are not sufficiently stable to justify retention of all these names but I see Hillier does retain the name **'Plumosa Compressa Aurea'** [Man. 475 (1971)] for a form distinguished by foliage gold tinged in summer.

This plant is susceptible to frost injury. In the spring as soon as new growth commences the dead growth should be cut out and usually a season's growth will close over the gap.

-- **'Plumosa Cristata'** (Onuma) Beissn. Mitt. d.d.d.G.**9:** 67 (1900); 'Plumosa Cristata' Bergm. Pl. and Gar. **21**, (1): 36 (1965). This seems the most distinctive of any of the group, if the picture and description given by Mrs. Bergman are maintained in other soils and situations. I saw a form when in the United States in 1970 that I think was this. I was told that it maintained the tight growth of the plants I saw and never misbehaved by producing loose growth. Young plants in the UK shew little or no improvement on 'Squarrosa Intermedia' in this respect.

-- **'Plumosa Flavescens'**. (de Vos) Beissn. (1884) 454 ; under *Retinospora* de Vos in Sempervirens **3:** 228 (1874). *Retinospora plumosa sulphurea* Hort. ex Den Ouden, Conif, 128 (1949). This is quite a distinct and attractive form which is seldom met with. It forms a conical or globose bush never over 1 m high and its cultivar name (which means 'yellowish' or 'turning yellow') is quite descriptive of the foliage. This is smaller in the leaf and neater than in any other of the 'Plumosa' varieties, and at first is sulphur yellow at the growing tips. It turns green by the autumn.

-- **'Plumosa Juniperoides'**. Welch, 183 (1979); 'Plumosa Aurea Compacta' Bergm. Pl. and Gar. **21**, (1): 35(1965); 'Plumosa Juniperoides Aurea' Hort Amer. This distinct form might, at a casual glance, be mistaken for 'Plumosa Compressa', having the same closely packed juvenile leaves and neat habit. It differs in being of much stronger growth and in holding its relatively broad leaves forwards and clasping the shoot (unlike the strongly decurved leaves of 'Plumosa Compressa'), especially on the young shoots, to which they impart a tetragonal effect that no doubt suggested the name. It is golden-yellow in early summer, turning green later.

-- 'Plumosa Nana' Hornb. Dw. Con. ed. 2, 74 (1939). This, according to Hornibrook, was a sub-globose form which has been lost to cultivation but **'Plumosa Nana Aurea'** [Hornb., l.c.] is quite a distinct and very uncommon variety which forms a low, compact, bun-shaped cushion. The leaves are much more of the 'Squarrosa' type than 'Plumosa' and are strongly recurved, a rich golden-yellow. A twig taken from the plant could be mistaken for 'Plumosa Aurea Compacta' or 'Plumosa Rogersii', but the low, slow growth and the bun-shaped habit of the plant make it distinct. It is less hardy than other clones in this group, and burns in summer sunlight.

-- **'Plumosa Pygmaea'**. Welch, Wansdyke Nurs.

Cat. 4 (1961) and Dw. Con. 150. (1966); *C. pisifera* 'Pygmaea' Hillier, Man. 476 (1971). On the rock garden in the National Botanic Garden at Glasnevin in Dublin I found a very old specimen about 2 m high bears this name, and although I have not been able to trace it in the literature, I am assuming that because of its age it must have been published at some time or other and therefore be legitimate. The plant could be roughly described as a green form of 'Plumosa Rogersii'. Young plants form the same cone-shaped bush and it is an attract-

Figure 135. *Chamaecyparis pisifera* 'Squarrosa'. A tree 12 m high. Such trees bear cones, despite the persistent juvenile foliage. (10)

Figure 136. *Chamaecyparis pisifera* 'Plumosa Aurea Nana'. Several dwarf selections are grown and others intermediate in vigour.

ive plant which should be better known, having the great advantage that it seems to be thoroughly wind resistant.

-- **'Plumosa Rogersii'**. Hornb. Dw. Con. ed. 2, 75 (1939); as 'Plumosa Aurea Compacta Rogersii' in Nurs. Cat. ± **1930**: 2. Dwarf, conical, to 1 m high; branches upright; leaves all juvenile, about 8 mm long, uniformly golden yellow during the whole winter. Raised at the nursery of W. H. Rogers Ltd, Southampton, England. This is a very popular form with golden-yellow foliage which is much nearer 'Squarrosa' than 'Plumosa'. It arose as as sport from 'Plumosa Aurea'. Unfortunately it is a little wind tender, especially in the spring, so should be given a sheltered spot.

-- **'Snow'**. Hillier [Man. 476. 1971] This is an introduction from Japan. Hillier describes it as a dwarf, bun-shaped bush with mossy, blue-grey foliage tipped with white, tending to burn in the sun or cold wind. It is, I think, a plant for the alpine house, like the similarly white-tipped *Cryptomeria japonica* 'Knaptonensis'. (See Illus. No. 140)

-- f. ***squarrosa*** (Zucc.) Beissn. & Hochst. ex Beissner, 13 (1887); *Retinospora squarrosa* Sieb. & Zucc. Fl. Japan **2:** 40 (1844). This is a botanical epithet covering *forms with persistent juvenile foliage not included in* f. *leptoclada. Leaves resemble the seed-leaves; long free standing, decurved; no distinction between facing and lateral pairs of leaves.*

-- **'Squarrosa'** (Zucc.) Beissn. & Hochst. in Gartenfl. 364 (1880); *Retinospora squarrosa* Zucc. in S. & Z. 40 (1844); *Ch. pisifera squarrosa veitchii* Hort. Juvenile form. Pyramidal, densely branched, loose, broad, 5–10 m; branches spreading, tops decurving; branchlets irregularly arranged, not frond-like, feathery, silvery grey, woolly or mossy in appearance, soft to the touch; leaves linear, about 4 mm long, flat, rather soft, glaucous. Introduced from Japan by Dr. von Siebold into Belgium (Gent) in 1843 and by J. G. Veitch into England in 1861. (See Illus. No. 135)

Several names ['Squarrosa Albospica' and 'Squarrosa Argentea' Hornibrook, 48 (1923); 'Squarrosa Variegata' Endlicher, 65 (1847)] were raised for now-unidentifable minor variants, and 'Squarrosa Argentea Compacta' Hort. ex Welch [151 (1966] has no validity, since plants thus described turn out in time to have the normal habit and rate of growth. The following are cultivars currently in cultivation.

-- **'Squarrosa Aurea'**. Nels. Pin. 77 (1866); Welch, Wansdyke Nurs. Cat. 4 (1962). This is an old variety, now seldom mentioned or planted. It has golden-yellow foliage and is a beautiful plant when happy, but as it dislikes sun, frost and wind it needs careful placing. There are trees several metres high in the Royal Botanic Garden, Edinburgh.

-- **'Squarrosa Dumosa'**. Beissn. Handb. 87 (1891); 'Plumosa Nana' Hort. This is a most attractive and truly dwarf form which forms a dense, globose bush with fairly large-leafed foliage which, in summer, is a quiet grey-green but which, in winter, takes on a most attractive appearance, giving the whole bush the suggestion of the beauty of old pewter. It does, however, tend to throw out coarse and only semi-juvenile growth. This should be at once cut away, since it is difficult to restore a plant once such unwelcome growth has taken hold. It was found in the Botanic Garden in Berlin.

-- **'Squarrosa Intermedia'**. Hornb. Dw. Con. 49 (1923); 'Dwarf Blue' Hort. Europe. This form is well known in England, where its easy rooting qualities endear it to nurserymen. At its best it is a low, rounded bush with tightly packed, congested foliage of the 'Squarrosa' type but smaller in leaf than in any of those forms. But it is prone to throw out longer shoots with more open foliage. If it is desired to keep the bush small and dense looking these should be cut away; they normally develop the congested type of foliage the following year. Sometimes strong coarse growth bearing more or less adult foliage appears and unless this is cut out at once the plant will end up as a loose conical tree to 2m tall, with a central leading shoot, and arching branches bearing intermediate green foliage, very similar to that of 'Plumosa'; and with only a few traces of the true juvenile type of foliage near the base of the tree. (See Illus. No. 137)

-- **'Squarrosa Lombarts'** is similar to 'Dumosa' but is a much more vigorous plant, attractively coloured copper-bronze in winter.

-- **'Squarrosa Lutea'**. Den Ouden, Naaml. 18 (1937) and Conif. 129 (1949) is a sport from 'Squarrosa' with smaller habit and yellowish-white foliage which arose in the nursery of Koster Bros., in Holland, and is distinct in colour habit from 'Squarrosa Aurea' already described, **'Squarrosa Aurea Nana'** listed by Hillier [Man. 476 (1971)] may well be the same, from the description given. I do not know it.

-- **'Squarrosa Minima'**. Hornibr. 49 (1923); 'Squarrosa Pygmaea' Hornibr. 72 (1932); *Ch. pisifera* var. *minima* Dall. & Jacks. 248 (1948);

Figure 137. *Chamaecyparis pisifera* 'Squarrosa Intermedia' as it should be seen, but adult foliage is prone to develop and soon 'takes over' if allowed to remain.

Figure 138. *Chamaecyparis pisifera* 'Compacta Variegata'. becomes a large, spreading bush. The variegation is in irregular patches of butter-yellow.

Figure 139. *Chamaecyparis pisifera* 'Filifera Nana'. The spreading habit distinguishes this dwarf form from the taller 'Filifera'.

Figure 140. *Chamaecyparis pisifera* 'Snow'.

Figure 141. *Chamaecyparis pisifera* 'Sungold'. Less prone to sunscorch than the older 'Golden Mop'.

'Squarrosa Argentea Pygmaea' Hillier, Dw. Con. 28 (1964) 'Argentea Pygmaea' Bergm. Pl. and Gar. **21**, (1): 35 (1965); 'Squarrosa Pygmaea' Spingarn. Bull. Amer. Rock Gar. Soc. **24**, (2): 41 (1966). I regard all these as the same as, or at the most as derivatives of, 'Squarrosa Intermedia.' Since any desirable differences in tightness and slowness of growth may be the result of regular pruning, mutation or years of careful selection of propagating material by particular nurserymen, its value to collectors and connoisseurs may be real enough but it is impossible to reduce to writing. **'Rucks Globe'**, introduced by J. W. Spingarn of Baldwin, Long Island, New York, originated from an extremely slow-growing mutation found by Wm. Ruck, nurseryman of Merrick, also of Long Island, New York.

Figure 142. *Chamaecyparis thyoides* 'Aurea' at the National Pinetum, Kent, U.K.

-- **'Squarrosa Sulphurea'**. Kent in Veitch, Man. ed. 2 1900. This, as its name indicates, is a variety with sulphur-yellow foliage. It forms a dense, conical plant and in spring and early summer is outstanding because of its colour, but this turns to green later in the year and by winter the plant has only its neat outline and soft looking foliage to distinguish it. **'Squarrosa Elegans'** was, according to Hornibrook [Dw. Con. ed. 2, 71 (1939)] a compact form with light yellow foliage not burning in the sun. There are certainly more than the single clone in cultivation, but they are difficult to separate.

-- **'Sungold'**. Grootendorst in Dendroflora **6:** 78 (1969). This clone slowly builds up into a dense 'mop-head' bush with tangled, overlapping threadleaf foliage. It does not sun-scorch and in this respect it is superior to 'Golden Mop'. It was selected from a batch of about 500 seedlings of *Ch. pisifera* 'Filifera Aurea' by Wm. Goddard of Floravista Nursery, Vancouver Island, BC., Canada. (See Illus. No. 141)

-- **'Tsukumo'**. A very dwarf clone of the 'Minima' type, forming a low bun of dense, light green foliage in fan-shaped sprays that are glaucous beneath. It is popular for bonsai, since it soon forms a stout little trunk.

Ch. sphaeroidea Spach, Hist. Nat. Veg. Phan. **11:** 331 (1842). Now known as *Ch. thyoides.*

Ch. thyoides, L.) Britton, Sterns & Poggenburg. Prelim. Cat. Pl. N.Y. City 71 (1888); *Cupressus thyoides* L. 1003 (1753); *Thuja sphaeroidea* Spreng., Syst. Veg. **3:** 889 (1826); *Ch. sphaeroidea* Spach, Hist. Nat. Vég. Phan. **11:** 331 (1842). Atlantic White Cedar, White Cedar.

Size and habit Conical tree to 20–30 m with narrow, spirelike crown.

Trunk Usually stout. *Bark* Grey- to reddish-brown with narrow, raised strips in spirals, coming away at the ends.

Branches Upright-spreading.

Branchlets Numerous, thin and slender; irregularly arranged at all angles round the shoot (ie. not in a flat plane as in the other species).

Leaves Very small but in dense sprays at the surface of the tree; acute, closely appressed, incurved or with apices free and with a prominent, oval gland; white beneath. The crushed leaves smell of ginger.

Mature cones Usually very numerous, in clusters on small branchlets; globose, 6 mm diam, bluish-purple with bloom, ripening to dark brown. *Scales*: 4–5 (rarely 6); acute, often with a reflexed boss.

Seeds 1–2 per scale, oblong, with a narrow wing, not wider than the seed.

Figure 143. *Chamaecyparis thyoides* 'Rubicon'.

Figure 144. *Chamaecyparis thyoides* 'Andelyensis Nana'.

Figure 145. *Chamaecyparis thyoides* 'Andelyensis'.

Habitat United States: Maine to Florida, West to the Mississippi.

Cultivation Introduced by P. Collinson into Britain in 1736. In nature it frequently is found growing in marshy ground or swamps, so in cultivation will presumably feel at home in such conditions. From other species of *Chamaecyparis* it can at once be distinguished by the foliage not being in flat sprays. From *Cupressus* it is distinct by the small cones. Zone 4. (See Illus. No. 96)

-- '**Andelyensis**'. (Carr.) Schn. 169 (1913); *Retinospora leptoclada* Sieb., Cat. Prodr. 6 (**1861**); under *Ch. sphaeroidea.* Carr. 123 (1867). This well known variety very slowly forms a tall, pointed column with upright branches bearing mainly adult foliage and coning freely. It was much planted years ago but now that we have other columnar forms available it is less in demand, which is rather a pity. Veitch in his first edition gives an account of this cultivar and its origin as a seedling in a Nursery at Andelys in France, its introduction at an exhibition in Paris and subsequent purchase by an English nurseryman. It is such an excellent account that it leaves Veitch with no excuse for adopting the purchaser's change to *leptoclada*, a name already in use for a form of *Chamaecyparis pisifera* (see above), and thereby introducing confusion into the nomenclature of both species. (See Illus. No. 145)

-- '**Andelyensis Nana**'. [*C. thyoides* var. *andelyensis nana* Hornb. Dw. Con. Ed. 2, 78. (1939)]. This form is identical with 'Andelyensis' in all respects except habit. It forms at first an upright rounded bush which tends to increase laterally more than in heigth with age. It is doubtless a cultivariant, produced from deliberate selection of weak-growing lateral growth for cuttings. It is therefore apt to throw up the strong vertical growth characteristic of 'Andelyensis' and this should be at once cut out if the dwarf characteristic is to survive. Hornibrook found it in the nursery of Détriché at Angers, France. (See Illus. No. 144)

-- '**Andelys Yellow**'. Den Ouden & Boom, 120 (1965). *Ch. thyoides* 'Andelyensis Aurea' Meyer, Plant Explorations 92 (1963), illeg. name. Similar to 'Andelyensis' but foliage yellowish and growth looser and less columnar. Origin unknown.

-- '**Aurea**'. *Ch. thyoides aurea* (Ottol.) Fitsch. (1930) *Ch. sphaeroidea aurea* Ottol. in Sempervirens **3:** 260 (1874). A regular conical form with yellow foliage in summer, fading to browish in winter, originated at the nursery of N. Gaujard, at Gent, Belgium, in 1872. (See Illus. No. 142)

-- **'Conica'**. den Ouden 131 (1949). A dwarf, juvenile, conical, dense, very slow-growing form; annual growth 2–3 cm; branches erect, stout; branchlets spreading; sprays densely set, tapering, tips nodding; leaves in threes, linear, slightly decurved, 6–8 mm long, flat above, bluish-green in summer, bluish-brown in winter, with 2 inconspicuous stomatic lines beneath; cones small, glossy crimson when young. Originated as a sport of 'Andelyensis' in the nursery of L. Konijn, Reeuwijk near Boskoop, Netherlands. Plants of about 10 years old are only 20–25 cm in height. It is, however, not quite stable and reverts occasionally to 'Andelyensis'.

-- **'Ericoides'**. Sudw. U.S. Dept. Agric. For. Serv. Bull. **14:** 79 (1897); *Retinospora ericoides* (Zucc) Gord. Pin. 294 (1858); *Chamaecyparis ericoides* Carr. Traité, 140 (1855). This is another early introduction from Japan and recorded as grown there under the name of Nazu (= dwarf). Its reputed appearance also in a French seed bed in 1840 as recorded by Den Ouden and Boom [Man. 120 (1965)] may be equally true, and this may account for the recognised variability of 'Ericoides' in the trade. It forms a regular, very close-growing, compact, pyramidal bush, eventually to about 1.5 m high. The wholly juvenile foliage is a dark grey-green in summer, turning to deep purplish or violet-brown in winter. It forms an attractive bush but is apt to be damaged by cold winds and dislikes dry conditions.

-- **'Glauca'**(Laws.) Sudw. 77 (1897); *Cupressus thyoides* 'Glauca' Laws. 63 (1852); 'Kewensis' Hort. Hoopes 190 (1866). Conical; branchlets densely set, short, reddish; leaves silvery glaucous green.

-- **'Heatherbun'**. Welch 187 (1979). I have received from several sources a very diminutive and slow-growing form with wholly juvenile foliage similar to that of 'Ericoides' but smaller in all its parts. The growth is consequently very tight and dense and it forms a round-topped little bush quite unlike the conical habit of 'Ericoides'. It changes colour like 'Ericoides', from which it may have originated as a bud sport, but it seems to be quite stable and reliable in cultivation. It is a most desirable form with the same rich, plum-purple winter colour. This clone is also in circulation in UK as 'Purple Heather'.

-- **'Rubicon'**. Found by H. J. Welch in a nursery in Tennessee, USA in 1970 and introduced into the trade in UK by Wansdyke Nurs. Devizes, Wilts. It forms an upright bush with dense juvenile foliage having much longer leaves than 'Heather Bun' but which turn a similar rich plum-purple colour in winter. It seems to be questionably hardy, but is worth a little cossetting. From the illustration given in Dendroflora **23:** 47 (1986), the plant there listed as 'Red Star', and stated to have been received in Holland from New Zealand would appear to be 'Rubicon'. (See Illus. No. 143)

-- **'Variegata'** *Ch. thyoides variegata* (Carr.) Sudw. 77 (1897); *Ch. sphaeroidea variegate* Carr. 133 (1855). Similar in growth to the typical form of the species, but less branched; leaves partly yellow-variegated. Loudon received a specimen from a nursery in Ireland in 1831. Rather common in cultivation.

CRYPTOMERIA

Taxodiaceae

Cryptomeria. D. Don in Trans. Linn. Soc. Lond. **18:** 166 (1841). A monotypic genus of forest tree with one recognised variety. From its nearest relative, another monotypic genus, *Sequoiadendron*, it differs in having bright green leaves, longer and wider-spreading and arranged in five ranks, rather than the three ranks and the dull grey-green colour of *Sequoiadendron*. The bark, also, is thinner and the Cryptomeria is usually a broader tree.

C. fortunei. Hooibrenk ex Otto & Dietrich, Allg. Gartenzeit. **2:** 21: 234 (1853) and of some present-day authorities. It is here retained as *C. japonica* var. *sinensis*, following Silba (1984).

C. japonica (Linn. f.) D. Don in Trans. Linn. Soc. Lond. **18:** 167 (1841). (See Illus. Nos. 146)

Figure 146. *Cryptomeria japonica.* Foliage with both the new and previous year's cones.

Size and habit Tall, conical, monoecious trees to 50 m, 2–6 m in girth, with a pointed or, sometimes, a broadly domed crown.

Trunk Straight, tapering, often ridged or well buttressed. *Bark*: reddish-brown, often dark-grey in old trees; thick and fibrous, often deeply grooved; peeling off in long shreds.

Branches Irregularly whorled, spreading and up-curved or pendent towards the tips.

Branchlets Horizontal or drooping, glabrous; green the second year.

Winter buds Minute, hidden by short leaves.

Leaves Awl-like, quadrangular in section but laterally compressed; deep glossy green on the two outer faces and with inconspicuous stomata on the inside. Spirally arranged in 5 vertical ranks, decurrent at the base, then directed forwards and curved inwards; as short as 6–8 mm or as long as 10–20 mm, the length often varying along a single season's growth. Leaves are retained for 4–5 years.

Flowers Male flowers borne axially, in clusters of 20 or more, sessile, oblong, forming short racemes at the ends of branchlets; female flowers globular and solitary, at the ends of branchlets.

Mature cones Sub-globose, at first green and leaflike, but ripening the first season to reddish-brown; hanging on the tree after shedding the seeds. *Scales:* 20–30, wedge-shaped, enlarged above into a disc with a recurved mucro in the middle and with 3–5 long, stiff teeth on the upper margin.

Seeds 3–5 on each scale, triangular-oblong and slightly compressed, narrowly winged.

Habitat Japan, China. Zone 5.

Cultivation Specimens of this species and of its varieties *chinensis* and *lobbii* (the latter is now regarded as a cultivar [See Mitchell, 89 (1972)] are commonly found in arboreta and large collections, but in gardens the genus is usually represented by one or more of the many cultivars that have been selected, many of them in Japan. As well as the snake-like forms that are planted for their interest as young plants (but outgrow their welcome in time) there are some truly diminutive plants that are a safe choice for even the smallest garden.

Figure 147. *Cryptomeria japonica* 'Lobbii'. Note the characteristic bunching of the foliage.

Figure 148. *Cryptomeria japonica*. Bark on main trunk. (10)

The nomenclature of the garden forms is complicated by the duplication of European and Japanese names. In Japanese horticulture names are often used more as descriptions than as names in the western sense and often with local variations in their application. The synonyms often given must therefore be treated with caution, since they may not be exact equivalents. But where they do relate to the same clones they may be regarded as 'commercial synonyms' in the two languages.

The word 'sugi' is the Japanese common name for the species. Under Recommendation 31A (j) it may not form part of a cultivar name introduced since 1st Jan. 1959, but in cultivar names published before that date it must be retained.

A number of recently claimed 'introductions' from Japan appear from descriptions to be merely re-introductions of well-known cultivars under different names. Where this turns out to be the case, such names must be dropped, as superfluous synonyms.

-- 'Alba Variegata'. See 'Argenteovariegata'.

-- 'Albo-spica'. See 'Argenteovariegata'.

-- f. ***araucarioides***. Henkel & Hochstetter 269 (1865). *All those forms characterised by rope-like or snake-like foliage, the result of strong main shoots and the paucity or irregular incidence of lateral shoots.*

The clones in cultivation in this peculiar group are quite impossible to relate to the names found 'in the literature'. This is partly because some early descriptions were evidently made from young plants whose potential growth was not known and which had not yet developed the features characteristic of the mature plant, and partly because (as Carrière pointed out) these bizarre forms turn up in the seedbeds from time to time, so we probably have many more clones (some of them very much alike) than we have names for them. Early descriptions of 'Araucarioides', 'Dacrydioides' and 'Lycopodioides' are adequate to distinguish, at least provisionally, these three groups of clones, and unless authentic living specimens of 'Athrotaxoides', 'Ha-o-sugi', 'Kewensis',

Figure 149. *Cryptomeria japonica* 'Cristata'. Detail of foliage. The coarse fasciations are sometimes as big as a man's hand.

Figure 150. *Cryptomeria japonica* 'Kilmacurragh'. The fasciations are quite distinct from those of 'Cristata'.

Figure 151. *Cryptomeria japonica* 'Araucarioides'. A group of clones with foliage mid- to deep-green, lacking 'tufts' of short branchlets.

'Mucronata', 'Selaginoides', 'Viminalis', 'Virgata' and other snake-like plants are available to provide good clonal descriptions I now suggest they should be discarded as being 'No longer identifiable' and that a little restrained selection of clearly distinguishable clones of real garden value for introduction as new cultivars might be a good thing. In any such case, as well as registering the proposed name, a description (protologue) of the plant being named with sufficient detail to ensure identification of the plant with the name chosen for it at any time in future, should be published and a copy sent to the World Conifer Data Pool to ensure adequate recognition of the new cultivar. Selection should be made only from mature trees and it should be followed by the distribution of authentic propagation material to selected centres throughout the world.

-- **'Araucarioides'**. Gordon, 21 (1862). A group of clones of f. *araucarioides* (above) based on the plant introduced by Dr von Siebold into Leyden from Japan in 1859 and which, eight years later, Seneclauze described 'as an outstanding ornamental tree, especially so as an isolated specimen, with branches horizontal, drooping and sparsely side-branched, and with leaves shorter (5–10 mm long), fewer, thicker and more incurved than normal, regularly arranged around the shoot and of a vivid green throughout the year.' The absence of any reference to 'tufts' of foliage characterises this group. (See Illus. No. 157)

-- **'Argenteovariegata'**. Sénéclauze, Conif. 62 (1868); 'Alba Variegata' Young in Gordon, 9 (1874); 'Albo-spica' De Vos 231 (1887); 'Argenteo-spicata' Beissner 151 (1906); 'Nana Albo-spicata' Beissner, 483 (1909); 'Nana Albospica' Den Ouden/Boom, 127 (1965). In the absence of clear evidence to the contrary I can only assume that the clone raised and named by Sénéclauze slowly made its way around the nurseries of Europe and was re-christened from time to time by well-meaning writers who had only seen small plants. Were Hornibrook able to see the present size of the plant on Isola Madre he would feel confirmed in his opinion [See 81 (1939)]. Sénéclauze describes his plant as 'A strong-growing dwarf variety that turned up on the nursery at Bourg-Argental, Loires, on which the extremities of the shoots and twigs are a pure, silvery white. Very pretty!'

-- 'Aurea' Beissner, 147 (1891). A clone with yellow foliage no longer identifiable, unless 'Ogon-sugi' is the same.

-- 'Athrotaxoides'. No longer identifiable in cultivation.

-- **'Aurescens'**. Den Ouden 133 (1949). Pyramidal, broad, dense; branches crowded; branchlets fan-shaped; leaves numerous, falcate, apex curved, 12–15 mm long, greenish-yellow in summer, yellowish-green in winter. Quite hardy. Originated in Blijdestein Pinetum, Hilversum, Netherlands, in 1937.

Figure 152. *Cryptomeria japonica* 'Lycopodioides'. A group of clones characterised by the development of foliage in 'tufts'.

Figure 153. *Cryptomeria japonica* 'Spiralis'. 'Granny's ringlets'. Slow growing, but it will eventually become a tree.

Figure 154. *Cryptomeria japonica* 'Spiraliter Falcata'. Similarly twisted foliage, but quite distinct from 'Spiralis'.

-- **'Bandai-sugi'**. Grootendorst ex Hornibrook, Dw. Con. ed. 2, 82 (1939). This is a popular garden variety. It is usually seen as a low globular bush, but with age it can reach to 2 m high. The foliage is very congested and irregular, all the branches being thick and the leaves thick, broad and tapered. The growth is most erratic, some of the branchlets being of fairly strong growth with stiff straight foliage pointing forward at an angle of 30°, very slightly incurved; other shoots being very short with short, closely set leaves, some not more than 3 mm long. The colour is green, turning a dull bronze in winter. (See Illus. No. 157)

-- 'Birodo-sugi'. This is an unsatisfactory name, having been applied at different times to several different cultivars. It should no longer be used in any clonal sense.

-- **'Broom'**. Kortman, J. P. in Dendroflora **24:** 13 (1987). This clone forms a globose to broadly conical tree 2–3 m high, with light green foliage in summer, darkening in winter. It was introduced by the nursery firm S. A. Minier of Angers in France, around 1968.

-- **'Compacta'**. Carr. in Rev. Hort. 60 (1878); 'Lobbii Compacta' Hort. Holl. Conical, compact, to 15 m; branches spreading, densely feathered near the top of the tree; branchlets 10–30 cm long, densely set, paniculate; leaves short, stiff, bluish-green. Found by Messrs Thibaut & Keteleer, horticulturists at Sceaux (near Paris) in 1877. Common in cultivation.

-- **'Compressa'**. Den Ouden, Conif. 134 (1949). This cultivar was imported from Japan and distributed by J. Blaauw & Co., Boskoop, in 1942. This is one of two pygmy globular forms very popular in gardens, the other form being 'Vilmoriniana' (described below). They both have short, densely set, neat, juvenile foliage and both form dense, globular bushes. They are of about equal garden value and are often confused with each other, but they differ in that the leading shoots in 'Compressa' never project above the general surface of the bush as they do in the case of 'Vilmoriniana' (giving that variety a somewhat less neat and tidy appearance); in that the leaves in 'Compressa' are slightly more decurved than they are in 'Vilmoriniana' (giving the shoot a more rosette-like appearance looked at from the end), and in that 'Compressa' colours to a rich red-bronze in the winter compared to the much less pronounced colouring of 'Vilmoriniana'. Finally, the foliage on the interior of the bush is blue-green in the case of 'Compressa' and yellow-green in 'Vilmoriniana'.

-- **'Cristata'**. Beissner in Mitt. d.d.d. Ges. **10:** 79 (1901); 'Sekka-sugi' Hort. Japan; 'Sekkwia-sugi' Dallimore and Jackson 184 (1923). Introduced from the Yokohama nurseries into Germany by L. Unger, ± 1900. There are several forms in cultivation characterised by abnormality in parts of the foliage. The form commonly grown under this name forms a fairly tall, upright bush or low tree to 3 m high with some of its foliage quite normal but developing monstrous foliage here and there. The cristations occasionally are as large as a man's hand on young plants, but become much smaller in maturity. The plant mentioned by Hornibrook as growing at Kilmacurragh, Rathdrum, Country Wicklow, is of quite a different type and is now called **'Kilmacurragh'** Welch, Dw. Con. 157 (1966). In this cultivar the foliage is juvenile and the whole bush is covered with fasciated shoots. The original plant is still at Kilmacurragh (having now reached the height of 6 m) to serve as our type plant and there are specimens in the National Botanic Garden at Glasnevin, Dublin and the Royal Botanic Garden, Edinburgh. (See Illus. Nos. 149, 150)

-- **'Dacrydioides'**. Carrière, ed. 2, 193 (1867); NON Mitchell 89 (1972), and other writers since 1960. Carrière describes this, without mentioning size or habit, as "a quite distinctive form with branches spreading and pendulous, leaves incurved, much smaller and thinner than normal to the species, of a rusty-brown as in some species of *Dacrydium*, and varying in size along the same shoot. This reference to the colour is repeated by later writers, down to about 1960, (although the time of year is never mentioned), so it is accepted here as the characteristic feature of this clone or, as is probably the case,

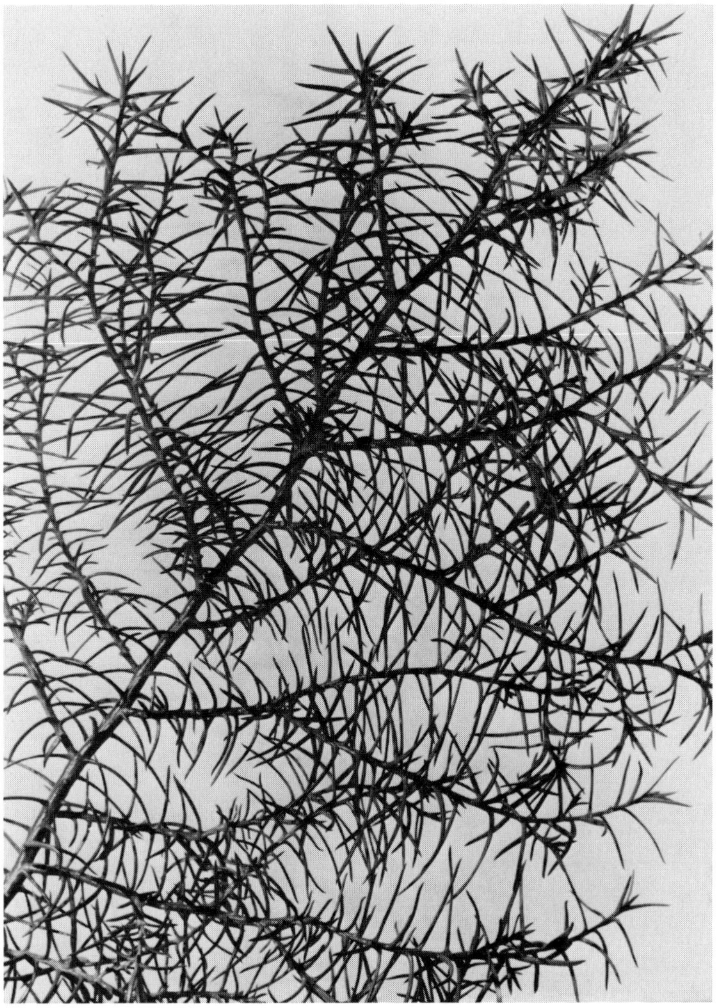

Figure 155. *Cryptomeria japonica* 'Elegans'. This is the 'fixed juvenile' form. It turns ruby-red in winter.

this group of clones.

Alan Mitchell tells me that the 7 m tree labelled 'Dacrydioides' at Bedgbury, Kent, UK, and described by him [89 (1972)] as "Branches spreading, upswept at ends, shoots long, unbranched, with distant, star-like whorls of unbranched minor shoots radiating in bunches of mixed lengths" is bright green, so he does not now regard it as having been correctly identified. The tufts or whorls of short branchlets that are so pronounced in that tree would suggest that it would fall within the group 'Lycopodioides' as defined below.

-- **'Elegans'**. Nelson, 60 (1866); *C. elegans* Makoy in Belg. Hortic. **14:** 356 (1864); *C. japonica elongata* Regel ex Gord. 72 (1875); *C. gracilis* Hort. A mop-headed tree 5–8 m tall; branches horizontal; branchlets densely set; sprays often spirally arranged; leaves soft, juvenile, widely set, long and sickle-shaped, green in summer, bright brownish-red in winter. First sent from Japan by Th. Lobb to the Veitch nurseries in 1854, but it presumably could turn up in a seed-bed at any time, since it seems merely a case of persistent juvenility. The tree is not well anchored and the foliage often suffers from wind so some very unsightly specimens can be found. (See Illus. No. 155)

-- **'Elegans Aurea'**. Harrison (NZ) 115 (1959). In

Figure 156. *Cryptomeria japonica* 'Jindai-sugi'. A medium grower, with regular and normal foliage.

Figure 157. *Cryptomeria japonica* 'Bandai-sugi'. Very slow-growing, with very irregular foliage.

Figure 158. *Cryptomeria japonica* 'Monstrosa Nana'. Foliage similar to but very much coarser than in 'Bandai-sugi'.

growth similar to 'Elegans', but retaining its colour (variously described as lemon-yellow, yellowish green or green) throughout the winter. A plant sent to me from New Zealand was green in summer and almost a true yellow during the winter. The seasonal change may explain the discrepancies in the records. It may well be identical with **'Elegans Viridis'** of Hornibrook ed. 2, 87 (1939).

-- 'Elegans Compacta'. Dallimore & Jackson, Handbook Conif. ed. 3, 255 (1948); Welch, Dwarf Conif. 157 (1966) and 191 (1979). Despite what I have regrettably written about this name I am now of the opinion that it is no more than a superfluous name for the cultivar 'Elegans Nana', a view supported by the virtually identical descriptions given us by Dallimore under the two names. The result of my re-examination of this difficult group of names is given in the next paragraph and under 'Nana'.

-- **'Elegans Nana'**. *C. elegans nana* Veitch in Man. Conif. 218 (1881); *C. j.* var. *elegans nana* Beissner in Mitt. d.d.d. Ges. 141 (1899) and

Hornibrook 58 (1923); 'Elegans Compacta' SENSU Welch, Dw. Conif. 157 (1966) NON 'Lobbii nana' l.c; 'Elegans compacta' SENSU Welch, Man. Dw. Conif. 191 (1979), NON 'Elegans Nana' loc. cit. Veitch describes it as "A low, dense, bushy shrub with leaves more crowded and slender than 'Elegans'. It retains in all its intensity the same glowing crimson hue during the winter months with the exception of the pendulous tips of the branchlets which remain green."

-- **'Elegans Variegata'**, van Geert, Nurs. Namebook, no. 540, 57, name only; *C. elegans fol. variegata*, de Vos, 230 (1887). Similar to 'Elegans', but beautifully white variegated, often more white than green; not strongly growing.

-- **'Enko-sugi'**. From the following description sent to me from Japan this would appear to be 'Lycopodioides' or a clone very close to it. This is a slow-growing, irregular shrub (2–4 m) without a distinct central leader, having long, stiff, coarse, whip-like main branches that at intervals along their length produce clusters of short, side branches. Needles differ in length according to the season of growth (forming branches of varying thickness) being usually long, thick, stiff, curved towards the branch and dark green. Old specimens may develop dense and constant tufts at the ends of the branches, with branchlets of nearly equal length; but young plants have few tufts and less branchlets."

'Fasciata'. Dall. & Jacks. Handb. 183, (1923). This is a collector's piece, more a curiosity than a thing of beauty, usually a low stunted bush with some healthy foliage but with the old wood studded with a curious, moss-like, leafy excrescence (one can hardly call it growth) which gives the bush an unhealthy appearance.

-- **'Globosa Nana'** Hornibrook 58 (1923); 'Lobbii Nana' Hillier, Man . 477 (1971) is an attractive dwarf form which bears normal foliage, but the trunk is entirely missing and it consequently forms a globose bush. It is densely set with

Figure 159. *Cryptomeria japonica* 'Globosa Nana'.

branches and the branchlets are pendulous. It is rather slow in getting going as a small plant, but eventually forms a bush 1–2 m high by as much or more across, and it would form an attractive lawn specimen. Hornibrook's choice of this name was rather unfortunate for it has left the way open for the name **'Globosa'** Lombart Nurs. Cat. 19 (1942/43); Den Ouden, Conif. 135 (1949) to be since used for a different cultivar. I have not seen large plants of 'Globosa' but it would appear that it is going to form a smaller plant than 'Globosa Nana'. The plants I have seen are low-growing, much wider than they are high. The juvenile foliage has straight, stiff, pointed leaves, similar to but shorter than in 'Nana' but the outstanding feature of 'Globosa' to my mind is the awful colour that it turns in the winter. This is a most unattractive bright red, the colour of a rusty tramp steamer. (See Illus. No. 159)

-- **'Ha-o-sugi'**. Dall. & Jacks. 184 (1923); (Hoo-o-sugi) Hort. Japan. A slow-growing small (2–3 m) tree, marked by the following features; branches snakelike, swung elegantly in different directions, needles very short, sometimes of scale-form, of nearly equal length, giving the impression of a thin braided rope, dark green. The name means 'Phoenix-Cryptomeria'; it is used for various forms, this being the form cultivated around Tokyo. They all, presumably, would be clones of f. *araucariodes*.

-- **'Ito-sugi'**. Krüssmann, 106 (1972). A form selected by Prof. G. Isa of Kyoto University, and therefore more exactly called 'Kyoto-ito-sugi', to distinguish it from other gardeners' forms: ('Ito' means 'thread'). It forms a small (2–3 m) tree of pyramid form, with a single trunk as central leader, branches regular but scarce and thin, more crowded near the top in older plants, with branchlets crowded at the tips of the branch, tips swung straight upwards, needles short, of nearly equal length (but of variable length on old plants), light (grass) green.

-- **'Jindai-sugi'** Hornb. Dw. Con. ed. 2, 83 (1939). This popular variety forms a compact, conical bush with erect and spreading branches. The foliage is very regular and dense, with short, stiff, straight leaves pointing forward at a narrow angle. The colour is a nice soft green which is retained all the year. What a pity a few more of these forms were not imported from Japan under their vernacular names! (See Illus. No. 156)

-- **'Knaptonensis'**. Lyttel in My Garden **7:** 360 (1934); Hornibrook, ed. 2, 83 (1939). Hornibrook tells us that this originated as a witch's broom on the plant of 'Argenteovariegata' on the Isola Madre in Lake Maggiore, and as one would therefore expect, when true to type it consists of a congested mass of irregular shoots. These are a glistening white, but it shows a tendency to revert to the looser, dirty creamy-white of the mother variety. It is an extremely attractive plant when well grown, but as it dislikes wind and frost it is really a plant for the cold greenhouse. (See Illus. No. 163)

-- **'Littleworth Dwarf'**, originated from a witch's broom found by H J Welch and W B Archer in a tree at Littleworth Cross, near Farnham, UK, and was introduced to the trade by the Wansdyke Nursery, Devizes, Wilts, UK. It forms a dense little plant of congested mid-green foliage, but is apt to develop loose, uncharacteristic growth in time. The illustration in the recent book *Conifers* is correct, but the name and caption are not.

-- **'Lobbii'**. Carr. 154 (1855). Although for long regarded as a botanical variety, the plant found by M. Lobb in a botanic garden in Java in 1953 was clearly a cultivar, so as such is here listed. But from the fact that so many trees in Britain are of this form it surely must come true from seed, so it may turn out to be entitled to botanical recognition after all. It forms a narrowly conical tree with each branch carrying a rather separate dense clump of foliage. The resulting irregularity in density gives the tree an uneven outline which is characteristic. The foliage otherwise is normal. (See Illus. No. 147)

-- 'Lobbii Nana' Dallimore & Jackson, Handbook Conif. 183 (1923) and ed. 3, 255 (1948); 'Pygmaea' and 'Hort. Kew' Hornibrook,

Figure 160. *Cryptomeria japonica* 'Vilmoriniana'. A well-tried favourite that remains reliably dwarf.

Dwarf Conif. ed. 2, 84 (1939); Welch, Dw. Conif. 159 (1966). This name has caused a lot of confusion. No cultivar entitled to this name either does or ever did exist. Dallimore & Jackson first describe it as 'Of dwarf habit'. When preparing for their third edition, Dallimore (or it may have been Jackson, or they could have been egging each other on) let himself really go and came up with 'Of dwarf habit with green leaves'. Neither description helps a great deal. The name originated with a plant so labelled in the Royal Botanic Gardens at Kew which Hornibrook incorrectly identified as 'Pygmaea', which Welch accepted at face value in 1966 but which turns out to be nothing more than a superfluous name for 'Nana'. The cultivar listed as 'Lobbii Nana' in Hillier's Manual is 'Globosa Nana'.

-- **'Lycopodioides'**. Carrière in Rev. Hort. 226 (1875); *C. j. lycopodiiformis* Beissner 147 (1891), altered to *C. j. lycopodioides* in ed. 2, 483 (1909) and to *C. j. viminalis* (*C. j. selaginoides* pro. syn.) in ed. 3, 461 (1930); *C. j. selaginoides* Rovelli ex Dallimore and Jackson ed. 3, 255 (1948); *C. j.* var. *viminalis* Hornibrook 56 (1923). Carrière describes this as "a very irregular shrub, TUFTED AS A RESULT OF ITS ABUNDANT BRANCHING. Main branches almost always devoid of branches or very irregularly branched, sometimes more than 1 m long and uniform in diameter throughout, looking like a long cord. In some places they are thinner, shewing some form of restriction, some are spreading and pendulous. The leaves are incurved, glossy dark green, lying close to the stem, drawn out to a point, very irregular in length, thus increasing the bizarre appearance." Carrière gives us his opinion that this variety is not very attractive but is curious, and certainly worthy of a place in every conifer collection. He records that this clone was imported direct from Japan by M. Mazel of Montsauve nursery near Anduze, France. Hornibrook and Bailey both mention the tufts of foliage, which must therefore be regarded as the characteristic of this group of cultivars.

I have no doubt that Rovelli Frères sent a

plant (as *viminalis*) to M. Hornibrook in 1909 as an as-yet-uncatalogued novelty which appeared later in their 1913 catalogue (shortly before the firm went out of business) as *selaginoides*. Hornibrook describes it as making a low, rather loose shrub sending out long, slender, snake-like branches with few, or no, side branches ... but at times crowned with a whorl of smaller branchlets. (See Ilus. No. 152)

-- **'Monstrosa'**. Beissner ed. 2, 482 (1909); 'Monstrosa' Hornb. Rep. Conif. Conf. 73 (1932), an upright, columnar or rounded bush to 2 m or more high.

Although the plant Beissner took his description form was columnar there seems no doubt that he was describing the form widely grown under this name today, but it usually forms a round-topped tree. Beissner accurately describes the coarse, irregular growth with partly coarse branches with long stiff pointed leaves and partly short branchlets crowded with similarly short leaves. It could very well be described as an exceedingly coarse, outsize version of 'Bandai-sugi'.

Figure 161. *Cryptomeria japonica* 'Archer's Greenbush' (= 'Nana' of Welch (1979) and Krüssmann (1979)).

-- **'Monstrosa Nana'**. Hornibrook, ed. 2, 83 (1939). This unfortunate plant has had much difficulty to establish an identity. The specimen described by Hornibrook is no longer to be found and his brief description does not altogether distinguish his plant from 'Nana', so the name 'Monstrosa Nana', amongst others, has been widely used for that other unfortunate plant. Specimens of 'Monstrosa Nana' that I have come across have all had the peculiar growth habit that is characteristic of both 'Bandai-sugi' and 'Monstrosa' but have been of a scale and vigour intermediate between those two forms. This agrees with Hornibrook's description. Growth characteristics and vigour can vary so greatly with the age of a plant and rootstock vigour that the existence of three clones different in scale but so alike otherwise needs some further verification, but in the meantime I suggest the use of the cultivar name 'Monstrosa Nana' be limited accordingly. It is probable that the name covers more than one clone, so the relationship of the Japanese names 'Mankichi-sugi', 'Mankitiana', 'Midare-sugi', 'Negire-sugi', 'Shishi-gashira' and possibly others to 'Monstrosa Nana' needs to be clarified by the study of living plants. (See Illus No. 000)

-- **'Nana'**. Hort. ex Knight and Perry, Synop. Conif. 22 (1850), Carrière, Traité Gén. 155 (1855), beissner 147 (1891) and Dallimore and Jackson, 183 (1923) – all of whom cite 'Pygmaea' H. B. ex Knight and Perry as a synonym; 'Nana' Fortune ex Gordon, 54 (1858); 'Pygmaea' London ex Hornibrook 84 (1939); NON 'Nana' SENSU Welch [161 (1966) and 194 (1979)]. The epithets *nana* and *pygmaea* first appeared within months of each other in the middle of the nineteenth century and Hornibrook convinced himself (See above) that 'Pygmaea' had priority over 'Nana' by a few months. But the fact that Lindley and Gordon's name was *sub nudum*, and the citing by Knight and Perry only a few months later of 'Hort.' as authority for the name 'Nana' (indicating that it was then in general acceptance in the

Figure 162. *Cryptomeria japonica* 'Nana'. (Misidentified by Welch (1979) as 'Elegans Nana'.)

trade) and its acceptance by all writers both in Britain and on the continent throughout the century that followed would establish a case for acceptance of the name 'Nana' under Art. 27 (c) of the *Cultivated Code* as "best preserving existing usage". This reduces the name 'Pygmaea' SENSU Knight and Perry to a superfluous synonym. 'Nana', then, is a widely grown bush to about 1.5 m, usually broader than high, with dense habit and stiff, straight juvenile leaves shorter than those in 'Elegans Nana' and never assuming the rich colouring of that clone in winter. On larger plants the top shoots terminate with closely packed clusters of male flowers in winter. (See Illus. Nos. 162, 164)

I much regret having generated much confusion by having misidentified in my previous books, two plants in cultivation as 'Nana' and 'Pygmaea', respectively, on the advice of those from whom I felt only able to learn at that time. I was similarly misled in the use of the names 'Elegans Compacta' and 'Elegans Nana' (aided considerably, no doubt, by the almost identical descriptions given in Dallimore and Jackson.) But one's knowledge must be permitted to grow in twenty-five years, and having

Figure 163. *Cryptomeria japonica* 'Knaptonensis'. Needs a sheltered situation to give of its best.

Figure 164. *Cryptomeria japonica* 'Nana'. The dense clusters of male flowers only develop on older plants.

here corrected the record on 'Nana' I suggest the new name '**Archer's Greenbush**' for the unfortunate plant to which I misapplied that name. Although no such cultivar as 'Pygmaea' "either does or did exist" the name is "occupied", so for the plant that is similar to 'Archer's Greenbush' except that it colours up in the winter I suggest the new name '**Archer's Redbush**'. It is uncommon in cultivation. (See Illus. No. 161)

-- '**Nana Albospica**'. Since my endorsement [194 (1979)] of the use of this undocumented name by Den Ouden and Boom in 1965 the use of this name has become generally accepted. As a young plant it is very slow-growing, but after what has happened to the one-time dwarf plant on Isola Madre on Lago Maggiore over the years it is more than likely that a hundred years from now its distinctiveness from 'Argenteovariegata' will be called in question.

-- 'Ogon-sugi'. See 'Aurea'.

-- 'Okina-sugi'. See 'Argenteovariegata'.

-- 'Plumosa'. A name listed by Duncan and Davies Ltd, New Plymouth, New Zealand in 1953. It is probably the same as 'Elegans Plumosa' described by R. E. Harrison in *Trees and Shrubs for the Southern Hemisphere* 115 (1959) as being "similar to 'Elegans' but with more attractive, 'plumey' foliage."

-- 'Pygmaea'. This name should be discarded. See under 'Nana'.

-- var. **radicans** Nakai. A vigorous, quick-growing selection widely used in forestry in Japan. It forms a straight trunk and any low branches root into the ground and form fresh trunks. Since, according to Krüssmann it is only propagated from cuttings it is difficult to understand its claim to any recognition botanically.

-- '**Rasen-sugi**'. Le Feber ex Grootendorst in Dendroflora **13–14:** 69 (1977). Indistinguishable from 'Spiraliter Falcata'.

-- '**Sekkan-sugi**'. A tree differing from the normal (and from 'Aurea') only in colour. This is a creamy-yellow, during the winter almost white.

-- 'Selaginoides'. A superfluous name for 'Lycopodoides'. See p. 276

-- var. ***sinensis***. Sieb. in S. & Z. 52 (1844); *C. fortunei* Hoorbr. ex Otto & Dietr. in Algem. Gartens. **21:** 234 (1853). This is a true geographical variant, being the form of the tree found in China. It was introduced into Britain nearly twenty years before the typical form from Japan, as a result of which all the oldest trees are of this variety, but they can only rank at the variety level because the Swedish traveller Thunberg had discovered the species in Japan in 1754. It is true that he wrote of it as *Cupressus japonica*, but the botanical rule is that the epithet '*japonica*' shall survive the transference of the species to the new genus *Cryptomeria*. The differences are not great: trees of var. *sinensis* are more open and the branches more slender and pendulous; the leaves are longer and thinner and the cones have fewer scales (seldom more than 20) and usually only two seeds per scale (instead of the 3–5 of the typical form).

-- '**Spiralis**'. Sieb., Cat. Prodr. 5, **1861**: name only; Sieb. in S. & Z. 52 (1844); NOT 'Spiraliter Falcata' Carrière, 340 (1876); 'Granny's Ringlets' Hort.] There are two quite different forms with the foliage twisted spirally round the branches. The present form is known in specimens of tree-like dimensions, but the garden variety (which may have been derived from the larger tree as a witch's broom) is normally seen as a low, congested little bush, unmistakable on account of the thick, incurved, and twisted foliage. This peculiarity varies with the season, being most noticeable in winter. It was introduced from Japan into the Netherlands (Leiden) by Dr. von Siebold, ± 1860. (See Illus. No. 153)

-- '**Spiraliter Falcata**'. Carr. Rev. Hort. 340 (1876); 'Kusari-sugi' Hort. Japan; 'Spiralis Elongata' Hort. This variety forms a much taller, upright bush and carries longer and thinner branches than in 'Spiralis' and they all tend to twist and curve. It bears much thinner, narrower leaves, rather irregular in size. These two spiral forms are quite distinct, even as small plants. As a young plant it is somewhat straggly. But with age it fills out and forms an attractive tree to 3 m. The forms 'Torta' and 'Yore-sugi' are indistinguishable, perhaps identical. (See Illus No. 154)

-- '**Tansu**' might well be described as a diminutive form of 'Bandai-sugi', the irregularity of growth being similar but the leaves and growth on a reduced scale. The foliage is a light green. Formerly distributed in U.S.A. as 'Yatsubusa'.

-- '**Tenzan**' is a very diminutive form, dense habit and light-green foliage. The needles are short and sharp.

-- '**Vilmoriniana**'. Hornb. The Garden, **87:** 37 (1924) and Dw. Con. ed. 2, 88 (1939). This is one of the most popular dwarf conifers. It forms a compact, roughly globular bush never seen more than 1 m high by as much through with very neat foliage and closely recurved leaves. The differences between this cultivar and 'Compressa', the only one likely to be confused with it, are given under that variety. Hornibrook found it in the garden of M. de Vilmorin at Verrières-le-Buisson, near Paris in 1923; it was sent from Japan by Philippe de Vilmorin as *Juniperus japonica*, about 1890. (See Illus. No. 152)

-- 'Viminalis'. A superfluous name for 'Lycopodioides'. See p. 276.

-- '**Yoshino**'. A clone long known in Japanese cultivation which forms a graceful small tree with light green foliage soft to the touch.

CUNNINGHAMIA

Taxodiaceae

Cunninghamia. R. Brown ex Richard, Mem. Conif. 149 (1826). A genus of two (doubtfully three) closely related species.

Size and habit Monoecious, tall evergreen tree, more or less columnar, with a domed crown.

Trunk Cylindrical, reducing sharply at each main branch. *Bark*: Bright red-brown; stringy; ridges formed by long, narrow, shallow fissures.

Branches Spaced widely apart; spreading.

Branchlets Sometimes pendulous; a pale shining green; opposite and in two ranks, the short lateral shoots being deciduous.

Winter buds Round, covered with leaf-like scales.

Leaves Linear-lanceolate, emerging spirally but often twisted into two ranks; decurrent; broadest at the base, tapering gradually to a sharp point; serrulate; remaining green for several years and even when dead hanging on the tree for a long time.

Flowers Male flowers in terminal clusters; female flowers also terminal, but solitary; globose.

Mature cones Spindle-shaped; long-pointed; 4 cm long. *Scales*: Long, thin and papery; sharply pointed, irregularly serrate.

Seeds Oblong, hard-coated; 3 per scale; with 2 narrow wings. (See Illus. Nos. 165, 166)

C. kawakamii Hayata in Ic. Plant. Formosa **5:** 207 (1915). This species was described as being intermediate between the two accepted species but is now regarded as indistinguishable from *C. konishii*.

C. konishii Hayata in Gard. Chron. 194 (1908). Found in Taiwan: Mt. Randai, 1300–2000 m. Zone 8. See under *C. lanceolata*.

C. lanceolata (Lambert) Hooker in Bot. Mag. **54:** t. 2743 (1827); *Pinus lanceolata* Lambert 53 (1803); *C. sinensis* R. Brown ex Richard, Mem. Conif. 80 (1826). Found in China, Yunnan to Chekiang at 700–2500 m. Zone 7.

Silba in *An International census of the Coniferae*, 21 (1984) quotes Professor Laubenfalls as stating "Many plants cultivated in the eastern United States as *C. lanceolata* are actually *C. konishii*". Cunninghamia seedlings take many years to develop their adult foliage and many cultivated trees in the eastern United

Figure 165. *Cunninghamia lanceolata*. Detail of foliage.

Figure 166. *Cunninghamia lanceolata*. Bark on trunk. (10)

States are still in a juvenile state. It is probable that the same situation prevails in Europe, since Mitchell remarks that young plants of the two species hardly differ.

In the circumstances it is not surprising that the text-book descriptions disagree. The best distinctions seem to be that *C. konishii* makes a larger (to 50m) and a somewhat more columnar tree and bears short (1.9–2.5 cm long), stiff and short-pointed leaves with stomata on both sides, pale green and inclined to bronze in winter; arranged rather higgledy-piggledy all round the shoot. *C. lanceolata* makes a smaller (30 m) tree with a rounded crown and with the bark peeling off in strips and it bears longer (3–7 cm) leaves with a long, pointed tip; dark, glossy green above, with two broad, noticeable bands of stomata on the lower side only and with most of the leaves twisted into a more or less pectinate arrangement.

-- **'Compacta'** Bano (1965). If this name was not validly in use before 1st January, 1959 the suggested name 'Bano' (See Roy. Hort. Soc. Journal 113 (1988) must be used.

-- **'Glauca'** Dallimore & Jackson 188 (1923) (Carrière) Fitschen 468 (1930); *C. sinensis* var. *glauca* Knight and Perry, Synop. Conif. 45 (1850), name only; Carrière 171 (1855). Leaves with a conspicuous bloom which imparts a silvery sheen to the tree. More hardly than the typical form.

× CUPRESSOCYPARIS

Cupressaceae

× ***Cupressocyparis*** Dallimore in Journ. Soc. For. of Gr. Brit. **11** (1): 3 (1937). The name embraces any bigeneric hybrid between the genera *Cupressus* and *Chamaecyparis.*

These interesting bi-generic hybrids (if that is what they really are) being of recent occurrence have been well documented. Of the several crosses that are on record *C. leylandii* is the most important. Its vigour, hardiness and rapid growth have swept it into popularity and, in the hands of innumerable householders seeking immediate shelter and privacy, have led to its having been planted in most unsuitable places. In almost every urban area in Britain a day of reckoning lies ahead when it is realised that those dear little trees that were so pleasing when first planted are forest trees demanding their right to adequate space and unwilling to take 'No!' for an answer. (See Illus. No. 167)

× ***C. leylandii*** (Jackson & Dallimore) Dallimore and Jackson in Kew Handl. Conif., ed. 4, 38 (1938); *Cupressus leylandii* Jackson & Dallimore in Kew Bull. 114 (1926). A hybrid: *Cupressus macrocarpa* × *Chamaecyparis nootkatensis*. Zone 7.

This interesting cross had spontaneously appeared at widely separated places in Britain at rare intervals over a long period, never more than a few plants (in some instances only a single seedling having survived on each occasion). Although some quite large trees had developed from the early crosses, some of them in well-known collections, interest in and

Figure 167. × *Cupressocyparis leylandii*. Group at the National Pinetum. Bedgebury, Kent, U.K.

Figure 168. × *Cupressocyparis leylandii* 'Haggerston Grey' (Clone 2). Detail of foliage.

appreciation of the valuable qualities of these trees was very slow to appear. Eventually Alan C. Mitchell set about tracing all the surviving trees and identifying the clones in use for propagation in the nursery trade. In 1964 four of these were given cultivar names. Since then, other clones from both the early crosses and more recent ones plus several colour mutations have been brought into circulation. A full account of this research will be found in the *Quarterly Journal of Forestry* **58** (1): 8–19 (1964), in the Forestry Commission Booklet No. 33 *Conifers in the British Isles* and in the *International Dendrology Society Year Book* 97–100 (1985).

All the clones form rather upright or columnar trees with a general resemblance to a 'Macrocarpa', with either steeply ascending or upswept branches according to the open-ness of the site. The foliage is very dense except towards the tip of the tree. The foliage is always intermediate between the two parents, some clones approximating to the flat-pinnate foliage of *Chamaecyparis nootkatensis* and others inclining to the angular-plumose leaf arrangement of *Cupressus macrocarpa*. The cross has been made in both directions, with apparently identical results. (See Illus. No. 167)

The following clones and mutations have been given cultivar names.

-- 'Belvoir Castle'. A superfluous name for 'Robinson's Gold'.

-- **'Castlewellan'**. (Plumose type). This very popular cultivar was one of the first of the golden-yellow forms to have been introduced. It originated at 'Castlewellan' in Co. Down, Northern Ireland when Mr McKeown, then head-gardener, sowed seeds taken from a *Cupressus macrocarpa* 'Lutea' standing next to a *Chamaecyparis nootkatensis* 'Lutea'. Despite the colour of both parents the golden colouration of the foliage is limited to the surface of the tree, the interior being variably dark green. It is hardy and both wind and smog tolerant, so is being widely planted, although it is not to everyone's taste when used as a hedge or shelter belt. It was at first known as *Cupressus macrocarpa* 'Keownii' and has also been distributed as 'Galway Gold', 'Mellow Yellow' and 'Castlewellan Gold'. It was introduced in the Slieve Donard Catalogue in 1974.

-- **'Clun'**. Mitchell in Int. Dendr. Soc. Yearbook 98 (1985). (Plumose type). Named for a tree in a garden at Clun in Shropshire, having a crown similar to 'Haggerston Grey' but with foliage quite unlike any other clone, being remote, slender, hard and dark grey-green with a tinge of blue. It probably dates from about 1900. The name 'Clun Rectory' first suggested was based on a misunderstanding.

-- 'Contorta'. An illegitimate name. See 'Picturesque'.

-- 'Galway Gold'. A superfluous name for 'Castlewellan'.

'Golconda'. Name registered by D. F. Wyant in 1977. It arose from a branch sport of 'Haggerston Grey' at Wyboston, Bedford and under test has turned out to be good grower with foliage a most brilliant golden-yellow throughout the tree.

-- 'Gold Cup'. A name listed in New Zealand.

-- **'Golden Sun'**. Mitchell l.c. (Plumose type). A branch sport on a tree of 'Haggerston Grey' at Barnham Nurseries, Bognor Regis, Sussex. It seems to be semi-dwarf and bushy but according to Mitchell is an exceptionally good bright gold and should be better known.

-- **'Goldness'**. Name registered by S. G. McMinn in 1976, and introduced to the trade by Ness Nursery, Londonderry, N.I. Of normal habit and rate of growth, but with sprays of yellow-variegated foliage alternating with the normal green.

-- **'Gold River'**. Van der Laar in Dendroflora **22:** 73 (1985). A bright golden-yellow mutation found on an unrecorded clone on a nursery at Boskoop, Holland. It has an open, horizontally branching habit, and is resistant to sunscorch.

Figure 169. × *Cupressocyperus leylandii* 'Leighton Green' (Clone 11). Detail of foliage, with cone.

-- **'Green Spire'**. Ovens, Bright and Mitchell in Quart. Journal For. **58** (1): 8 (1964). (Plumose type). It forms a very dense and narrow-columnar tree; the shoot with the main axis often losing dominance; branchlets irregularly spaced and arising at various angles around the shoot, lying forward in a flat spray; foliage bright green. It was named for Clone No 1 at Haggerston Castle. It has not been widely propagated.

-- **'Haggerston Grey'**. Ovens, Bright and Mitchell 1964, l.c. (Plumose type). This is much the commonest clone in England, probably because it is one of the easiest to propagate from cuttings. It forms a tree that is columnar, but rather more open than 'Green Spire'. The foliage is long, narrow, asymmetrical and uneven, with the final sprays rising and lying in two planes at right angles, often in opposite or decussate pairs with the tips rounded. The colour of the leaves is a slightly greyish-green, the shoots being bright yellow. It was at first designated 'Clone 2'. (See Illus. No. 168)

-- **'Harlequin'**. Mitchell l.c. (Plumose type). This is one of several mutations that were detected at Weston Park, Shifnal in Shropshire, seat of Lord Bradford. This branch sport on 'Haggerston Grey' has proved very consistent in its creamy-white variegation. Mitchell writes "It is the plumose version of 'Silver Dust', with all the peculiar vagaries of colour distribution of that form. Indeed, for the first few years until the foliage develops, they are not easy to distinguish and have been found mixed in one nursery. Close comparison shows the more sparse side-shoot in 'Harlequin', lacking the more solid, diamondshaped shoot of 'Silver Dust', and the white of 'Harlequin' is generally a clearer, more ivory-white, and there is less of the random, scattered little outbreak of white."

-- **'Hillspire'**. A listed name in U.S.A.

-- **'Hyde Hall'**. Name registered by R. H. M. Robinson in 1976. This originated from a witch's broom that appeared in a private garden in 1973. It develops into a rather uninteresting bush difficult to distinguish from *Chamaecyparis nootkatensis* 'Compacta'.

-- **'Jubilee'**. A variegated sport that arose on the Wansdyke Nursery at Devizes in Wiltshire U.K. I shewed it at one of the Royal Horticultural Society Shows in London but it was not sufficiently distinctive to receive an award and I ceased propagating it.

-- **'Leighton Green'**. This clone is usually seen as a narrowly columnar tree, but it is outstanding as a specimen, where in isolation it forms a broader tree. The branchlets are lanceolate in outline, symmetrical and even; the final leaf sprays lying predominantly quite flat with the tips flattened. The shoot is pale yellow. It cones frequently and was at first designated 'Clone No. 11'. (See Illus. No. 169)

-- 'Mellow Yellow'. A superfluous name for Castlewellan Gold'.

-- 'Picturesque'. A name registered in UK for a plant already in circulation in USA which I understand is objected to by the originator and may therefore not be validly published. Unless the right to register a name is restricted to the originator, or at least to persons having some claim to the plant or its introduction, and unless applicants for the registration of a name are required to shew that they have made some real attempt to trace the ultimate origin of the cultivar they wish to name and establish that it is not already in circulation under a different name, cases of this kind are liable to arise and bring the whole idea of names registration into disrepute.

-- **'Robinson's Gold'**. (Flat-pinnate type). Name registered by the Dept. Agric. For. Service, N. Ireland in 1977. It was apparently a natural hybrid and has an interesting history. Alan Mitchell tells us "Mr Robinson, Head Gardener at Belvoir Castle, Co. Down, N. Ireland, noticed a minute seedling among some rhododendron jungle that was being cleared. He took it home and grew it for a while in a 'Kit-e-Kat' tin on his kitchen window-sill before planting it in his garden." It is now a fine tree, 36 ft × 3 ft in 1982. It has solid foliage compared with 'Castlewellan' and is really to all appearance a golden 'Leighton Green' and equally vigorous and strongly branched. The colour varies with season and with year but is usually brighter and more consistently yellow than 'Castlewellan' although it can at times be a mossy kind of yellow-green. 'Belvoir Castle' is a superfluous name.

-- **'Rostrevor'**. Mitchell l.c. ('Flat-pinnate type'). This rather mysterious clone may well have been the very first Leyland Cypress. Three trees seen by Alan Mitchell in 1976 at Castlewellan, and there known as 'The Rostrevor Leyland', can be traced back to cuttings taken from a tree at Rostrevor in Co. Down, N. Ireland that an eye-witness account records as having been 'enormous' before the First World War of 1914–1918. There is no record of when that tree was planted nor of its parentage – it was a seedling given by an unknown stranger to the gardener at Rostrevor – but that it predated the crosses at Leighton Hall must be indisputable.

-- **'Rua'**. A clone introduced in New Zealand about 1977. It carries corded foliage, a lighter green than normal, and the stems are a purplish shade.

-- **'Silver Dust'**. Name registered by S. G. March, U.S. Nat. Arb., Washington DC in 1976. (Flat-pinnate type). It arose in 1960 as a branch sport on 'Leighton Green' at the National Arboretum, Washington DC, USA. About 20% of its foliage is a clear creamy-white in oval patches. The distribution of the variegation is irregular. Some main branch-tips grow out white bearing solid white side-shoots and the main branch-axis remains as vigorous as the green branches but the white side-shoots are small. The main axis turns green again and bears normal, vigorous side-shoots, so there is a sector some 15 cm long of small, white shoots between normal, big, green ones. On less strong branches, fully white side-shoots grow at random but sparsely. Many short sectors of minor branches are white on one side only, bearing small white lesser shoots on that side. Some small shoots, themselves clad in green scales, bear very small white shoots each side, giving a pretty, speckled effect but only 3 cm long. An occasional ultimate shoot on a fully green spray has a white tip. See also 'Harlequin' above.

-- 'Smith's Gold'. A superfluous name for 'Spring Gold'.

-- **'Spring Gold'**. Mitchell l.c. (Plumose type). This golden-yellow branch sport was found on a 'Haggerston Grey' tree at Pickering, Yorkshire by a Mr Stephen Smith. 'Smith's Gold' is a superfluous synonym.

-- **'Sparkler'**. (Plumose type). A yellow or white-tipped sport found by Laurence C. Hatch on a tree in a shelter-belt of 'Haggerston Grey' on the North Carolina U.S.A. State Nursery University Arboretum. All growing tips are brightly coloured throughout the growing season in contrast to the green older leaves, so the plant appears to sparkle in sunlight. It is resistant to sun-scorch.

-- **'Stapehill 20'**. Ovens, Bright and Mitchell. Quart. Jour. For. **58** (1): 8 (1964) (Flat-pinnate, type, open and rough). About 1940, seed was collected from a tree of *Cupressus macrocarpa* growing in a nearby garden and sown on the nursery of M. Barthelemy at Stapehill, Ferndown, Dorset. Two of the resulting plants developed abnormal foliage, suggestive of hybridity. Because of the lack of any record of any *Chamaecyparis nootkatensis* tree in the vicinity there has been speculation as to the male parent, but from the foliar characteristics these are now both regarded as Leyland cypresses. Of the two clones 'Stapehill 20' by its behaviour least supports this conclusion. It is susceptible to drought and the foliage it has abandoned to assist its recovery hangs on and is most unsightly. It forms a much more open tree. It received a limited circulation, but is no longer being propagated.

-- **'Stapehill 21'**. Ovens, Bright and Mitchell l.c. (Flat-pinnate type). This clone is luxuriantly furnished, very much like 'Leighton Green' and marginally even more vigorous. It propagates easily from cuttings.

Two other species in this hybrid Genus have been recorded. This adds to doubt as to how *Chamaecyparis nootkatensis* should be classified, because of its evident affinities with *Cupressus*.

× ***C. notabilis*** Mitchell, Conif. Brit. Is. 95 (1972). This hybrid between *Cupressus glabra* and *Chamaecyparis nootkatensis*, raised in 1956, also originates from Leighton Hall. It makes a narrowly conical tree with characteristics intermediate between the parents. It is interesting botanically but is of no particular horticultural value. Zone 8.

× ***C. ovensii*** Mitchell, Conif. Brit, Is. 96 (1972). This is a hybrid between *Cupressus lusitanica* and *Chamaecyparis nootkatensis* raised in 1961 by Mr Howard Ovens from seed collected at the Westonbirt Arboretum in Gloucestershire. It makes a vigorous, upright, sturdy plant with habit and foliage characteristic leaning towards its Nootka Cypress parent. Zone 8.

CUPRESSUS

Cupressaceae

A genus of 13 species or more (the number depending upon which authority you are following), occupying a wide range of soils and curiously scattered habitats. In the Old World they occur from Morocco to Iran and from the Himalayas to Southwest and Central China. In the New World they occur from the extreme Southwest United States, all through Mexico and on to Southern Guatemala and Northwest Honduras. But despite this wide range they are all remarkably similar, a fact that is reflected in the numerous attempts that have been made over the years to deal with the botany of the group. These attempts have resulted in much re-ranking and transferences to and fro that are very puzzling. Here I follow John Silba in *A revision of Cupressus* L. [Phytologia **49** (5): 416–420 (1981) and a further *Revision* in the same Journal **52** (5): 349–361 (1983) in which papers he recognises 13 species and 12 varieties. But almost certainly a cloud of uncertainty still hangs over the Cupressus Genus.

Although the cypresses as a group are hardy, disease-free and long-lived, they do not have the appeal for garden culture of some other genera such as *Chamaecyparis*. In result, they are not (save for a number of cultivars) widely found outside arboreta and specialised collections (and some of the species are but rarely represented even in such favoured spots). From *Chamaecyparis* they are distinguished by their larger cones composed of 6 to 14 scales, each with numerous seeds.

Size and habit Evergreen shrubs or trees 10–60m tall, columnar or conical even in maturity.

Trunk Usually single. *Bark*: Usually splitting, in long strands.

Branches Ascending or horizontally spreading.

Branchlets Usually more or less evenly disposed around the branch and either cylindrical or quadrangular in section, clothed with leaves. The ultimate shoots are held erect, unlike those in *Chamaecyparis* which always nod, more or less.

Winter buds Hidden.

Leaves Foliage the same colour on both sides. Adult leaves are scale-like, 1–3 mm long, usually closely appressed to the twig, decussate, minutely toothed overlapping, triangular in section, bluntly acute, ridged on the back and with a narrow, transparent edge. Sometimes glandular.

Flowers Borne on separate branchlets. The male cones, on the ends of short branchlets are cylindrical to oblong, 3–7 mm long, often brightly coloured; the female flowers are borne erect.

Mature cones Large; globose to ovoid, woody, formed of close-fitting, peltate scales, each with a central umbo, ripening to dark brown second year but hanging on the tree, sometimes for years.

Seeds numerous on each scale; compressed and angled, with two narrow wings.

Cultivation All the species are reasonably hardy, given suitable positions. The Cypresses are not fastidious as to soil, tolerating even a chalk sub-soil. They succeed in dry conditions but are not happy in damp situations. They are intolerant of root disturbance, so must be given permanent homes when very young. Reject pot-bound plants. Propagation is from seed which germinates readily.

The guide on pp. 290–291 is not a complete key, but used in conjunction with the genus and species descriptions, it will aid identification.

C. abramsiana Wolf, El Aliso **I**, 215 (1948). See *C. goveniana* var. *abramsiana.*

C. arizonica Greene in Bull. Torr. Bot. Club **9:** 64 (1882); *C. benthamii* var. *arizonica* (Greene) Mast. 340 (1896). Rough-bark Arizona Cypress.

Size and habit Narrowly conical tree 20–25m with a dense, rounded crown.

Trunk Thick. *Bark:* Greenish-brown, separating in long, regular fissures, slightly stringy. On old trees, grey and furrowed.

Branches Short, dense, horizontal.

Branchlets Rather stout, quadrangular; irregularly arranged around the shoot, standing off at ± 80°, light green.

Winter buds Hidden.

Leaves All scale-like (adult); thick, sharply pointed, keeled, almost or entirely without, white resin sports; pale green.

Mature cones On stalks, globose or oblong, 1.5–2.5cm long; glaucous blue ripening to dark reddish-brown and remaining closed. *Scales:* 6–10; flat or slightly depressed on back, each scale with a small but prominent mucro (point).

Seeds 8–10 per scale; oblong or irregular, dark brown. *Wings:* Two, narrow, thin. Cotyledons 3–5.

Habitat U.S.A.: S.W. Arizona, New Mexico, Mexico.

Cultivation The type (i.e. var. *arizonica*) is rare in cultivation in Britain, most of the cultivars being, in a strict sense, selections from var. *glabra*. Zone 7.

A number of minor variants, mostly from very limited areas, are given specific status by some authorities; others treat them as varieties of *A. arizona*. Here I take the latter path except for var. *glauca* which I prefer to leave as *C. glabra* Sudworth. This apparently inconsistent treatment best preserves widespread usage in a large, widely planted group of cultivars, and avoids the confusion that would result from acceptance of a swing in botanical opinion that may not even be final.

-- **'Aurea'**. A seedling raised at the Woods and Forests Department Nursery, National Park, Belair, South Australia in 1955 and introduced by Kemps Nurs. South Australia. It makes a small broadly conical tree 10–12 m high, with prolific branching resulting in a dense compact crown of almost horizontal branches, with golden-yellow foliage, the intensity of which varies from year to year, and in any year can fade to green by winter. To be consistent, this should probably here be listed under *C. glabra*.

-- var. *bonita*. Lemmon, 76 (1895). Here listed as *C.glabra*.

-- **'Clemson Greenspire'**. A name registered under this species by the Dept. of Forestry, Clemson University, Carolina, USA, in 1980 for a fast-growing, easily rooted clone with very full, lacy, bright green foliage. It was raised from seed by E. Wilson of Columbia, South Carolina, in 1962.

-- var. *glabra* (Sudworth) Little (1966). See *Cupressus glabra*, and Note above.

-- 'Janice'. Name registered in 1984 under this

GUIDES TO IDENTIFICATION OF CUPRESSUS SPECIES

	FOLIAGE in flattened sprays	in 3-dimensional sprays	HABIT of growth	COLOUR of first-year shoot	FOLIAGE colour	FOLIAGE characteristics	RESIN SPOTS present	Not present	Aroma of crushed foliage □	Remarks
C. arizonica		X	Broad conical	Orange-brown	Rich green		X			Bark thick, regularly fissured
var. stephensonii		X	Fastigiate tree	Yellow-green to red	Pale grey-green		X		Lemon verbena	
C. bakeri		X	Conical tree		Dark grey-green	Very Acute tip	X		Sage or cigars	In 4 rows; conspicuously glandular
C. chengiana	X				Grey-green	1–1.5 mm Acute tip, incurved				Flat, open sprays. Glandular
C. corneyana	X					1–2 mm incurved				In 4 rows. Sharp yellow tip
C. duclouxiana		X	Columnar	Yellow-brown	Blue-green	Incurved, acute tips		X	Resin (slight)	Terete
C. funebris See CHAMAECYPARIS funebris										
C. glabra		X	Upright conical	Orange-brown	Pale blue-grey	Sprays upright	X		Grape-fruit	Bark smooth. Leaves glandular. Many cvv.
C. goveniana		X	Branches ascending	Purple to brown	Dark to v.d. green	Obtuse, incurved		X	Petrol or citrinella	Shoots long and slender
var. abrahamsiana		X	Vigorous upright	Purple to red	Dark green	Tips free,		X	Soap or lemon	Bark furrowed; leaves thick, rounded
C. guadalupensis		X	Upright branches	Yellow-green to brown	Sea-green (variable)	Pointed, terete		X	Resinous (Feint)	Long slender sprays
var. forbesii		X	Open; irregular	Dark chestnut-red	Bright green	Acute, finely tipped		X	Sweetish: petrol	Leaves terete with whitened edges, sharply acute
C. lusitanica	X		Upright	Pink to dark red	Dark to grey-green	Acute, tips free		X	None	Bark furrowed, peeling in strips
C. macnabiana	X		Broad, bushy widespread	Bright brown	Dark green	Very fine; appressed swollen ends	X		Lantana or bergamot	Blunt tips sprays; glandular

□ According to Mitchell or other writers.

GUIDES TO IDENTIFICATION OF CUPRESSUS SPECIES

	FOLIAGE in flattened sprays	in 3-dimensional sprays	HABIT of growth	COLOUR of first-year shoot	FOLIAGE colour	FOLIAGE characteristics	RESIN SPOTS present	Not present	Aroma of crushed foliage □	Remarks
C. macrocarpa		X	Tall tree Brs. ascendg.	Pink to dark brown	Deep green	Terete, tips thickened		X	Lemon-verbena	Leaves with incurved margins. Large fruits
C. sempervirens		X	Fastigiate	Copper-brown to red-green	Dull green	Terete, 1.0 mm rounded		X	None	Leaves sharply tipped
C. torulosa	X		Rounded, pendulous	Orange to dark grey	Yellowish green	1.25 mm incurved		X	New mown grass	Bark peeling. Leaves glandular
var. cashmeriana	X		Pendulous	Red-brown	Pale blue, whitish	Sharply pendulous sprays	X			Not hardy; long pendulous sprays

□ According to Mitchell or other writers.

Figure 170. *Cupressus bakeri*. Detail of foliage. (10)

species for a seedling four years old. To attempt to select a new cultivar at this age, before its mature behaviour can possibly be predicted, seems to me to be quite absurd, and to register a name for it, only little less so. It may be that the absence of any connection between the registration of a name, on the one hand, and the securing of plant variety rights or a plant patent, on the other, is not always understood.

-- var. ***montana*** (Wiggins) Little in Madrono **18:** 163 (1966); *C. montana* Wiggins, Contrib. Dudl. Herb. **1:** 161 (1933). A tree to 20 m, with grey, furrowed bark and dark green leaves, prominently glandular. The cones open when ripe.

-- var. ***nevadensis*** (Abrams) Little in Madrono **18:** 164 (1966); *C. nevadensis* Abrams in Torreya **19:** 92 (1919). Piute Cypress. A tree to 20 m, with grey, furrowed bark and light grey-green

Figure 171. *Cupressus (now Chamaecyparis) funebris.* (5)

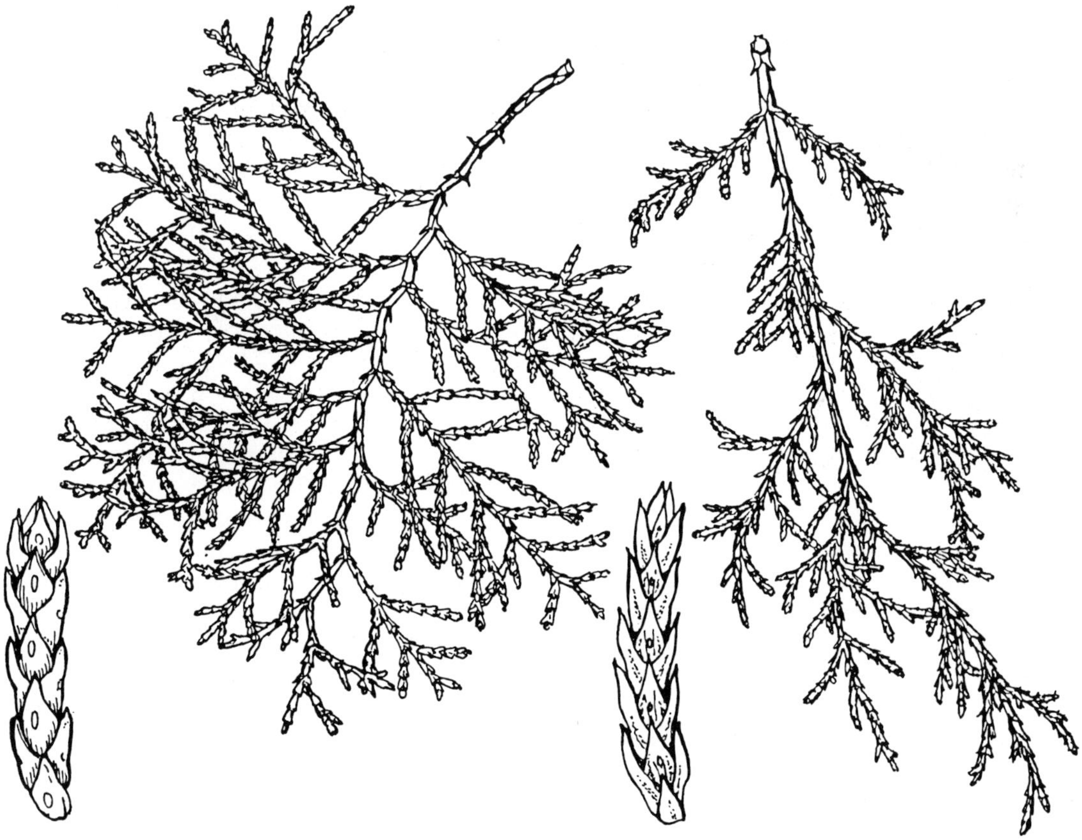

Figure 172. (Right) *Cupressus lusitanica* and (Left) var. *benthamii*. Detail of foliage. (5)

leaves, prominently glandular, the cones remaining closed when ripe, as is normal in this species.

-- var. *revealiana* Silba in Phytologia **49** (4): 393 (1981). A minor variant no longer distinguished. Probably not in cultivation.

-- var. ***stevensonii*** (Wolf) Little in Phytologia **18:** 164 (1966); *C. stephensonii* Wolf in El Aliso **I**, 125 (1948). A little known local form, intermediate between *C. arizonica* and its var. *glabra*.

C. bakeri Jepson, Fl. Calif. **I:** 61 (1909); *C. macnabiana* var. *bakeri* (Jeps.) Jeps., Man. Fl. Pl. Calif. 58 (1923); *C. bakeri* ssp. *matthewsii* Wolf in Aliso **1:** 83 (1948). Modoc Cypress.

Tree 10–30m with an open crown; *Bark:* thin, reddish, peeling in thin strips; branch system irregular. *Branchlets:* quadrangular, slender. *Leaves:* dark or bright green, acute, keeled, with a conspicuous gland marked by a white resin dot, lemon scented. *Cones:* sub-globose, green, ripening the second year to brown or silvery-grey. *Scales:* 4–8, abruptly contracted into a short umbo. *Seeds:* about 50 to each cone, 3–5 mm long, with narrow wings. *Habitat:* California and Oregon on Pacific coast. Zone 6. Sent to the Arnold Arboretum in 1917, but still rare in cultivation. A sub. sp. ***matthewsii*** Wolf in El Aliso **1:** 83 (1948) - but questionably distinct - is even rarer. (Illus. No. 170)

C. cashmeriana Royle ex Carrière, 161 (1867). Now known as *C. torulosa* var. *cashemeriana*.

C. chengiana Hu, Taiwania **10:** 57 (1964). A columnar tree to 30 m; *Bark:* brown-grey, fissured; *Leaves:* in four rows, grey-green, obtuse, glands distinct. *Cones:* globose, 15–25 mm. *Scales:* 8–10. *Seeds:* 4–5. *Habitat:* China, Szechuan. Zone 6. Probably not yet in cultivation.

C. corneyana Carrière (1855). The Bhutan Cypress. This name is used for a species that reaches to 40m in the wild, which shews a strong affinity with *C. torulosa*. Carrière, under the general heading '*Espèces peu connues*', gives a recognisable description, but the reference in his synonymy to Knight is based on nothing more than the guesswork that was prevalent amongst writers in the nineteenth century. Knight and Perry, to whom the tree was even more '*peu connu*', introduce it as " 'Mr Corney's Cypress' with a possible origin of Japan or North China". Prof. Henry, in *Trees of Gt. Britain and Ireland* cites a letter from Knight and Perry from which it seems that the species was raised on the Chelsea nursery from seed received from a Mr. Corney, a merchant in China, who occasionally sent seeds to them, and was named by them in his honour. There is no suggestion that the tree was regarded otherwise than a species, so no use of the word as a cultivar name is justified.

Much confusion over the years has resulted from a determined impulse to identify a particular tree felt by someone having access only to an inadequate library (often only a single book) not covering all the possibilities. Eliminating all the improbables at his disposal one by one, such an enquirer is apt to reach the conclusion that his tree MUST BE the only entry finally available to him. This may well not be the case. When it is not, a mistake has been made, and who can say what trouble it may cause?

C. duclouxiana Hickel in Camus, Les Cypres 91 (1914); *C. sempervirens* Franchet 263 (1899), not Linnaeus; *C. torulosa* Rehder and Wilson 54 (1916), not D. Don. Yunnan Cypress. A species related to *C. sempervirens*, but the cones are larger, the foliage finer and noticeably more grey in tone. *Tree* to 25–50 m with dark red-brown bark, finely furrowed; *Branches:* wide-spreading and upswept. *Branchlets:* very thin, 0.8 mm across, quadrangular, ascending; *Leaves:* 1–1.5 mm long, apex bluntly pointed, bluish-green with grey margin, faintly resinous scented. *Cones:* sub-globose, 1.5–3 cm, shining red-brown. *Scales:* 6–8, rather flat but with a small, thick umbo; *Seeds:* 5mm long with a narrow wing. Very rare in cultivation. Zone 8.

C. dupreziana Camus (1926). Now known as *C. sempervirens* var. *dupreziana*.

C. forbesii Jepson, (1922). Now known as *C. guadalupensis* var. *forbesii.*

C. funebris Endlicher (1847). Chinese Weeping Cypress. Now known as *CHAMAECYPARIS funebris.* (See p. 196)

C. gigantea. Cheng et Fu, Acta Phytotax. Sin. **13:** 85 (1975). Giant or Tsangpo Cypress. A newly described species, which, according to Silba is: *Tree:* to 45m. *Bark:* grey-purple-brown, fissured. *Leaves:* blue-grey to green, blunt-tipped; glands noticeable. *Cones:* 16–20 mm. *Scales:* 12, umbos prominent. *Habitat:* given as China: Tibet; Tsango River, Linchi westwards. Zone 7.

C. glabra Sudworth in Amer. For. **16:** 88 (1910); *C. arizonica* var. *glabra* (Sudworth) Little in Madrono **18:** 162 (1966); as var. *bonita* Lemmon, W. Amer. Cone-bearers 76 (1895). The Smooth Bark Arizona Cypress. This quite distinct tree was originally introduced and planted along with the typical form. When distinguished, it was given specific status but is here treated as a botanical variety, following Silba [22 (1984)]. Apart from the smoothness of the bark it differs in making a smaller tree, in the bark being orange-brown to purplish-brown and in bearing dense foliage with leaves that are distinctly blue-grey in colour and noticeably glandular, this showing as a white resin dot on each leaf. The bruised foliage smells of grapefruit. It carries its yellow flowers in profusion during the winter. It is reliably hardy and is tolerant of alkaline soils and drought. Zone 7. (See Illus. No. 173)

-- **'Arctic'**. A selection listed in New Zealand that makes an open, upright tree with emerald green foliage with glaucous blue young growth in the Spring.

-- **'Blue Ice'**. A New Zealand selection, also making an upright tree, but with blue-grey foliage.

-- **'Blue Pyramid'**. A medium size conical form with very pale blue-grey foliage, almost white in overall effect. Received by Duncan and Davies, New Plymouth, New Zealand from Ken Bruns, Timaru. This rapid grower retains a compact, symmetrical habit for some years, but eventually becomes more broad and open.

-- **'Blue Spire'**. Found and introduced by P. C. Nitschke, S Australia in a garden at Tanunda, S. Australia. It has good blue foliage and branches that hang vertically down the trunk. Annual growth approximately 90 cm.

-- 'Blue Streak'. I understand that this was an early favourite in New Zealand that has been superseded by later selections. The relationship of the recent introductions in New Zealand to the older clones in Europe is not clear. Horticultural nomenclature being essentially global, only action on a worldwide basis whenever a new cultivar name is raised can prevent the risk of confusion arising from a proliferation of names on a regional basis. The international registration of the proposed name must be accompanied by the simultaneous publication of an adequate 'protologue' (See p. 54), with a copy deposited with the World Conifer Data Pool for general reference. A counsel of perfection would also include the distribution of clonal material to selected institutions throughout the world.

-- **'Canny'**. This clone has the same origin as 'Aurea'. 'Canny' is slower growing and less compact than 'Aurea', with a more open habit and branches that upturn at their tips, resulting in an irregular outline to the tree. The golden foliage on both cultivars is replaced by varying amounts of dusky greenish foliage in regions that have cool, wet winters. In warm summers the golden colour becomes dominant. It is resistant to sun-scorch. Named for J. Canny who discovered these two plants. ('Canny's Golden'). (See p. 289)

-- **'Carolina Sapphire'**. Name registered in 1987 by the Department of Forestry of Clemson University, Clemson. SC, USA, for a clone raised from seed in 1961. It forms a conical tree broader in habit than 'Blue Pyramid', with stringy but dense steel-blue foliage. It is claimed that it roots well from cuttings.

Figure 173. *Cupressus glabra (C. arizonica).* Detail of foliage and cones.

Figure 174. *Cupressus glabra* 'Pyramidalis'.

Figure 175. *Cupressus glabra* 'Compacta'. A reliable dwarf form.

-- 'Clemson Greenspire'. Name registered under *CUPRESSUS arizonica*. (See p. 289)

-- **'Compacta'**. Schneider in Silva Tarouca 174, Ill. 178, (1913). An attractive dwarf, forming a globose bush with congested, grey-green, adult-leaf foliage and red-brown branchlets. (See Illus. No. 175)

-- cv. **'Conica'**. den Ouden 140 (1949). Conical; branches ascending; branchlets crowded, widely set, short, stiff, bluntly-pointed; leaves tightly appressed, conspicuously blue-bloomy. Imported from France, origin unknown. It is probably the clone most widely distributed in U.K.

-- **'Crowborough'**. Welch, 196 (1979). This is a compact form with congested but rather stout foliage that was raised from seed by H. E. Bawden, then living at Crowborough in Kent, UK. After ten years the mother plant was a round-topped bush a little under 2m high. It has no connection with Australia.

-- **'Fastigiata'**. Krüssmann, 116 (1972). Growth narrowly upright but compact; foliage blue-grey. Origin is given as Gino Bartolini. Considered by Krüssmann to be an improvement on 'Conica'.

-- **'Fastigiata Aurea'**. Krüssmann, (1983). Of conical, upright habit, rather open, branchlets sulphur-yellow. Of presumed French origin. In both this and 'Fastigiata', legitimacy depends upon whether publication before 1st January 1959 can be established.

-- **'Garee'**. A clone selected by Mr. Garee, Noble Nurs, Noble, Oklahoma in the early 1960's. Introduced by Monrovia Nursery, Azusa, Cal. U.S.A., 1973. Selected for hardiness and the rich silvery blue-green foliage. The habit is conical and the branches have a distinctly whipcord texture. It is said to be drought resistant.

-- **'Glauca'**. Woodall in Garden **88:** 16 (1916). This name was published for a clone that was conical and regularly branched; with juvenile, bright bloomy leaves; slightly hardier than 'Conica'.

A new name will need to be found for the columnar plant raised from seed in Canberra, Australia in the late 1950s. This clone has a graceful habit and dense bluish-green foliage, and makes an outstanding specimen tree. It was introduced into the USA and listed in the Monrovia Nursery Cat. 34 (1978). Other clones in cultivation may be in the same situation.

-- '**Golden Pyramid**. A strong-growing golden clone, reported by Duncan and Davies of New Zealand as having been received from Vic. Levy of Brisbane, Queensland, Australia.

-- '**Greenwood**'. Raised at the USDA Field Sta., Woodward, Oklahoma. Selected for its good green colour by Mr. Garee and named for Greenwood County, Kansas, his boyhood home.

-- '**Hodginsii**'. Lord, E. E., Trees Austral. Gard. 1948. A tree to 10 m, erect habit, compact growth, foliage a deep blue-grey, conspicuously resin-spotted, strong smelling if bruised, rough to the touch. It was raised from seed on Hodgins Nurs. Essendon, Victoria, Australia and introduced by them in 1936.

-- 'Janice'. Name registered under *CUPRESSUS arizonica*. (See p. 289)

-- '**Lakeview**'. A very slow-growing plant introduced by E. H. Lohbrunner of Victoria, BC., Canada. It makes a grey mound 20 cm high and 30 cm across after ten years. It is difficult to propagate.

-- '**Pyramidalis**'. Considered by many to be an improvement on 'Conica', making a dense, compact tree of moderate size with bright blue-grey foliage, each leaf with a prominent white dot (resin.). It carries the usual crop of bright yellow male flowers in the winter and cones freely. Introduced by Hillier Nurs. Winchester, Hants. in 1918. (See Illus. No. 174)

-- '**Silver Smoke**'. Listed Duncan and Davies, New Zealand, 1984. Described as compact, conical, with bright silvery-grey foliage.

-- '**Sulfurea**'. Compact, upright habit, foliage yellow-grey. It is said to have originated in France, but from the description and Mr van Hoey-Smith's picture in the book *Conifers* it would seem to be very near to 'Fastigata Aurea'.

-- '**Variegata**'. A slow-growing conical tree with blue-green foliage nicely interspersed with creamy-white variegation. Effective, but requires a sheltered location. It was introduced on the Minier Nurs, Angers, France.

C. ***goveniana*** Gordon ex Lindley in Journ. Soc. Lond. **4:** 295 (1849); *C. pygmaea* (Lemmon) Sargent in Bot. Gaz. **31:** 239 (1901). Gowen Cypress. Columnar tree to 20m. or more, with a conical, or sometimes broad, open crown. *Bark:* Smooth, dark reddish-brown, pale grey-brown on old trees, peeling in strips. *Branches:* spreading, *Branchlets:* irregularly set, quadrangular, about 1 mm broad, odorous when rubbed. Leaves: ovoid, 1–1.3 mm long, acute or sharply acuminate, densely appressed, glossy dark- to vary dark-green, paler at edges, some whitish bloom; with an inconspicuous dorsal gland. It is free flowering. *Cones:* Globose or short-oblong 15–18 mm broad, rough, glossy dark brown, persisting many years after shedding the seeds. *Scales:* 6–8 with a rounded or pointed mucro in the middle; *Seeds:* 10–12 on each scale, 3–4 mm long, glossy brown or nearly black, with narrow wings. Mitchell likens the scent of the crushed foliage to oil of citronella or, in some trees, petrol. This, he adds, at once distinguishes this species from *C. macrocarpa* in which the foliage is similar in appearance. It is uncommon in Britain. *Habitat:* California, coast regions on Monterey and Mendolina. Discovered and introduced in 1848 by Hartweg, who sent seeds to the Hort. Soc. at Chiswick, Britain. Zone 8 (See Illus. No. 176).

-- var. ***abramsiana***. (Wolf) Little in Phytologia **20:** 435 (1970); *C. abramsiana* Wolf, El Aliso **1:** 215 (1948). Santa Cruz Cypress. *Tree* to 18 m. *Bark:* Purple or brown, finely ridged. Branching irregular. *Leaves:* bright green in broad sprays, bluntly acute. *Cones:* Ovoid, 20–30 mm long, green, strongly lemon-scented when crushed, ripening to grey-brown. *Scales:* 8–10, with a stumpy umbo. *Seeds:* Non-warty, flattened, often glaucous. *Habitat:* Santa Cruz County, California.

-- **'Pendula'**. Henry 1172 (1910); *C. california* Carr. 127 (1855). Rather dense, graceful tree; branches long, pendulous; leaves often spreading, apex sharply pointed.

-- var. ***pygmaea*** Lemmon, Handb. West Amer. Cone-bearers, ed. 3, 77 (1895). An unfortunate name for a tree that is, if anything, more vigorous than the type variety, differing otherwise mainly in its dull, blackish-green leaves and its usually glossy seeds. Like all trees it responds by slower growth in uncongenial conditions and, in impoverished soil, might possibly live down to its name! Mendocino County, California.

C. guadalupensis S. Watson in Proc. Amer. Acad. **14:** 300 (1879); *C. macrocarpa* var. *guadalupensis* (S. Wats.) Mast. in Gard. Chron. Guadalupe Cypress. 62 (1895). *Tree:* 12m or more with wide-spreading but usually dense crown. *Bark:* Grey-brown to deep red, shining, cracking off in patches, showing grey beneath; *Branchlets:* slender. *Leaves:* long, slender, variably green to blue-green, acute or acutish, obscurely glandular, *Cones* irregular but more or less globose 25–40 mm diam. *Scales:* 6–8, very thick, light grey, flattened, umbos inconspicuous. *Habitat:* Guadalupe, Islands, southern and Lower California. Introduced into Europe in about 1880. Not hardy. An endangered species. (See Illus No. 178)

-- var. ***forbesii*** (Jeps) Little in Phytologia **20:** 435 (1970); *C. forbesii* Jepson in Madrono **1:** 75 (1922). *Tree:* to 10m, with a spreading, irregular crown. *Bark:* Dark red, scaling off and exposing lighter colour beneath. *Branchlets:* Irregularly arranged. *Leaves:* 1.5 cm long, bright green with a whitened margin, acute with a short sharp tip. *Cones:* 20–30mm diam, grey-brown. *Scales:* 6–10, thick with an inconspicuous micro. *Seeds:* Brown, non-glaucous. Very rare in cultivation. *Habitat:* South-West California; Mexico.

C. lambertiana. The origin of this name will be found in Gordon, Pinetum, 65 (1858). The

Figure 176. *Cupressus goveniana*. Detail of foliage and cones.

Figure 177. *Cupressus lusitanica.* A young tree on the Minier Nursery, Angers, France.

Figure 178. *Cupressus quadalupensis.* Detail of foliage.

name, although widely used, was never validly published, so it had to give way to *CUPRESSUS macrocarpa* under the rules, but not before the name (and presumably some seeds) had reached Australia. Early clones distributed under the wrong species name were 'Horizontalis' and 'Horizontalis Aurea'.

C. lusitanica Mill. no. 3 (1768). Mexican Cypress. *Tree:* 30 m. *Bark:* Reddish-brown, furrowed and peeling in long, narrow strips. *Branches:* widely spreading – ascending or pendulous. *Branchlets:* short, with red bark, quadrangular, irregularly disposed, more or less pendulous, not flattened. *Leaves:* 1.5–2 mm long, tightly appressed, ovoid, glaucous grey-green, apex often long-pointed with tips free. *Cones:* small, about 15 mm broad, globular, at first whitish green, ripening to dark brown and opening immediately. *Scales:* 6–10, usually with a hooked micro. *Seeds:* 8–10 on each scale. *Habitat:* Mexico and Guatemala. Probably introduced from Mexico by the Spanish or Portugese in ± 1500; known in English gardens since 1682, Chelsea. Only hardy in the warmer temperate regions, rather commonly planted in S. and W. France. (See Illus No. 172)

-- **'Adelaide Gold'**. T. R. Lothian in Adelaide Bot. Gar. Journ. **1**(3): 193 (1978). Growth habit and morphology similar to the species but foliage and branchlets near growing points golden yellow.

-- var. ***benthamii*** *C. lusitanica* var. *benthamii* (Endl.) Carrière 155 (1867); *C. benthamii* Endl. 59 (1847); *C. lusitanica* var. *skinnei* Henry 1177 (1910). Bentham Cypress. Narrow, pyramidal tree to 20m, branches ascending and regularly spaced; foliage dimorphic, ultimate branchlets flattened and compressed. *Leaves:* pale green, 1.5 mm long, glandular, acute, with incurved tip, set in even fernlike pinnate sprays; slightly resin scented. *Cones:* Sub-globose, brown, opening as soon as they ripen. *Habitat:* Mexico at 1800–2400m and doubtless sporadic throughout the whole range of the species. (See Illus No. 172)

-- **'Glauca'**. Henry 1178 (1910). Blue Mexican Cypress. *Branchlets:* not flat; *Leaves:* very bluish-glaucous, glands distinct; found at Montserrat, Portugal.

-- **'Glauca Pendula'**. Hillier, Cat. **36T:** 10 (1925). A beautiful glaucous drooping tree. If grown as a bush by stopping leader growth at 1–1.5m it will develop like a giant-size 'Pfitzer Glauca' Juniper. In an open spot with plenty (repeat-'plenty') of space to develop and be seen from a distance it makes a spectacular lawn specimen.

-- **'Nana'**. A very slow-growing, congested 'dwarf conifer' of rounded, dome-like habit and blue-green foliage of which I have not been able to trace the origin.

C. macnabiana A. Murray in Edinb. New Phil. Journ. **1:** 293 (1855). MacNab Cypress. *Tree:* or shrub 10–25 m, with a broad, bushy crown. *Bark:* dark reddish-brown and scaling – grey-brown on old trees. *Branches:* numerous, ascending. *Branchlets:* fine, terete, evenly spread in flattened sprays; *Leaves:* 1.2–1.5 mm long, tightly appressed, ovoid, obtuse, thickened and blunted at the apex, shiny dark- to glaucous-green with a distinct gland shewing as a white resin spot. *Cones:* in clusters, oblong-globular, 2–2.5cm diam, grey-green, ripening to dark, glossy reddish-brown or grey-brown. *Scales:* usually 6, with a conical, curved mucro. *Seeds:* small, angular, very numerous; winged. *Habitat:* California; Introduced 1854 by W. Lobb, who sent it to the Veitch nurseries. UK. (See Illus No. 182)

-- **'Fastigiata'**. Distributed by Wansdyke Nurs. Devizes U.K. This clone originated from a tall tree of unrecorded origin at the Westonbirt Arboretum near Tetbury. Glos. which has since died. It is not reliably hardy in UK, but is a spectacular columnar tree.

-- **'Lilah'**. Regd. by A. Stewart in 1984. From seed sown in 1980, this plant has branchlets tinged red and a regular, persistent variegation. But my remarks above, under *C. arizonica* 'Janice' apply here also.

-- **'Sulphurea'**. Berckmans ex Rehd. 209 (1923). Branchlets yellow-tipped.

C. macrocarpa Hartw. in Journ. Hort. Soc. Lond. **2:** 187 (1847); *C. hartwegii* Carr. in Rev. Hort 233 (1855); *C. lambertiana* Carr. 124 (1855); *C. macrocarpa* var *lambertiana* (Carr.) Mast. 343 (1896). Monterey Cypress.

Size and habit Tree 20–25 m, columnar when young, later broadening (rather variably from tree to tree) and developing a rounded crown, often becoming flat-topped in old age.

Trunk Single and stout; sometimes multiple. *Bark:* On young trees pale brown or grey, dividing into broad, scaly ridges at random angles.

Branches Horizontal or upswept; varying from tree to tree; often of uneven lengths with the longer branches protruding.

Branchlets Short, irregularly spreading all round the branch and pointing forwards at 45–60°; 1–1.5 mm broad; dull- to dark grey-brown. Foliage is dense, in forward-pointing plumes.

Leaves Scale-like, to 1.8 mm long, in decussate, opposite pairs; rhombic-ovate, appressed except towards the broad and swollen tip; dark green with a paler margin, faintly bloomed white all over some leaves; sometimes with an inconspicuous gland not central on the leaf.

Mature cones Short-stalked, solitary or two or three together on short branchlets, globose or broadly ellipsoid, 25-40mm across; shiny chestnut-brown to dark grey. *Scales:* 8–10mm, varying in size, wrinkled, convex or slightly indented with a waved margin and a short central hooked boss. *Seeds:* Numerous, often as many

Figure 179. *Cupressus macrocarpa*. Detail of foliage, with cones.

as 20 to each scale; small, warty with long, narrow, brown wings.

Habitat U.S.A. Monterey County, Point Cypress, nearly at sea level. Zone 7.

Cultivation Introduced in 1839 by Lambert who sent seeds to the Hort. Soc. at Chiswick, England. One of the hardiest species. This species, well known as the fast-growing 'Cypress' (at one time so popular for hedges and shelter planting but now ousted by the bigeneric hybrid *Cupressocyparis leylandii*) has given rise to a number of seedling variants, particularly golden forms. Many of these have originated in Australia or New Zealand where they are very popular. Mr. Davies of the nursery firm of Duncan & Davies of New Plymouth, New Zealand once told me that in that country they never grow 'Macrocarpa' in pots, a practice he thinks may largely contribute to their reputation for instability in this country.

There is also a small group of very diminutive forms. These are not garden plants at all. In some cases they are not hardy out of doors in a hard winter, but they are sought after by collectors and are usually a point of interest to visitors, partly, I think, because most of them know the species well in its usual, arboreal form and are so surprised to see such a tiny representative thereof a few centimetres high growing in a pot, in such striking contrast to all the other *Cupressus macrocarpa* trees they know.

This is the Cypress that is most widely planted for ornament in gardens today. It is also one of the best species of tree for use near the sea, the reason being that its own very restricted native habitat is almost at sea level.

Surprisingly, the golden-yellow cultivars behave in maritime situations even better than the green form typical of the species. (See Illus. Nos. 179, 180, 181)

-- '**Arctic**'. A selection from seed made by Ewen B. Perrott, Tanilba Nurs, Pukeatua, New Zealand and introduced to the trade in 1975. It is marked by the young foliage being intensely 'frosted' with pale cream, up to 80%. In young trees especially, the branches are fourway horizontally opposed. Considering the lack of chlorophyll, the growth is quite vigorous.

Figure 180. *Cupressus macrocarpa*. Trunk of an old tree at Montacute, Somerset, U.K. (10)

'**Antarctic**' is a similar selection on the same nursery and considered by Mr. Perrott to be the better of the two.

-- '**Aurea**'. Certainty can only be reached by some serious delving into early records in Australia itself, but I am reasonably satisfied that the tree traced by Den Ouden and Boom [140 (1965)] to the entry *C. macrocarpa lambertiana aurea* in Duncan and Davies' 1953/4 catalogue, has the prior name 'Horizontalis Aurea' (see below). But it is now so widely known as 'Aurea' that that name must at least be accepted as a 'commercial synonym'. It is the Australian equivalent of the equally venerable selection 'Lutea' in Europe.

-- '**Aurea Saligna**'. It would seem that a strict application of the rules would require the name 'Coneybearii Aurea' to be used. But 'Aurea Saligna' is in such general use as to justify its acceptance as a 'commercial synonym'.

Figure 181. *Cupressus macrocarpa*. Note the tree's 'spiky' silhouette. (10)

-- **'Berrigold'**. S. Austral. Nurs. Assoc. Bull. 30: (2): 7 (1972). A golden form at first distributed by Lasscock Nurseries Ltd, Lockleys, S. Australia either as 'Raywood Gold' or as 'Aurea' (Berri Strain).

-- **'Brunniana Aurea'**. R. E. Harrison, Trees & Shrubs 117 (1959). A clone described as conical in habit, retaining its golden colour well throughout the year. It was probably raised by Geo. Brunning and Sons Nurs, Victoria,

Figure 182. *Cupressus macnabiana*. Detail of foliage.

Australia and was listed by them in 1915.

-- **'Cascade'**. A clone introduced by Cedar Lodge Nurseries, Egmont Road, New Plymouth, New Zealand in 1988. It has foliage resembling *Chamaecyparis lawsoniana* 'Imbricata Pendula', but forms a wider plant.

-- **'Charles Détriché**. A more manageable name suggested by Krüssmann [117 (1983)] for a very old cultivar formerly listed as 'Souvenir de Charles Détriché'. It is of conical form, more columnar than 'Lutea', and carries delicately branched golden-yellow foliage.

-- **'Coneybearii Aurea'**. Welch, 197 (1979); *C. macrocarpa* 'Aurea Saligna' Lord, Trees and Shrubs, (1948); *Cupressus coneybearii aurea* Hazlewood Bros, Epping, N. S. W. Nurs, Cat. 1933. Raised from seed at Woolrich Nurs. Olinda, Victoria, in the mid 1920's. A clone with golden threadlike foliage. It requires stem-training, but a 3–4m specimen with a dense thatch of weeping branches is very impressive. Such a tree, unless 'top worked' requires permanent staking since neither the root-system nor the trunk will support such a heavy top.

-- **'Crippsii'**. R. Smith 16 (1872); Gordon 93 (1875). An old clone with short, stout, stiff, widespreading branches, numerous branchlets bearing juvenile foliage, and with the young growth creamy-white. Still worth planting. It was raised from imported seed on the Cripps

Nursery, Tunbridge Wells, U.K. about 1850.

-- **'Donard Gold'**. Dunster in Gard. Chron. 28 (1946)(I). Leaves deep golden-yellow, it tends to be a broader and more shapely tree than usual. Raised in the Slieve Donard nursery, N. Ireland.

-- **'Erecta Aurea'**. Harrison 117 (1959). An erect-growing, golden-foliaged tree introduced by Hazlewood Bros, Nursery, Sydney, Australia. Said to stand clipping well.

-- **'Fastigiata'**. Carr. in Jacques & Hérincq, Man. Pl. **4:** 326 (1857); *C. lambertiana* var. *fastigiata* Carr. 124 (1855). Columnar or narrow-pyramidal; branches upright. Probably introduced by Lambert in 1838 together with the typical form. Probably no longer identifiable.

-- **'Fine Gold'**. This clone was raised on the South Taranaki Nurseries, Hawera, New Zealand about 1927 from seed collected from a tree in New South Wales being grown as *C. macrocarpa* 'Lambertiana Aurea'. Out of a batch of ten seedlings (all 'yellows') this one was outstanding, even when small, for its noticeably finer foliage. The rate of growth is about half the norm and also the branches are shorter and the foliage more tufty. An outstanding quality that endears 'Fine Gold' to the nursery trade is that it roots readily from cuttings.

-- **'Globe'**. Den Ouden and Boom 141 (1965). *C. macrocarpa* 'Globosa' Hillier, Dw. Con. 32 (1964); var. *compacta* Hornb. Dw. Con. 59 (1923). (Nom. illegit.) is a moderately dwarf, globose bush with a dense habit and scale-like leaves. Its normal growth is compact, but if any loose shoots appear they should be cut out. At the time Hornibrook suggested a name it had not been propagated so was not, technically, a cultivar. So any name given to it in 1923 is rejected under Article 42 of the *Cultivated Code* because at that time the cultivar 'neither does nor did exist'.

-- **'Goldcrest'**. A medium size tree of narrowly columnar habit with dense rich golden foliage and one that retains its juvenile character for some years. It was raised on the Treseder Nurs. Truro, U.K. about 1948. It roots readily from cuttings of juvenile wood and is now one of the most popular forms in U.K.

-- **'Gold Cone'**. A strong-growing form of very compact habit, with ascending branches, the shape of a conventional candle-flame and with adult foliage of a good colour; raised on the Barthelemy Nurs. Wimborne. U.K.

-- **'Golden Halo'**. Plant Variety Rights in New Zealand were granted to D. J. Liddle in 1984. I understand it was developed from a close-growing mutation on 'Gold Spread', but it is reported to be liable to reversion.

-- **'Golden Pillar'**. Jackson, Planters' Handbook. An extremely fastigiate clone of a very good colour. During the 1939–45 war Rowland Jackman rescued a tree about 2m high that had been a victim of land clearance for food production on a nursery at Southampton. As it had hardly any surviving roots he divided it up for cuttings. These developed very slowly so it was not until 1955 that he was able to introduce it. This slowness to build up size is characteristic of 'Golden Pillar', but it should always be propagated from cuttings, since grafted plants lose the slim outline which is its chief attraction.

-- **'Goldspire'**. New cv. This is a recent introduction in U.K., raised from seed by J. W. Archer. It propagates easily, readily forming a good leader and is of good colour. 'Warnham Gold' appears to be a superfluous synonym.

-- **'Gold Spread'**. Welch 197 (1979). Two vigorous, widespreading, low-growing forms with golden yellow foliage have been received in U.K. from New Zealand. 'Gold Spread' has made a plant 3 m across in 10 yr and appears to be stable. Given plenty of space it makes a fine lawn specimen. The other plant is faster-growing and less stable, being apt to throw up a leader (consistently doing so if grafted). See 'Horizontalis Aurea' below.

-- **'Greenstead Magnificent'**. Hall in Seed and

Nursery Trader, Jan. (1970) as 'Greenstead Magnifica'. This interesting variant occurred as a bud mutation 8 metres up on a 40 year old specimen of 'Horizontalis' growing near a vineyard at Hope Valley, South Australia. After 8 years, propagations have formed plants 1 m wide by 30 cm high, so it would seem to be reliably dwarf. It forms a dense, low-growing plant, almost prostrate in habit, with blue-grey foliage pendulous at the tips, the blue colour being more intense when growing in shade. It was introduced to the trade by Mr. Roger Hall of C. F. Newman & Sons, Tea Tree Gully, South Australia, who first described it in an Australian trade journal and in *Australian Home Gardener* May 1970. It is only hardy in the most favoured parts of U.K., but will make a spectacular plant given winter protection.

-- **'Hodginsii'**. Lord, Trees Austral. Gard. 1948; Hodgins Nurs. Cat. 19 (1898) under *C. lambertiana*. It is a clone that originated in Australia having a horizontal branching system and foliage with a cream variegation.

-- **'Horizontalis'**. E. E. Lord, Trees Austral, Gard. 1948; Newman, Adelaide, Cat. 1894, under *C. lambertiana*. This is a very old cultivar recorded as being widely used in Australia for shelter-belts, being very hardy and salt tolerant. The distinctive feature is the markedly horizontal branching system. G. Brunning and Sons, St Kilda Nurseries, Melbourne, Victoria, in their 1898 catalogue state, "It is now more than a quarter of a century since this fine tree was introduced to Victoria and there are many fine specimens in the public parks and gardens around Melbourne".

-- **'Horizontalis Aurea'**. E. E. Lord, loc. cit.; Brunning Cat. 19 (1898) under *C. lambertiana*. A golden counterpart of 'Horizontalis'. Plants raised from cuttings at first display this horizontal habit and, unless a leader is encouraged to develop, remain widespreading plants difficult to distinguish from 'Golden Spreader'. I do not think there is much doubt but that the spreading forms are merely cultivariants, 'Golden Spreader' perhaps having arisen from material selected from slightly denser growth, the subject possibly of a minor mutation.

-- **'Lohbrunner'**. This slow-growing dwarf makes a dense globose little bush with attractive light green foliage. It was introduced by E. H. Lohbrunner of Victoria, BC., Canada. Ed. writes "About 20 years ago I saw growing in a garden in Victoria a compact, rounded form of *C. macrocarpa*. The owner told me that about eighteen years previously she had raised a number of Monterey Cypress for a hedge. These had grown quite large, but most of them were killed during the unusual freeze in 1955. The dwarf had survived and was then about 30 cm by 30 cm and she was disappointed that it would not grow. I offered her three dozen choice rose bushes for the plant: she said I must be insane and readily accepted. I find it quite easy from cuttings". After ten years it will reach 25 cm high by 30 cm across, but any tendency to put out vigorous shoots must at once be checked.

-- **'Lutea'**. Webster 43 (1896). Golden Monterey Cypress. Pyramidal, narrow; leaves bright yellow in the first year, green later; cones yellow. Raised just prior to 1895 in Dickson's Nursery, Chester, England. Although perhaps the first golden selection, this clone is still of value, especially for planting close to the sea. The branching system is horizontal.

-- **'Minima'**. Corley, Gar. Chron. **151, 18:** 326 (1962); 'Minimax' Den Ouden and Boom, Man. 141 (1965); 'Nisbets Gem' Welch. Dw. Con. 165 (1966).

We now come to a group of very diminutive forms that have long served as 'collectors' items in UK. 'Minima' is the hardiest of these. It was raised many years ago by R. Menzies at the Golden Gate Park, San Francisco, California, U.S.A., and was sent by the late James Noble to A. H. Nisbet who had made a notable collection at Gosport in England. It forms at first a low, rounded bush with neat, congested and mainly juvenile foliage and as it is much easier to propagate than the other forms it will always be in easier supply. In time it will form a rather untidy, sprawling bush, so it pays for some cosmetic pruning to be done from time to time. **'Pygmaea'** A. B. Jackson in

Gar. Chron. **151** (15): 267 (1938); Hornibrook, 90 (1939). This is rather slower-growing and soon develops adult foliage. The original plant at over thirty years of age was a spreading little tree about 45cm high and wide. It is much less hardy than 'Minima', and **'Woking'** requires winter protection under glass. It may now be lost to cultivation.

-- **'Pendula'**. Hillier, Cat. No. 49: 172 (1952). Branchlets drooping. Originated in a private garden in Ireland. There is also a very fine specimen in a house garden in Hillworth Road, Devizes, not far from the Wansdyke Nursery. Given time and space it could make a spectacular specimen.

-- 'Saligna Aurea'. A mistake. See 'Coneybearii Aurea'.

-- **'Sulphur Tip'**. This was found as a stray seedling by D. W. Hatch, Chantry Nurs. Honiton, Devon, U.K. when visiting the Knaphill Nurs. Woking, Surrey in company with J. W. Archer during the early 1960s. It is a slow-growing form that builds up into an upright bush with attractive foliage that is grey-green but with all growing tips a pale, sulphur-yellow. The colour is brightest during the summer.

-- **'Variegata'**. Nels. 74 (1866); as 'Lebretoni' Hillier, Cat. No. 36: 10 (1925). Branchlets irregularly white variegated.

C. pygmaea Wolf and Wagener, El Aliso **1:** 195 (1948). Now merged with *C. goveniana*.

C. sargentii Jepson, Fl. Calif. **1:** 61 (1909). This is a little-known species described as a tree 10–35 m. *Bark:* Brown, grey or almost black, furrowed. *Leaves:* Foliage stiff and hard to the touch; bluntly acute, grey-green; glands noticeable. *Cones:* 15–30mm across. *Scales:* 4–10, seeds glaucous, brown. It has a very limited natural distribution in California, U.S.A. It was named in honour of Dr. Charles Sprague Sargent (1841–1927), Director of the Arnold Arboretum form 1872 until his death.

C. sempervirens L. 1002 (1753). *C. horizontalis* Miller, Gard. Dict. ed 8, n. 2 (1768). The nomenclature of this species presents us with rather a conundrum. It is a common tree all round the Mediterranean area and into Asia. It is usually found as a broadly conical, later, a widespreading tree, sometimes in old age with a wide crown as spectacular as an old *Cedrus libani*, but occasionally it occurs as a fastigate, almost columnar tree. History records that it was introduced into Italy at a very early date. Presumably, since as the columnar form is still widely used for its architectural value in that country today, this was the form in garden cultivation. It would seem that Linnaeus described the species from this cultivated material, but there is no possible doubt that this was a selection uncharacteristic of the heterogenous wild population, rather than that the spreading trees are variants of a characteristically fastigiate species.

The following description relates to the species in the wild, differences to be expected in trees in cultivation being referred to under var. *stricta*, below.

Size and habit Tree to 25m, very variable, usually a broadly conical, eventually wide-spreading tree often open and flat-topped in old age, but occasionally narrowly columnar trees occur.

Trunk Often divided, *Bark:* Thin, glabrous, pinkish-brown often discoloured by grey algae; with broad, shallow fissures and flaky, often spiralled ridges; fissured and ridged on old trees.

Branches Horizontally spreading or ascending.

Branchlets Coppery-brown turning to brownish-purple. Shoots 'wiry'-looking, terete or somewhat quadrangular, about 1mm thick tapering towards the tip. The youngest shoots are curved, in flat, unsymmetrical sprays, held strongly forwards all round the shoot.

Winter buds Hidden.

Leaves Small (to 'very small' at the tip of the shoot); imbricate; dull dark green, acute, with blunt, incurved tips and margins minutely

Figure 183. *Cupressus sempervirens.* Detail of foliage, with cones.

Figure 184. *Cupressus sempervirens*. The selected fastigiate form usual in cultivation. (10)

toothed. The crushed foliage has little or no scent.

Mature cones Ovoid, often in clusters on short stalks; 2–3cm across, shining green, ripening to dark brown in late Autumn; retained on the tree, dull grey-brown the second year. *Scales:* 8–14; smooth, convex or slightly sunk, with a small central spike.

Seeds: 8–20 to each scale, ovoid or narrowly oblong, 3–6mm long, with narrow wing.

Habitat Southern Europe, the Mediterranean area and Western Asia.

Cultivation Despite its Mediterranean origin, the species seems reliably hardy. Zone 7. (see Illus. Nos. 183, 184)

Apart from the fastigiate habit of all specimens seen in cultivation, this species can be distinguished from the only other similar species, *C. macrocarpa*, by the lack of the strong lemon scent of the crushed foliage of that species, by the darker colour of the leaves and their more finely pointed tips and the tail-like tips of the shoots, and (if cones are available) by their dull grey colour and occurrence in clusters.

-- var. ***atlantica*** (Gaussen) Silba Phytologia **49** (4): 398 (1981); *C. atlantica* Gaussen, Le Monde des Plantes **45:** 55 (1950). This, and var. ***dupreziana*** Camus in Bull. Mus. Hist. Nat. Paris **32:** 101 (1926), are merely geographical forms from N. Africa which vary only in minor respects from the type variety. Neither is of other than minor botanical interest.

-- var. *horizontalis* (Miller) Loudon, Hort. Brit. 588 (1830); *C. horizontalis* Miller, No 2 (1768). This is here recognised as the typical form of the species, so would today be distinguished as var. ***sempervirens.***

-- var. ***indica*** Royle ex Gord. 72 (1858); *C. whitleyana* Carr. 128 (1855). Pyramidal and compact; cones globose; scales mucronate, acute. Southern Himalaya Region. Introduced into England in 1852, but probably no longer recognised in cultivation.

Having in this work accepted the one-time var. *horizontalis* as the typical form of the species, to be logical we must also accept var. ***stricta*** as the botanical home for any more-or-less fastigiate specimens ever to turn up, including all the trees in cultivation. But seen through horticultural eyes these are by no means a homogeneous unit. Mitchell [107 (1972)] distinguishes two distinct groups, one with multiple stems producing a columnar tree with a flattened or irregular top and the other maintaining a strong central leader and resulting in a thin, spire-like tree, and for this latter group he suggests the use of 'Stricta' as a cultivar name. Whether the difference in habit is genetic or no more than the effect of environmental factors in early life I am not clear, but there is much to be gained by limiting cultivars to a single clone, and (following the general plan of this book) I advocate the use of the word **'Stricta'** in a horticultural context only in a collective sense or as a group name, finding new names for sufficiently distinctive clones, one such being **'Totem Pole'**.

-- **'Swane's Golden'**. Swane Bros. Nurs. Cat. 1961. This is a bright golden-yellow small tree of compact and very fastigiate habit that arose as a seedling on the nursery of Swane Bros. Pty. Ltd., of Sydney, N.S.W. from seed collected in Canberra. Plants in New Zealand reach 3 m high and 1 m across in 15 years. Like all golden conifers it requires a position in full sun and unfortunately it is not reliably hardy in Britain. **'Gold Rocket'** is very similar but with a more intense pale gold at certain times of the year.

C. torulosa D. Don ex Lambert, Genus Pinus ed. 1 (2): 18 (1824); Don in Prod. Fl. Nepal 55 (1925) Himalayan Cypress.

Size and habit Broadly conical tree to 45m with a rounded crown.

Trunk Usually straight. *Bark:* 12mm thick, dull brown, peeling off in long, narrow fibrous strips; on old trees sometimes grey and spiralled.

Branches Horizontal; ascending at the tips; dull orange turning purplish-grey, terete or nearly so.

Branchlets Long, thin, sparsely branched; pendulous.

Winter buds Hidden.

Leaves In dense, irregular, hanging bunches; 1–5mm long, narrow, incurved, with blunt tips, obtuse, bright yellowish- or dull, glaucous-green, with slight whitish bloom; grooved above, with pale, finealy toothed margins.

Mature cones Freely borne; nearly sessile, on short recurved branchlets, globose or elliptic, 12–20mm broad, whitish-green ripening to dark red-brown and glaucous violet. *Scales:* 10–12, indented but with a small bent round spine. *Seeds:* 6–8 or more per scale, 4mm, reddish-brown; round or somewhat flattened; smooth, with narrow wing, small mucro. *Habitat:* W. Himalayas and W. Schechuan.

Cultivation Introduced by Dr. Wallich into Britain in 1824. Only reliably hardy in warmer temperate regions, so is uncommon in Britain. (See Illus. No. 185)

Figure 185. *Cupressus torulosa.* Detail of foliage. (5)

-- var. **cashmeriana**. (Carr.) Kent in Veitch Man. Conif. ed. 2, 234 (1900); *C. cashmeriana* Carr. ed. 2, 161 (1867). Kashmir Cypress. This, perhaps the most elegant and beautiful tree in the world, is something of a conundrum. It was recorded as having been introduced from Tibet in 1862, but cannot now be traced 'in the wild', so its botanical status is difficult to determine. It closely resembles *C. torulosa* in its cones, leading many authorities, beginning with Carrière himself, to regard it as a form of that species, but since it develops cones and grows to a great size with no alteration in the foliage, other botanists would give it the benefit of the doubt and have us regard it as a species.

A very graceful small (eventually enormous) tree with ascending branches arching over and ending in systems of long, remarkably pendulous branchlets; pink-brown darkening to reddish-brown; bare of leaves by the third year; conspicuously flattened. *Leaves:* 2mm long, in opposite pairs with lancelike, spreading tips; soft grey-blue above, darker below with pale margins and both sides having a whitish bloom that disappears when the foliage is wet. *Cones:* Globose 10–13mm diam, glaucous greenish-yellow, ripening to dark brown. *Scales:* 10, each with a central depression and an acute, angular depression. *Seeds:* About 10 per scale, short-winged. (See Illus. No. 186)

-- 'Corneyana'. See *C. corneyana.* There is no cultivar entitled to this name.

-- '**Majestica**'. Knight ex Carr. 118 (1855). More robust, taller and hardier; *Branches:* yellow, somewhat thicker than usual. *Branchlets:* de-

Figure 186. *Cupressus torulosa* 'Cashmeriana' at Isola Madre, Italy.

curved, thicker. *Leaves:* scale-like, 4-seriate, tightly imbricate, sometimes slightly spreading at the apex, with a longitudinal groove on the back.

DACRYCARPUS

Podocarpaceae

Dacrycarpus (Endlicher) De Laubenfels in Journ. Arn. Arb. **50**: 315 (1969); Endlicher 221 (1847). A genus first mentioned by Endlicher but only recently studied. It now consists of nine species. It is closely related to *Dacrydium* but differs in its flattened branchlets and by slight differences in the cones. The habitats range from Burma to New Zealand, so some species will not be hardy in cold-or temparate regions.

The following species are now recognised in this Genus; all formerly listed under *Podocarpus*. There are no recorded cultivars.

DACRYCARPUS cinctus (Pilger) De Laubenfels (1969); ***D. cumingii*** (Parlatore) De Laubenfels (1969); ***D. dacrydioides*** (Richard) De Laubenfels (1969); ***D. expansus*** De Laubenfels (1969); ***D. imbricatus*** (Blume) De Laubenfels (1969), var. ***curvulus*** (Miquel) De Laubenfels (1969), var. ***patulus*** De Laubenfels (1969) and var. ***robustus*** De Laubenfels (1969); ***D. kinabaluensis*** (Wasscher) De

Figure 187. *Dacrydium franklinii*. Detail of foliage.

Figure 188. *Dacrydium laxifolium* in the Royal Botanic Gardens, Edinburgh.

Laubenfels (1969); ***D. steupii*** (Wasscher) De Laubenfels (1969); ***D. viellardii*** (Parlatore) De Laubenfels (1969).

DACRYDIUM

Podocarpaceae

Dacrydium Solander ex. Forster in De Pl. Escul. Ins. Ocean. Austral. Comm. Bot. 80 (1786) is a genus of 25 species ranging from evergreen tall trees to quite low shrubs; usually dioecious (except for all the New Zealand species). *Leaves*: Juvenile, usually large, linear, spreading, adult leaves scale-like, closely imbricate, appressed and sharply keeled as in *Cupressus* but spirally arranged. Female flowers usually solitary, borne erect, usually brightly coloured but inconspicuous. *Cones*: In axils of upper leaves, oblong-cylindrical, erect, surrounded at the base by a cup or disc; *Seed*: Ovoid, enclosed within a fleshy outer shell and a more or less hard seed coat. *Habitat*: Subtropical regions of S.E. Asia, in Australasia, Tasmania and in W. South America. Few of the species have been introduced into Britain and none is reliably hardy save in the more sheltered areas. As with several of the tropical genera in this part of the world the botany does not yet seem to have reached finality. A proposal by Quinn in the Australian Journal of Botany **30**: 317 (1982) to transfer three species found in New Zealand to a new genus *HALOCARPUS* is accepted here, following Silba (1984), but Quinn's further depletion of the genus *DACRYDIUM* by transfers to a formerly recognised genus *LEPIDOTHAMNUS* and a proposed new one *LAGAROSTROBUS* have not been universally accepted, so are not here followed.

D. bidwillii Hooker f. ex T. Kirk. Now known as *Halocarpus bidwillii.*

D. biforme (Hooker) Pilger ex Engelmann. Now known as *Halocarpus biformis.*

D. colensoi Hooker Ic. Pl. **6**: t. 548 (1843); *D. westlandicum* Kirk; *Lagarostrobus colensoi* (Hooker) Quinn in Austral. J. Bot. **30**: 317 (1982). A conical tree to 15m with a straight, clean trunk and bearing almost wholly adult foliage. Found on the west coast of South Island, New Zealand. It is very rare in Britain.

Figure 189. *Fitzroya cupressoides*. Detail of foliage.

Figure 190. *Fitzroya cupressoides.* A tree in the National Pinetum, Bedgebury, Kent, U.K.

Zone 8.

D. cupressinum Solander ex Forster. (1786). Rimu. Tree 15–45m in the wild, conical when young, becoming round-headed in maturity; with a straight trunk and dark red bark. Juvenile leaves awl-like, dark green, linear, 4–6 mm long, wide-spreading; adult leaves scale-like, sharply pointed, yellow green, predominating on older plants. A native of New Zealand (both islands), and there an important timber tree. Young plants in cultivation make very attractive specimens with the ultimate branchlets long and pendulous. Unfortunately, it is only marginally hardy in even the most favoured areas in Britain. Zone 9.

D. franklinii Hooker f. Lond. Journ. Bot. **4**: 152, t.4 (1845); *Lepidothamnus intermedius* (Kirk) Quinn in Austral. J. Bot. **30**: 316 (1982). Huon Pine. A conical tree to 30m in the wild, with a dense crown of fine, slender, pendent branchlets. It bears wholly adult foliage with leaves about 1mm long, blunt and concave, dark green with scattered stomata; closely pressed to the stem like *Cupressus*, sharply keeled. Cones very small, borne at the tips of the branchlets. The terminal growths are long, arching over and more or less pendulous. This is another species that is a valuable timber tree in its native Tasmania, but, in Britain is seldom seen as more than a large bush, being only marginally hardy, even in sheltered parts of the country. Zone 8. (See Illus No. 000)

D. laxifolium Hooker f. ex Hooker in Lond. Journ. Bot. **4**: 143 (1845); *Lepidothamnus laxifolius* (Hooker f.) Quinn in Austral. J. Bot. **30**: 316 (1982). A prostrate or widespreading shrub, not only an anomaly in a genus mostly consisting of forest trees, but unique in being the smallest conifer in existence. Branches long and slender, branchlets fine and filiform; juvenile leaves lax, spreading, narrowly linear, acute, curved, 3–7mm long; adult leaves scale-like, linear-oblong, blunt at tips, 1–1.5mm long, green, reddish inthe winter. With this growth characteristic it has neither economic nor ornamental value, but a well-grown specimen in a large pan (it needs frequent cutting back of the trailing main shoots) can make a point of interest in a display of alpine plants. Zone 8. (See Illus. No. 188)

-- **'Blue Gem'** Welch, Man. Dw. Conif. 200 (1979). This is a very good form with blue foliage selected in the wild and introduced to cultivation by the Otari Plant Museum, Wellington, New Zealand. The name 'Blue' under which it was at first distributed contravenes the Recommendations of the *Cultivated Code*.

DECUSSOCARPUS

Podocarpaceae

A genus recently separated from *Podocarpus* embracing 12 species with pine-cone-like male flowers. All are from sub-tropical and high rainfall areas and so are never found in cool temperate zones.

DISELMA

Cupressaceae

Diselma Hooker fil. A monotypic form found only

in mountainous areas of Australia and W. Tasmania.

D. archeri Hooker fil. Fl. Tasm. **1**: 353 t. 98 (1857). An erect or spreading, sometimes straggly tree to 6m. *Bark*: grey-brown or dark grey. *Branchlets*: much ramified. *Leaves*: Exceedingly short, 1mm long, closely packed, imbricate, opposite or decussate producing a quadrangular shoot; obtuse, keeled; rich green (at first reddish green), with white stomata beneath. *Cones*: Globose, 8–10mm across, with few scales.

EPHEDRA

Ephedraceae

Ephedra L. A genus of about 40 species of mostly rather straggly bushes of no garden value. They are strictly *Gymnosperms* but are certainly not conifers in the minds of most gardeners.

FITZROYA

Cupressaceae

Fitzroya Hooker fil. ex Lindley, Journ. Bot. Soc. Lond. **6**: 264 (1851). A monotypic genus found only in southern Chile, named in honour of Capt. Fitz-Roy, who discovered it.

F. cupressoides (Molina) Johnson in Contr. Gray Herb. new ser. **70**: 9 (1924); *Pinus cupressoides* Molina, Sagg. Stor. Nat. Chile 168 (1782); *F. patagonica* Hooker f. ex Lindley (1851). Monoeceous or dioeceous tree to 35 m, 3.5–5.5 m in girth, shrubby at higher elevations; *Bark*: Reddish, furrowed, peeling off in long strips; *Branchlets* pendulous, slender, flexible, dark green when young, becoming reddish with age; *Winter-buds*: Ovoid or globose; *leaves*: In alternating whorls of 3, oblong or lanceolate, about 3.5mm long, minutely incurved, pointed, free part spreading, base flattened and attached to the branchlets, concave with 2 sunken white stomatic bands above, convex with a broad green midrib and 2 white bands beneath; *Cones*: 6–9mm across, often freely borne on quite young trees, apex terminating in gland-like, resin-secreting bodies of about 4mm length, which exhale a fragrant odour, ripening the first season. *Habitat*: Chile and N. Patagonia, from the coast-range north of Valdivia to the Central Corderilla of the Andes; introduced into cultivation in 1849 by W. Lobb and again in 1859 by R. Pearce; both collectors sent seeds to the Veitch nurseries in England. Not quite hardy in colder regions. The combination of the leaves being in whorls of three and dark green in colour, distinguishes this genus from any other with which it might be confused. (See Illus Nos. 189, 190)

FOKIENIA

Cupressaceae

Fokienia Henry and Thomas in Gar. Chron. ser. 3. **49**: 66 (1911). A genus from China, presently regarded as monotypic.

F. hodginsii Henry and Thomas loc. cit., t. 32–33; *F. kawaii* Hayata in Bot. Mag. Tokyo **31**: 116 (1917). A monoecious small tree with foliage in broad, fan-shaped, tri-pinnate and very much flattened sprays. Juvenile leaves (which are all that the few small trees in Britain have managed to produce) are sharp-pointed, decussate, the lateral pair much larger than the facing pair (as in *Calocedrus*) and have glistening white bands on both sides, especially beneath; adult leaves much smaller; buds covered by the terminal leaves which, being little or less than other leaves give a squared off look to each shoot. It is very rare in Britain, where it is only hardy in sheltered conditions. It should not be confused with the much more common *Thujopsis dolabrata*.

GINKGO

Ginkgoaceae

Ginkgo L. Mant. Pl. **2**: 313 (1771). This tree, although still very much alive with us today, is interesting as being the sole survivor of a class of plants that in early geological times dominated the forests of the world. Not surprisingly it finds itself rather isolated in any present-day botanical classification, so not only

Figure 191. *Ginkgo biloba*. Detail of foliage and fruit. (10)

is the Genus itself monotypic, the Ginkgo is the sole occupant of the Family *Ginkgoaceae*, of the Order *Ginkgoales* and of the Class *Ginkgoopsida*. It differs from the *Coniferales* by its broad, fan-shaped leaves and by its method of reproduction (which it shares with the ferns) and so finds a place in a book on conifers more by tradition than as of right.

G. biloba L. Mant. Pl. **2**: 313 (1771); *Salisburia adiantifolia* Sir James Smith. Maidenhair Tree.

Size and habit A deciduous tree to 30m, usually monoecious; habit variable – either a conical tree with a straight trunk and a single leader or else with the trunk dividing early, resulting in a more spreading, rounded tree. *Bark*: dull grey, deeply furrowed.

Branches Branch pattern variable, from ascending to spreading, with laterals more or less pendulous.

Branchlets Of two kinds (as in *Cedrus* and *Larix*) – (i) Short, stout spurs, with leaves in clusters and numerous leaf scars, or (ii) long, free-growing shoots with leaves spirally arranged.

Winter buds Broad, flat, conical, short pointed. *Scales*: brown; 6mm long.

Leaves Held on long (20–50mm) thin stalks; fan-shaped, widening from the narrow base to an irregularly jagged, lobed or sometimes notched top edge; 50–75mm long and as much wide; at first a bright yellowish green with a few stomata, mainly beneath, darkening by the

Figure 192. *Ginkgo biloba* in summer. The Ginkgo is one of the very few deciduous conifers. (10)

Figure 193. *Ginkgo biloba*. Detail of bark on trunk.

Summer and turning bright butter-yellow for a short period before falling in the Autumn: glabrous, without a mid-rib but with parallel veins running longitudinally, branching as the leaf widens.

Flowers Borne on the spurs; the male flowers in cylindrical, short-stalked pale yellow catkins about 20mm long, consisting of green stamens appearing before the leaves are fully developed; the female flowers solitary or in pairs on a stalk 35–50mm long.

Fruit A yellowish green fruit like a bullace plum, 25–40mm long, ripening the first year to pale orange consisting of a long stem carrying a nut surrounded by a fleshy layer that, when the fruit falls, splits open and gives off a most unpleasant, rancid stench, despite which the kernel is well flavoured and edible.

Habitat In Japan, and in its more probable home-China, it is highly regarded for religious and sentimental reasons, so has been widely planted for centuries past. For a long time it seemed uncertain whether or not the Ginkgo still existed in the wild, but it is now known to exist in the wild in Chekiang Province, China.

Cultivation The Ginkgo responds to good growing conditions but will not prosper in shallow, dry soils,. Grown well it makes an excellent specimen tree but, if used as an avenue, careful selection of clonal stock is advisable, since as a species it is very variable. At one time it was believed that the conical trees were the males and the lower, spreading type of tree the females. This may have arisen through the vegetative propagation of proven clones (a conical habit and inability to produce fruit both being desirable attributes), but it is not true in any biological sense.

The Ginkgo, with its interesting and unusual leaf-shape, the dense, green summer foliage and the abrupt change to golden-yellow in Autumn and then the attractive branch-pattern during the winter months makes a good tree for planting in parks, highway verges and private gardens and even in cities, provided that sufficient space is available. But it is very variable from seed and this, together with the need to avoid the planting of female trees makes vegetative propagation of selected clones necessary. In 1951 the Saratoga Horticultural Foundation, of Saratoga, California, USA, a non-profit corporation dedicated to studies in trees and shrubs, commenced a search, both in America and in Europe, for shapely and distinctive staminate (male) clones, several of which are now in production. (See illus. Nos. 191, 192, 193)

-- 'Argentea' (the silvery), 'Aurea' (the golden-variegated), 'Macrophylla' (the very-large--leaved), 'Microphylla' (a smaller-leaved variety) and other names consisting of a latin descriptive adjective were listed by Nelson [Pinaceae 164 (1866)] and other early writers, for clones that cannot now be identified in cultivation. The names should not be applied to newly selected clones answering to such brief descriptions. The re-typifying of old names is impermissible.

-- 'Aurea Variegata' is an illegitimate name for a plant received by J. R. P. van Hoey-Smith from an American nursery in 1967. Unless its identity with 'Variegata' (below) or Nelson's clone 'Aurea' can be established (which is extremely unlikely) a new name must be found.

-- 'Autumn Glory' See Dendroflora **9**: 61 (1982). In the absence of a description distinguishing this clone from many others, this name does not qualify for acceptance as legitimate under the *Cultivated Code*. The proliferation of unrecorded and undocumented cultivar names in the past is the prime source of the confusion in the conifer world today, so its continuation must be strongly deplored.

-- **'Autumn Gold'** M. van Renslaar in Journ. Calif. Hort. Soc. **17**: 100 (1956). This was the first selected clone distributed by the Saratoga Horticultural Foundation. It is a male clone with a broadly conical habit. At the date of selection the mother tree was 12m high, 6m across and of very regular growth. The autumn colour is of an especially bright golden yellow.

-- 'Epiphylla'. A mere curiosity, distinctive only by its fruit stalks being broad and wing-like on the apical portion, adnate to the leaf petiole.

-- **'Fairmount'**. A vigorous male clone of narrowly upright habit introduced into cultivation by the Saratoga Horticultural Foundation. Although maintaining a single leader it is more open in growth than 'Autumn Gold'. The mother tree is in Fairmount Park, Philadelphia, PA, U.S.A., planted at the time of the 1876 Centennial Exhibition. In 1973 this tree was 20–25m high with a crown 10.5m and a trunk diam. of 1m at 1.2m from the ground.

-- **'Fastigiata'**. Masters in Kew Hand-list Conif. 19 (1896). Narrow-conical to columnar habit. (See *Amer. Nurs.* 1st June 1965.) Mr. van Hoey-Smith of Rotterdam says that the plant he grows under this name originated with a tree in Hortus Botanicus in Leiden that holds all its branches up at 45°. But unless its identity with the clone originally given the name can be established, a new name should be found for it.

-- 'Horizontalis'. Also emanating from Leiden Botanical Garden where there is a very low horizontally-growing specimen. This name also is doubtfully legitimate.

-- **'Laciniata'** Carrière, 504 (1855), under *Salisburia*, in Revue Hort. 412 (1854). Vigorous growing, conical habit; leaves very large 20–30cm wide, kidney-shaped, divided into 2–5 main lobes frequently sub-divided, venations distinctly raised. Originated as a seedling in the Reynier Nursery, Avignon, France, before 1850. Van Hoey-Smith's plant came from the Cologne Botanic Gardens, so may well be authentic.

-- **'Lakeview'**. See Dendroflora **19**: 61 (1982). A narrowly conical male form suitable for street and landscape use.

-- **'Mayfield'**. Male form, quite narrowly columnar. Illustrated in Proc. Amer. Plant Prop. Soc. 28 (1951). Selected Ohio about 1948.

-- 'Ohazuki' From the only description available this is probably 'Epiphylla'.

-- **'Palo Alto'**. See Dendroflora **19**: 62 (1982). Of normal colouring but with a broad, spreading crown. Raised E. H. Scanlon before 1962?

-- **'Pendula'**. Carrière, ed. 2, 713 (1867). Small tree with a broad umbrella form or semi-spherical crown, branches spreading or more or less nodding. It has to be propagated by 'top-working', i.e. grafting onto a tall stem, and seldom makes an attractive specimen.

-- **'Princeton Sentry'**. See Dendroflora **19**: 62 (1982). A very narrowly conical form introduced by the Princeton Nurs, NJ, U.S.A. in 1967, propagated from a specimen in the Roosevelt Boulevard in Philadelphia.

-- **'Saint Cloud'**. Meyer, Plant Explor. 135 (1959). A form with a clean trunk and good leader, the main branches widely spaced or ascending, rarely with secondary branches but covered along their whole length with very short side shoots densely clothed with foliage giving the tree an unusual, skeletal aspect. The original tree was found in Jardin Kahn, St. Cloud-sur-Seine, France.

-- **'Saratoga'**. A male clone selected by the Saratoga Horticultural Foundation which forms a small conical tree, the mother plant having reached 6m × 2.5m in 12 years. The main branches are ascending, the laterals numerous and short, densely clothed with foliage, making a full, dense specimen. The leaves are distinctive and characteristic, being variable in shape but usually with long, narrow lobes.

-- **'Tit'**. See Dendroflora **15–16**: 54 (1979). A dense, congested dwarf form received in Holland from Japan and presumably originating from a witch's broom. Leaves sprout from swellings on the stems.

-- **'Tremonia'**. Krüssmann. Handb. Conif. 133 (1972). A narrow, columnar form 12m high by 80cm across in 1983. Like a Lombardy poplar. Raised from seed in 1930 and introduced in 1970 by the Dortmund Botanical Garden.

-- **'Variegata'**. Carrière ed. 2, 712 (1867). A somewhat less hardy clone, distinctive by the colour of the leaves which are irregularly striped ped yellow. It arose as a seedling on the M. A. Leroy nursery at Angers, France.

GLYPTOSTROBUS

Taxodiaceae

Glyptostrobus. Endlicher, Synop. Conif. 69 (1847). A (probably) monotypic genus consisting of a shrub or small, fastigiate tree sharing the deciduous branchlets and other characteristics of *Taxodium* but differing in the pear-shaped, long-stalked cones, obovate scales and small-winged seeds.

G. lineatus (Poiret) Druce. According to Cheng & Fu (1978) this was a mistake, being based on a cultivated specimen of *Taxodium distichum* var. *nutans*.

G. pensilis (Staunton) K. Koch,Dendr. **2** (2): 191 (1873); *G. heterophyllus* Endlicher, Synop. Conif. 70 (1847). Chinese Swamp Cypress. This species, although long known in the literature, was not introduced into Britain until the present century and Mitchell [116 (1972)] states that only a very few trees, all with juvenile foliage, are found in this country, older trees so labelled turn out to be *Taxodium distichum* var. *nutans*. He further states that the leaves of *Glyptostrobus* are in three ranks, very slender, 1.5–2.5cm long, soft to the touch, pale green on both sides; on slender, bright green shoots 10cm long which arch down from the branches.

HALOCARPUS

Podocarpaceae

Halocarpus Quinn in Austral. J. Bot. **30**: 317 (1982). A genus recently separated from *Dacrydium*. It consists of three species of evergreen shrubs or small trees of which two only are hardy in the cool-temperate region. Buds inconspicuous. Juvenile leaves are linear, flattened, distichous, in two ranks as in *Podocarpus*. Adult leaves emerging spirally, small, scale-like, imbricate. Zone 8. New Zealand, South Is.

H. bidwillii. (Hooker fil. ex Kirk) Quinn in Austral. J. Bot. **30**: 317 (1982). Tarwood. A closely branched large bush bearing both juvenile and adult foliage; the former flat, linear, abruptly pointed, 8–10mm long, mostly in the lower part of the bush, the latter, borne on upright shoots, small, closely appressed, triangular in section, blunt and thick. Both leaves and shoots are yellowish-green. Very uncommon in Britain, but apparently quite hardy. Habitat: South Island and Stewart Island, New Zealand.

H. biformis (Hooker) Quinn in Austral. J. Bot. **30**: 318 (1982). Formerly *Dacrydium biforme* (Hooker) Pilger, (1903).

-- **'Copper Glow'** A recent selection in New Zealand made (presumably) on account of the colour of the foliage.

H. Kirkii (F. Mueller) Quinn. This species makes a medium-size tree.

JUNIPERUS

Cupressaceae

Juniperus L. 1038 (1753), pro parte; ***Sabina*** Miller, Gar. Dict. 3 (1754). A genus of about 60 species that is so extremely varied that it is not possible

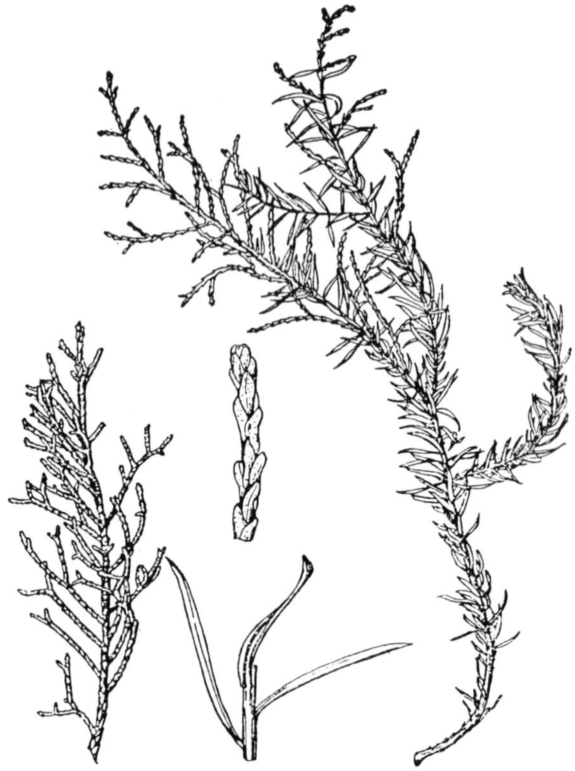

Figure 194. *Halocarpous bidwilii* (formerly known as *Dacrydium bidwillii*). Detail of foliage.

to produce a general description that is not so hedged about with alternatives as to be useless. The only features common to all species are the flowering parts. The male cones are yellowish, ovoud or oblong with many scales on a central axis each carrying pollen sacs. The female cones have six coalescent fleshy scales which develop to form a berry-like fruit (hardly a 'cone' in the usual sense) which remains closed even when ripe. In a few species the foliage is so indistinguishable from *Cupressus* that the presence of fruit is essential to identification.

Junipers like a well-drained soil and some species are drought-resistant. They are mostly tolerant of lime, even chalk. All species are readily raised from seed, which may need a pre-sowing treatment to break dormancy which otherwise can delay germination for a year. Most cultivars propagate readily from cuttings, a few requiring to be grafted. A general compatibility within the genus exists, but the vigour of the scion should influence the choice of understock, especially with the very slow-growing selections.

With the Junipers we come to one of the most interesting, and certainly the most useful to gardeners of all the genera of conifer. It is also perhaps the most widely spread, at least in the northern hemisphere, to every corner of which it has penetrated.

There is good reason for this. During the geological ages during which the conifers were evolving their present forms and colonising the surface of the earth, the Junipers succeeded where other genera failed because they were more enterprising and adaptable. Did other genera dislike a taste of lime in the soil and turn back, not so the Junipers, they pushed on and developed a tolerance of, perhaps even a preference for, alkaline soil. Mountains; the dry summers of continental climates; salt-laden gales near the coast; every barrier to colonisation except crossing the equator they accepted as a challenge to be overcome. Even the development of the berry-like fruit was a smart move. It might not represent the ultimate in juiciness but it was at least edible and many a hungry bird would have realised this and, later, dropped the seed miles away.

This inherent adaptability that enabled some species to settle down and find a home, whatever the conditions, has remained with some species as an inherent variability, and, as befits the genus with the largest distribution throughout the Northern Hemisphere, it has given us many useful garden forms. These come in every conceivable size, colour and texture, and in shape they range from the completely prostrate form of *Juniperus horizontalis*, through spreading, globose and pyramidal forms to perhaps the most spectacularly needle-like plant known, *Juniperus scopulorum* 'Sky Rocket'. As well as green forms of every hue there are golden, glaucous, blue-grey forms and variegated forms and some of the plants take on rich purple hues in the winter.

It is a genus that must have sadly tried the patience of botanists because, particularly on account of the habit that many of the forms have of clothing themselves wholly or mainly with juvenile foliage when young and changing over gradually and leisurely (sometimes incompletely, and in some cases not at all) to adult foliage in the mature plant, and its other habit – in species that are characteristically dioecious – of not running invariably true to type, it does not lend itself to being studied by the normal methods of orthodox botany, which rely mainly upon herbarium specimens and little if at all upon the knowledge and study of living plants.

There are several species of Juniper which never become large trees and these can therefore be used in gardens equally as well as the dwarf forms of taller-growing species. The persistent juvenile habit of others is an invitation to make use of them at this stage of their life, since many of these are most attractive both in foliage and in habit, which in the early years is often regularly conical.

Since the Junipers tolerate calcareous soils and dry soil conditions (quite a different thing to needing such conditions) better than many other genera, those whose lot it is to garden on limestone or chalk may regard themselves fortunate that the Junipers cover such a wide range in colour, habit and texture.

Guides to species identification

A complete Key to the species in so large a genus would be both unwieldy and of very little practical use, since quite a number of them are rare in cultivation and such as are widely planted are usually represented by one of the numerous cultivars that have been selected and named. So, following the plan of this book, only simple listings are here given, helping a searcher towards final identification from a study of the full species and cultivar descriptions that are given later. (See pages 324, 325)

GUIDES TO IDENTIFICATION OF JUNIPERUS SPECIES

1. Foliage wholly juvenile, i.e. all leaves awl-like (acicular).

(i) Leaves in threes and jointed at the base, winter buds visible.

Shoots triangular; sharply pointed	J. communis (and numerous cvv)*
Prostrate; foliage soft	J. conferta
Leaves and fruits very large	J. drupacea
	J. oxycedrus
Pendulous tree; Leaves very prickly	J. rigida*
Species rare in cultivation	J. brevifolia, J. cedrus J. formosana, J. taxifolia

(ii) Leaves always in threes, long-tapering, not jointed at the base, decurrent, pointing forward. Winter buds hidden. Cones ripen to black the 2nd year.

Pendulous tree	J. coxii
	J. morrisonicola
	J. recurva (and cvv)*
Small tree or shrub	J. pingii (and cvv)*
	J. squamata (and cvv)*
Prostrate shrub	J. procumbens

2. Both juvenile and also small, adult leaves are usually present; the former in 2's or 3's, the latter in decussate pairs as *Cupressus*.

(i) Leaves with entire (i.e. not toothed) margins (Use a hand-lens)

Small to medium tree, or shrub	J. chinensis (and many cvv)*
Not fully hardy	J. excelsa
Foetid smell	J. foetidissima
Low-growing shrub	J. horizontalis (and many cvv)
Spreading shrub; faint aroma	J. × media (and numerous cvv)*
Low shrub; acrid smell	J. sabina (and numerous cvv)
Prostrate shrub	J. sargentii
Small upright tree	J. scopulorum (and numerous cvv)*
Musty smell	J. thurifera
Upright tree	J. virginiana (and numerous cvv)*
Species rare in cultivation	J. barbadensis, J. bermudiana J. blancoi, J. centrasiatica, J. gaussenii, J. komarovii, J. macropoda, J. procera J. semi-globosa, J. silicicola, J. tibetica

(ii) Leaves with finely toothed margins (Use a hand lens)

J. deppeana (and vars.)
J. indica
J. flaccida
J. wallichiana

(Guides to identification of Juniperus species, contd.)

J. durangensis, J. erythrocarpa,
J. gamboana, J. jaliscana.
J. martinezii, J. monosperma,
J. occidentalis, J. osteosperma,
J. phoenicea, J. pinchotii,
J. przelwalskii, J. pseudosabina,
J. sabinoides, J. saltuaria,
J. standleyii, J. saltillensis

3. Low-growing, spreading or prostrate shrubs

J. conferta
J. davurica
J. horizontalis
J. procumbens
J. sabina
J. sargentii
And, some cultivars of all the above species marked*

J. ashei Buchholz in Bot. Gaz. **90**: 329 (1930): *C. sabinoides* H.B.K., Nov. Gen. Sp. **2**: 3 (1817); *J. mexicana* Sprengel, Syst. Veg. **3**: 309 (1826); *J. sabinoides* (H.B.K.) Nees in Linnaea **19**: 706 (1847), not Grisebach (1845); *J. monticola* Martinez in Mex. Univ. Nac. An. Inst. Biol. **17**: 79 (1946). Mountain Cedar.

Dioecious shrub or small tree slow-growing, 4–6m; trunk short, crooked, divided near the ground into large, ascending branches; bark thin, ashen-grey, splitting into long, narrow, persistent, shreddy scales; twigs rust-brown, slender, 4-angled, becoming round and ashen-grey with age; winter-buds naked, leaf-covered; leaves on old twigs decussate, scale-like, 1–1.8 mm long, ovate, acute, on vigorous shoots ternate, lanceolate, 2–8 mm long, rigid, sharp-pointed; foliage of young plants and terminal twigs yellowish-green to light brown, of the mature branches dark green, becoming black upon drying; cones glaucous, deep blue, subglobose, 6–8 mm in diameter, maturing in one season; flesh thin, pungent, sweet; skin thick; seeds 1–2 in each cone, 5 mm in length, ovoid, acute, slightly grooved at apex.

S. Missouri, N. Arkansas, Colorado, Arizona, Central Texas, Mexico at altitudes from 180–600 m, in association with *J. virginiana*. Introduced in 1926 by Dr. Buchholz into cultivation in U.S.A., but little planted. Zone 8.

J. barbadensis L. Sp. Pl. 1039 (1753); *J. ekmanii* Florin, Acta Hort. Gothoburg. 3 (1928); *J. gracilior* Pilger in Symb. Ant. **7**: 481 (1913); *J. lucayana* Britton in N. Amer. Trees 121 (1908); *J. saxicola* Britton et Wilson in Bull. Torr. Cl. **50**: 35 (1926); *J. virginiana barbadensis* Gordon, Pin. 114 (1858); *J. gossainthanea* Lodgiges cat. A tree to 15 m with an open crown and thin, pendulous branchlets. Adult foliage light green, opposite, imbricate, 1.0mm long, 0.7mm wide. Known in Britain in 1759, but is still extremely rare in cultivation as, also, it now is in the wild. It should not be confused with *J. bermudiana*, which has thick and conspicuously quadrangular branches or with *J. silicola* which bears oval fruits. Bahamas; Cuba; Haiti and Barbados. Zone 9.

J. bermudiana L. Sp. Pl. 1039 (1753). Bermuda Juniper. A conical tree 12–15m tall, with thick, noticeably quadrangular branches: Adult leaves in four ranks, imbricate, blunt, incurved; a few juvenile leaves sometimes present. It was cultivated in Britain as early as 1684, but is very rare today, not being hardy. It is also now rare in the wild as the result of blight. Zone 9.

J. blancoi Martinez in Ann. Inst. Biol. Univ. Mexico **17** (1): 73/4 (1946). A rare shrub or small tree 2.5–15m high usually with an erect trunk and rounded to conical crown and pendulous branches. Adult leaves opposite, closely adpressed. Mexico. Zone 8.

J. brevifolia Antoine 16 (1857); *J. oxycedrus* var. *brevifolia* Hochstetter. in Seubert. Fl. Azor. 26 1844). Azores Juniper. Differs from *J. oxycedrus* in having shorter (6–9 mm long) leaves, the upper side with a midrib not continuing to the top and the smaller (± 8 mm across) cones.

Azores at an altitude of ± 1500 m. Zone 9.

J. californica Carrière in Rev. Hort. 352 **(1854:).** Californian Juniper. Shrub, sometimes a tree, 14 m, conical; bark ashy-grey, thin, separating in long strands; branchlets stout, flexible, often glaucous sprays moderately thick, stiff, 1–1.3 mm broad; leaves appressed, usually ternate, yellowish green, rounded or acute, prominently toothed on the margins, usually with conspicuous glands on the backs; cones ripening in the second season, short ovoid, 12–20 mm long, reddish-brown, bloomy, pulp thick, dry, sweetish; seeds 1–2, large, angled, apex sharppointed.

From S. Oregon to Lower California. Introduced into Britain by W. Lobb in 1853. Zone 9.

J. cedrus Webb & Berthelot, Phyt. Canar. **3**: 267 (1836–1840). Canary Islands Juniper. Closely related to *J. oxycedrus* L., differs from it by its drooping branchlets and its closer set, not pungent leaves. Not hardy in Britain.

Canary Islands, on Palma, nearly extinct in the wild. Zone 9.

J. centrasiatica Komarov, Notulae Syst. **5,2**: 28 (1924). Tree, 12 m; crown dense; branches erect; leaves scale-like, greyish-green not glandular; cones globose or ovoid, about 10 mm in diameter, dark red, slightly bloomy; seeds solitary, oval, flat-keeled.

Asia on the Kuen-luen at an altitude of 3000–4000 m. Zone 5.

J. chinensis L. Mant. Pl. **1**: 127 (1769): *J. sphaerica* Lindley in Paxton, Fl. Gar. **1**: 58 (1850): *J. sheppardi* Veitch, Man. 290 (1881); *J. sphaerica glauca* Fortune ex Gordon 122 (1858) NOM; *J. sheppardii* (Veitch) Van Melle, Phytol. **2** (6): 187 (1946). The Chinese Juniper. China, except far south and west; Japan. Zone 3. (See Illus. No. 195)

As currently accepted; A conical tree to 20–25m with an ascending branch system, or sometimes a large, spreading bush. Leaves of two kinds, the adult form scale-like, narrowly rhombic, closely appressed, obtuse, in decussate pairs 1.5mm long, the back convex and glandular; the juvenile form leaves awl-shaped, in pairs or whorls of three, 12mm long, spiny pointed, with 2 white bands above. The cone irregularly globose or oblong 5–7mm across, outline of scales often distinguishable, at first whitish-blue, ripening to dark brown, often with a whitish bloom. Seeds: 1–4, usually 2–3 oblong-obtuse or nearly 3-angled, glossy brown.

This is a very difficult and perplexingly complex species with many nomenclatural problems. These spring from the inherent behaviour of this variable species (or group of species, if that is what it finally turns out to be). Within a single plant the foliage can be wholly juvenile with awl-like leaves (either in pairs or set in threes in different parts of the plant); wholly adult with scale-like leaves in pairs, or a mixture of both. A plant may carry solely male, or solely female flowers or both, changing its behaviour from time to time. The habit of growth as each plant matures may change and the normal takeover of adult foliage may proceed erratically or not at all. Added to this, individual clones may depart from the usual upright conical shape and form spreading or horizontally-branched plants. There is also a considerable variation between clones of the same species in the size of the fruits.

Clearly all this variability made any satisfactory classification of this important group by the normal methods of herbarium botany impossible, so Mr P. J. Van Melle or Poughkeepsie, New York made a study of this and certain allied Junipers and, in 1946, he published his findings in the American publication *Phytologia*. The following year these were collected and republished in book form entitled *Review of Juniperus Chinensis*. This was published by the New York Botanical Garden.

Briefly, Van Melle complained that the original species *J. chinensis* L. was inadequately defined by

Linnaeus when he named it, that *J. sphaerica* Lindl. was subsequently quite unjustifiably reduced to a variety of *J. chinensis* and that *J. chinensis* var. *sheppardii* should have been recognised as a distinct species and that in result (what with all the garden forms that have appeared and a whole group of natural and garden hybrids with the prostrate species *J. sabina*) the name *J. chinensis* had come to include, as he puts it, 'everything but the kitchen stove – a loose aggregate, incapable of definition in terms of a species'.

To overcome this he proposed to limit the use of *J. chinensis*, to resuscitate *J. sphaerica* and to raise *J. sheppardii* to specific status. He also proposed a new hybrid species *J.* × *media* to contain all the more or less bush-like forms in which *J. sabina* blood – or should it be sap? – was discernible by the characteristic savin odour of the crushed foliage. He also separated certain garden forms being grown as cultivars of *J. chinensis*, regarding them as belonging to another species *J. davurica* and he regarded *J. chinensis* 'Sargentii' as a separate species.

In support of all these suggestions he clearly did an immense amount of work both amongst the literature and herbarium specimens of all these Junipers and amongst the living plants, but his choice of principal criterion – habit – was unfortunate (just think how readily any variable species, such as *Chamaecyparis lawsoniana*, could be dismembered by grouping forms of similar habit into separate 'species'), and like persons always tend to do once they have found a tenable theory, he disregarded every contribution from earlier botanical work that did not support his proposals. That many of these early writers were often mere copyists or at best were writing with but limited knowledge of the Junipers of which they wrote is only too clear (e.g., many of them as young plants bearing predominantly juvenile foliage are quite unlike the same trees as adults) so he had good grounds for exasperation over much of the literature he was trying to sort out, but Van Melles's work has not been accepted in the botanical world. Rehder wrote [Bibl. 649. (1949)] 'An evaluation of the classification of these Junipers as proposed by Van Melle is not possible without further study and must be left to a future monographer of this group'.

His proposals have also only received a limited acceptance in horticulture [See Welch, 203 (1979) and most later writers]. But in 1979 I was able to interest Dr. Laurence Gough, a scientist at the Polytechnic of the South Bank, Battersea, London (who had used the method to clear up with me a long-standing conundrum regarding the plant now known as *Platycladus orientalis* 'Sanderi') in exploring the possibility of clearing up at least some of the problems in this complex group by means of gas-chromatographic analysis of leaf extracts.

We worked together on this (he doing the scientific work, I supplying plant material and encouragement) until about 1982. He came to some valuable conclusions which he passed on to me, but about then his health, never good, deteriorated and, sadly, he passed away in 1985 before being able to complete the work and publish the results.

Dr. John Mills, Technical Adviser to the National Gallery (who is acting as Gough's scientific executor) tells me that he doubts the feasibility of ever publishing Gough's conclusions in a technical paper, but I feel that he is at least entitled to be credited with a valuable contribution, so I give here (as a 'personal communication') a summary of his latest findings. Gough was conscious of how poor was his access to material of known wild origin and also of his own lack of any formal botanical training. His conclusions therefore do not, standing alone, justify any taxonomic changes, but they will doubtless be taken seriously into account by some future investigator.

Briefly, Gough based his conclusions on the analysis by gas chromatography of leaf extracts of over 70 specimens, nearly all of them in cultivation in England, many being recognised cultivars, others not.

Comparing his results with those of Van Melle, Gough found no correlation between the habit variations upon which Van Melle's system was largely based and the conclusions he arrived at by his own technique, which did fully support the earlier worker's recognition of *J. sargentii* (Henry) Takeda ex Nakai as a separate species. Gough also accepted Van Melle's hybrid species *J.* × *media* as a practical expedient, although he disagreed as to the parentage involved. Beyond that he found little with which he could fully concur.

Gough found a large group (approaching 40 specimens) which was acceptably coherent. It included plants of both Chinese and of Japanese origin which between them covered a wide range of habits. Plants in this group are predominantly, although not exclusively dioecious. In general, mature plants carry mainly adult (scale-like) leaves, and they are mostly bright- to dull-green in colour, although glaucous, yellow-diffused and variegated selections are known. Awl-like leaves are mainly confined to the lower part of the plant and are in threes with usually a substantial, sometimes a predominant number paired.

This group includes plants currently listed under *J. davurica* and some that are currently (following Van Melle) included in *J.* × *media*.

Gough accepts the name *J. chinensis* L. (1867) for this group and here lists the following cultivars.

'Arbuscula'
'Aurea'
'Blaauw'
'Columnaris Glauca'
'Expansa Aureospica'
'Expansa Variegata'
'Globosa'
'Globosa Cinerea'
'Kaizuka'
'Mas'
'Mordigan'
'Olympia'
'Parsonsii'
'Pendula'
'Plumosa'
'Plumosa Argenteo-variegata'
'Plumosa Aurea'
'Plumosa Aureo-variegata'
'Robusta Glauca'
'Rosefield'
'San José'
'Sheppardi'
'Shimpaku'
'Tama Ibuki'
'Tremonia'
'Variegata'

Gough identified a second group, differing quantitatively from the first but not definitely distinct therefrom, and possibly divisible into two sub-groups, the first comprised of plants shewing rather persistent juvenility, even mature trees carrying a substantial proportion of awl-like leaves in all parts of the crown. These plants are often of a blue-green colour, with stout awl-like leaves in threes, rarely in pairs. Adult branches may be thicker and shorter than in the plants in the first group (above). Plants in the other sub-group, although giving similar analyses, is comprised of plants in which juvenility is not notably persistent and which are not usually of bluish shades. Their awl-like leaves are paired or in threes and are moderately slender, while adult branchlets tend to be thinner than with plants in the first group. Gough points out that some of these differences may be merely clonal distinctions. Pending further investigation, he leaves this group also within *J. chinensis* L. (1867), but suggests that a future worker might consider separating this group and referring it to *J. sphaerica* Lindley (1850). He lists the following clones.

'Blue Alps'
'Columnaris'
'Jacobiana'
'Echiniformis'
'Keteleerii'
'Luptonii'
'Maney'
'Pyramidalis'
'Story'
'Stricta'

Further and most importantly, Gough makes out a case for recognising *J. japonica* Carrière (1855), despite Van Melle having convinced himself that the title could be dropped entirely. This group is both qualitatively and quantitatively distinct from the other three groups in its analyses, and in some morphological respects too. It includes some monoecious plants. Its most distinctive feature is that some awl-like leaves are wide spread, very stiff and very prickly, in pairs or in threes. Adult branchlets seem to be slightly more slender than in Group 1. Branch tips are often at least slightly drooping.

Here are listed:-

'Japonica'
'Japonica Variegata'
'Kuriwao Gold'
'Kuriwao Sunbeam'
'Neaboriensis'
'Oblonga'

The Table on p. 329 sets out a comparison between Van Melle's proposals and those made by Gough. The latter makes several suggestions for new infra-specific combinations. I know that it was his intention to enlist the help of a taxonomic botanist and validly publish these new names, but his failing health having presented this they can only be regarded as 'Provisional', as permitted in Article 34 of ICBN. Nevertheless his conclusions commend themselves to me and are incorporated in this book.

So there the matter rests. The full text of Gough's last Report is available to any research worker who cares to write to me through the Publisher.

No doubt a future monographic study will clear the matter up finally. In the meantime, for the purposes of this book I accept Gough's return to *J. chinensis* of the 'Plumosa' group of cultivars transferred by van Melle to *J.* × *media* and also the var. *parsonsii* group of cultivars transferred by him to *J. davurica*. I otherwise concur in his retention of *J.* × *media*, on the grounds of practical expediency, and continue to recognise the independence of *J. sargentii*.

A comparison of van Melle's proposals with those made by Gough.

Van Melle 1947		Gough 1985		Cultivars examined by Gough
J. chinensis	var. *chinensis*	*J. chinensis* L.	sensu stricto	'Aurea', 'Columnaris Glauca', 'Mas', 'Mordigan', 'Robusta Glauca', 'Variegata', etc.
do. in the synonymy –				
	oblonga Bobbink & Atkins	*J. japonica* Carrière	'Oblonga'	
	Sylvestris Hort	do.	'Japonica'	
J. sphaerica Lindl.	var. *sphaerica*	*J. chinensis* L.		
	var. *pseudo-mas*	*J. chinensis* L.		
	var. *dioica*	Uncertain		
	var. *neaboriensis*	*J. japonica* Carrière		'Japonica', 'Japonica Variegata', 'Kuriwao Gold', 'Kuriwao Sunbeam', 'Neaborensis', 'Oblonga'.
	var. *pendula*	*J. chinensis* L.	f. *pendula*	'Pendula' (of Franchet, not of other authors)
	var. *keteleeri*	do.	'Keteleeri'	
	var. *luptonii*	do.	'Luptonii'	
J. sheppardii (Veitch) Van Melle		*J. chinensis* L.		
	var. sheppardii		f. *sheppardii* (in part)	'Arbuscula', 'Rosefield', 'Sheppardii'
	var. *torulosa*	do.	f. *torulosa**	'Blaauw', 'Olympia', 'Shimpaku', 'Tama Ibuki', 'Tremonia'
	f. *albo-notata*	do.	'Variegated Kaizuka'	
	f. *aureo-variegata*	*J. japonica* Carrière	'Japonica Variegated'	
	var. *pyramidalis*	J. *chinensis* L.		'Jacobiana', 'Keteleeri', 'Luptonii', 'Obelisk', 'Pyramidalis', 'Story', 'Blue Alps', 'Echiniformis', 'Maney'
	f. *variegata*	do.	'Variegata'	
J. x *media* Van Melle	var. *pfitzeriana*	*J.* x *media* Van Melle	'Pfitzeriana'	Numerous cultivars in the 'Pfitzer' group
	var. *arbuscula*	*J. chinensis* L.	f. *sheppardii**	
	var. *globosa*	do.	'Globosa'	And colour forms.
	var. *plumosa*	do.	'Plumosa'	And colour forms.
J. davurica Pall.	var. *parsonii*	do.	f. *parsonsii**	'Expansa' group.
J. sargentii (Henry)	Takeda ex Nakai	*J. sargentii* (Henry) Takeda ex Nakai		

* Names suggested by Gough provisionally. See ICBN. Article 34.

-- '**Ames**'. Haber and Lantz, Iowa Farm Sc. 2 **11**: 10. 1948. The late Prof. T. J. Maney, sometime of Iowa State College, introduced four interesting new Junipers (the others are 'Maney', 'Iowa' and 'Story'), selections from a large number of seedlings raised by Dr. Vernon Stoutemeyer from Japanese seed (reputedly of an upright form of *J. sargentii* but evidently from a glaucous plant of *J. chinensis*) sown about 1934. 'Ames' makes a small, compact upright tree well clothed with foliage, giving a very dense appearance. The colour is blue-green and it makes an attractive specimen. It is very hardy.

-- 'Arctic'. See *J.* × *media* 'Arctic'.

-- 'Armstrongii'. See *J.* × *media* 'Armstrongii'.

-- 'Armstrong Gold'. See *J.* × *media* 'Gold Coast'.

-- '**Arbuscula**'. (Van Melle) Cornman in Gentes Herb. **8**: 79 (1949); *J. chinensis* var. *smithii* Slavin 103 (1932) not Loudon (1838): *J.* × *media* var. *arbuscula* Van Melle, 189 (1946). A female clone of unrecorded origin that forms a small tree or erect bush with several ascending, mostly undivided branches but with all the terminal growths nodding. The foliage is

Figure 195. *Juniperus chinensis.* Chinese juniper. (10)

Figure 196. *Juniperus chinensis* 'Obelisk'.

mainly adult, a light bright- to dull-green, any juvenile leaves present being either short, and soft as in *J. sabina* or longer and more stiff.

-- **'Aurea'** R. Smith, Pl. Fir Tribe, ed. 3, 17 (1865); Young, Nurs. Cat. 3. 1875; f. *aurea* (Young) Beissn. in Rehd. 60. (1949). Golden Chinese Juniper. This in time grows too large to be regarded as a dwarf conifer, but being of slow growth it should be retained as long as possible on account of its wonderful colour, this being a brilliant golden-yellow during the spring and summer if the plant is grown in full sunlight, but it becomes less outstanding by winter. It is a male clone originating as a branch sport on a normal tree in M. Young's Nursery at Milford, Godalming, in Surrey, and introduced by him to the trade. It bears juvenile and adult foliage in approximately equal amounts.

-- 'Berry Hill'. See *J.* × *media* 'Berry Hill'.

-- **'Blaauw'**. Den Ouden and Boom, Man. 149. (1965) Blaauw's Varietät Grootn. Mit. d. d. d. G. **53**: 206. (1940). This is a dense growing and upright female cultivar in which the differentiation in vigour between the leading shoots and the sub-laterals is very pronounced

Figure 197. *Juniperus chinensis* 'Olympia'.

Figure 198. *Juniperus chinensis* 'Stricta'.

so that the relatively few, strongly growing main branches striking up at a steep angle give a characteristically rugged outline to the plant which is most attractive. The colour of the foliage, a rich blue-grey, is an additional attraction. It is quite distinct from 'Globosa Cinerea' in habit although somewhat similar in colour. Introduced from Japan by the firm J. Blaauw & Co, Boskoop, Netherlands, in ± 1924.

-- '**Blue Alps**'. A vigorous, spreading large shrub with ascending main branches and gracefully overhanging growth tips. The juvenile leaves are 7–12mm long, 1.5–2mm wide, very sharp-pointed, blue-green with two whitish bands of stomata above, giving an overall silver-blue colour that is the brightest blue in any juniper. The young shoots have a brown tinge. It was discovered by H. J. Welch on 4th November 1968 as a plant in Alpengarten Frohnleiten in Austria, raised from seed of unrecorded origin and there being grown as *J. communis nana*. It had a limited distribution from the Wansdyke Nurs. Devizes, U.K. as 'Willow Pattern', but the stock was later taken over by the Grootendorst and Sons Nurs. of Boskoop, Holland. At the suggestion of Hermann Grootendorst the name was changed to 'Blue Alps' and this name was registered with COPF (Canada) in 1981 and with the International Registration Authority in 1983.

-- 'Blue and Gold'. See *J.* × *media* 'Blue and Gold'.

Figure 199. *Juniperus chinensis* 'Plumosa'. (Wrongly listed by Van Melle under *Juniperus* × *media*.)

-- 'Blue Cloud'. See *J. virginiana* 'Blue Cloud'.

-- **'Blue Point'**. Monrovia Nurs. Cat. 56 (1983). This clone builds up into a symmetrical cone the shape of a conventional tear-drop, with extremely dense blue-grey foliage.

-- 'Blue Vase'. See *J.* × *media* 'Blue Vase'.

-- **'Columnaris'**. Fairchild in U.S. Dept. Agric. Ann. List New Pl. Intr. **9**: 22 (1920) under no. 18577; *J. chinensis pyramidalis* Kumlien 164 (1946); *J. chinensis columnaris viridis* Hort. Green Columnar Chinese Juniper. Narrowly columnar, 8m, loose; leaves awl-shaped, sharp pointed, green; selected from an introduction of F. N. Meyer from Hupeh (C. China) sent by him to the U.S. Dept. of Agriculture at Chico, Calif. in 1905. The nice hedge from which this and the following cv. were selected is still present in the Plant Introduction Garden at Glenn Dale, Md. U.S.A.

-- **'Columnaris Glauca'**. den Ouden 152 (1949); *J. chinensis pyramidalis glauca* Slavin (1932). Blue Columnar Chinese Juniper. Similar to the preceding cv. but foliage silvery grey and with the same history.

-- 'Daub's Frosted'. See *J.* × *media* 'Daub's Frosted'.

-- 'Den Boer'. See *J.* × *media* 'Den Boer'.

-- 'Densa Erecta'. Now known as 'Spartan'.

-- **'Dropmore'**. Krüssmann, 136 (1972). A form collected by Frank L. Skinner at Dropmore, Manitoba, Canada, is an extremely slow-growing bun-forming plant with a very dense habit and highly ramified branching habit and a dense, neat thatch of very decussate adult or semi-adult foliage, in which the very small (to 3 mm long) leaves are held closely forward so that the plant entirely lacks the starry-eyed look

Figure 200. *Juniperus chinensis* 'Parsonii'. (Wrongly listed by Van Mell under *Juniperus* × *media.*)

of 'Echiniformis'.

-- **'Echiniformis'**. *J. oxycedrus echiniformis* Knight, Syn Conif. 11 (1850) *J. hemisphaerica* Gord. (NON Presl) Pin. 96. (1858) and *J. echinoformis* Rinz; *J. oxycedrus echinoformis* Van Houtte (pro syn.); *J. communis echinoformis* Beissn. Conif. Ben. 41 (1887) and Handb. 137. (1891) This curious little plant has suffered much from the coiners of names. Certainly it has no affinity with either *J. communis* or *J. oxycedrus* so has now found a more likely resting place here. It forms a tiny little flattened or rounded bush with short, irregular, crowded branches and very tiny leaves. Its slow rate of growth and apparent weakness of constitution in Britain may be due to a virus, for whilst it seems hardy, quite established plants suddenly may die back. It is a great favourite with owners of alpine houses, under which conditions very fine specimens can be grown, given enough patience. Gordon and some others confused it with *J. communis hemisphaerica* Presl and so the description of the plant and its origin by these writers are a useless hotchpotch. It seems reasonably certain that 'Echiniformis' was first sent out by a nurseryman Rinz of Frankfort. Whence he received it is not known, but it would hardly have been a wildling found on Mt. Etna. It seems to be a much better 'doer' in the United States than in Britain.

-- 'Expansa', 'Expansa Aureospicata' and 'Expansa Variegata'. See 'Parsonsii', p. 337.

-- **'Fairview'**. F. J. Grootendorst & Sons, Cat. 59 1958: II; *J. chinensis 'Hetz'* Cat. Fairview Nurs. 4 1942; *J. chinensis hetzii* Kruessm. 129 (1955); *J. chinensis 'Hetz's Columnaris'* Fairview nurs., Cat. 1960: 8; Wyman in Amer. Nurs. 117 (1): 115 (1963). Narrow-pyramidal and strong-growing; leaves mostly subulate, on some branchlets scale-like, bright green. Originated as a seedling in the Fairview Evergreen Nurseries, Fairview, Pa, U.S.A. during the early 1930's.

-- **'Femina'**. (Carr.) Beissn. in Mitt. d. d. d. G. **10**: 75 (1901); *J. chinensis (femina)* Carr. 32 (1855); *J. reevesiana* Hort.; *J. sphaerica* var. *dioica* van Melle 186 (1946). Conical, 3–5 m tall, free-fruiting; branches spreading, thin, weak; sprays crowded, thin; leaves scale-like, appressed, blue-bloomy, awl-shaped leaves rarely present; cones globose or oblongoid or irregularly globose, 5–7 mm across, rarely thicker, at first whitish-blue bloomy, dark blue when ripe. Probably introduced by R. Fortune into Britain in 1861.

-- 'Fruitland'. See *J.* × *media* 'Fruitland'.

-- **'Globosa'**. Hornb. Dw. Con. 62, 1923; *J.* × *media* var. *globosa* Van Melle, Review, 29, 1947 (excl. synonymy); *J. virginalis globosa* Yokohama Nurs. Cat. 33 1921/22 (and earlier?); *J. virginiana globosa* (Hort. Eng.); *J. japonica Bandai-sugi* Hort. Amer. (in part.). 'Globosa' and its colour forms have had an interesting and amusing history. They first came on to the western scene as importations into Britain and the United States from a Japanese nursery of plants labelled '*Juniperus virginalis globosa*'. These were certainly nice, globose plants and came in three colours – green, gold and grey – and they apparently sold well. After a while the fact that there was no '*J. virginalis*' known to botanists was noticed and it was thereupon assumed that the correct name should have been '*J. virginiana globosa*', so the next few customers received plants under this name and thus labelled they can still be found. Later it was seen that they were not a *J. virginiana* form at all but were cultivars of *J. chinensis*, so Hornibrook in his second edition described them as '*J.c. globosa*', '*J.c. globosa aurea*' and '*J.c. globosa cinerea*', respectively.

Finally, it was found that they had not even any moral right to the designation '*globosa*', for the plants remaining unsold, deprived of the careful attention from the knife to which they had been accustomed in their native Yokohama, were beginning to develop a habit that was by no means globose.

The present position, therefore, is that 'Globosa' differs from 'Plumosa' only in minor respects. It forms a more symmetrical plant, the branches are much shorter and stand at wide angles all over the plant and the foliage is a lighter shade of green. The true 'Globosa' is rare in cultivation, but the difference between the forms is slight and both are only needed in a very complete collection. 'Globosa' is a female form.

-- 'Globosa Aurea'. Lost to cultivation. See 'Plumosa Aurea' below.

-- **'Globosa Cinerea'**. Hillier, (1964) and Man. 488 (1971) *J. chinensis* var. *globosa cinerea* Hornb. Dw. Con. Ed. 2, 95. (1938). This is very close to the cultivar 'Blaauw'; so much so that, especially in young plants, they are difficult to tell apart. 'Globosa Cinerea' does not, however, throw up quite such strong leader growth and the angle of divergence of side shoots is greater, so it forms a much broader plant of less rugged outline; also the foliage, especially in wintertime, is more a grey-green than the blue-grey of 'Blaauw'. As in the case of the green forms, these two grey varieties are too close together for both to be needed in the average garden, and my choice would always rest with 'Blaauw' (see also 'Shimpaku' below).

-- 'Gold Coast', 'Golden Saucer', 'Goldkissen', 'Gold Star', 'Green Jungle' and 'Hill's Blue' are all cultivars of *J.* × *media*.

-- 'Helle'. A superfluous synonym for 'Spartan'.

-- 'Hetz Columnaris'. Now known as 'Fairview'.

-- 'Hetzii' (= 'Hetz' Glauca'). This is now listed under *J. virginiana*, but it is possibly a hybrid.

-- **'Iowa'**. Haber & Lantz in Iowa Farm Sc. **2,11**: 11 (1948); Wyman in Amer. Nurs. **117,1**: 115 (1963). Similar to 'Ames' but a little more spreading and not quite so compact; leaves acicular and scale-like, green with a bluish cast; it is a free fruiting form; origin see 'Ames'.

-- **'Japonica'** [*J. japonica* Carr. Traite, 33. (1855) (excl. synonymy)*; *J. chinensis* f. *japonica* (Carr.) Lipa in Rehd. Bibl. 61. (1949); var. *nana* Hochst. Conif. 90. (1882) *J. japonica* Carr. Traité. Ed. 2, 32. (1867), (incl. var. pyramidalis); *J. chinensis* 'Nana' Welch, Dw. Con. 172. (1966) f. *oblonga* Slavin, Rep. Conif. Conf. 102. (1932) var. *Veitchii* Hornb. Dw. Con. Ed. 2, 101. (1938): NON var. *japonica* Vilm. Hort. Vilm. 58. (1906); NON *japonica* (Vilm.) Elwes & Henry, Trees **6**: 1432 (1912) Hornibrook seems to have made a valiant attempt to unravel what is usually referred to as 'The Japonica Tangle', this being the confusion amongst several *J. chinensis* forms that resulted from the free and light-hearted use made of the word '*japonica*' both as a specific and as a varietal name by nineteenth century writers on conifers. But in spite of all his trouble I, personally, remain

*See p. 328 for an account of the suggested re-establishment of this specific name by the late Dr L.J. Gough.

uncertain as to just how many forms he was trying to leave with us. Van Melle, on the other hand, convinced himself that the title could be dropped entirely. I find it impossible to make head or tail of his synonymy. In his zeal to eliminate the epithet '*japonica*' entirely he found no place for the plant here listed as *J. chinensis* 'Japonica', so he had to put undoubted references to our plant where best he could.

Van Melle was such a stickler for taking into account the differences in habit and foliar characteristics of these junipers at varying ages and the importance of studying trees at all stages of growth that it is strange that he did not also appreciate the importance of following the changes as they took place in particular plants over a period of years. Had he done so he would have been saved a curious mistake. Referring to Veitch's description "It is a dwarf bush, sometimes with only one stem but oftener with a divided one; in the latter case the branches or divisions of the stem assume a procumbent habit", Van Melle says "There is no juniper...which behaves in so variable a manner". But in this he was wrong.

I originally obtained propagating material of 'Japonica' from two sources, one a spreading form bearing mostly juvenile foliage, the other form an upright plant with a good proportion of adult foliage. Apart from this the two samples were indistinguishable. As often happens on a busy nursery there was difficulty in keeping the two forms separate – the prostrate form and the upright form would persistently get mixed – or so I thought until I noticed that one by one the spreading plants were changing sides. They did this not by throwing up a leader but by pulling up their main branches, at first to a steep angle, eventually to the vertical. Eliminating double leaders being second nature to a nurseryman when on his rounds I never secured plants with Veitch's 'divided leader' but have no doubt some of the plants would have so developed, left to themselves, and of course a couple of snips with secateurs would have kept any young plant in procumbent shape indefinitely.

With this growth characteristic known there is no difficulty in accepting apparently divergent accounts that neither Hornibrook nor Van Melle could reconcile. I have no doubt that Carrière, Gordon, Veitch, Slavin and Nicholson were all writing about the same plant but at different stages of growth. When preparing his second edition Carrière was fooled by the behaviour of some of his plants into treating them as var. *pyramidalis* and also there seems to have been widespread uncertainty about the relationship to *J. procumbens*. Otherwise the name serves us well as a cultivar name for a plant which usually starts life as a fairly low bush with two or three ascending main branches bearing adult foliage at their tips but densely covered throughout the rest of their length with juvenile foliage which is very densely set and extremely prickly. This prickliness is one of the main characteristics and is so pronounced that it is advisable to wear gloves when handling the plant. Eventually, however, left to itself, it becomes a small, more or less upright little tree, the proportion of adult to juvenile foliage increasing as time goes on, and often varying considerably on different sides of the same tree. The colour (especially in the adult foliage) is an attractive, rich green.

-- **'Japonica Variegata'**. *J. japonica variegata* Carr. Traité. Ed. 2, 31. (1867) *J. chinensis japonica aureo-variegata* Veitch Conif. 288 (1881); The plant tallies with Carrière's description, and it is quite clearly a variegated form of 'Japonica' with no affinity whatever with 'Kaizuka', although often grown as 'Kaizuka Variegated' in Europe.

-- **'Kaizuka'**. Yokohama Nurs. Cat. 1920; *J. chinensis* var. *torulosa* Bailey 196 (1933); *J. sheppardi* var. *torulosa*. (Eastwood) Van Melle, Review, 28. (1947;) 'Hollywood Variety' Hort. Amer. (in part); 'Torulosa' Hort. Amer. (in part). This attractive form will grow eventually to tree-size but it is slow-growing and for many years in its early stages is a most desirable plant, having a picturesque, rugged outline with a tendency for its trunk or main branches to be slightly out of upright and this character should not be modified by pruning or staking.

Part of the attraction of this plant is the picturesque effect given by the irregular crowding of the branchlets into dense mop-like clusters which often form cylindrical effects (hence the name 'Torulosa' – not to be confused with the word 'tortuosa' which means twisted) along the branches. The foliage is a bright, rich green which will hold its own with the green of *Chamaecyparis obtusa* 'Nana Gracilis'.

-- 'Kaizuka Variegated'. See 'Variegated Kaizuka'.

-- **'Keteleeri'**. Beissn. in Mitt. d. d. d. G **19**: 140 (1910); under *J. virginiana* Rehd, 205 (1923); *J. sphaerica* var. *keteleeri* (Beissn.) van Melle 187 (1946). Narrowly pyramidal, dense, 10–12 m, free fruiting; branches ascending; branchlets crowded, short, very thin; leaves scale-like, very pointed, green, slightly glaucous bloomy; cones with a recurved stalk, regularly globose, 12–15 mm across, distinctly meally, finally glossy light green; seeds to 7 mm long. Probably raised by Keteleer, nurseryman at Brussels, Belgium. The cones are probably the largest of any cultivar in the species.

-- **'Kohankie's Compact'**. Wyman in Amer. Nurs. **117** (2): 64 (1963). Very dense and almost globose, bluish-green in colour; an unauthenticated sport that arose at the former Kohankie Nurs, Painsville, Ohio, U.S.A. Not recommended by Wyman, and possibly not in cultivation.

-- **'Kuriwao Gold'**. Griffin, B.A. (1975). This originated as a seedling raised by Ross Stuart of Clinton, South Otago, New Zealand from seed collected from a plant of 'Keteleeri' on the Ces. Wallis Nursery at Mosgiel, growing near a plant of 'Pfitzeriana Aurea'. Of several interesting seedlings that resulted, this was the pick and it has fulfilled its early promise. It makes a plant 1.25 m high and almost as wide at 14 years: a rounded little barrel with prickly, juvenile foliage that is a bright gold through the summer and autumn, fading somewhat with the reduced sunlight during the winter, but in spring again its soft new growth is a most attractive greeny-gold. It must be planted in full sun and colours best in poor soil. It was introduced in 1975. The mixed parentage (if the species *J.* × *media* is accepted) raises a delicate question for the purists as to where this group of seedlings should be listed, but they all seem to side with Mother so here is where they seem to belong.

-- **'Kuriwao Mist'**. has the same origin as 'Kuriwao Gold'. It makes a much taller plant and is of much looser growth and takes an upright oval shape. In form and foliage it is much nearer the parent 'Keteleeri', but its most interesting characteristic are the seasonal colour changes; from misty white when the new spring growth appears, to blue and finally bluey-green later in the autumn.

-- **'Kuriwao Sunbeam'**. also has the same origin as 'Kuriwao Gold'. It was given as seedling by Ross Stuart to Dennis Hughes of Blue Mountain Nursery and was so named because of its brilliance when the sun shines on it. (The names 'Sunburst' and 'Sunshine' are incorrect.) It is a rather more vigorous and upright plant than 'Kuriwao Gold' and is a paler colour. It is a male plant.

-- 'Lemon Hill'. See *J.* × *media* 'Lemon Hill'.

-- **'Maney'**. Haber & Lantz in Iowa Farm Farm Sc., **2**, 11: 10 (1948). This is the third of Professor Maney's introductions the history of which is given under 'Ames' above. 'Maney' forms a semi-erect but spreading bushy plant with a blue cast to the foliage. It thus fills a useful place between the upright form normal to *J. chinensis* and the *J.* × *media* spreading forms. It looks well either massed or grown as a specimen in a small garden.

-- **'Mas'**. Gord. 176 (1858); *J. chinensis mascula* Carr. 30 (1867). An upright and quite loose male tree, to 20 m; leaves scale-like and awl-shaped, the former yellowish-green, the latter distinctly white-striped above, green beneath, with a pungent apex; smothered with yellow male flowers in spring.

-- 'Mathot', 'Milky Way' and 'Mint Julep'. See under *J.* × *media*.

-- **'Mission Spire'**. A vigorous, upright-growing selection with glossy, bluish-green foliage that has distinctive lilac-mauve cast during the winter months. It was introduced by the Mission Gardens, Techny, Ill., U.S.A. It would appear to be very similar to 'Iowa'.

-- **'Monarch'**. Grootn. in Dendroflora **2**: 47 (1965). An upright-conical plant with an erect branching system and strong growth, with an irregular outline due to the development of several sub-leaders. The foliage is wholly juvenile, the leaves being of varying sizes, the free-standing part up to 1 cm long on the main branches, less on laterals and sub-laterals. It was raised on the Grootendorst Nurs. from seed received from Japan.

-- 'Moraine'. See *J.* × *media* 'Moraine'.

-- **'Mordigan'**. An irregular bush with mainly juvenile foliage and sharply pointed silvery-blue needles when young and an increasing amount of green, adult foliage as the plant ages. It was raised from seed from 'Columnaris' and was registered with A.A.N in 1951.

-- 'Mordigan Aurea'. See *J.* × *media* 'Mordigan Gold'.

-- **'Mountbatten'**. Ashworth in Amer. Nurs. **88** (10): 42 (1948). A dense, narrowly conical form to 4 m, similar to *J. communis* 'Hibernica'. It is a female clone and fruits freely; foliage greyish-green, predominantly juvenile. It originated as a seedling on the Sheridan Nurs, Etobicoke, Ont., Canada, and was named in honour of Lord Mountbatten when he visited Toronto in 1948 to open the Canadian National Exhibition.

-- 'Nick's Compact'. Now known as *J.* × *media* 'Pfitzeriana Compacta'.

-- **'Obelisk'**.F. J. Grootendorst & Sons, Cat. **1946**: 12; den Ouden 154 (1949). Narrow, columnar, slender, erect topped; branches spreading, short, stout; branchlets short, stout, bluish-green; sprays numerous, densely set, short; leaves awl-shaped, directed forward, slightly decurved, apex very sharply pungent, narrow at base, 10–15 mm long, underside glaucous bloomy. The upper side of the leaves turning outwards gives the plant a distinct glaucous-white appearance; keeps its habit without trimming. A slight slope to the leader seems to be characteristic of this variety and for formal situations would be a defect. Raised by F. J. Grootendorst & Sons, Boskoop, Netherlands from seed imported about 1930 from the Chugai Nursery Co., Yamamota, Japan. (See Illus. No. 196)

-- **'Oblonga'**. Bobbink & Atkins ex Slavin 102 (1932). One stemmed, subglobose, semi-dwarf, with a single, erect leader bearing bright green scale-like leaves, the rest of the plant with dark green, awl-shaped leaves, very prickly like those of *'Japonica'*, from which it is a sport. Originated in the nursery of Bobbink & Atkins, Rutherford, N.J., U.S.A.

-- **'Olympia'**. F. J. Grootendorst & Sons, Cat 10; **1957**: 'Chugai', Hort. Columnar, slender; branches ascending; branchlets short; leaves of 2 kinds, the scale-like ones closely appressed, bluish; the awl-shaped leaves to 12 mm long, sharp, spiny-pointed, with 2 white bands above, greenish-blue beneath. Raised from seed imported about 1930 from the Chugai nursery Co., Yamamota, Japan, and distributed by the firm F. J. Grootendorst & Sons, Boskoop, Netherlands in 1956. (See Illus. No. 197)

-- **'Parsonsii'**. *J. chinensis* var. parsonsii Hornibrook, 96 (1939); *J. squamata* var. *parsonsii* Bailey, Hortus Second 400 (1940); *J. chinensis* var. *expansa* Grootendorst in Mitt. d. d. d. Ges. **53**: 207 (1940); *J. davurica* var. *parsonsii* Van Melle, 192 (1946). Apart from any mention of the name occurring in earlier Parsons Nursery catalogues, this name has clear priority over 'Expansa'. The transference of this cultivar to the little known species *Juniperus davurica* Pall. by Van Melle seems to have had no justification since the late Laurence Gough was unable to distinguish this plant and its variegated forms chemically from *J. chinensis*. It is a female clone, bearing berries that are more or less globose, 7 mm diam, containing three seeds and it is said to have been sent by Dr. Hall to the Parsons Nurseries, Flushing, N.Y., U.S.A., in

1862.

It is not a prostrate juniper. Individual plants in open, uncrowded situations develop stout and very rigid, horizontally-spreading primary branches which do not rest upon the ground but extend themselves horizontally just clear of it. Because of this characteristic, it is important not to plant this juniper too deeply and I have noticed that when set so that the main branches are touching the ground, the plant does not prosper and the growth becomes uncharacteristic, lacking the sturdy, horizontal-branching habit that it normally has.

The foliage is light green, somewhat glaucous, the leaves mostly acicular in twos or threes, but the long main branches carry very small scale-like leaves giving them a filiform appearance. There are two variegated forms. '***Expansa Aureospicata***' Grootendorst in Mitt. d. d. d. Ges. **53**: 207 (1940) is a form with butter-yellow variegation, somewhat less vigorous than the green form and carrying mainly juvenile foliage. In '**Expansa Variegata**' Hornibrook, 94 (1939) the variegation consists of bold creamy-white splashes on both juvenile and adult foliage. This variegation is apt to suffer the winter, but it appears again with the new season's growth. With time it makes a striking plant 2 m across, and the variegation is very effective in shade. (See Illus. No. 200)

-- '**Pendula**'. Franchet 101 (1884), not of de Vos 1887 and other authors; *J. sphaerica* var. *pendula* (Franch.) van Melle 186 (1946). A conical tree with an irregular outline; branches few and widely ascending, pendulous at their tips. Introduced by Father A. David into France in ± 1882 from Shenai, W. China. Probably no longer in cultivation.

-- 'Pfitzer', 'Pfitzeriana' and several names derived therefrom, see *J.* × *media*.

-- '**Plumosa**'. Hornb. Dw. Con. 66. (1923) and ed. 2, 102 (1938); *J.* × *media* var. *plumosa* Van Melle, Review, 30 (1947). Van Melle having established his *J.* × *media* as a resting place for all hybrids between *J. chinensis* and *J. sabina* had no justification whatsoever for having included therein quite a number of clones of *J. chinensis* which were under no suspicion of any such mixed parentage merely because they were shrubby in habit. These, including 'Plumosa' and its colour forms are in this book returned to *J. chinensis* where they rightly belong.

'Plumosa' is a broad, spreading shrub, seldom over 1.5 m high with long main branches and numerous short side branches all pointing strongly forwards, the sprays being crowded and plume-like, with all growing tips curved gracefully over. There is usually a small amount of juvenile foliage near the base of the plant. Unless otherwise trained in its early days it limits itself to relatively few main branches and when one of these predominates the plant has a lop-sided development which may or may not be an advantage according to the situation. The colour is deep green. It is a male clone. (See Illus. No. 199)

-- '**Plumosa Albovariegata**'. Hornb. Dw. Con. ed. 2, 106 (1939); *J.* × *media* var. *plumosa* f. *albovariegata* Van Melle, Review, 30 (1947) is similar to 'Plumosa' with a somewhat spreading habit and with its foliage liberally peppered with a small creamy-white variegation.

-- '**Plumosa Aurea**'. Den Ouden and Boom, Man. 157 (1965); *J. chinensis* var. *globosa aurea* Hornb. Dw. Con. 63 (1923) and ed. 2, 95. (1939); *J. chinensis* f. *aureo-globosa* Van Melle, Review, 29 (1947); *J. virginalis globosa* (Yellow), Yokohama Nurs. Cat. (1910?); *J. japonica globosa aurea* Bobbink and Atkins Nurs. Cat. (c. 1916); NON *J. chinensis* var. *plumosa aurea* Hornb. Dw. Con. 66 (1923) and ed. 2, 103 (1938); NON *J.* × *media* var. *plumosa* f. *aurata* Van Melle, Review, 30 (1947). This is the most decorative colour form of this group. It forms an attractive flat-topped, spreading bush wider than high with mainly adult foliage in gracefully arching sprays, the foliage and branchlets yellow-green in spring, the colour becoming more pronounced through the year until by winter it is a rich golden-bronze. In my previous book I stated that there is no golden form of 'Globosa', since fresh importations from Japan always turn out to be the cultivar 'Plumosa Aurea' of the trade. I have since seen a form with the same colouring but having the erect habit mentioned by Hornibrook as characteristic of 'Plumosa Aurea' as he used the name in 1939. There has clearly been a name switch here since 1939 and as the more spreading form has become so widely known as 'Plumosa Aurea', common sense suggests retaining this usage (as permitted by Article 27c of the Cultivated Code – despite its posing a problem with the synonymy) and accepting the name 'Tremonia' (see below) for the form distinguished by the more upright habit.

-- **'Plumosa Aureovariegata'**. (Otto) Hornibr. 105 (1939); *J. japonica aureovariegata* Otto in Hamb. Gart. u. Bl. Zeit. **29**: 203 (1873); *J. chinensis japonica aureovariegata* Mast. ex Henry 1432; (1912); *J. media plumosa aureovariegata* (Otto) van Melle 192 (1946). A slower-growing form with its deep green foliage carrying a deep yellow variegation, usually somewhat irregularly.

Where there is room for such a group to be seen from a little distance away, a remarkable colour and texture effect could doubtless be obtained by planting a large group of these 'Plumosa' forms, including 'Blaauw' and 'Globosa Cinerea' for their contribution in colour, the rich green of the taller-growing clone 'Kaizuka' at the back or centre of the group.

-- **'Pyramidalis'**. [*J. japonica* SENSU (Carr.) Sargent Gar. and For. **10**: 42 (1897) (NON Carr.); *J. chinensis* f. *pyramidalis* Beissn. Handb. 120. (1891) *J. chinensis pyramidalis* f. *glauca* Slavin in Report. Conif. Conf. 103 (1932). The correct (even if misleading) name of a plant Sargent described as "of distinct compact habit with many erect branches with bluish-green juvenile foliage...it retains its peculiar compact juvenile habit". Beissner tells us that it forms a dense, slim blue-green column bearing at first acicular leaves and squammiform leaves later, and is a male clone. Den Ouden adds that the leaves are sharp, and that it was introduced from Japan into Belgium by Dr. von Siebold in 1845. It is significant that neither the word 'pyramidal' nor 'conical' appears in any of these descriptions. Nor is there any mention of a strong central leader. Instead we read of a compact column of a distinct or a peculiar shape and of many erect branches, and one account notices the sharpness of the leaves. These clearly distinguish this form from 'Stricta' below, with which this cultivar is frequently confused. The foliage of 'Pyramidalis' is relatively soft to the touch when stroked upwards.

-- 'Ramlosa'. See under *J.* × *media*. See also *J. ramlosa*.

-- 'Richeson'. See under *J.* × *media*.

-- **'Robusta Green'**. H. J. van der Laar in Dendroflora **11–12**: 58 (1975). A slow-growing, medium-size and rather informal tree with ascending branches and blue-green foliage carried in dense tufts, giving an irregular outline to the tree. Very popular in U.S.A. where it originated.

-- 'Rockery Gem'. See under *J. sabina*.

-- **'San José'**. W. B. Clarke in Amer. Nurs. **88,7**: 40 (1948); *J. japonia 'San José'* Clarke & Co, Cat. **1935**. San José Juniper. Dwarf shrub, lies flat on the ground; branches irregularly spreading, in all directions; foliage greyish-green, mostly acicular, especially so in young plants.

-- var. *sargentii*. See *J. sargentii* (p. 000).

-- 'Sea Green'. See *J.* × *media* 'Sea Green'.

-- **'Shimpaku'**. Welch in Dw. Conif. 198. (1966). This is a very slow-growing and delightful dwarf form of the Plumosa group, with dull, greyish-green foliage. It is a form much used by the 'manufacturers' of Bonsai trees, by whom it is frequently misnamed Sargent's juniper (see below), but when it becomes freely available it should become very popular as a garden plant or for use in troughs.

Herman J. Grootendorst of Boskoop, Holland, considered it to be identical with 'Globosa Cinerea', 'Shimpaku' being merely that variety growing 'on its own roots'. A golden form which I have received as **'Shimpaku Gold'** appears to be a diminutive 'Plumosa Aurea'. Since the latter form is another that is usually supplied as a grafted plant, this could well confirm Mr. Grootendorst's statement regarding 'Shimpaku', but the special requirements of the Bonsai cult justify retention of the separate name for commercial reasons. See also under *J. sargentii*.

-- **'Shoosmith'**. Den Ouden and Boom 159 (1965); *J. chinensis nana Shoosmith* Southside Nurs. Cat. ± 1930. Dwarf bush, globular to pyramidal, very compact, resembling the box in growth; leaves awl-shaped, 5–10 mm long, 0.6–0.8 mm broad, very thin, spiny-pointed, with 2 inconspicuous bands above, dark green beneath. Originated in the Southside Nurseries, Pa., U.S.A., and distributed by Fred Shoosmith, during the 1930 s.

-- **'Spartan'**. *J. chinensis densaerecta 'Spartan'* Monrovia Nurs. Cat. **1961/62**: 6. A fast dense grower with a compact pyramid habit of growth; foliage rich green. Found in the Monrovia nurseries, Azusa, Calif. U.S.A.

-- **'Story'**. Haber & Lantz in Iowa Farm Sc. **2,11**: 11 (1948). Male, symmetrical, slender, with a dominant central leader; branches horizontal; foliage adult, dark green. Same origin as *'Ames'*.

-- **'Stricta'**. den Ouden 156 (1949); *J. excelsa stricta* Hort. amer., not of Sénéclauze; *J. procera* var. *africana* Hort. N.Z. (in part); *J. squamata campbellii* Krüssm. (1960). Conical, erect, mostly pointed, dense bush; branches ascending, regularly borne; branchlets regularly erect, rather short; leaves strictly awl-shaped, soft to the touch, somewhat glaucous above, equally blue-bloomy beneath, steel-blue in winter. Very common in cultivation.

'Stricta' becomes too large in time but it serves us well as a garden plant, for in its young stages it forms a regularly conical bushlet with ascending branches and wholly juvenile foliage of a very beautiful, light blue-grey. Brushed with the hand, the foliage feels distinctly prickly. It remains an attractive plant for some years but eventually its habit of holding on to its dead foliage for several years gives the plant an unhealthy appearance, by which time it is usually best to replace it. (See Illus. No. 198)

-- **'Titlis'**. [Draijer. Nurs. Cat. M. 1972/73] is a new form introduced by the nursery firm Draijer B. V. of Heemstede, Holland, and described as a somewhat irregular and very compact fastigiate form with silvery-blue foliage, very sharp to the touch.

-- **'Tremonia'**. Welch, 231 (1979); *J. chinensis* var. *plumosa aurea* Hornb. Dw. Con. 66 (1923) and Ed. 2, 103 (1939). Under 'Plumosa Aurea' above I state a case for retaining that name for the wide spreading plant for which it is in general use. The more erect form which Hornibrook doubtless had in mind was first brought to my notice in the Dortmund-Brunninghausen Botanic Garden by G. Krüssman who has suggested the name. 'Tremonia' is identical with 'Plumosa Aurea' in colour but its habit is that of 'Blaauw'.

-- **'Variegata'**. R. Smith, ed. 2, 22 (1864); *J. chinensis albovariegata* Veitch 288 (1881); *J. chinensis argenteovariegata* Sieb., Cat. prodr. 5, **1861**: name only; Beissn. 120 (1891); *J. sheppardii* var. *pyramidalis* f. *variegata* (Carr.) van Melle 81 (1947). This clone, which received its present name some years before Veitch suggested the name *'albovariegata'* quoted by Hornibrook, forms a slow-growing but eventually large, conical bush which is worthy of a better position than the rough corner assigned to it by Hornibrook, as its blue-green foliage (very similar in colour to 'Stricta') with its combination of juvenile and adult foliage, both types being irregularly but often liberally splashed with a creamy-white variegation, can be very effective when seen from close at hand.

-- **'Variegated Kaizuka'**. A nicely variegated mutation from 'Kaizuka'. Sometimes confused with 'Japonica Variegata'.

-- **'Wilson's Weeper'**. A bowl-shaped plant produced by removing the leader from a normal plant of *J. chinensis*. Introduced in 1934 by W. B. Clarke Nurs., San Francisco, California. It is merely a cultivariant.

J. comitana Martinez in A. Inst. Biol. Mexico **15**: 12–13 (1944). An uncommon species allied to *J. monosperma*, from which it differs only in minor respects.

J. communis L. 1040 (1753). Common juniper. Despite the variation in growth and habit, both under cultivation and 'in the wild', this versatile species can be clearly distinguished from almost all other commonly seen species by the leaves. They are invariably awl-like and grow straight out at a wide angle from the shoot, never clasping it as in *J. sabina* and many other species. The branchlets are triangular with projecting ridges and the leaves are in whorls of three. The leaves are stalkless and taper from a swollen base to a sharp point, their upper surface having a single broad band of stomata (rarely divided by a faint mid-rib), the lower surface being convex and bluntly keeled. The species is dioecious and bears solitary ovoid or globose fruits, green at first, blue the second year, ripening to black the third season. Zone 2.

J. communis is truly circumpolar, being widely distributed throughout Europe, Asia and North America and, although usually a small upright tree, it also occurs widely as a low-growing or prostrate shrub. This variability has long been recognised and numerous attempts have been made by botanical writers to deal with the taxonomy, but the absence of any clear distribution pattern and the sporadic appearance of the deviant plants makes any satisfactory classification almost impossible. In the literature', *alpina, cambricus, canadensis, communis, davurica, densa, hemispherica, intermedia, nana, pygmaea, rebunensis*, and *sibirica* appear as specific epithets, and in various infra-specific ranks can be found *alpina*, β, γ, *canadensis, communis, dealbata, depressa, hemisphaerica, hornibrookii, intermedia, minor, montana, nana, nana compacta, prostrata, saxatilis* and *sibirica*.

The nomenclatural pendulum has over the years swung through all the available categories, (species, sub-species, *varietas* and *forma*) but the general consensus of opinion has been to treat these interesting Junipers as varieties of *J. communis*. I have elsewhere [Gar. Chron. **166**] (5): 14 (1969) argued the case for continuing this practice. The theory that the selection of appropriate rank for any taxon is a matter of botanical judgment and that each worker has unfettered choice is stoutly maintained by botanists, but is is not always so, since sometimes quite different considerations override. So my lack of deference to so eminentan authority as *Flora Europaea* is probably more apparent than real since, as a matter of editorial policy, the only infra-specific rank recognised in that monumental work is subspecies. These low-growing forms therefore had to be recognised at that rank or NOT AT ALL. Editorial limitations, like Railway Regulations, can put an author in a similar predicament to the ticket collector who told the old lady 'Cats is dawgs and parrots is dawgs, but your canary is a hinsect'.

I have previously suggested [*Manual of Dwarf Conifers*, (1979)] a workable way out of these nomenclatural problems would be to retain only the published vars. *depressa* Pursh, *hemispherica* (Presl) Parl., *hornibrookii* Hornibrook and *montana* Aiton; to accept at the rank *forma* certain other local variants that have been given botanical recognition in the past and deal with all the selections that have been brought into cultivation as cultivars.

Every seedling or wildling brought into cultivation must fall into one or other 'var', so no objection can be raised to labels such as '*J. communis* var. *montana* Ait'. Of the many dozens of the selected wildlings only a handful have been named (or are worth naming) as cultivars and if the clonal name is unknown (or is non-existent) that is as far as the identification needs to go – or, indeed, can go. A plant so labelled would quite properly have to share its identity with other plants botanically within the same *varietas*, even although to a gardener's eye they may look different.

-- var. ***communis***. When you partition off one part of a room it is undeniable that you are at the same time partitioning off the remainder of that room, and this kind of varietal name (which merely repeats the specific epithet – the technical term is 'autonym') owes it existence to a useful botanical rule which says, roughly speaking, that whenever a varietal name is established, automatically the name for what remains the species will be an autonym. So since *J. communis* var. *communis* has this rump-like character it is a 'hold-all' for all examples not falling within any of the previously mentioned varieties.

A large number of selections have been made 'in the wild' and there are many more clones in cultivation than there are valid names but many of them are too much alike to justify retention. There is a need for a study of this group in a comprehensive reference collection of living plants, leading to a restrained selection of a limited range of clones of proven garden value and the distribution of clonal material to representative horticultural centres for wider circulation.

-- var. ***depressa*** Pursh, Fl. Am. Sept. **2**: 646 (1814); Rehd. Bibl. 57 (1949); var. *canadensis* Loud. **4**: 2490 (1838). *The low-growing forms of J. communis* L. *found in eastern North America. A prostrate low shrub with branches ascending at tips; leaves upturned, rarely almost spreading, linear, acuminate-subulate, to 15 × 1.6 mm, with a narrow (as broad as, or less than each green margin) glaucous (stomatal rows distinct) band above.*

This is the common wild juniper of eastern North America. It is a low-spreading shrub never more than 1 m high and occurs sometimes in large groups, there being a very good illustration in *Cultivated Conifers in North America* by L. H. Bailey, 1933, Plate XXXIX. It thrives in dry, sandy, stony or gravelly soil, is extremely hardy and is a rapid, vigorous grower. It is very useful for massing in front of trees or in the wild or moorland garden. The leaves are 8–10 mm by 1.5 mm, oval to lanceolate, with a long and finely drawn out transparent tip. These make the plant prickly to the touch. The lower side of the leaf is prominently keeled, yellowish to brownish-green in summer, sometimes touched with darker brown near the tips, and turns brown in winter, unless grown in full shade. The leaves are in whorls of three and each leaf is sharply curved near its base and twisted so as to face the ground, consequently the white stomatic band 1/2 to 2/3 the total width of leaf (not quite reaching to its base) on the upper side of the leaf is not seen. The leaves at the strong growing tips are strongly appressed to the stem and the growth is in irregular, flat sprays.

-- var. ***hemisphaerica*** (Presl) Parl. Fl. Ital. **4**: 83 (1867); *J. communis* var. *saxatilis* (Attrib. to Pall. Fl. Ross. **1,2**: 12. and t. 54. 1788 by) Willd. Sp. Pl. **4,4**: Ed. 2, 854 (1806) (pro syn.); Suring. Mitt. d. d. d.. G. **38**: 330 (1927), nom. illegit.; *J. hemisphaerica* Presl, Del. Prag; 142 (1822); Endl. Syn. Conif 12. (1847); *J. communis* ssp. *hemisphaerica* (Presl) Nym. Consp. Fl. Europ. 676 (1881); *J. nana A. montana* Endl. Syn. Conif. 14 (1847); *J. communis* var. *montana* Hornb. Dw. Con. Ed. 2, 110 (1939); NON *J. hemisphaerica* (Presl) Gord. Pin. 96 (1858). *The low-growing forms of J. communis L. found in the high mountainous regions of Europe, Asia and northern Africa and distinguished by their bushy habit and densely set, broad, thick, usually tapered leaves, sharply up-curved at the base and thence straight, ending in a long prickly tip.*

This has the reputation of not being hardy in Britain and only one selection, to my knowledge, is in cultivation. There is no objection to using the cultivar name for this clone.

-- var. ***hornibrookii*** Grootn. ex Hornb. Dw. Con. Ed. 2, 111 (1939); *J. communis* var. *prostrata* Hornb. Dw. Con. 68 (1923); NON Beissn. Mitt. d. d. d. G. **5**: 55 (1896). *The low-growing forms of J. communis L. distinguished by their bushy, prostrate or mat-forming habit and by their very small, straight, widely spreading and loosely set leaves with long, sharp tips, with stomata above in distinct rows not embedded in a waxy coat.*

Although it is arguable that var. *hornibrookii* is botanically not distinguishable from vars. *communis* and *montana* because of the infinitude of intermediate forms, they constitute a well understood horticultural group.

-- var. ***montana*** Aiton, Hort. Kew. **3**: 414. (1789); *J. minor montana* Bauh. Pinax 489 (1596); *J. communis* γ L. Sp. Pl. 1040 (1753); *J. communis* Lamarck, Encyc. Meth. Bot. **2** 625 (1788); *J. nana* Willd. Sp. Pl. **4,2**: 854 (1806) in par; *J. communis alpina* Gaud. Fl. Helv. **6**: 301 (1830);

Figure 201. *Juniperus communis* 'Canadensis Aurea'. The Golden Canadian juniper.

J. communis 3 *nana* Loud. Arb. Frut. Brit. **4**: 2489 and Fig. 2344 (1838); *J. nana alpina* Endl. Syn. Conif. 14 (1847); *J. communis* var. *nana prostrata* Hornb. Dw. Con. Ed. 2, 110 (1938); *J. communis* ssp. *nana* Syme in Sowerby, Engl. Bot. Ed. 3, **8**: 275 (1868); Franco in Bot. Soc. Brot. **36**: 112 (1962); Flora Europaea **1**: 38 (1964); *J. sibirica* Loddiges ex Burgsd., Anleit. Sich. Erzieh. Holzart **2**: 127 (1805); *J. communis* var. *jackii* Rehd. Mitt. d. d. d. G. **1907** (16): 70 (1908); *J. communis* var. *minima* Grootn. Mitt. d. d. d. G. **53**: 200 (1940). *The low-growing forms of J. communis L. found in Greenland, the British Isles, and through Europe, the Himalayas, India and to Japan and in western North America and distinguished by the prostrate, often mat-forming habit and thick, broad and relatively short, boat-shaped, upcurved leaves which because they are densely set (especially so on the young growing shoots) are frequently subimbricate. The leaves carry a broad white (in the American forms often conspicuously so) stomatal band more than twice the width of each green margin, are bluntly keeled below and terminate in a point which varies in shape considerably (often in different parts of the same plant) but is always obtuse relative to the long fine point normal to the species, but is sometimes drawn out at the tip to a short point.*

The following *formae* and cultivars are in cultivation.

-- **'Arnold'**. Grootn. in Dendroflora **5**: 33 (1968). A moderately quick-growing, narrowly columnar form with very short leaves, similar in colour to 'Hibernica'. It came from Germany.

-- **'Ashfordii'** Kammerer in Bull. Morton Arb. **7**: 29 (1932). Ashford Common Juniper. Columnar, narrow or spire-like; branches erect; branchlets erect, thin, slender; leaves densely set, 5–9 mm long, 0.8–1 mm broad, short-pointed, with a dull bluish broad band above, margins very narrow, green, with a raised midrib and green beneath.

-- f. *aurea* (Carr.) Rehder Bibl. 57 (1949); excluding synonymy. A botanical designation of doubtful application which at least serves to cover any upright plants with yellow foliage that there may be that lack cultivar names of their own.

-- f. *aureo-spica* Rehder in Bailey, Cult. Everg. 199 (1923). This is the golden variegated form of var. *depressa* Pursh, now known as 'Canadensis Aurea'.

-- **'Aureovariegata'**. Beissner 138 (1981). Similar to typical *J. communis*, but young shoots and leaves part golden-yellow, part green in summer. Not a strong grower, and possibly no longer in cultivation. A similar plant was discovered by Tage Lundell in Sweden in 1982.

-- **'Bakony'**. E. Barabits in Magyar Fenyoudon-

Figure 202. *Juniperus communis* 'Depressed Star' in the Proefstation in Boskoop, Holland.

sagok 64 (1965). As seen at Kamoni Arboretum in Nov. 1968, this was an upright, narrowly-oval tree 2.5 m tall, 0.5 m wide. It is claimed to be very hardy.

-- **'Berkshire '**. Welch, 213 (1979). This is a very attractive and extremely slow-growing, bun-forming plant. It has very small, dark-green leaves with prominent, silvery-white stomatic bands. I admired it in the Arnold Arboretum growing under accession number 814/166 whilst being shewn round by Alfred Fordham in 1970, and he named it after the area where it had been found.

-- **'Bruns'**. Timm & Co., Cat. (1953/54): 139. Strictly erect and vigorous shrub like '*Suecica*', to 2.50 m high, rather loosely branched; leaves bluish-green, 15 mm × 1.5 mm. Found in Sweden and introduced into cultivation because of its putative resistance against rust, by Timm & Co., Elmshorn, Germany.

-- **'Canadensis Aurea'**. Beissner, 133 (1891), under *J. nana*; 'Aurea' Bailey in Stand. Cycl. Hort. **3**: 1727 (1915) NON Nicholson, **2**: 212 (1884); 'Aureo-spica' Rehder in Cult. Evergreens 199 (1923). This is the 'correct' (as well as the logical name for a golden variant of var. *depressa* Pursh, having the same wide-spreading habit and origin in the wild. I have not seen it in cultivation in Europe. The use by Hornibrook [109 (1939)] of the name 'var. *depressa aurea* Hort. Amer.' for a European selection of var. *montana*, a plant having no connection whatsoever with var. *depressa* Pursh, has long caused confusion. (See Illus. No. 201)

-- var. ***communis***. See p. 341.

-- f. ***columnaris*** Hornibrook in Gar. Chron. 50 (1929), Dwarf Conif. ed. 2, 113 (1939) as 'var.' Hornibrook described this as "a narrow, fastigiate form, an outsize *compressa*" and leaves open the question (his p. 118) of its possible identity with 'Suecica Nana' which Bailey [Cult. Conif. 189 (1933)] described as "a dwarf form of 'Suecica". Although that name is nowadays of rather loose application it is always associated with a description that stops short of fastigiate. So in *Manual of Dwarf Conifers* I suggested limiting the use of 'Suecica Nana' to the fastigate-but-flat-topped clone for which it is in general use in Britain. I also suggested treating var. *columnaris* as 'Lost to cultivation', since the particular clone Hornibrook wrote about can no longer identified. But there are now several cultivars within the description "narrowly fastigiate plants with a looser, freer growth than 'Compressa'", so as a collective or sub-group name it still has its use. Here would come 'Sentinel' (See p. 352) already widely known, and two recent selections. **'Sieben Steinhauser'**, introduced from West Germany, in 10 years will form a dense, narrow column 80 cm tall and 20 cm diam, with bright blue-grey foliage and leaves averaging 9 mm long. **'Norwegen'**, found (as the name suggests) in the wild in Norway, also forms a narrow column, much stronger-growing than 'Compressa' but reliably winter-hardy.

-- f. ***compressa*** Carrière. Traité. 22 (1855); *J. hibernica compressa* Hort. Although I do not for one moment believe Veitch's statement (Man. 275. (1881) repeated by Nicholson [Dict. Gar. **2**: 212. (1884)] that this dainty and rather delicate little treasure is found on the Pyrenees at 5000 feet of elevation yet it lends an air of respectability to its claim for recognition as a botanical *forma* and this suits us very well horticulturally, for there is evidently more than one clone in cultivation under this name, although they are virtually impossible to tell apart. The name used collectively should be restricted to narrowly columnar dwarf forms and, ideally, the cultivar name **'Compressa'** should be limited to the well known, minute, columnar plant so universally popular, with its steeply ascending branches, pointed apex and tiny leaves, about 5 mm by 1 mm, and an average annual growth of 5 cm or less – a tiny replica of the Irish juniper. Unfortunately it is not an easy plant, being very prone to damage by wind, frost and red spider.

-- **'Constance Franklin'**. Name registered by E. C. Franklin in 1982 for a sport from 'Hibernica' with an attractive creamy-white variegation well distributed over the entire plant. It is a most attractive garden plant, deserving to be better known.

-- **'Corielagen'**. Dendroflora **21**: 6 (1984). This is an attractive and unusual mat-forming clone. It carries needles 8–12 mm long that are bronze beneath but the upper sides carry a noticeable white stripe, the overall effect being an attractive bluish-grey. Collected by Dr. Coombes on the west coast of Scotland before 1975, it was introduced by P. Foley of Bolton, Lancs.

-- **'Cracovia'**. Carrière, 21 (1855) Columnar, narrow, dense, to 2.5 m rather slow-growing; branches ascending, covered by branchlets; branchlets erect to spreading, light brown, tips nodding; leaves 10 mm long, 1.7 mm broad at base, with a bluish-green midrib, with bluish-white bands laterally, margins narrow, green above, glossy green and with a conspicuous midrib beneath. The clone in cultivation was found near Krakow in Poland.

-- var. ***depressa*** Pursh. See p. 342.

-- **'Depressa Aurea'**. Hornibrook, 109 (1939). In coining this name Hornibrook was under the impression that he was dealing with the American clone 'Canadensis Aurea', but he inaugurated the unfortunate practice in Europe of using the term 'Depressa' for clones that are properly to be listed under var. *montana*. The plant widely grown as 'Depressa Aurea' is an at attractive low shrub that is at its best in late May, when its golden-yellow colour is at its peak. Later it becomes a dull bronze and by winter, almost green. It makes a sturdy bush, with its branches much more ascending than (as Hornibrook doubtless intended to say) those of 'Nana Aurea'. It is also a much better 'doer'.

-- **'Depressed Star'**. Den Ouden & Boom, 164 (1965). Name raised for a broad, wide-spreading shrub with a mainly horizontal branching system and leaves broad and decurved, with a silvery band above but held so that only the green undersides are seen, changing to brown

Figure 203. *Juniperus communis* 'Suecica Nana'. The blunt top is characteristic.

Figure 204. *Juniperus communis* 'Suecica Nana' shewing coarse, reverting growth.

in late autumn growing at the Proefstation at Boskoop in Holland. It is a selection from a low-growing form of var. *montana* common in northern Europe in which the characteristic broad leaves bronze during the winter. Probably as a result of Hornibrook's confusion over the plant he named 'Depressa Aurea' this European group is frequently confused with the American var. *depressa* with which it has no connection. Other that have selected and named vary mainly in habit. **'Dumosa'**. *J. communis depressa dumosa* Den Ouden, Conif. 160. (1949), **'Effusa'** Den Ouden, Conif. 160 (1949), **'Gimborn'**. *J. communis depressa* 'Gimborn' Den Ouden, Conif. 160 (1949) and **'Vase'**. *J. communis* 'Vase Shaped' Kuml. Friendly Evergreens, 173 (1946) are shrubby forms, increasingly upright in the order given, the first three having been seedlings selected on the Gimborn estate, Doorn, Holland, and the third being an American selection introduced by D. Hill Nursery Co., Dundee, Illinois, U.S.A. 'Vase' colours to a rich chocolate brown in winter and, especially as a young plant, is an

Figure 205. *Juniperus communis* 'Hibernica'.

attractive form. These shrubby forms all become more widespreading as they age.

Of these, I understand that the European selections are no longer in production, plants labelled 'Effusa' usually being 'Repanda' or a clone difficult to distinguish therefrom.

Figure 206. *Juniperus conferta*. Shore juniper. Will cover several metres in time.

-- **'Gew Graze'**. Welch, 215 (1979) is an interesting and attractive new introduction not unlike 'Repanda' but with much larger (10 mm × 2 mm) leaves but slower growth and less prostrate habit. The foliage remains green during the winter. It was selected by Dr. D. E. Coombe of Cambridge University 'in the wild' near the Lizard, in Cornwall, where a once quite extensive local population is in danger of extinction by heath fires. It does not seem sufficiently distinctive for recognition botanically, but it makes an attractive and quite 'different' garden plant and could become popular when better known. 'Gew Graze' is a female clone. **'Soapstone'** (Welch, loc. cit.) is the male clone from the same collection by Dr. Coombe. As a young plant in the Pygmy Pinetum it is distinct only by fractionally shorter leaves, but it turns red-bronze during the winter, at which time the two forms are easily distinguished, even in non-flowering plants.

-- **'Gold Beach'**. Wyman in Arnoldia 19 (1968). A prostrate form with dense, widespreading growth reported as having been found on the west coast (U.S.A.). In the early Spring the growing tips are bright yellow, but they turn later to the usual green. The mother plant in the Arnold Arboretum when 160 cm across was only 15 cm high.

-- **'Gold Cone'**. This is a very attractive clone

having the habit of 'Hibernica' but with bright golden-yellow foliage, the colour well maintained throughout the season. It was introduced by the Kordes Nursery, Bilsen/Holstein, West Germany. In Britain it was at first incorrectly distributed as 'Suecica Aurea'. Another recent introduction of similar habit with golden young foliage but reputedly with larger leaves is 'Schneverdinger Goldmachangel'. (Goldmachandel?)

-- **'Graciosa'**. A shrubby form raised from seed by J. Konijn and Co. Reeuwijk, Holland. Freely branching, it has light brown shoots and narrow, sharply pointed leaves 5–10 mm long, light green with the usual stomatic lines above.

-- **'Green Carpet'**. This was discovered by Mr Arne Bryne on a cliffside near Stavanger, Norway. It makes a ground-hugging plant well clothed with foliage, dark green in winter, brighter green in summer, and being very slow-growing it is a safe choice for even the smallest garden.

-- **'Haverbeck'**. A very slow-growing form, possibly from a witch's broom, named for the village in Germany where it was found. The foliage is congested and irregular; the needles around the (relatively) large needles are very short (3 mm); elsewhere they can reach to 14 mm. After ten years a plant will be 35 cm high and 20 cm across. It is similar to, but a better 'doer' than *J. chinensis* 'Echiniformis'.

-- **'Heidegeist'**. A fast-growing, open-columnar clone with silvery-grey appearance, found in the wild on the Luneberger Heath in Germany, hence the name. (= *Heath Ghost*.)

-- var. ***hemispherica***. See p. 342.

-- **'Hemispherica'**. This, as a cultivar name, can be properly used for the only clone that, so far as I am aware, is in cultivation in Britain. I received it at Wansdyke Nursery from the National Botanic Garden, Glasnevin, Dublin, under the name *montana*. It has no connection whatsoever with *J. chinensis* 'Echiniformis'.

'Hibernica'. Gordon, Pin. 94 (1858); *J. hibernica* Loddiges Nurs. Cat. 1836 and Carrière [with *J. stricta* pro syn] 22 (1855); var. *fastigiata* Bean, Trees and Shrubs 671 (1914.) (See Illus. No. 205). *J. communis* is one of several conifer species that come either as more or less columnar trees or as low-growing or spreading shrubs. The latter on the whole are more useful in our gardens, but the upright kind (probably more typical of the species) have also given us several good garden forms. These all occur within var. *communis* as defined above and constitute no real botanical entity, so I consider the advantage lies in treating them as cultivars. The rules for deciding the priority of varietal epithets when botanical names are changed sometimes bring to the surface a long forgotten and almost unused name but the rule for cultivars is more simple – basically, we retain the first name used for that cultivar. And fortunately, as in this case, it is usually the name in general use.

'Hibernica', therefore, is the well-known 'Irish Juniper' forming a tall, narrow column reaching 3–5 m with tightly packed vertical branches with tips stiffly erect, needles 5–7 mm long by 1 mm wide, sharply pointed but less prickly than in 'Suecica', bluish-white to near the apex above, with a narrow margin; bluish-green below. Several other rather similar selections have been named. **'Bakony'** a Hungarian selection is claimed to be hardier in winter; **'Kiyonoi'** from an American source is said to be a dark-green form which withstands summer-heat better and **'Wilseder Berg'**, a recent selection made in Germany is claimed to be very hardy and of a good colour.

-- **'Hils Freiburg'**. Hesse ex Meyer, Plant Explorations 120 (1963). Low and slow-growing, tips of the new growths ascending. A selection found in the H. A. Hesse nurseries, Weener on Ems, Germany, in 1952.

-- var. ***hornibrookii***. See p. 342.

-- **'Hornibrookii'**. Grootn. ex Hornb. Dw. Con. Ed. 2, ttt (1939); *J. communis* var. *prostrata* Hornb. Dw. Con. 68 (1923), in part. NON

Beissner. This is the clone found by Murray Hornibrook in County Galway in Ireland. It was there a perfectly prostrate mat, flowing down over rocks and absolutely following their contours. It is very popular for rockeries because of its neat habit and dainty leaves, but over the years other more or less similar forms have crept into cultivation. The true plant is a very prostrate and *slow-growing* male form with leaves 5 by 1 mm (rarely to 8 mm by 1.25 mm) and an annual growth of 50 to 80 mm. In this form the older leaves are often twisted, so that the wide (always more than half the total leaf width) stomatic band shews and this gives a somewhat grey look to the plant as a whole. There is no tendency to bronze during the winter. (See Illus. No. 208).

The clone grown in Britain as 'Prostrata' has the same horizontal habit and the same small leaf as 'Hornibrookii' but it is much more vigorous and strong-growing. This makes it quite distinctive, but I regret that my attempt in an earlier book to isolate the clones 'Edgbaston' and 'Inverleith' was premature, since with age neither plant has maintained sufficient distinctiveness to justify retention. 'Prostrata' should not be grown as 'Hornibrookii' nor be confused with 'Depressed Star' which falls in the 'Montana' Group.

-- '**Hornibrook's Gold**'. Grootn. in Dendroflora **15–16**: 63 (1979). An alleged sport from 'Hornibrookii' (more probably the more vigorous clone 'Prostrata'), its foliage turning bronze yellow-gold in winter.

-- '**Hulkjaerhus**'. Thomsen, Cat. 1962. A columnar form, like '*Suecica*' but shorter and a little broader in youth, more green and less prickly, older plants have almost the same narrow shape as '*Suecica*'. Found in the Thomsen nurseries, Skalborg, Denmark.

-- f. ***jackii*** (Rehder) Welch 216 (1979); Rehder in Mitt., d. d. d. G. **16**: 70 (1907) and Man. 10 (1927) as 'var.' This is a local variant occasionally met with in Oregon and northern California with long, whip-like main branches, often reaching over 1 m in length unbranched, save for occasional clusters of short branchlets. The plant bears leaves characteristic of var. *montana* and, consistently with the general treatment in this book, has no claim to varietal status.

-- '**Jensen**'. A. M Jensen, Cat. **1944**. According to the firm Asger M. Jensen this form differs from 'Pendula': pendulous, strong-growing; leaves shorter and set widely apart in whorls. Raised in the nursery of the firm Asger M. Jensen, Holmstrup, Denmark.

-- '**Loensgrab**'. Krussm. 144 (1979) broad form, 65 cm high and 50 cm wide, branching obliquely upwards, twigs numerous, very dense, pendulous, totally covering the branches; needles light green, white stomatal bands, 6 mm long.

-- '**Mayer**'. Den Ouden & Boom, 168 (1965); 'Suecica Major' Hort. Holl. A form selected by Erich Mayer of Barmstadt, West Germany in about 1945, and introduced by J. Timm and Co. Elmshorn, Germany. It makes a very sturdy upright-conical plant in all respects more vigorous than is normal to the species. Is hardy and very accommodating as to soil and situation.

-- '**Minima**'. Grootn. Mitt. d. d. d. G. **53**: 200. (1940) var. *nana prostrata* Hornb. Dw. Con. Ed. 2, 110. (1939) 'Silver Lining' Welch, 182 (1966). I am now satisfied that the clone named by Herman J. Grootendorst in 1940 and my plant are one and the same, so – used as a cultivar name – his name has priority and my own rather more euphonious effort is redundant or, since it is now widely in use, reduced to the level of a 'commercial synonym'. It forms a very prostrate plant which is hard and rough to the touch but not prickly, with strong leading growths which nod at the growing tips. Young twigs are distinctly triangular, pink above, green below; older branches brownish-purple. Entre-noeud up to 10 mm, less towards tips of shoot. Leaves 5–8 mm by 1.5–2 mm, upcurved, but narrow at the base and bluntly pointed; each leaf viewed from above or the side being boat-shaped, rather like a canoe. The leaves are shiny green below but the whole leaf is covered with a heavy glaucous bloom (easily

Figure 207. *Juniperus communis* 'Repanda'. Name loosely in use for several rather similar procumbent clones.

rubbed off) giving the plant a grey-green appearance which is maintained throughout the year. Above, the leaves carry a very wide band of minute stomata covered with a chalky white bloom and as many of the leaves are twisted so that this whiteness is seen especially near the centre of the plant, the general effect is most attractive. It is a fairly strong grower, but is a poor transplanter and is slow to establish itself.

-- var. ***montana***. See p. 342.

-- ssp. *nana* Syme and var. *nana* (Willdenow) Baumgarten. See var. *montana*, p. 342.

-- **'Nana Aurea'** [*J. communis nana aurea* Beissn. Zierg. Ed. 2, 461. (1884) *aurea* (Hort. Holl.) Beissn. Handb. Ed. 2, 625. (1909) NON *aureus* Nich. Dict. Gar. Suppl. 455. (1900)]. This is a low, flat-spreading form with main branches nodding at the tips. The leaves are silvery-white, striped with green edges above, but golden-yellow to brown-yellow below, this colour being more or less retained throughout the year. It is very easily distinguished from all other forms, but it is not a very good doer, so is seldom planted. Since it roots poorly from cuttings it is usually grafted.

-- **'Norwegen'**. A clone that is similar to although somewhat faster-growing and more open than 'Compressa', but also very much hardier. Leaves short, 5–7 mm, very close.

-- 'Obergaertner Bruns'. See 'Bruns'.

-- **'Oblonga Pendula'**. Loud. 2490 (1838); *J. communis* f. *oblongopendula* (Loud.) Beissn. (1887) 19. Broad Weeping Common Juniper. Erect, rather broad shrub, 3–3.5 m; branches nearly horizontal, tops at first curved slightly upwards, tips gracefully drooping; branchlets nearly pendulous, 10–20 cm long, brown when young, darker brown in winter; leaves 15–18 mm long, thin, stout, apex very sharp-prickly

Figure 208. *Juniperus communis* 'Hornibrookii'. Name loosely in use for several rather similar prostrate clones.

pointed; on sprays the leaves are much shorter, with an inconspicuous band above, margins green, green beneath, brown in winter. **'Horstmann'** is a recent selection from W. Germany which will, but only if stem-trained when young, form a very vigorous and elegant tree with the same pendulous habit.

-- var. *prostrata*. Beissner in Mitt. d. d. d. Ges. **5**: 55 (1896). Now included in var. *montana*. See also under 'Depressed Star', p. 345.

-- **'Prostrata'**. Beissner, NON Hornibrook 68 (1923). A prostrate, widespreading selection differing from the clone 'Hornibrookii' only by its very much greater vigour. It remains green in winter.

-- **'Repanda'**. Grootn. Mitt. d. d. d. G. **53**: 200 (1940). A very dense, prostrate clone (or group of clones) always forming a neat, circular plant, very useful as a ground cover. The whole plant is very soft to the touch with no trace of prickliness. The twigs are very supple. Leading shoots send out their laterals in regular flat sprays but the branching towards the centre of the plant is irregular and twigs stand up at all angles. The twigs are light brown. The leaves vary in size, up to 10 mm by 1.5 mm, and are *upcurved so that they lie close to the shoot*, the underside is dark, dull green, prominently keeled but with a sunk mid-rib, the upper side with an ill defined, broad stomatic band, light grey in colour but largely hidden from view because of the way the leaves are held. It was in the first instance found in Ireland by Maurice Prichard of the firm M. Prichard & Sons, Riversley Nurseries, Christchurch, Hants, England, and distributed by the firm M. Koster & Son, Boskoop, Netherlands, in 1934, but it is certain that more than one clone is now in circulation under this name, the difference between them being mainly the habit. One, that is somewhat less prostrate, circulates in Europe as 'Effusa'. **'Derrynane'** originated with a berry-bearing plant found on a headland at Derrynane in Co. Kerry, Eire. It is not worth distinguishing as a separate cultivar. (See Illus. No. 207)

-- **'Ruud'**. Named for a plant found in Norway and registered by E. Ruud in 1977. Unless and until such a findling is propagated it is not, technically, a cultivar.

-- var. *saxatilis* Pallas. See var. *montana*, p. 000. I have elsewhere shewn [Taxon **17,5**: 545. (1968)] that this much-used name was never validly published until it had become redundant, so it should no longer be used in any combination. Plants labelled *J. communis saxatilis* in gardens are usually var. *montana* Ait., or one of its named clones.

-- var. *sibirica* Burgsd. See var. *montana* p. 342.

-- **'Sentinel'**. F. J. Grootend. in de Boomkw. **16**: 140 (1961); *J. communis 'Pencil Point'* Sheridan nurseries Cat. 19 **1963**. An extremely narrow columnar, almost pointed form similar in outline to 'Compressa', but very much larger in every respect. It originated on the Sheridan Nurseries, Etobicoke, Ontario, Canada.

-- var. *suecica*. (Miller). Beissner, Aiton **3**: 414 (1789); *J. suecica* Miller, No. 2 (1768). Swedish Juniper. A minor variant found in Scandinavia and East Russia which includes different habit forms. Some plants form strictly columnar trees 8–12 m high, others are broader and lower, but always the branches are ascending with shoot-tips nodding. Leaves blue-green, cones oblong. The cultivar name **'Suecica'** is in use for a widely cultivated clone that is almost as columnar as 'Hibernica' but differs in that the top few centimetres of each shoot arch over to + 10°. Also the foliage is less prickly to the touch and shews more of the stomatic bands, so the whole plant has a greyish cast. Leaves 8 mm long, 1 mm broad.

-- 'Suecica Aurea'. See 'Gold Cone'.

-- **'Suecica Nana'**. Bailey, Cult. Conif. ed. 2, 189 (1933). It being described by Bailey as 'a dwarf columnar form of var. *suecica*, usually not exceeding 1 metre' is unlikely to mean that this clone originated in Sweden, and in Britain this name is used for a looser and freer-growing edition of 'Compressa'. Although the branching of both are equally fastigiate, each little shoot of 'Suecica Nana', if closely examined, will be seen to have much less leader dominance, so even as a young plant it is usually blunt-topped by comparison with 'Compressa', which usually terminates in a single shoot, this resulting in a pointed little tree. It will make a plant 1.5 m high by not more than 30 cm across, but without some assistance from the knife such old plants usually become a bit irregular in outline towards the top. It seems much more wind resistant and hardy than 'Compressa' and is, of course, quicker growing, but 'Suecica Nana' is apt to throw out loose, strong growth. This should be cut out as soon as it appears. See Illus. Nos. 203, 204.

I do not feel able to be definite on the point, but in my experience the true 'Compressa' does not make this coarse growth. If this be correct, the presence of such growth on any plant would indicate that it is 'Suecica Nana' rather than 'Compressa'. The coarse growth, if propagated, gives plants of the normal 'Hibernica' type. It might very well be Hornibrook's Columnaris.

-- **'Tage Lundell'**. Kruessmann, 146 (1983). A dwarf form found by Tage Lundell of Helsingborg, Sweden. Makes a columnar plant 1 m by 20 cm in ten years. Some leaves are green and others are either wholly or partly yellow.

-- **'Taurifera'**. Listed Mitsch Nurs. Cat. 6 (1977). A compact and very slow-growing selection similar to the well-known 'Compressa'.

-- **'Tempelhof'**. Dendroflora **21**: 9 (1984). A slow-growing form introduced by the Konijn Nurs, Reeuwijk, Holland, about 1966, which forms a low, round-topped bush with very dense foliage that bronzes in winter but is otherwise as the type.

-- **'Velebit'**. Dendroflora **21**: 32 (1984). This was found on Mount Velebit near Rijeka in Jugoslavia by M. Frei. It forms a very slow-growing low bush to 25 cm, of fairly open growth; leaves 10–13 mm long with noticeable stomatic bands, bronze-green, turning bronze in winter.

-- **'Volcano'**. Van der Laar 218 (1986). A dwarf form that forms a vase-shaped bush with silvery-grey foliage. It was found in a wild park in Sweden and was introduced by the Draijer

Nursery, Heemstede, Holland.

-- **'Wallis'**. Draijer Nursery Catalogue (1980). Another selection from the same nursery. It makes a densely clothed bush of upright habit with main branches arching over and with its terminal shoots pendulous.

-- **'Windsor Gem'**. Welch, 218 (1979). This name I suggested for a distinctive, diminutive and attractive form of which there are several well established plants in the Saville Gardens, Windsor Great Park, the origin not being on record. It has foliage and habit similar to 'Minima' but with the leaf size and rate of growth very much reduced.

-- **'Zeal'**. Welch, 218 (1979). This is another selection with quite a different habit. It was found in a garden in Devon, U.K. It is very slow-growing, and the upright leaves shew both the dark green undersides and the white stomatic band on the upper-sides of the foliage, giving a dark blue-green overall appearance.

J. conferta. Parl., Conif. Nov. **1**: (1863); 481 (1868); *J. litoralis* Maxim. in Bull. Acad. Sci. St. Petersb. sér. 3, **12**: 230 (1868). Shore Juniper. Dense, bushy, procumbent shrub; bark brownish-red; branches thick, widely spreading over the ground; branchlets densely set, spreading, tips striking upwards, long, thick, grey-brown in the second season; leaves awl-shaped, in whorls of 3, directed forwards, scarsely spreading, covering the branchlets, 10–15 mm long, 1 mm broad at base, straight, apex gradually acuminate, not prickly, deeply grooved, with a white band and without a greenish midrib above, margin partly green; sometimes also grooved and green beneath; cones subglobose, flat at base, 8–12 mm across, dark blue or bluish-black, bloomy at maturity; seeds 3, 3-angled-ovate, with longitudinal grooves on back, acuminate. Japan and Sachalin, on sandy sea-shores; Sent by E. H. Wilson to the Arnold Arboretum, in 1915.

It has become very popular as a ground-cover plant and several selections have been given cultivar names. In its native habitat it grows on the seashore, so presumably would be a valuable plant for coastal gardens. Zone 6. (See Illus. No. 206)

-- **'Akebono'**. A selection that produces new growth at the tips of the branches that is creamy-white with greenflecks, but the leaves become totally green by late Summer. Imported into U.S.A. from Japan. See Arnoldia **43** (4): 10 (1983) and the Nihon Kaki Nursery, Angio, Japan Nursery Cat., Spring 1982.

-- **'Blue Pacific'**. A selection with a trailing habit of growth, needles 10–12 mm long with noticeable white bands of stomata and a blue-green tinge to the foliage to which it owes the choice of name. Listed Monrovia Nursery 1975.

-- **'Emerald Sea'**. This selection, which differs from the norm by its more prostrate habit and a much reduced tendency to build up hummocks of foliage, originated as cuttings collected by Dr. John L. Creech, then Director U.S. National Arboretum in 1967 from low sand dunes along the shore at Ashiarai, Ibaraki Prefecture, Honshu, Japan in 1967 and was released to the trade in 1972. Its habit and salt-tolerance make it a valuable ground cover and stabilisation planting in maritime areas. The leaves are 10–15 mm long, an emerald colour which changes to yellowish-green in winter.

-- var. *maritima* Wilson ex Nakai. Now known as *J. taxifolia* var. *luchuensis*.

-- **'Silver Mist'**. (= 'Shiro Tosho', 'Blue Tosho'). A selection introduced from the Shimabichi Nursery Nurs., Kawaguchi City, Japan. The habit is similar to 'Blue Pacific', but the growth is denser and the shorter leaves, (1 cm long, compared with an average length of 1.5 cm in 'Blue Pacific') and a more distinctly grey colour of the foliage distinguish this new cultivar. See Arnoldia **43**: 10 (1983).

J. convallium Rehder and Wilson in Sargent, Pl. Wils. **2**: 62 (1914); *J. mekongensis* Komarov in Not. Syst. Herb. Hort. Bot. Reip. Ross. **5**: 29 (1924); *J. ramulosa* Florin in Acta Hort. Goth. **3**: 5, t. 2 (1927). Upright tree, 5–12 m with a dense crown. *Bark*: Grey, smooth, peel-

Figure 209. *Juniperus coxii* (by some authorities listed under *Juniperus recurva*). Detail of foliage and clones.

ing in long strips. *Branches*: Short, stout, ascending or spreading. *Branchlets*: Thin, cylindrical, pendent or twisted. *Leaves*: Scale-like, opposite, appressed, ovoid, 1 mm long, apex blunt and incurved, grey-green; back rounded with a long narrow groove and a single gland. *Berries*: sub-globose or ovoid, 6–8 mm long with a single whorl of scales, ripening to purple-brown or black-brown, somewhat glaucous. *Seeds*: One per cone, 5–7 mm long.

-- var. ***microsperma*** (Cheng & Fu) Silba, 1984. A tree with branches less twisted than the type. Differing in having smaller and more acutely pointed leaves and in other minor respects.

J. coxii A. B. Jackson in New Flora and Silva **5**: 31 (1932); *J. recurva* var. *coxii* (Jackson) Melville in Kew Bull. 533 (1958). Coffin Juniper. It is difficult to understand the reluctance there is to allow this magnificent tree – described by Dallimore and Jackson as "the finest of all junipers and one of the most beautiful of conifers" – to retain its specific status. When publishing the name, Jackson wrote "It differs from *J. recurva* in the much more pendulous habit, the longer and more tapering leaves which are marked on the upper surface by two greenish-white bands of stomata and by the fact that it shows none of the habit of *J. recurva* of producing multiple stems. To this we might add that it grows much more rapidly and to an enormous size, that the final shoots hang gracefully at –90° (compared to the –75/80° of *J. recurva)*, that the leaves are straight and much more widely spaced and are widest at the base (of the free-standing part) and thence taper uniformly to the tip without any trace of a boatlike broadening amidships. The distribution pattern has been destroyed through the demand for the aromatic timber for the making of coffins, a sad reflection on man's avarice and pride. The leaves are all juvenile, borne in threes, a rich mid-green, the stomatic bands yellowish, rather than white. It is monoecious, and the fruits are egg-shaped, 10 mm long, dark purplish-brown and contain a single seed.

The leader needs tying up for several years and the lower branches must be cut away and the foliage thinned drastically as the tree grows, to bare the trunk and display the graceful habit. (See Illus. No. 209)

J. davurica Pall., Fl. Ross. **1,2**: 13 (1788); Loudon, Arb. Frut. Brit. **4**. 2500 (1838); Loddiges, Cat. 42 (1826). Daurian Juniper. An early discovered but still imperfectly known species that forms a low or trailing shrub with very thick ascending branches. The foliage is either juvenile with spreading leaves 5–9 mm long, or adult with scale-like, imbricate, backs convex, bluntly acute, glandular leaves 1–3 mm long in four ranks giving a quadrangular shape to the shoot. The disposition of the two types of foliage is erratic and it is (improbably) recorded that the male flowers are predominantly borne on the adult shoots and the female flowers on the juvenile. These latter occur singly, are carried on short stalks and the berries ripen to black or dark brown, covered with a white mealy bloom and containing usually three seeds. Specimens were in the Loddiges collection in Hackney, near London, by 1826, where doubtless Loudon made the description and illustrations that appear in *Arboretum et Fruticosum Britannicum*, since he used that collection extensively. It has an extensive distribution, from Lake Baikal eastwards right across Asia, probably as far as Japan. Van Melle no doubt rightly supposed that with so wide a distribution, both geographical variations and the opportunity for hybridisation with other species would occur, and he also felt that the material used by Pallas, being from the extreme westerly limit might well not be typical of the species as a whole, so he transferred to *J. davurica* a plant (of unrecorded origin) that was in cultivation in America under the name *J. chinensis* var. *Parsonsii*. But Dr. Laurence Gough (See p. 328) included this clone in his analyses of foliage extracts by gaschromatography and is satisfied that Van Melle was mistaken, so (along with its two variegated forms) it will be found described on p. 337 under *J. chinensis*.

J. deppeana. Steudel, Nomencl. ed. 2, **1**: 835 (1840); *J. mexicana* Schiede & Deppe in Schlecht. & Cham. in Linnaea **5**: 77 (1830), not Spreng.

Figure 210. *Juniperus drupacea*. A male tree at Wakehurst Place, Sussex, U.K.

(1826); *J. pachyphloea* Torr. in U.S. Rep. Expl. Surv. Pacif. Miss. **4**: 142 (1857). Alligator Juniper. Dioecious tree, 20 m, broadly pyramidal; trunk 1–1.5 m in diameter; bark grey, deeply split, scaling off in flat plates, very solid; branches spreading, long, thick; branchlets slender, quadrangular, 1 mm thick; leaves scale-like, ovate-rhomboid, apex sharp-pointed. 1–1.2 mm long, bluish-green, back keeled, with a gland at base; on young plants awl-shaped, opposite or in whorls of 2 or 3, apex sharp-pointed, concave and whitened above, keeled

Figure 211. *Juniperus drupacea*. A fruiting tree at La Fosse in France.

beneath; cones globose or subglobose, 10–12 mm across, reddish-brown, more or less bloomy, pulp thick, dry mealy, ripening in the second season; seeds normally 4, angled, shining brown. Arizona and New Mexico to southwestern Texas and Central Mexico. Introduced by C. A. Purpus to the botanic garden of Darmstadt, Germany in 1904. Not hardy in the colder regions of the Continent.

-- var. ***pachyphlaea***. (Torrey) Martinez, An. Inst. Biol. Mexico **17**: 47 (1946). Alligator Juniper. A distinct variety with finer foliage than normal; also the foliage is more glandular, marked by dots of white resin. The popular name refers to the bark which is deeply cracked into square plates, suggestive of an alligator's back.

-- **'Silver Spire'** This selection takes the normal narrow-upright form, but is distinguished by the bright silvery leaves, especially noticeable in summer and on young plants. It is difficult to root from cuttings, so is usually propagated

Figure 212. *Juniperus drupacea*. Detail of foliage.

by grafting.

J. distans Florin in Medd. Göteb. Bot. Trädg. **3**: 6 (1927). A small to medium-size conical tree to 4 m, with thin, papery, light brown bark and often a twisted trunk and tortuous branches. Juvenile leaves (on young plants and fruiting branches) are in whorls of three, 7–8 mm long, narrowly lanceolate with a pointed prickly tip, light green beneath with two bluish bands above; adult leaves towards the top of the tree are scale-like, tightly appressed, 1.5–3 mm long, triangular ovoid. Leaves intermediate in character also occur. Berries ovoid, 8–12 mm long, 7–8 mm across, reddish-brown, scales free. One large seed. Western Szechuan (S.W. China). Introduced into Britain in 1926 but still very rare. Mitchell records that the foliage is strongly scented of soap.

J. drupacea Labillardiere, Icon. Pl. Syria **2**: 14 (1791); *Arceuthos drupacea* Antoine et Kotschy, Oestr. Bot. Wochenbl. **4**: 249 (1854). Syrian Juniper. A dioecious species forming a broadly conical tree with a pointed crown (all the recorded trees in Britain are male and almost columnar) with orange-brown bark becoming ashy-grey on old plants, peeling in strips; *Branches*: Ascending or spreading, smooth; *Branchlets*: triangular, with prominent ridges; *Winter-buds*: 3 mm long; *Scales*: small, apex sharp-pointed; *Leaves*: In whorls of three, spreading, linear-lanceolate, stiff, apex spiny-pointed, 15–25 mm long, 2.5–3 mm broad (by far the broadest of all junipers), shorter and more closely set at the top of branchlets, scarsely grooved, a bright glossy green outside with 2 white bands and with a broad greenish midrib above, keeled below; *Cones*: globose or ovate, composed of usually 6–9 fleshy scales, which are often pointed at the apex, 20–25 mm across, bluish-black or brown, glaucous edible, ripening in the second season; *Seeds* 3, fused into a three-celled horny stone. *Habitat*: Asia Minor, Syria and Greece at elevations of 600–1500 m, in Europe confined to Mount Malevo (Peleponesus). It was already known to Clusius in 1600. Introduced into Britain in 1820. Zone 8. As a tree, it is quite the finest of all Junipers. Zone 8. (See Illus. Nos. 210, 211, 212)

J. durangensis Martinez in An. Inst. Biol. Mex. **17**: 94 (1946). Durango Juniper. A shrub or small tree 2.5–5 m. tall, to 25–35 cm. in girth. *Bark*: Grey, fibrous, exfoliating in thin strips. *Branches*: Twisted, irregularly arranged. *Branchlets*: Quadrangular, rather dense with numerous short compact side branchlets, to 1.5 mm. in diameter. Branchlets spreading out at wide angles. *Leaves*: Scale-like, obtuse, broadly ovate, dull yellowish-green, 1–2 mm long, glandular, margin finely toothed. Male cone subglobose to oblong-ovoid, to 2 mm. long, with 12 scales. *Fruits*: Irregular, subglobose to oblong, 5 mm. wide, orange to reddish with a glaucous bloom. Seeds mostly 2–3 per cone, narrow, ovoid-acuminate, curved, striated, 4 mm. long by 2–2.5 mm. wide, dark reddish-brown. *Habitat*: Mexico: Sonora to S.W. Zacatecas; N. Jalisco; Aguascalienties, 1600–2900 m. Zone 8.

J. ekmannii. Florin. Now merged in *J. barbadensis*.

J. excelsa. Bieb., Beschr. Länd. Flüss. Terek. Kur. **204** (1800); Fl. Taur. Cauc. **2**: 425 (1808). Greek Juniper. Monoecious tree, 14–20 m, pyramidal, crown finally rounded; *Bark*: Ashy-grey, peeling off in strips; *Branches*: Upright or spreading, round, dark grey; *Branchlets*: Slender; sprays thin, sometimes scarsely 1 mm broad; *Leaves*: On older trees mostly all scale-like, appressed, rhombic-ovate, with free incurved acute apex, glandular on back, dark or bluish-green; solitary on short branchlets, crowded, composed of 4–6 scales, each marked by a minute point, globose, 9–12 mm across, dark purplish-brown, bluish-bloomy; Seeds: 4–6, oblong. *Habitat*: Asia Minor forming extensive woods in mountainous regions, also in the Balkan (Macedonia), Armenia and the Caucasus to Persia. Not hardy on the Continent. Introduced into Britain by Dr. J. Banks in 1806, reintroduced into France in 1830.

-- var. ***polycarpos***. (C. Koch) Silba, 34 (1984). A dioecious variant that occurs sporadically in Turkey, U.S.S.R., Iran, Afghanistan and Pakistan. It makes a smaller tree or a low bush, with short, stout branchlets.

-- **'Stricta'**. Sénéclauze, 90 (1868); (Rollisson) Gordon, 144 (1875). The description by Sénéclauze, was "a dense, vigorous plant with both awl-like and squammiform leaves, differing from the normal only by all parts being a pale glaucous green". Whether this was the same plant described by Gordon as a "beautiful variety forming a tall, dense, narrow conical head, tapering gradually from the ground to a sharp terminal point, and of a fine silvery, glaucous colour, originating in the nursery of Messrs. Rollisson at Tooting" is now difficult to say, but what is certain is that this name has been widely misapplied to a slow-growing juniper with juvenile foliage properly known as *J. chinensis* 'Stricta'. See p. 340.

J. flaccida. Schlechtendal in Linnaea **12**: 495 (1838). Mexican Weeping Juniper.

Size and habit Tree to 8–12 m, at first narrowly conical with a rounded crown, broadening with age.

Bark Reddish grey-brown, stringy, peeling off in long strips, exposing the orange-brown colour beneath.

Branches Widespreading, irregularly arranged.

Branchlets Long; pendulous; pale pink-brown.

Foliage For some years wholly juvenile with leaves in whorls of three or in opposite pairs, widely spreading, 8 mm long, gradually tapering to a horn-like tip; outer surface a bright, rather grey-green, dusted with stomata; inner surface with two broad grey-green bands. With age, adult foliage appears in patches, eventually becoming predominant. The opposite, scale-like adult leaves are oval-lanceolate, 1.5–2 mm long, free but held closely to the shoot, with a spiny tip, dark in the centre, glandular, with pale margins. Mitchell [122 (1972)] states that the crushed foliage has a soapy, resinous scent.

Figure 213. *Juniperus flaccida*. Detail of foliage and fruit.

Fruit Maturing the second year, elongated-globose; glaucous green, ripening to red-brown with a waxycoating; 1 cm long with 6–8 scales, each with a minute flat green spine in the centre.

Habitat Mexico; Texas, U.S.A.

Cultivation Although apparently hardy, it is seldom seen in Britain. Zone 8.

-- var. ***poblana*** Martinez in Anal. Inst. Biol. Mex. **17**: 31 (1946). A regional variant that makes a shrub 4–6 m high with the foliage noticeably distichous, as in the genus *Thuja*, and with much larger cones.

J. foetidissima. Willdenow, Sp. Plant. **4** (2): 853 (1806). Stinking Juniper. A conical tree 5–10 m high (old specimens much higher) akin to *J. excelsa*, with similar grey-green foliage, differing mainly in having much thicker shoots. The specific name refers to the foetid smell of the foliage when bruised. Mountains of Greece, Albania, Jugoslavia, Asia Minor to Transcaucasica, often in rocky areas. Known in cultivation since 1910, but rarely seen. Zone 9.

J. formosana. Hayata in Gard. Chron. series 3, 4 (1908). A small tree 8–12 m high, similar to *J. rigida*, with wide-spreading branches and pendulous, 3-angled branchlets. *Leaves*: In whorls of three, needle-like, pointed forward, with 2 white stomatic bands and broad green margins above. *Fruits* Are on short stalks and mature 2nd year to red- or orange-brown, grooved. China. Zone 9.

J. gamboana. Martinez in Anal. Biol. Mex. **15**: 7,8

Figure 214. *Juniperus horizontalis* 'Andorra Compact'. (The word 'Plumosa' is not now acceptable in a cultivar name.)

Figure 215. *Juniperus horizontalis* 'Douglasii'. The Waukegan juniper.

Figure 216. *Juniperus horizontalis* 'Wiltonii'. Probably the best of the neat, silvery-blue clones.

(1944). A rare tree, forming a species with a limited distribution in S. Mexico and Guatemala. It resembles *J. deppeana* but has the single seed of the *J. monosperma* group of junipers.

J. gaussenii. Cheng, 1940. Now merged in *J. chinensis*.

J. glaucescens. Florin, 1927. Now merged with *J. komarovii*.

J. gracilior. Pilger, 1913. Now known as *J. barbadensis*.

J. gymnocarpa. (Lemmon) Cory, 1936. Now

J. horizontalis. Moench, Meth. Pl. 699 (1794); *J. prostrata* Persoon, Syn. pl. **2**: 632 (1807); *J. sabina* var. *prostrata* (Michaux) Louden 2499 (1838); *J. sabina* var. *humilis* Hooker, Fl. Bor. Amer. **2**: 166 (1839); *J. virginiana* var. *prostrata* Torr., Fl. New York **2**: 235 (1843). Creeping Juniper. This is a widely distributed and variable species in which hybridisation long ago with other species is evident. Numerous clones have been selected in the wild and brought into cultivation. They are all more or less prostrate and many are vigorous plants capable of covering a large area, so are valuable for groundcover. Having been regarded at one time as a prostrate variety of *J. sabina*, at another a variety of *J. virginiana* and even, by one authority, simply as *J. prostrata*, plants can still be found incorrectly labelled.

The main branches are long, the branchlets very numerous, short and dense. The leaves on the cultivated forms are mainly needle-shaped, often in whorls of three, green or blue-green, frequently colouring to blue or mauve tones in the winter, 2–6 mm long and standing free at the tips, with very sharp points; convex on the underside, with a gland. On mature foliage the leaves are scale-like and loosely imbricate. The species is dioecious, so only female clones carry berries.

Like all junipers, this species and its cultivars withstand hot, dry situations, growing under town conditions and tolerating moderately alkaline soils. They grow slowly at first and can therefore be disappointing for a couple of years, but eventually produce excellent ground cover, especially useful in covering steep, rough banks. They are usually much improved by pinching back the leader growths during the summer for two or three seasons after planting to induce a dense branch system, and thereafter by an occasional hard pruning. A large number of clones have been selected and named, especially in the United States of recent years, and the number now available in the trade is, surely, many more than is necessary.

-- '**Admirabilis**'. Is similar to 'Emerson' but is a male clone. It has prostrate main branches, with branches and branchlets held at a steep angle. The foliage is dense. The leaves are all juvenile, a yellowish-green with a heavy grey bloom, especially on the upper sides of the leaves. It is a vigorous plant reaching to 30 cm high, but spreading to several metres. It is one of the cultivars introduced by the Plumfield Nurseries, Fremont, Nebraska, U.S.A. in 1936.

-- '**Adpressa**'. Introduced by Plumfield Nurs. Fremont, Nebraska, U.S.A. (Nurs Cat. 1936) as *adpressus*. A quick-growing selection with a dense, mat-forming habit of growth only 10–15 cm high. Foliage green with very conspicuous whitish tips on the young growth. See also Morton Arb. Bull. Pop. Inf. **35** (1): 3 (1960); Welch 220 (1979). A number of selections made on this nursery were given names that were descriptive adjectives taken from a Latin dictionary. The rule in Latin is that adjectives must agree in gender with the noun they qualify, so since the noun *Juniperus* is feminine these adjectives would have to be the same. So in this case *adpressus* should have been *adpressa*. Several similar cases occur below.

-- f. **alpina**. (Loud) Rehd. Journ. Arn. Arb. **6**: 203. (1925); *J. sabina* var. *alpina* Loudon. 2499 (1838). Although Rehder's description was based on a single clone, this is a botanical form found in the Manitoba province of Canada, although possibly only a single male clone is in cultivation. It is a rather atypical form which carries its one-year branches nearly erect, becoming horizontal by the second year. By this means it spreads over a large area, never reaching over 50–60 cm high. The leaves are exclusively awl-shaped, 3–4 mm long, more or less bluish or greyish changing to a dull purple in autumn. It is a rather rank grower and is therefore not a good ground cover plant. It seems to need yearly snow-cover to flatten the current year's growth. Known in cultivation in England since 1836.

-- '**Andorra Compact**'. Krüssmann, Handbuch Nadel. 149 (1972). This name has had to replace the illegitimate but descriptive name 'Plumosa Compacta' under which it was at first widely circulated. The foliage is similar to 'Plumosa' but it has denser growth and a less spreading habit; the branches rising at a steeper angle, forming a cushion-shaped plant 30 cm high, 1 m across. Several further selections have been named. (See Illus. No. 214)

-- '**Argentea**'. Plumfield Nurs. Cat. 1936. Another early Plumfield introduction; a plant with a mat-like habit of growth, not over 20 cm high and very dense. The smooth foliage has a silvery-grey colour.

-- '**Aunt Jemima**'. Wyman in Amer. Nurs. 120 (1963); Welch 224 (1979). (= 'Plumosa A. J.' of some nurseries.) A selection made in 1957 and introduced by the D. Hill Nurs. Dundee, Illinois, U.S.A. The growth is very low and spreading, with a blue-green colour retained throughout the growing season. It will make a densely clothed bush 1.5 m across by only 20 cm high.

-- '**Bar Harbor**'. Teuscher in New Flora and Silva **8**: 194 (1936); Hornb. Dw. Con. Ed. 2, 120 (1938). This form comes from Mt. Desert Island, Maine, where it grows in crevices on the rocky coast and may frequently be found well within reach of the salt spray. It is a ground hugging form with its matted growth covering large areas and rooting from the long branches. In colour it is a pleasant grey-green resulting from a white glaucous bloom over its true green colour. In autumn and winter it turns a delicate mauve-purple. Being a population in the wild several clones may have found their way into cultivation. These differ slightly, but some plants labelled 'Bar Harbor' are certainly no such thing. Bar Harbor being a place name the spelling should not be altered.

-- 'Black Hills Creeper'. Now known as 'Emerson'.

-- **'Blue Acres'**. A selection raised by Professor C. E. McNinch and Joerg Leiss from seed collected in Alberta, Canada. It was registered in 1971 and introduced by the Sheridan Nurs. Oakville, Ontario, Canada in 1973. It forms a prostrate plant with young growth at first blue in colour and standing upright, later turning green and trailing. Its ultimate height is 30 cm and it is said to propagate easily.

-- **'Blue Chip'**. (= 'Blue Moon' of some nurseries in error.) It has dense foliage of a beautiful, soft, silver-blue on a low (to 15 cm) spreading and mounding plant with all its branchlets standing up at 30°. This cultivar was selected in 1940 on the Asger Jensen Planteskole, Denmark by Mr. Thomsen of Thomsens Planteskole, Skalborg, Denmark and introduced by the latter as 'Blue Moon' about 1970. A stock had, however, earlier been sent as an unnamed form to D. Hill Nurs., Illinois, U.S.A. and there had been put into circulation as 'Blue Chip'. Mr. Thomsen was already concerned about possible confusion of the use already of the name 'Blue Moon' for a cultivar of *J. scopulorum* so he withdrew his own suggestion and allowed the name 'Blue Chip' to stand. The provision of Art. 41 of the *Cultivated Code* which lays down that publication of a name is not valid if against the expressed wish of the originator must surely allow a change of mind in such circumstances, so the name 'Blue Moon' should no longer be applied to this clone. Nor is it available for re-use for a fresh selection.

-- **'Blue Horizon'**. Welch, 221 (1979). This clone makes a very low-growing, creeping plant that retains this characteristic, never mounding up in the centre as does 'Wiltonii', which it otherwise resembles.

-- **'Blue Mat'**. Mitsch Nurs. Cat. 1973. A very compact, blue green selection.

-- 'Blue Moon'. See 'Blue Chip'.

-- **'Blue Pearl'**. Has the same origin as 'Blue Chip'. This clone carries numerous upright shoots all of similar length, forming a low bun-shaped plant about 25 cm high, with juvenile foliage. It is not unlike a miniature replica of 'Alpina'.

-- '*Blue Rug*'. See 'Wiltonii'.

-- '*Blue Wilton*'. See 'Wiltonii'.

-- **'Cabot Trail'**. A very compact form, about 10 cm high, found by Professor Robert B. Livingston on Cabot Trail, Highlands National Park, Nova Scotia.

-- **'Clearwater Blue'**. Originated from a plant collected in the wild. It has a low spreading habit, to 10 cm. Growth laterally is rapid, the foliage being silvery-blue, becoming purple by winter. It was registered by T. H. Machin in 1979.

-- **'Coast of Maine'**. Hillier 487 (1971). This would appear to be a superfluous name for 'Grey Carpet', an example of what can happen if plant material is distributed before a name has been registered or published.

-- **'Douglasii'**. Rehd. in Mitt. D. D. G. **24**: 214 (1916); *J. horizontalis glauca major* Grootend. in Mitt. D. D. G. **53**: 212 (1940). Waukegan Juniper. This is one of the earliest selections (1855) from the one-time Douglas Nurseries, Waukegan, Illinois (hence it is sometimes listed as the Waukegan juniper.) It is a form with very small leaves tightly appressed. Main branches horizontal or nearly so, all branchlets ascending at a steep angle. Side branches turning upwards so as to produce a V-shaped spray (i.e., not flat.) The foliage is grass green but covered all over with a glaucous bloom giving a rich grey-green appearance which changes to a soft greyish mauve in winter. Shoots are orange-brown at the growing tips. Teuscher [New Flora and Silva **8**: 195. (1936)] tells us 'To see this Juniper as it grows naturally is a revelation. Its main branches are entirely buried in the sand, only the branchlets coming above the surface, where they form a dense, soft carpet which follows in every detail the outline of the dune... the natural habitat suggests the proper treatment of this juniper in cultivation... not very well suited for use in the rock garden, but for sandy gardens, especially for dune gardens it offers one of the finest lawn substitutes'. (See Illus. No. 215)

-- **'Dunvegan Blue'**. Welch, 221 (1979). This clone was collected from the north bank of the Peace River by John A. Wallace in 1959 and introduced by the Beaverlodge Nurs., Beaverlodge, Alberta, Canada. It makes a dense, spreading plant, reaching 4.5 m across, 15 cm high in five years. The colour is steel-blue in early summer, turning to purple tones in late Autumn, and the foliage is soft and persistently juvenile.

-- **'Emerald Spreader'**. Monrovia Nurs. Cat. 6 (1973). This is an exceedingly low, ground-hugging selection introduced by the Monrovia Nurs., Azusa, California, U.S.A. It is well furnished with branchlets densely covered with emerald-green juvenile foliage. It is an excellent ground-cover plant and is becoming widely planted in both N. America and Europe.

-- **'Emerson'**. Marshall ex Wyman in Amer. Nurs. **117,1**: 120 (1963); *J. horizontalis 'Marshall'* Marshall, Cat. **1929**: More in Green Thumb **13** (3): 32 (1956); *J. horizontalis 'Black Hill's Creeper'* Hort. A dense, dwarf shrub; branches spreading, finally creeping, 3–5 m in diam., tops fast growing; leaves both acicular and scale-like, bluish, holding this colour throughout the winter. A female form found in 1915 in the Black Hills of S. Dakota by G. A. Marshall of the Marshall nurseries, Arlington, Nebraska and by Prof. Emerson of the University of Nebraska. At first it was distributed as 'Black Hills Creeper' but was given its present name in honour of one of its finders. The plant is very prostrate, seldom exceeding 30 cm. It is slow-growing and drought resistant, but is seldom planted because of the availability of improved selections, such as e.g. 'Hughes'.

-- **'Eximia'**. As seen at Phytotektor Nursery, Winchester, Tennessee, U.S.A., in 1970, this appeared somewhat more upright than 'Emerson' and was a lighter green.

-- **'Filicina'**. [*J. horizontalis filicinus* Plumfield Nurs. Cat. 1936; More, Green Thumb **13,3**: 36. (1956) and **15,3**: 93. (1958); *filicinus minimus* Hort. Amer.] This is a very prostrate, slow-growing and dense form with delicate green foliage in fern-shaped sprays. A female clone – and a very dainty, lady-like looking plant. It deserves to be better known.

-- **'Fountain'**. Welch, 224 (1979). This selection, introduced by the D. Hill Nurs., Dundee, Illinois, U.S.A. is of fairly vigorous growth and a habit that is intermediate between 'Plumosa' and 'Aunt Jemima', a flat-topped plant to 40 cm high.

-- 'Glauca'. This name has been loosely applied (as is the case in other species in which such forms turn up) for so many glaucous selections as to have lost any practical value, so it should not be used in any clonal sense. Whether the clone sent by Professor Sargent from the Arnold Arboretum to M. Hornibrook before 1923, but not referred to by him as 'var. *glauca*' until his second edition in 1939, (and described in that edition as having been selected by the Professor as "the best coloured form of this juniper") is the clone 'Wiltonii' (recorded as having been found by Jacob C. van Heiningen on an island off the coast of Maine) or one of similar description, is an academic point not worth pursuing.

-- **'Glenmore'**. More [Green Thumb. **13,3**: 32. (1956)] is a dwarf form, one of the lowest and slowest of the species, main branches horizontal, held just clear of the ground, branchlets thin, a dull light brown; leaf sprays nearly upright; leaves all juvenile, very closely set, medium green, tips brownish in winter. A female form, selected by Robert E. More, in Wyoming. A neat little plant, but because of the way the leaf sprays are held, of less value for ground cover than the more rampant forms.

-- f. ***glomerata***. Rehder in Journal Arn. Arb. **6**: 202 (1925). Although Rehder records that this form had been found more than once in the wild it is probable that only a single clone is in cultivation and for this the cultivar name **'Glomerata'** is in order. It is a most distinctive variety carrying long, prostrate, often twisted, pinkish-brown main branches; the branchlets rising at a steep angle and bearing juvenile leaves nearly appressed, and carrying very dense, rich green foliage and tiny leaves 2 mm long, glaucous inside in flat sprays almost suggestive of one of the tiny forms of *Chamaecyparis obtusa*. If the strong-growing leaders are regularly pinched out this type of foliage will slowly develop into a dense mat only 2 cm in height. The colour turns rather dull in winter.

-- **'Gracilis'**. Listed Woodland Nurs. ex Plant Buyers' Guide 1958; selected in the Brooklyn Botanic Garden about 1937. It is not particularly graceful and not common in the trade.

-- **'Green Acres'**. A fast-growing, prostrate selection by Prof. C. E. McNinch of Guelph University and released through the Sheridan Nurs. in 1971. It is similar to 'Blue Acres' save for foliage colour, which is an attractive dark green, especially so in winter.

-- **'Grey Carpet'**. Kammerer in Bull. Morton Arb. **6**: 40 (1931). Creeping, with long, trailing but well furnished branches forming low cushions. It differs from 'Bar Harbor' by its somewhat more green foliage. Found along the coast in Maine, U.S.A.

-- **'Grey Pearl'**. Dendroflora **21**: 15 (1984). A selection with much the same habit as 'Alpina'. Introduced by the Thomsen Planteskole, Skalborg, Denmark, in about 1975. It makes a shrub of unusual habit but of no great beauty.

-- **'Hermit'**. A.A.B.G.A. Bulletin **7**: 51 (1973). This form was found growing on Hermit Island, Maine, by Dr. R. B. Livingston, University of Massachusetts, Amherst, Massachusetts, and was registered at the Arnold Arboretum by N. E. Pellett in 1972, as a possible hybrid *J. virginiana* × *J. horizontalis* but Dr. Livingston writes (30 May 1974) that it is a *J. horizontalis* form, found many miles outside the range of *J. virginiana*. It is described as a horizontally spreading shrub 3 times wider than high, vigorous but of compact habit: leaves mostly acicular, deep green in summer, silvery purple in winter. It was at first distributed as 'Livingston No. 11'.

-- **'Hughes'**. Dirr. Man. Woody. Plts. 241 (1975). A strong-growing new form selected by the Cedar Rapids Nursery, Iowa, U.S.A. It has a wide-spreading but ascending branch system and as a young plant has very much the habit of *J. virginana* 'Grey Owl', but the foliage is grey-green. Useful for ground cover in a rough spot, or under trees. Illustration 235 in Welch, Manual of Dwarf Conifers, is incorrect, being of a much taller-growing clone found, incorrectly, in the trade under this name.

-- **'Huntington Blue'**. A name listed in Monrovia Cat. 58 (1983). Described as a widely spreading, densely branching plant covered by very intense blue-grey foliage. Its matted growth becomes silvery-plum colour in Winter.

-- **'Jade River'**. A vigorous, dense, mat-forming selection from the Reimer Nursery, Yarrow, Canada. The adult leaves are small and closely held, juvenile leaves are only 3–4 mm long, a silvery-blue at its best during the winter months. A valuable ground-cover plant. If **'Jade Spreader'**, listed by Monrovia Nurs. Azusa, U.S.A. is the same cultivar, whichever name can claim priority should be retained and the other dropped; if they are distinctive the second name should be replaced by a name not likely to be confused with the accepted name. See Recommendation 31A (h) of the *Cultivated Code*.

-- **'Limeglow'**. Reported L. C. Hatch, 21 Oct. 1984. A mutation from 'Youngstown' with chartreuse-yellow new growth.It retains the colour for some time, but by autumn has faded to the normal colour.

-- **'Livida'**. Plumfield Nurs. Cat. 1941 (as *lividus*.) Forms a dense mat of yellow-brown main branches and many side branches pointing forwards, all completely prostrate to the tips. Branches are numerous and very short, seldom rising to 45° even at the centre of the plant (at least when young – I have not seen an old specimen). Foliage is all juvenile, the leaves are very small except on strong-growing main shoots, much resembling *J. sabina* 'Tamariscifolia'. Colour is actually a deep grass green, but a heavy grey bloom all over the plant gives it a greyish appearance, especially so during the winter.

-- **'Livingston'**. Hebb, Arnoldia **33**: 205. (1973). Another of Dr. R. B. Livingston's finds on Hermit Island, Maine; introduced by the University of Vermont. It is described as 'a procumbent plant generally 15–20 cm high; foliage steel-blue in summer, bluish-green in winter, mostly scale-like, minute; occasional fruit, light blue, with bloom. The plant is much branched and growth is dense'. The name was registered by N. E. Pellett in 1972 with the Arnold Arboretum.

-- **'Magic Carpet'**. Found as a mutation on 'Wiltonii' with a general cream-coloured variegation, and registered by F. A. Menser in 1983. Introduced by Cedar Pond Nursery, Rock Hill, South Carolina, U.S.A.

-- **'Marcella'**. *J. horizontalis marcellus*. Plumfield Cat. 1940. An old selection, but still one of the best, that makes a very low, ground-covering plant never over 13 cm high with long, prostrate main branches and numerous ascending laterals, densely clothed with juvenile foliage giving a soft appearance. The colour is light green, shading to deep blue-green in the interior of the plant. A female clone.

-- 'Marshall'. See 'Emerson'.

-- **'Petraea'**. *J. horizontalis petraeus* Plumfield Cat. This is another early selection which should give way to later cultivars. It forms a dainty little spreading plant about 20 cm high with ascending main branches tending to become bare at the base, so it is insufficiently dense to make good ground-cover. The leaves are green, turning dull mauve in winter, but the glaucousness on both sides give a greyish-green appearance. A female clone.

-- **'Planifolia'**. *J. horizontalis planifolius* Plumfield Nurs. Cat. (1941). This clone is a rapid grower with long, feathery, silver-blue leaves. It reaches to 25 cm in height and makes good solid ground-cover. A useful plant for quick cover in a rough spot.

-- f. ***plumosa***. Rehder in Journ. Arn. Arb. **6**: 204. (1925). A botanical term, if one be required, that embraces all the clones relating to the Andorra juniper.

-- **'Plumosa'**. Hornb. Dw. Con. Ed. 2, 121 (1938) *J. communis depressa plumosa* Andorra Nurs. Cat. 8. (1919) (Spring). This, the Andorra juniper, forms a flat-topped, open spreading shrub, never over 50 cm high but spreading laterally by means of main branches radiating from the centre of the plant. The branches rise at an angle of about 45° (reminiscent of the Pfitzer juniper) and bear dense, awl-like foliage loosely appressed and arranged in plumes. The somewhat prickly foliage is a light grey-green, becoming purplish-plum colour in the winter. It was discovered in 1907 and introduced by the Andorra Nurseries of Philadelphia, Pennsylvania. The name is, however, sometimes loosely used to cover a number of sports or selections that have been named in the United States which are better garden plants. (See 'Andorra Compact'.)

-- **'Prince of Wales'**. Wyman, in Arnoldia **29** (1): 3 (1969). This is possibly the best selection yet made of this species. It forms a dense, low, prosperous-looking plant 10–15 cm high with foliage a rich mid-green, tinged purple during the autumn and winter. Young foliage has a bluish tinge due to a waxy bloom. Plants 18 years old in the arboretum at the Agriculture Research Station, Morden, Manitoba, Canada carry both juvenile and adult foliage. The original plant was found growing amongst others of the species along Pekisko Creek on the property of the E. P. Ranch near High River, Alberta, by Mr. Wellesley White, a cattle breeder and farmer in the Morden district. In 1931 Mr. White brought rooted layers back to Dr. W. R. Leslie, formerly Superintendent of the Morden Station. The name was registered by W. A. Cummery in 1968.

-- **'Procumbens'**. *J. horizontalis procumbens* Slavin, Rep. Conif. Conf. 106 (1932). A misidentification. As well as recording a description unlike any form of *J. horizontalis*, Slavin tells us that his plant had been grown under many different names. I have no doubt but that it was some form of *J. procumbens.*

-- **'Prostrata'**. Fulling, Bull. New York Bot. Gar. **14**, 118. (1929); Grootn. Mitt. d. d. d. G. **53**: 212 (1940), *J. sabina* var. *prostrata*, Loudon 2499 (1838). As I know it, this is a prostrate, dense and mat-forming variety, making a plant up to 30 cm high by 3–4 m across. Main branches are long, stout and prostrate but with tips slightly turned up, the branchlets are numerous and densely set, covering the ground; the sprays are blue-grey with the tips purplish; the leaves are mostly awl-shaped and thin but towards the growing tips are scale-like, a glaucous green. It is one of the best prostrate forms and one that has been in cultivation a long time, but often found wrongly named. Plants supplied in Britain as *J. horizontalis* are usually of this cultivar.

-- **'Pulchella'**. More, Green Thumb, **16**: 32. (1959) is a very slow-growing and dense, compact variety, forming a symmetrical mat-like plant seldom over 10–15 cm high but more than 1 m across, with trailing branches and greenish-grey, acicular leaves borne on numerous, almost vertical branchlets in fern-like sprays. Claimed by the late Mr. Robert E. More to be an exceptionally hardy form, withstanding sun, shade and drought.

-- **'Repens'**. Grootend. in Mitt. d. d. d. G. **53**: 212 (1940); *J. sabina repens* Hort. Prostrate, mat-shaped, irregular and less dense than 'Prostrata', 3 m broad, 0.2–0.3 m high; branches spreading mostly in one direction; branchlets directed forward; leaves scale-like, greyish-green, ash-coloured in winter. Not a very distinct form, often forming open plants.

-- **'Schoodic Point'**. A very low-growing form found by Professor R. B. Livingston of the Massachusetts University near the shore at Schoodic Point, Maine. It is very slow-growing, quite flat and flexible, creeping down over any obstacle. It should be an excellent choice for a small rockery or trough garden.

-- **'Sunspot'**. Listed Girard Nurs. Geneva, Ohio, U.S.A. 1970. A mutation on 'Douglasii', spotted yellow. Hardly distinct from 'Variegata', below.

-- **'Turquoise Spreader'**. Monrovia Nurs. Azusa, California, U.S.A. Nurs. Cat. 6 (1973). A strong-growing ground-cover plant with a wide-spreading horizontal branching habit which has no tendency to mound up at the centre and is densely covered with soft sprays of juvenile foliage of (with a little imagination) turquoise green. A tendency to become 'open' at the centre of the plant should be checked by pruning back of the stronger branches.

-- **'Variegata'**. Slavin 106 (1932). Prostrate, 2 m broad, 0.35 m high; branches spreading, decurving; branchlets erect, tips creamy white; leaves awl-shaped, 2–4 mm long, glaucous green.

-- **'Venusta'**. Raraflora Nurs. Cat. 1966 is a ground hugging, dense form of a much, darker shade of bluish-grey, but otherwise similar to 'Wiltonii'.

-- **'Viridis'**. *J. horizontalis* var. *viridis* Grootn. Mitt. d. d. d. G. **53**. 213. 1940 is a prostrate, densely branched, broad shrub, 50–60 cm high; branches horizontal, branchlets spreading, covering the branches; leaves for the greater part awl-shaped with pointed apex, light green above, sometimes slightly blue bloomy beneath. It is a female form and distinctive by its green colour in summer, but is otherwise not outstanding.

-- **'Wapiti'**. Buckley, Amer. Nurs. **161,2**: 14. (1967). This interesting form was collected by John A. Wallace, Head Gardener at the Beaverlodge Research Station, Alberta, Manitoba, Canada, in an area overlooking the Wapiti River to the north in 1960. It is extremely hardy; young plants are upright but soon form a solid, dense, flat mass of fine textured, lustrous dark green foliage turning to dull purple during the winter months. It reaches a uniform height of 30–35 cm, but will spread to several metres.

-- **'Watnong'**. A very low-growing female clone of good foliage colour that berries well. Registered by Don Smith, Watnong Nurs. Morris Plains N.J. in 1967.

-- **'Webberi'**. Sherwood Nurs. Cat. (1971) (or earlier?). A new form which makes an extremely low, mat-like spreading plant. The foliage is green with a glaucous blue sheen.

-- **'Wilms'**. This is a clone of the 'Plumosa' group, somewhat dwarfer then 'Andorraa Compact' and with a finer texture. It carries adult foliage of a greyish-green. It was raised by J. Paul Wilms and introduced by the Gwenn Gary Nurs. Columbiana, Ohio, U.S.A. in 1972. The name was registered in 1975.

-- **'Wiltonii'**. Vermeulen Nurs. Cat. 4. (1953); listed also as Wilton Creeping Juniper, 4. 1955; as Wilton Carpet Juniper, 15. (1957) and as 'Blue Rug', 11. (1964) *J. horizontalis* 'Blue Wilton' More, Green Thumb. **13,3**: 32. (1956) and **15,3**: 93. (1958); *J. horizontalis* 'Glauca' Hort. Europe. It is still perhaps the best blue clone for ground-cover that is available. The colour is bright blue, well maintained throughout the year. It was found on Vinalhaven, an island off the coast of Maine, by Jacob C. van Heiningen of the South Wilton Nurseries, Wilton, Connecticut, and was grown by him as '*J. horizontalis* var. *glauca*'. Under this unfortunate name the F. J. Grootendorst & Sons Nurseries imported it into Holland around 1933 and introduced it just before the war. Being one of the finest blue carpeting Junipers it caught on and is now grown by tens of thousands in Continental Europe, and as always happens, the unsatisfactory name continues to be used by some producers. As the name 'Glauca' was (in the U.S.A.) being loosely used for several other collected blue leaved forms John Vermeulen listed it as 'Wiltonii' after the town where Jacob van Heiningen lived. (See Illus. No. 216)

-- **'Winter Blue'**. Heasley Cat. 8 (1970). Another American selection similar to 'Andorra Compact', distinctive for its bright blue colour in winter.

-- **'Youngstown'**. Monrovia cat. 1973. Yet another form of 'Andorra Compact'. It is described as having a low habit, tightly knit branches that maintain a 'full' look to the plant even as it ages. The colour of the foliage is bright green.

-- **'Yukon Belle'**. Monrovia Nurs. Cat. 58 (1976). A very hardy selection from the Plumfield Nurs, Fremont, Nebraska, U.S.A. making a widespreading ground-cover plant with mainly adult foliage and very short leaves of a bright, silvery blue.

J. indica. Bertoloni in Mem. Accad. Sci. Bologna, ser. 2, **1**: 228 (1862). Black Juniper. A low, spreading shrubby, autumn-flowering species found along the dry inner valleys of the Himalayas; for many years confused (and by some authorities still confused) with the little-known *J. pseudosabina* from central Asia and with the (usually) tree-forming species *J. wallichiana*. A very clear account that should disperse all uncertainty will be found in the recent volume *Conifers* by Keith Rushforth (Christopher Helm, London.) Rushforth states "Black Juniper occurs as high as 5,200 m and makes a dark patch on the distant hillside, hence the name. It is sometimes confused with *J. wallichiana* from which it is distinguished by the smaller cones and different flowering period." He considers that this confusion may have been largely due to the fact that as dried herbarium specimens the foliage of both of these species is very similar, but he tells us that 'in the field' both in growth habit and in flowering period the two species are unmistakable.

J. jaliscana Martinez, An. Inst. Biol. Mexico **17**: 69 (1946). A small tree, 5–10 m, with an erect trunk and a dense, broadly conical or irregularly rounded crown. *Bark*: Soft and spongy, peeling in strips, cinnamon-brown in colour. *Branchlets*: Scabrous, reddish-brown, the ultimate shoots more or less quadrangular, directed up at wide angles. *Leaves*: Small, scale-like, dark green, 0.7–1 mm long, opposite or ternate, ovate, obtuse, free at the tip with a short pointed tip and with a narrow dorsal gland. *Fruits*: On a short stem, very irregularly globular, 7–8 mm across, yellowish-green, ripening to reddish-brown; fleshy and sweet. *Seeds*: 6–7, irregularly arranged, angular, grooved, 4 mm long, 2 mm across. Found in the mountain-areas of Durango and Jalisco, Mexico. Very rare in European cultivation.

J. kanitzii Csató in Mag., Növéngt. Lapok **10**: 145 (1886), *J. communis* × *J. sabina*, according to Wettstein in Sitzb. Kais. Ak. Wiss. **96**, 1e Abt: 334 (1887). Differs from *J. sabina 'Tamariscifolia'* by microscopical leaf characters only. Found by Csató near Remeta, Hungary, in 1886.

J. komarovii Flor. in Medd. Göteb. Bot. Trädg. **3**: 3 (1927); *J. glaucescens* Florin, l. c., 5 (1927). A dense, small monoecious tree to 8 m; crown dense; *Branches*: Ascending or pendulous; sprays erect, straight, round at the top of the branches; juvenile leaves in threes, broadly triangular, 6 mm long; adult leaves scale-like, appressed, triangular-ovoid, 1–2.5 mm long, bluntly acute, blunt-keeled, often with a conspicuous gland at the base. *Cones*: More or less globose, 9–10 mm long, dark brown to blackish-blue, bloomy; *Seeds*: Solitary, 6–7.5 mm long, slightly flattened on both sides, apex rounded, with four longitudinal grooves. N.W. Szechuan, China. Not in cultivation.

J. lucayana Brittain. See *J. barbadensis*.

J. macrocarpa Sibthorp and Smith. See *J. oxycedrus*.

J. macropoda Boissier, Not. Syst. Herb. Hort. Petrop. **2**: 28 (1924). A widespread species from Iran, Afghanistan and Baluchistan that is probably not in cultivation. It seems to be intermediate between *J. chinensis* and *J. excelsa*.

J.* × *media Van Melle in Phytologia **2** (6): 189 (1947). "*Juniperus* × *media (J. sabina* × *J. sphaerica)*. Frutices plerumque habitu *J. sabina* similis certe in plantarum juvenis. Foliis acicularibus interdum dimorphis: his brevibus textura mollibus ut in *J. sphaerica*. Floribus dioicis. 'Pedunculis' galborum matutorum semper nutantibus. Galbulis maturis plusmin usve distincte bilobatis." (In this book *J. sphaerica* is accepted as included in *J. chinensis* L., so the parentage given above must be construed accordingly.)

The Botanical Code does not control botanical classification (taxonomy), it regulates nomenclature. In simple terms it is not concerned with what (in this context) you do, only with how you do it. There is nothing to stop any botanist studying some group of plants and putting forward whatever suggestions he thinks fit for changes and/or extensions in their classification – provided he correctly follows the Rules in the names he finally elects to use. The only thing that deters him from acting irresponsibly and suggesting unwarranted changes is the knowledge that any such contribution would not be accepted but just be ignored throughout the botanical world. Even when this does not occur, a later, fuller study published by another worker can challenge the currently accepted taxonomy of the group in question. This then, technically, leaves everyone free to follow whichever system he himself prefers. This is the reason why different authorities sometimes disagree, and it results in the confusing situation that a name is 'correct' only in the context of the classification being followed by the current writer. This is why some conifer books list certain juniper cultivars under *J.* × *media*, whereas others still use *J. chinensis*.

Reference has already been made (See p. 326) to the work done by Van Melle in the *J. chinensis* complex. The Botanical Code regulates the naming

Figure 217. *Juniperus* × *media* 'Pfitzeriana'. The 'type' of this hybrid species. It can reach to several metres in time.

of hybrids and includes provision whereby when the projeny of the same two parents has become a group of sufficient importance or complexity to justify it, a 'hybrid species' can be created. Van Melle considered the large and horticulturally important group of widely planted junipers known in garden parlance as 'the Pfitzers' justified such action and transferred (to *J.* × *media*) a number of these shrubby forms.

This raises the inevitable problem – whom shall we follow?

In this book by no means all of Van Melle's suggestions are accepted, but this particular group of junipers certainly forms a very distinct unit, so to have a designation for them is of great practical advantage: moreover *J.* × *media* has been widely accepted in the horticultural world, so it is here retained.

But the published diagnosis is clear: the 'species' consists of (and only of) all hybrids between *J. sabina* L. and *J. sphaerica* Lindley (emend. Van Melle) wheresoever or whensoever they might have arisen, or may yet arise (each clone receiving its own cultivar name).

Dr. Laurence Gough (See p. 328) has pointed out that Van Melle transferred to his new species a number of cultivars (the 'Plumosa/Globosa' group) having no connection whatsoever with *J. sabina*. These have been returned to *J. chinensis*, where they have always rightly belonged.

All the clones of *J.* × *media* in cultivation derive in some way from *J.* × *media* 'Pfitzeriana', which serves as the 'type' of the hybrid species, so each conforms more or less to the description given on p. 373.

-- **'Arbuscula'**. *J.* × *media* var. *arbuscula* Van Melle, Phytologia 189 (1946); *J. chinensis Smithii* Slavin in Rep. Conif. Conf. 103 (1932). A rather mysterious product of Van Melle's proposals. It is (or was) an erect bush, lacking a main stem but with main branches emerging at a narrow angle; a female clone with fruit on a slender stalk as in *J. sabina* but larger, and the crushed foliage gave out a less unpleasant smell than it does in that species. (Is it still identifiable or in cultivation?)

-- **'Arctic'**. Hill Nurs. Cat. 5. (1972). This is a selection which makes a wide-spreading plant 2 m across by no more than 50 cm high. It grows rapidly and is reputedly very much hardier than most of the Pfitzeriana group. The colour is bluish-green. It is said to be very hardy.

-- **'Armstrongii'**. *J. chinensis Armstrongii* Armstrong Nurs. Cat. 42. (1932) 'var.' Bailey, Hortus Suppl. (1935) *J.* × *media* var. *Pfitzeriana* f. *Armstrongii* (Bailey) Van Melle, Review, 29. (1947) *J. pfitzeriana nana* Hort. Amer. A low-growing shrub of the 'Pfitzeriana' type but of more compact and lower growth, seldom reaching much beyond 1 m high. The leaves are mainly scale-like except at the base of the bush. Foliage is soft to the touch, of a bright yellowish-green colour. A sport of 'Pfitzer' introduced by the Armstrong Nurseries, Ontario, California, U.S.A., in 1932. The name 'Armstrong Gold' is a misnomer for 'Gold Coast'.

-- **'Berry Hill'**. Minier Nurs. Cat. 1976. A sport from 'Pfitzeriana', differing only in lower and more spreading habit of growth. Introduced by Minier Nurs., Angers, France.

-- Blound. The 'name' under which the cultivar 'Gold Sovereign' was granted U.K. Plant Breeders' Rights.

-- **'Blue and Gold'**. Recommendation 31A (g) of the *Cultivated Code* discourages the use of an 'attribute' (such as colour) as a cultivar name. But in this case the combination of the two colours is distinctive so is acceptable. It is also quite descriptive of a mutation in the Pfitzer group that is grey- to blue-green with patches of bright yellow foliage. This makes it a noticeable plant, but one not to everyone's taste. The yellow foliage is liable to be damaged in hard weather, but reappears with the new growth in the Spring.

-- **'Den Boer'**. *J. chinensis* 'Pfitzeriana Den Boer'. AAN Regd. No. 258, (1951). Compact and spreading to 25 cm by 1 m diam, with green foliage. Selected in 1930 by A. F. den Boer of Des Moines, Iowa, U.S.A. and introduced by the Heard's Landscape Nurs. at Des Moines.

-- **'Fruitland'**. Monrovia Nurs. Cat. 55. (1977). A vigorous, semi-prostrate, spreading form with a compact branch system and dense, neat, silvery green all juvenile foliage. Claimed to be an improvement on 'Pfitzeriana Compacta', but the branchlets are stiff, with no trace of droop at the growing tips, so its placement in this species is suspect.

-- 'Globosa' and colour forms. Now listed under *J. chinensis*.

-- **'Gold Coast'**. Sakiyama Nurs. Cat. (1965). This arose as a mutation on a plant of 'Old Gold' on the Sakiyama Nurs. of California U.S.A. Growth is more compact and the colour of the young growth is a somewhat deeper chrome yellow, especially so during the winter.

-- **'Goldkissen'**. Oltmans ex van der Laar in Dendroflorea **20**: 70 (1983). A compact selection from 'Pfitzeriana Aurea' that carries bright yellow adult foliage at the growing tips and blue-green juvenile leaves in the interior of the plant.

-- **'Gold Sovereign'**. This arose as another mutation on 'Old Gold' that occurred on the Bressingham Gardens Nursery in Norfolk, U.K. It forms a very much slower growing plant and with a good gold colour well maintained throughout the year. It is the plant protected by Plant Breeders' rights in U.K. under the designation 'Blound'.

-- **'Golden Saucer'**. A very compact mutation from 'Pfitzeriana Aurea' introduced by W. M. Van Nierop of Boskoop, Holland. It is a bright golden-yellow, at its best during the winter.

-- **'Gold Star'**. A mutation from 'Pfitzeriana Aurea' with compact habit and golden-yellow, all juvenile foliage. It was found by David Bakker in 1961, introduced and the name registered by Bakker and Sons Nurs. St. Catherines, Ontario, Canada in 1971 under *Juniperus chinensis*.

-- 'Hetz'. See under *J. virginiana*.

-- **'Hills Blue'**. *J. chinensis* 'Pfitzeriana Blue' D. Hill, Nurs. Cat. 5. (1939). This is close to 'Pfitzeriana Glauca' – perhaps in colour lying between that and *J. virginiana* 'Grey Owl', turning less purple in the winter, and making a lower and flatter plant.

-- **'Kallay's Compact'**. This is one of a number of compact forms of the Pfitzeriana juniper, all arising no doubt from bud mutations. This one, which originated in the Kallay Nursery, Painesville, Ohio, U.S.A., is one of the most popular. It is of a deep green colour and carries a preponderance of juvenile foliage. It makes a flat-topped plant which seldom exceeds 80 cm but which will spread to nearly 2 metres. It is sometimes sold as 'Pfitzeriana Compacta' – a name belonging to the 'Nicks Compact' of American nurseries.

-- *'Kosteri'*. See *J. virginiana* 'Kosteri'.

-- **'Mathot'**. *J. chinensis* 'Mathot' Den Ouden, Conif. 154. (1949) Meyer, Pl. Expl. **2**: 121. (1963). This is very similar to 'Pfitzeriana Compacta' (Nicks' Compact) the leaves are all juvenile, in pairs, to 7.5 mm long, with bluish-green bands above, green below. As the upper side of the leaves are turned outwards the plant has a generally glaucous appearance. It originated on the Mathot Nursery, Boskoop, Holland, in 1940.

-- **'Milky Way'**. Listed Konijn 1968. A mutation of the 'Pfitzer' type, with very grey foliage and a nice, creamy white variegation. Not a strong grower.

-- **'Mint Julep'**. *J. chinensis* 'Mint Julep' Monrovia Nurs. Cat. 5 (1960). This is a valuable addition because of its colour, which is a rich mid-green, the best green yet. It makes a rather flat spreading bush, fairly compact but with long, arching, main branches. It is said to be a female clone. This would preclude the possibility of its origination as a mutation from 'Pfitzeriana'.

-- **'Moraine'**. A slow-growing selection with a compact habit and blue-green, entirely juvenile foliage. The name was registered by J. B. Siebenthaler in 1951, but the plant does not now seem to be in circulation.

-- **'Mordigan'**. A shrub with an irregular growth habit and mixed juvenile and adult foliage with a bluish cast that lessens with age. It was registered by P. Mordigan in 1951.

-- **'Mordigan Gold'**. Hillside Nurs. Cat. (1970). A form with a low, spreading, graceful habit; main branches at +30°, tips stiff or only slightly nodding and juvenile foliage, golden yellow, the colour well maintained through the year. The name 'Mordigan Aurea' is illegitimate.

-- 'Nick's Compact'. A superfluous name for 'Pfitzeriana Compacta'.

-- **'Old Gold'**. [*J. chinensis pfitzeriana* 'Old Gold' Grootn. Nurs. Cat. 11. (1958/59). This is a very attractive form, descriptively named. It forms a much more compact plant than the typical 'Pfitzeriana' and the lovely bronzy-golden colour goes right through the plant. This clearly distinguishes it from 'Pfitzeriana Aurea', in which the golden colour at its best is limited to the young growth. The colour is retained throughout the winter. It arose as a sporting branch in the F. J. Grootendorst & Sons Nursery in Boskoop, Holland. (See Illus. No. 218)

-- **'Ozark Compact'**. Watnong Nurs. Cat. **6**: 1964. Similar to 'Nicks Compact' but bluer colour and less vigorous growth.

-- **'Pfitzeriana'** *J. chinensis pendula* Späth, Mitt. d. d. d. G. **5**: 28. (1896) NON Franchet, (1884); *J. chinensis Pfitzeriana* Späth, Nurs. Cat. 104. (1899–1900); *J.* × *media* var. *Pfitzeriana* Van Melle, Review, 19. (1947). This is a good example of the growth in popularity of a good plant. When Hornibrook wrote his second edition in 1938 he evidently (from the paucity of his description) had not seen a plant, but now it is one of the most widely planted conifers and is so well known as not to need a description here. It is planted by the thousand each year, mainly I expect because of its picturesque outline as a young plant. One excellent use for it is to mask the presence of a manhole-cover on a lawn. This, its vigour and characteristic habit of growth enables it to do effectively in a short while and yet permit free access to the drains beneath on lawful occasions, but I wonder how many planters of this excellent juniper bear in mind its ultimate size. In the Arboretum Trompenburg in Rotterdam, for instance, there is a specimen large enough to afford complete shelter for quite a number of persons during a shower.

Its picturesque habit of growth is due largely to the preponderance of a semi-juvenile type

Figure 218. *Juniperus × media* 'old Gold'. A popular selection, appropriately named.

of foliage and its habit of throwing up its strong main branches at an angle of 45°, from which angle the tips of the leader and the principal laterals arch over so that all the growing tips are pendulous. The foliage is a light yellowish-green but owing to the fact that the juvenile foliage (of which the centre of the bush mainly consists) shows the white uppersides, the general effect is that of a grey-green colour. (See Illus. No. 217)

An engraving of the mother plant in the Späth nursery shows a broad, pyramidal tree with gracefully arching branches. Sceptics who find it hard to believe that 'Pfitzer' will submit to such treatment will be amused by a photograph in *Die Gartenwelt* 403. (1900–01) which clearly shows the leader of this same tree, although several metres high, still trained up to a bamboo cane! It was this practice in the nursery that fooled Beissner and other early writers into using for it the epithet *pendula*.

Van Melle [Review, 81, (1947)] traces the history and probable introduction of this juniper, which was known in France and Belgium before Späth renamed it for his nurseryman friend W. Pfitzer of Stuttgart. Späth's plant is a male form, so if it is the origin of the ever-increasing number of variants they must all have arisen as bud sports. As Van Melle points out, this means that the mutations are themselves mutable (i.e. subject to further change, or to reversion) so careful selection of propagating material is important.

This might be a good place to explain about the spelling. In Latin it has always been usual to create adjectives by tacking a syllable or two onto a noun, as we have done in English with the words *sheeplike, chancy* and *fanciful*. There are no rules: what is 'correct' is what is customary among educated people. So we do not say *sheepy, chanceful or fancilike* even if our meaning might still be clear.

The suffix *-ianus* has always been used in Latin to convey the sense 'belonging to'. In Roman times no personal names ending in *er* existed, but when in the middle ages many such names came into use the same suffix was used, producing words like *pfitzerianus*. No one queried this until an unfortunate entry in the *International Code of Botanical Nomenclature* of 1905 introduced the name *Verbena* *hasslerana*. To a trained botanist the *Code* has absolute authority, binding the faithful down even to the spelling he uses, so this became a precedent and botanical and horticultural writers had to learn to omit the 'i' whenever a personal name ending in 'er' was involved. This not only challenged the custom of centuries, it created discrepancy with other, non-botanical, users of Latin, so at the International Botanical Congress held in Sydney in August 1981 a decision was made to amend the *Code* and reinstate the original spelling, any deviation meanwhile being treated as an 'orthographic error' ('spelling mistake' to most of us), to be corrected.

A very thorough account was published by Dr. W. T. Stearn (author of the book *Botanical Latin*) in the Journal of the Royal Horticultural Society, Vol. **110** (10): in October 1985.

One point remains. In Latin, adjectives as well as nouns have genders, and three of them, at that. In our case *ianus* is the masculine form; *iana* is the feminine and *ianum* the neuter equivalent. When the epithet in honour of W. Pfitzer of Stuttgart was invented the Botanical Code was considered to be involved, so since *Juniperus* is a feminine noun the form *Pfitzeriana* was appropriately chosen. Now that the plant in question is regarded as a cultivar, we may retain the epithet as a cultivar name. But 'Pfitzeriana' it must still be.

-- **'Pfitzeriana Aurea'**. *J. chinensis pfitzeriana aurea* D. Hill. Nurs. Cat. 4, (1937) var. *pfitzeriana aurea* Grootn. Mitt. d. d. d. G. **53**: 207. (1940) has the typical Pfitzeriana growth but with a tendency to a flatter habit, the branches being held at an angle of 30°. The branchlets and leaves are a bright golden-yellow in summer but become yellowish-green by the winter.

-- **'Pfitzeriana Compacta'**. *J. chinensis* 'Nicks Compact' Wyman, Amer. Nurs. **117,1**: 116. (1963). This is another sport from the Pfitzer juniper which Wyman onsiders one of the best of the compact junipers. It forms a flat-topped plant to nearly 2 metres across but only 30 cm high, of excellent compact form with stiff, thick branches and very prickly foliage. It was introduced by Bobbink and Atkins Nursery, Rutherford, New Jersey, U.S.A., in 1930 as *J. chinensis pfitzeriana compacta*. It carries mainly juvenile leaves, the white upper sides giving quite a greyish cast to the green colour. It is very difficult to distinguish from the clone 'Mathot'.

-- **'Pfitzeriana Glauca'**. *J. chinensis pfitzeriana glauca* Sherman, Texas Nurs. Cat. (1940) *J.* × *media* var. *pfitzeriana* f. *glauca* Van Melle, Review, 29. (1947). This is another sport with its foliage varying between silvery-blue and a dull greyish-blue. The foliage of 'Pfitzeriana Glauca' is very prickly, which distinguishes it at once from 'Hetzii', which is quite soft to the touch. 'Pfitzeriana Glauca' forms a taller plant than 'Hills Blue' and colours more noticeably in winter. Several variegated sports have also appeared. Of these, one named **'Silver Tip'** is attractively speckled with creamy-white similarly to *J. chinensis* 'Plumosa Albovariegata'.

-- 'Plumosa' and colour forms. Now listed under *J. chinensis*.

-- **'Ramlosa'**. A selection made on the H. Jensen nursery, Helsingborg, Sweden, before 1973. It is distinctive only in its very low and spreading habit, the branches being almost horizontal. This cultivar should not be confused with the species *Juniperus ramlosa*.

-- **'Richeson'**. *J. chinensis 'Richeson'* Armstrong nurs., Cat. **1946**; Wyman in Amer. Nurs. **117,1**: 117 (1963). Compact and low-growing, to 30 cm high and 1 m diam; foliage partly short-acicular, bluish-green. A sport of '*Pfitzeriana*' found by L. J. Richeson, Ontario, Calif. in 1941, distributed by the Armstrong nurseries in 1946.

-- **'Saybrook Gold'**. A selection introduced by the Girard nursery, Geneva, Ohio, U.S.A. in 1980. It forms a strong-growing, low plant with foliage that is fresh green in Spring, darkening during the Summer and turning a bronze-yellow by the Winter.

-- **'Sea Green'**. A name in use in U.S.A. for a selection that is indistinguishable from, and possibly identical with, 'Mint Julep'.

-- **'Sea Spray'**. A very hardy, vigorous procumbent selection with mixed juvenile and adult foliage, never over 20 cm tall, soon spreading symmetrically to 2 m but remaining densely covered at the centre. It has a distinctive sea-green colour. It was found by Frank J. Sherpa of Fremont, Cal. in 1963 and was registered by Hines Nurs., Santa Ana, Cal., U.S.A. in 1970, under *Juniperus chinensis*.

-- 'Shimpaku' Now listed under *J. chinensis*.

-- **'Winter Surprise'**. Van Klaveren, Boskoop in Dendroflora **21**: 23 (1984). A yellow-flecked, low-growing sport from 'Pfitzeriana Aurea'. It differs from but is questionably an improvement on 'Blue and Gold'.

J. mekongensis Komarov in Not. Syst. Herb. Hort. Petrop. **2**: 28 (1924). Now merged in *J. convallium*.

J. monosperma. (Engelm.) Sarg., Silva of N. Amer. **10**: 89 (1896); *J. occidentalis* var. *monosperma* Engelm. in Trans. Acad. Sci. St. Louis **3**: 590 (1877); *J. erythrocarpa* Cory in Rhodora **38**: 186 (1936); *J. texensis* van Melle 26 (1952). Cherrystone Juniper. A monoecious shrub or tree to 12–18 m, with a short, thick trunk and a dense, spreading crown. *Bark*: Fibrous, shredding in long strips, light reddish-brown or grey. *Branches*: Stout, thick, curving upwards, branchlets about 1 mm thick, quadrangular. *Leaves*: Mostly adult, scale-like, opposite, closely appressed, apex sharply long-pointed, often slightly spreading at the tip, rounded on the back, conspicuously glandular; glossy greyish-blue. Juvenile leaves in 3s, 10–12 mm long. *Cones*: Irregularly globose or ovoid, 5–7 mm long, dark blue with resin or a glaucous bloom, pulp firm, thick. *Seeds*: Solitary (rarely 2 or 3), broadly ovoid, often quadrangular, dull chestnut brown, apex blunt, sometimes visible between the scales. *Habitat*: Colorado to Westtern Texas, New Mexico, Nevada and North Mexico: Introduced in about 1900. Zone 7.

J. monticola. Martinez in An. Inst. Biol. Mexico **17**: 79 (1946); *J. sabinoides* Humboldt, Bonpland and Kunth. A shrubby species found in some mountainous areas of Mexico. Rare in the wild, and not in cultivation.

J. morrisonicola. Hayata in Gar. Chron. series 3, **43**: 194 (1908). A species that in cultivation makes an attractive large shrub or small tree with outstanding, silvery-glaucous juvenile foliage. By some authorities it is regarded as being a variety of *J. squamata*.

J. occidentalis. Hook., Fl. Bor. Amer. **2**: 166 (1838). Western Juniper. A monoecious (occasionally dioecious) tree 12–20 m with a tall, straight trunk and a dense crown, (Sometimes seen as a large shrub with several upright stems). *Bark*: Thick, scaling, bright cinnamon-red. *Branches*: Widely spreading. *Branchlets*: Stout, short, nearly round, 1.5 mm diam., yellowish-brown. *Leaves*: On old trees predominantly scale-like, mostly in whorls of three, 1.5–3 mm long, closely appressed, ovate, bluntly acute, greyish-green; back conspicuously glandular. A few awl-shaped juvenile leaves sometimes present. *Cones*: Held upright on short branchlets, solitary, crowded, sub-globose or oblong, 7–9 mm long, 5–8 mm across, composed of 6, rarely 9 scales, ripening 2nd year to bluish-black with whitish glaucescence. *Seeds*: 2–3, moderately large, ovoid, back deeply grooved or pitted. Edible. Young plants of *J. occidentalis*, both female and male, are conical in form with arching pendulous branchlets and awl-shaped distinctly bluish leaves. *Habitat*: Mountains of Washington and Western Idaho to Southern California; Introduced into England about 1840. Zone 7.

-- **'Sierra Silver'**. Monrovia Nurs., Cat. **1960/61**: 6. Habit twisted; foliage very bold, of a lavish grey-green colour. Found in the high Sierras of Colorado.

J. osteosperma. (Torrey) Little in Leafl. West. Bot. **5**: 125 (1948); *J. tetragona* var. *osteosperma* Torrey in U.S. Rep. Expl. Survey. Miss. Pacif. **4**: 141 (1857); *J. utahensis* (Engelmann) Lemmon in Calif. State Board For. Bein. Rep. 183 (1890). Utah juniper. A tree to 12 m with a

Figure 219. *Juniperus oxycedrus.* Detail of foliage.

dense, rounded crown. The ultimate shoots are somewhat quadrangular, due to the acute, slightly toothed scale-like leaves being in pairs, rarely in 3s; tightly appressed, usually with a dorsal resin gland, pale yellow green. 3 mm long. Cone globular or nearly so, glaucous, ripening 2nd year to red-brown, flesh dry, thin. This species is very near *J. californica*, differing mainly in the leaves being in pairs, not in 3s as in that species.

J. oxycedrus. 1038 (1753); *J. rufescens* Link in Endlicher, Synop. Conif. 11 (1847). Prickly Juniper. Dioecious shrub of small tree to about 12 m. *Branchlets*: Triangular and ridged in cross section slender, erect. *Leaves*: Widely spaced, awl-shaped, in whorls of three, 12–16 mm long, springing directly from a swollen base and tapering to a long, prickly point; flat, upper surface grooved with a narrow green midrib, 2 wide white bands and narrow green margins; strongly keeled below. *Cones*: Solitary in the leaf axils on very short stalks, globose to obovoid, 10–12 mm across, composed of 3–6 scales; yellowish, ripening the 2nd year to glossy red or brown. *Seeds*: Usually three, oblong, triangular, ridged, with resin-glands at base. *Habitat*: Mediterranean region, eastwards to Persia.

This is a Juniper with economic value for etherial oil, but without decorative value. It is extremely rare in cultivation. Zone 9. (See Illus No. 219)

-- subsp. **macrocarpa**. (Sibthorp and Smith) Beil 1868; *J. macrocarpa* Sibthorp and Smith in Fl. Graeca Prod. **2**: 263 (1813). This differs from the type only by its larger leaves, 16–25 mm long and larger fruits, 15–12 mm long and almost as much wide, so hardly justifies recognition as a sub-species. It occurs mainly in the coastal areas.

-- var. ***transtagna***. (Franco) Silba in Census Conif. 35 (1984); as 'subsp.' Franco, Repr. Nov. Spec. Regni. Veg. **68**: 166 (1963). A fastigiate shrub found only in Portugal, between Azeito and Coina, in maritime areas, mostly south of the River Tejo.

J. phoenicea. 1040 (1753); *J. oophora* Kuntze, Flora 637 (1846); *J. lycia* L. 1039 (1753). Monoecious shrub or small tree, usually monoecious, 6–12 m tall with ascending branch system and a dense, rounded or bluntly conical crown. *Bark*: On old branches dark brown. *Branchlets*: Light cinnamon-red; twigs round, scarcely 1 mm in diam. *Leaves*: Juvenile leaves (rare on mature trees) in whorls of 3, awl-shaped, 6 mm long, decurrent, apex acute. Adult leaves scale-like, opposite or in whorls of 3, rhombic-ovoid, closely appressed, about 1 mm long, apex sharp-pointed but turned inwards and therefore apparently blunt, dull- often bluish-green, margins finely toothed; the back convex and often with one gland. *Cones*: Very variable in size and shape; solitary on a curved stalk 5 mm long, sub-globose, 8–10 mm diam, composed of 3–4 pairs of scales, ripening the 2nd year to a glossy yellow-green or reddish-brown. The yellowish pulp is fibrous, dry and adherent. *Seeds*: 3–9, ovate, roughly triangular, apex blunt, brown. *Habitat*: Canary Islands and Mediterranean region. Not hardy in Central Europe. It was in cultivation in England as early as 1683, but is still very rare. Formerly a distinction was made between the shrubs with small fruits, occurring in rocky situations and those with large globose fruits, growing on the maritime sands, the latter being named *J. lycia* L., or *J. phoenicea lycia* Lois. Zone 8. (See Illus. No. 232)

J. pinchotii Sudworth in For. and Irrig. **9**: 204 (1905). Differs from *J. osteosperma* by its more bushy habit, its glandular leaves and its smaller (6–9 mm across), globose, 1–2 seeded cones. Central to northwestern and Trans Pecos, Texas, also in western Oklahoma and reported from southwestern New Mexico. Not hardy in colder regions. *J. pinchotii* is named after Gifford Pinchot, distinguished American forester and Statesman and first chief of the U.S. Forest Service. Zone 8.

J. pingii. Cheng in Trav. Lab. For. Toulouse **5,1** (2): 93 (1939). Low shrub or tree to 10–30 m high, with grey-brown, fissured bark; branchlets quadrangular, threadlike, tips nodding or pendent. Leaves ovate-lenceolate, broad, imbricate, slightly keeled, 3–5 mm long; decurrent with tips free but incurved; yellow-to mid-green, silvery white above. Cone on short peduncle, ovoid, 7–9 mm long, one-seeded, blackish, shiny, with 6 scales 8 mm long by 4 mm wide. *J. pingii* was by many authorities for long included in *Juniperus squamata*, but its recognition by botanists as a separate species now seems to be accepted. This, as is unfortunately always the result of a botanical name-change, requires the transfer of several cultivar names. China: Yunnan to Szechuan at 2600–3000 m.

-- f. ***wilsonii*** (Rehder) Silba; *J. squamata* f. *wilsonii* Rehder in Journ. Arn. Arb. **1**: 191 (1920). Among several consignments of seed sent home from China by E. H. Wilson as *Juniperus squamata* was one from his collection No. 985 which, in U.S.A. and in Britain, gave rise to an interesting assortment of shrubs or small trees, several of which have been recognised as cultivars. These Professor Rehder lumped together so it is not surprising that Van Melle regarded f. *Wilsonii* as 'something of a loose number'. It serves us correspondingly well as a group name for a number of clones in cultivation well separated from each other in garden use by their habit.

The name *J. squamata* var. *fargesii* is often misapplied in cultivation to specimens of *J. pingii*. This is without any apparent justification since although that tree might possibly on casual inspection be mistaken for *J. coxii* it bears no resemblance whatever to any form of *J. pingii*.

-- **'Forrestii'**. Welch, 240 (1979). This is described as being intermediate in habit between 'Wilsonii' and 'Loderi'.

-- **'Glassell'**. Ashberry, Min. Trees and Shrubs, 82. (1958). This is a dwarf and extremely slow-growing form with branches ascending at steep, picturesque angles and with short, curved branchlets. Foliage is as in 'Wilsonii', but is much shorter and denser. It is an ideal plant for a trough garden, as its branching habit gives a 'Ye olde gnarled apple tree' appearance to the tiniest plant, especially if all its foliage is nipped off except at the growing tips. It should appeal to Bonsai addicts.

-- **'Loderi'**. *J. squamata* var. *Loderi* Hornb. Rep. Conif. Conf. 74 (1932); *J. wallichiana* var. *loderi* Hornb. Gar. Chron. **85**: 50. (1929). Although not one of the Wilson seedlings this also seems to have come from a stray seed in a mixed packet and it certainly is one of our present group, differing from the other cultivars only in habit. 'Loderi' forms a dense, columnar bush of the habit of *J. communis* 'Compressa', for a giant specimen of which it can be mistaken at a distance. Even in this fastigiate form the characteristic nodding at the growing tips is just discernible. It may well turn out to be merely an arboreal form typical of the species. (See Illus. No. 220)

-- **'Prostrata'**. Under *J. squamata* 'Prostrata' I hope to have finally laid to rest any claim to this name by the plants so labelled in America and elsewhere which invariably turn out to be a form of *J. procumbens*. But, from the fact that Hornibrook claims to have raised the plant he named *J. squamata* var. *prostrata* from the famous seed packet No. 985 and from his description we can identify a low-growing plant with typical f. *wilsonii* foliage that, although uncommon, is in cultivation in Britain. Despite Hornibrook's statement that the branchlets point forward and do not droop, the growing tips arch over in the way characteristic of f. *wilsonii*.

Another clone cultivated in Britain, **'Pygmaea'**, so named because of its smaller leaves and neat habit when small, has the same vertical branching system but develops in time into a rounded, spreading bush.

-- **'Wilsonii'**. The clone in cultivation under this name (presumably that from which Professor Rehder took his description) forms a rounded bush about as broad as high. It has the short, broad, boat-shaped leaf that characterises the whole group, but these stand off at 80–90°, exposing the white stomatic bands, and in this clone all the growing tips arch over noticeably. (See Illus. No. 222)

J. procera. Hochstetter ex Endlicher, Synop. Conif.

26 (1847). East African Juniper. A tender species from equatorial East Africa, allied to *J. excelsa*. It makes a tree to 40 m. with peeling bark, and yellow-green foliage in thin shoots. Leaves mainly adult, long-pointed, grooved, glandular; occasional juvenile leaves in 3s, 10 mm long. Fruit glaucous violet-brown, globose, smaller than in *J. excelsa*. East Africa. Zone 10.

-- var. *africana*. Name at one time used mistakenly for *J. chinensis* 'Stricta'.

J. procumbens. (Endlicher) Miquel in Siebold & Zuccarini & Z. 59 (1870); *J. chinensis* var. *procumbens* Endl. 21 (1847). *J. horizontalis* var. *chinensis* Slavin in Rep. Conif. Conf. 106 (1932); *J. japonica* Hort ex Carrière, 34 (1855). Creeping Juniper (Gr. Brit.); Japgarden Juniper (Amer.). Dwarf shrub, low-spreading, 50–60 cm high, more than 2 m broad; branches stout, stiff, spreading, turning upwards at the ends, spreading and closely packed in the centre of the plant; branchlets with glaucous ridges; leaves in threes, linear-lanceolate, spiny-pointed, 6–8 mm long, concave above and glaucous with a green midrib toward the apex, bluish with 2 white spots near the base below, from which 2 glaucous lines run down the edges of the pulvini. Mountains of Japan. Sent by Teysman to Dr. von Siebold, Leiden, in 1843 and shown at an exhibition in the Netherlands in 1850 as *J. procumbens*, but seems not to have become widely established in European cultivation at that time. Reintroduced by M. Vilmorin into France in 1903 and distributed as *J. japonica*. Zone 4.

This unfortunate species has had much more difficulty in establishing its identity than such a useful and unassuming plant deserves. It was first recorded by Siebold and Zuccarini in 1844. The word *procumbens* being an adjective it soon became reduced to a varietal epithet and so the plant was pushed at one time or another into *J. chinensis*, into *J. horizontalis* and even (in confusion for quite another plant) into *J. squamata*.

But this was not all. A fresh introduction into Europe from Japan never could have hoped to escape becoming known as *J. japonica*. A slightly

Figure 220. *Juniperus pingii* 'Loderi'.

Figure 221. *Juniperus pingii*. Detail of foliage. (This species was, until recently, included in *Juniperus squamata*.)

Figure 222. *Juniperus pingii* 'Wilsonii'.

Figure 223. *Juniperus squamata* var. *fargesii* in the U.S. National Arboretum, Washington, D.C., U.S.A.

glaucous clone was early introduced into America where (since other, larger, junipers also supposedly from Japan were also known as *J. japonica*), this smaller plant became *J. japonica nana*, under which name it appeared in the D. Hill Nursery Catalogues from 1904 until 1942. Inevitably, when its true specific allegiance was recognised, this clone became *J. procumbens nana*.

This raises a problem, for where there is a 'Nana' form normal plants should also exist and for a long time the whereabouts of these was not known. A rather stronger-growing plant, a mid-green form collected many years ago 'in the wild', was introduced into U.K. as **'Bonin Isles'** It never fruits, but Rushforth [132 (1987)] states that the species is found in Southern Japan along the seashore on the Bonin Islands and on Kyushu, reaches to 2m across by 0.75m high, bears predominantly juvenile foliage and has sub-globose fruits ripening to brown or black, containing 2–3 seeds. So it may be that 'Bonin Isles' fairly represents the species in cultivation but, being a male plant, is condemned in isolation to a life of celibacy.

-- **'Nana'**. *J. procumbens* var. *nana* Grootend. in Hornibr. 122 (1939); *J. japonica nana* D. Hill, Cat. **1942**; *J. squamata* var. *prostrata* van Melle (1947) plate XII (bottom). Dwarf, smaller than the type form, forming a compact mat; branchlets vary in length spreading out as a compact mass of sprays; leaves small, bluish-green, slightly purplish in winter. It forms a dense mat and is particularly useful for ground cover or planting so as to trail downhill over rocks or a tree root. (See Illus. No. 224)

First claim to the name 'Nana' must surely go to the plant brought home from Japan by Arthur Hill about the turn of the century and listed by the D. Hill Nursery Company as *J. japonica nana* from 1904 onwards until 1942, since when it has been listed as *J. procumbens* 'Nana'. It makes a very dense plant and is of a distinctly blue-green colour. How the other form in American cultivation reached that country I cannot say. This is mid-green in colour and is, I think, the form generally known in Europe, where the blue-green form tends to attract the name 'Nana Glauca'. There is a form – **'Bonin Isles'**, which Will Ingwersen

Figure 224. *Juniperus procumbens* 'Nana' at Wakehurst Place, Surrey, U.K.

tells me was given to his father by a friend who collected it (presumably in the Bonin Isles south of Japan) many years ago, exact records not being available. It has the same dense, mat-forming habit as 'Nana' but is a much stronger grower. See also p. 380.

J. przewalskii. Komar., nonn. Asiat. **2**: 28 (1924). Tree 12 m; branchlets spreading to upright; leaves scale-like, spreading or appressed, about 3 mm long, greyish-green, back tubercled; cones 10–15 mm across, reddish-brown; seeds obcordate or irregularly globose, about 10 mm long, wrinkled on both sides, with 2 narrow keels.

China, in the eastern part of the Kansu province. Zone 7. Not in cultivation.

J. pseudosabina. Fisch. & Mey., Ind. Sem. Hort. Petrop. **8**: 15 (1842). Upright shrub; dark ashy-grey; branches erect, round; branchlets short, thick, quadrangular, to 2 mm broad; leaves scale-like, opposite, closely appressed, broad, ovoid-rhomboid, apex blunt, 1–3 mm long, back convex or nearly blunt keeled, with a gland, greyish-green, with awl-shaped leaves in threes on young plants and often partly on older ones; cones on short decurved branchlets, ovoid, 10–12 mm long, 6–8 mm across, nearly black, glossy, composed of 4 scales; seeds solitary, large, ovate.

Altai and Targabatai mountains of Turkestan. Not in cultivation. Zone 6.

J. ramulosa Florin. See *J. convallium*.

J. recurva Buchanan - Hamilton ex D. Don, Prodr. Fl. Nepal. 55 (1825). Drooping or Sacred

Juniper (Gr. Brit.); Himalayan Juniper (Amer.). Shrub or small tree to 10 m, crown mostly broadly conical. *Bark*: Dull grey-brown, peeling off in broad strips. *Branches*: Spreading, upswept, the bark reddish. *Branchlets*: Drooping. *Leaves*: Awl-shaped, densely imbricate, in threes, lower part appressed, upper part somewhat boat-shaped, pointing forward but free, 8–6 mm long, 0.5–1 mm broad, glaucous or bright greyish-green with light margins; apex long and sharply-pointed; with two white bands of stomata above, grooved beneath from the base to the middle. The withered leaves are retained for several years and rustle when disturbed. *Cones*: ovoid, 7–10 mm long, 3.5–4 mm across, sometimes longer and thicker, composed of 3–6 scales, ripening the 2nd season to a glossy purplish-brown or black. *Seeds*: Solitary, large, ovoid with blunt apex; pitted.

Eastern Himalaya to Yunnan: Introduced into France in ± 1822. Not hardy in Central Europe. Zone 7. (See Illus. No. 226)

-- **'Castlewellan'**. Den Ouden and Boom, 186 (1965). A clone with very threadlike foliage. The very large tree at Castlewellan, N. Ireland is spectacular, but as a young plant it can be disappointing. (See Illus. No. 225)

-- var. *coxii*. See *J. coxii*.

-- **'Densa'**. *J. recurva* var. *densa* Carr. Traité, 27 (1855) Gord. Pin. 106 (1858); *J. densa* Gord. Pin. Suppl. 32 (1862), with *J. recurva nana* pro syn.; *J. recurva* var. *squamata* Veitch Man. Ed. 2, 185 (1900) *J. recurva* 'Nana' Welch, Dw. Con. 202 (1966). Several clones are here involved. They are all low, spreading, rather attractive plants seldom exceeding 60 cm high, with main branches held at +45° but with all growing tips strongly recurved. The leaves are very densely set, 4 mm long, parallel-sided and blunt ended but with a narrow, sharp-pointed tip. The colour is green, but two bands of stomata (separated by a narrow mid-rib) give a greyish tone. The crushed foliage has a strong smell reminiscent of turpentine.

-- **'Embley Park'**. Welch in Wansdyke Nurs. Cat. 6 (1961); *J. recurva* var. *viridis* Hillier, Dw. Con. 39 (1964). This plant is an attractive garden variety making a wide-spreading bush with densely packed, rich, deep green foliage. It was first noticed in a derelict rock garden at Embley Park, near Romsey, Hampshire (at one time the home of Florence Nightingale and now a boy's school), and it was named by Mr. Hillier var. *viridis* on the assumption that it had been sent home by the plant collector George Forrest and so was established in the wild. The foliage is in all respects akin to *J. coxii*, to which species this clone may belong. It needs shelter from high winds and is sometimes more successful in partial shade, but when it is doing well it is a beautiful plant.

J. rigida Siebold & Zuccarini (1846) 233. Temple Juniper (Gr. Brit.), Needle Juniper (Amer.). A dioecious shrub or tree to 10m. with an open, conical crown. *Bark*: Brown or grey, thin; on old trees furrowed or peeling in long strips. *Branches*: Spreading, upswept. *Branchlets*: Long, pendulous thin, triangular, grooved, brownish-yellow, the buds noticeable. *Leaves*: Awl-like in whorls of three, rather widely set (but the foliage is irregularly collected into clusters), spreading at a wide angle or directed slightly forward, narrow-linear, 15–25 mm long, ± 1 mm wide, tapering from the middle towards a spiny (and very sharp) point; C-shaped in cross section with a narrow blue-white band inside (above) with narrow green margins; prominently keeled below; glossy green, becoming yellow towards the tip. This gives a yellowish cast to the foliage viewed in the mass. *Cones*: Solitary, in rows along the shoot; globose, on very short stalks, 6–8 mm across; composed of 6–9 scales; glaucous green, ripening the second season to a burnished, glossy brown-black. *Seeds*: 2–3, oblong, angular, marked near the base with 3–4 resinous pits. This species should not be confused with the rather similar *J. oxycedrus* which is usually a more open and pendulous tree with the leaves held more definitely forward, nor with the very uncommon *J. communis* 'Oblonga Pendula'. It is found in N.E. China; Korea and Japan, Zone 6. (See Illus. Nos. 227, 228)

Figure 225. *Juniperus recurva* 'Castlewellan'. A thread-leaf clone, beautiful as a mature specimen.

-- var. ***nipponica***. *J. rigida* ssp. *nipponica* (Maxim.) Franco in Bol. Soc. Brot. **36**: 119 (1962); *J. nipponica* Maxim., Mél. biol. **6**: 374 (1868); *J. communis* var. *nipponica* (Maxim.) Wils 81. (1916). Nippon Juniper. Prostrate Shrub; leaves closely set, 6–10 mm long, deeply grooved, with a narrow stomatal band above, distinctly keeled below; seeds solitary. High mountains of northern Japan, sent by E. H. Wilson to the Arnold Arboretum, in 1915. Uncommon in cultivation.

J. sabina L. 1039 (1753). Savin Gr. Brit.); Savin Juniper (Amer.) Monoecious or dioecious shrub 5 m, spreading or procumbent, seldom

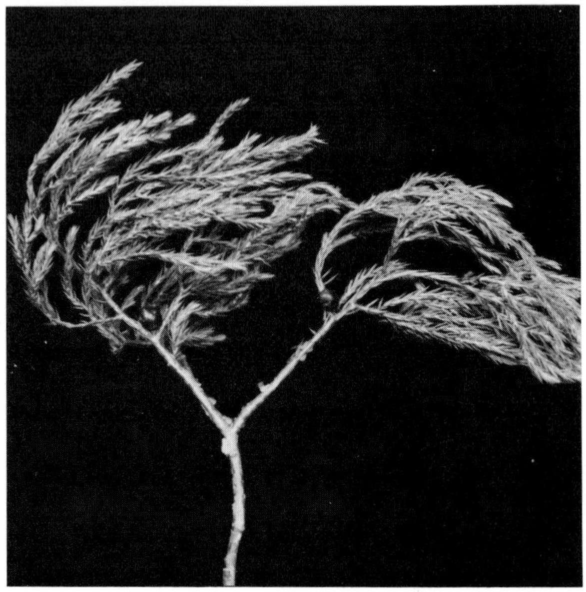

Figure 226. *Juniperus recurva*. Detail of foliage.

with an erect stem; bark on older branches reddish-brown, peeling off; branches procumbent or ascending, tops curving upward; branchlets densely set, rather slender, thin, scarcely 1 mm thick; leaves on young plants and often on some branchlets on older trees awl-shaped, spreading, straight, 4 mm long, apex sharply long-pointed, bluish-green and with an conspicuous midrib above; scale-like leaves opposite, ovoid, 1 mm long, entire, apex blunt (on main branches to 3 mm long, apex sharply long-pointed), back convex, mostly with a gland; cones on recurved stalks, globose to ovoid, 5–7 mm across, bluish-black, bloomy, composed of 4–6 scales, ripening in the first season of in the spring of the second season; seeds 1–3, ovoid, furrowed.

Mountains of central and southern Europe, Western Asia, Sibiria and Caucasus.

The foliage of *J. sabina* emits a disagreeable odour when bruised and has a bitter taste. In polite language – it smells of 'cats.' Cultivated since ancient times; first mentioned in Capitulare de Villis of Charlemaigne (812), but many cultivars are of recent introduction. Zone 4. (See Illus. No. 230)

-- '**Arcadia**'. Grootn. ex Den Ouden, 175. (1949);

Figure 227. *Juniperus rigida* in the Royal Botanic garden, Edinburgh.

Figure 228. *Juniperus rigida*. Detail of foliage and fruit.

Figure 229. *Juniperus sabina* 'Tamariscifolia'. Only blight-resistant clones should now be used.

Figure 230. *Juniperus sabina*. Detail of foliage and fruit.

'Russian Savin No. 3' (Hort Amer.) is one of four hardy selections (the others being 'Broadmoor', 'Buffalo' and 'Skandia') made from many thousands of seedlings imported by the D. Hill Nursery Co., Dundee, Illinois, U.S.A., from a government forestry station in Russia in 1934. All are proving resistant to the juniper blight which is seriously troubling most plants of *J. sabina* in the mid-west and so they are proving popular in the United States. 'Arcadia' is a grass green colour and its habit resembles a diminutive plant 'Tamariscifolia', but the leaves are predominantly scale-like and small, giving a lace-like texture. It will make a dainty looking plant 1.25 m across and 50 cm high.

-- **'Blau Danau'**. Usually translated to 'Blue Danube'.

-- **'Blue Danube'**. Blaauw, Nurs. Cat. 1956; 'Blaue Donau' Hesse Nurs. Cat. 41. (1961/62) forms a low shrub with wide-spreading branches having their tops curved upwards. The branchlets are crowded, the leaves mainly scale-like but the awl-shaped leaves inside the plant are a light greyish-blue. It is not in my opinion a particularly valuable form, but being very hardy, is widely planted on the continent. This clone was at first distributed as an unnamed seedling by L. Visser, Preszbaum, Austria.

-- **'Broadmoor'**. More, Green Thumb. **13,3**: 33 (1956). Russian Savin No. 4 (Hort. Amer.) This is dwarf, low, staminate form, wider spreading and stronger than 'Arcadia'. It looks when young like a neat version of 'Tamariscifolia' but the plant tends to build up at the centre with age. The main branches are strong and spreading horizontally; the branchlets short and reaching upwards; the sprays very short and occurring mainly on the upper side of the branchlets. The foliage is bright green. It is reported to do well in coastal situations.

-- **'Buffalo'**. More Green Thumb **13,3**: 33 (1956). This is another of the Russian seedlings. (see 'Arcadia' above.) It is a wide-spreading form which stays lower than 'Tamariscifolia'. Bright, deep green, bearing berries when quite young, very hardy, and tolerating coastal conditions.

-- **'Calgary Carpet'**. Monrovia Nurs. Cat. 61 (1978). A selection from 'Arcadia' making a much lower and more spreading plant; with soft green foliage.

-- f. *cupressifolia*. Aiton, Hort. Kew. **3**: 414 (1789). This is an early varietal name of uncertain application. It was described as being a low-growing, berry bearing shrub with bluish-green, wholly adult foliage. The name is now valueless, unless as a group heading.

-- **'Erecta'**. *J. sabina* Hort. Holl.; *J. sabina cupressifolia* Hort. NON Ait. Dutch Savin juniper. This form approaches the normal form of the species. It is a tall, loose-growing shrub, with a predominance of adult foliage but with branches all ascending at a steep angle. It is not a particularly good garden plant and is hardly worth distinguishing from *J. sabina* in its typical form.

-- **'Fastigiata'**. (Beissner) Boom, 99 (1942); *J. sabina fastigiata* Beissn. 111 (1891), pro parte; *J. sabina pyramidalis* Hort. A form similar to, if not identical with, 'Erecta'. Growth is dense, top many headed, 5–6 m tall; branches ascending; branchlets crowded, erect; leaves chiefly scale-like, dark green, glaucous bloomy. With early training it can form a small, upright tree.

-- 'Femina' and 'Mas' are names once given to 'sexed' clones otherwise undistinguished and no longer identifiable.

-- **'Hicksii'**. Grootn. Mitt. d. d. d. G. **53**: 210 (1940); Shrub, rather strongly growing; branches at first ascending, becoming more widespread as the plant ages. Leaves mostly acicular, distinctly greyish-blue, with a mauve tinge in winter. It was introduced by I. Hicks & Son, nurserymen at Westbury, N.Y., U.S.A. as an improvement on the type, but is now seldom planted.

-- **'Jade'**. Krüssmann, Nadel. ed. 3, 126 (1979) A selection made on the Fairview Nursery, U.S.A. with rather fine, blue green foliage.

-- **'Knap Hill'**. (At one time distributed by the

nursery of that name near Woking, Surrey) I am unable to distinguish from *J.* × media 'Pfitzeriana'. It is still listed by one nursery in England.

-- '**Musgrave**'. A low, spreading form which closeely covers the ground, carrying mostly tightly appressed adult foliage, but awl-shaped leaves occur here and there over the bush and in the centre of the plant. The foliage is glaucous. It was found in a garden in Surrey and was introduced by the Ingwersen Nursery, East Grinstead, Sussex, who still list it.

-- '**Rockery Gem**'. Welch, 237 (1979). This variety arose as a seedling selected on the nursery of Le Feber and Co., Boskoop, Holland. In 1969 the mother plant there was 1.25 m across on a short stem, plants in the nursery rows being much lower. The main branches are horizontal with numerous branchlets densely set at the sides and top; foliage is all acicular leaves 5 mm long, green below and with a wide glaucous band above. The leaves on weak shoots in the centre of the plant stand off at 45° shewing the glaucescence clearly; leaves on strong-growing shoots lie closer and give a green appearance, although the general effect, seen from a little distance, is grey-green. Although at first listed under *J. chinensis*, the appearance of the plant and the characteristic *sabina* pungence of the foliage indicate its affinity with the latter species.

-- '**Skandia**'. Grootn. Nurs. Cat. 9 (1953/54); 'Scandens' Hillier, Dw. Con. 40 (1964). This is the last of the four Russian seedlings (see 'Arcadia').

It is a very low-growing form of dense habit. The foliage is predominantly juvenile, dark green but with a glaucous bloom which gives the plant a greyish overall appearance.

-- f. ***tamariscifolia***. Aiton, Hort. Kew. **3**: 414 (1789). Spanish Savin Juniper (Gr. Brit.); Tamarix Juniper (Amer.). This is found growing wild on the mountains of southern Europe so it is clearly entitled to botanical recognition, but since the cultivated plants are a single clone (or a group of clones so similar as to be indistinguishable) the use of the name cultivarwise is justifiable. It makes a low spreading shrub, 1.5–2 m broad, 0.8–1 m high; branches horizontal, heaping up; branchlets crowded, spreading, short; leaves awl-shaped, very short, nearly appressed, apex pointed, bluish-green, scale-like near the rop of branchlets, ovoid, sharply-pointed, with a gland on the back. A rather variable variety, occurring in Southern Europe and according to Carrière 35 (1855), already known in the first century (Pliny).

-- '**Tamariscifolia**', then, is a low, spreading shrub carrying mostly juvenile, pointed leaves in opposite pairs, small and loosely appressed. It has been widely planted in the past, having a happy knack of fitting into the garden planting scheme at several points. Unfortunately, the clone thus widely used in the past has become very prone to juniper blight, caused by a fungus known as *Phomopsis juniperovora* which causes die-back of branches and even the death of the tree. The blight is almost impossible to control even under nursery conditions so the continued planting of this popular nurseryman's standby has become impossible. This has stimulated search for resistant clones, several of which may now be in cultivation. One, '**Tam No Blight**', introduced by Monrovia Nursery, Azusa, U.S.A., about 1970 is much less prone to this trouble, but as young plants they are almost impossible to distinguish between the two clones. The name 'New Blue' under which it was first distributed, and the names 'Tamariscifolia Compacta' and 'Tamariscifolia No Blight' should not be used. (See Illus. No. 229)

-- '**Thomsen**'. Thomsen, Nurs. Cat. **1964**: Extremely compact, very bushy shrub, 10–20 cm high; branches spreading or pprostrate; branchlets thin, but stout, greyish-brown; sprays numerous, 5–6 cm long, whipcord-like, densely feathered, those on the top of the sprays nearly all uniformly rectilinear streaked; leaves scale-like, very closely appressed, very small, dark green. Raised by Thomsen, nurseryman, Skalborg, Denmark.

-- '**Variegata**'. F. G. Hayne Dendrol. Flora, 204

Figure 231. *Juniperus sargentii* 'Glauca'. The normal green form is more prostrate.

(1822); *Juniperus-Sabina sabina* var. *variegata* Weston, Fl. Angl. 19 (1775); *J. sabina* var. *tamariscifolia variegata* Nels. 151 (1866). Dwarf shrub, 1–1.5 m broad, 0.8–1 m high; branches spreading, tips decurving; sprays white-variegated; leaves scale-like, nearly appressed. It makes an interesting small variegated bush which might well be more widely planted.

-- **'Von Ehren'**. L. L. Kumlien in Hill's Book of Evergreens, 225 (1936); An upright, spreading form with light green foliage, with very small, adult leaves. Left to itself it forms a rather loose, open plant with no particular attraction (at least when young), but in America I understand it is in much demand sheared into a dense, globose shrub.

J. sabinioides. (H.B.K.) Nees; *J. monticola* Martinez (1946). Mountain Juniper. A rare, usually shrubby species found in high mountainous regions in Mexico. It is not in cultivation.

J. saltuaria. Rehd. & Wils. 61 (1914), Blackseed Juniper. Monoecious tree, 15 m, very densely branched; crown pyramidal; bark greyish-brown or cinnamonous; branches ascending or spreading; branchlets crowded, short, quadrangular, about 1 mm broad, slightly curved; leaves scale-like, opposite, tightly appressed, 3-angled-ovoid, apex short-acuminate or obtuse, slightly incurved, with an inconspicuous gland, 1–2 mm long, light green, ternate, on terminal branchlets, 2–4 mm long, cones erect, ovoid or nearly globose, about 6 mm across, across, shining black, not bloomy; seeds solitary, ovoid, about 3 mm long, nearly glabrous.

China, in the province Kanzu on the Peling-shan and in northern Shechuan on the mountains north of Sung-pan-Ting at altitudes between 3300–4000 m, forming extensive woods;

Very rare in cultivation. Zone 6.

J. sargentii. Takeda ex Koidzumi (1919); *J. chinensis* var. *Sargentii* Henry Trees Gt. Brit. Ire. **6**: 1432 (1912), NON *J. chinensis sargentii* SENSU Newsum, Dwarfed Tree Manual 74. (1960). NEC Yashiroda, Plants & Gardens, **22,2**: *Front cover illus.* and 45. (1966); NON *J. sargentiae* SENSU Yashiroda, Bonsai Plants. 1,22 and 48, (1960) (and other writers on Bonsai trees). Sargent's juniper is a prostrate shrub discovered by Dr. Sargent on the coast of northern Japan, very useful for ground cover as it is hardy, very accommodating as to soil and situation and remarkably disease free. (The plant so much used by Bonsai addicts is *J. chinensis* 'Shimpaku'.)

It has been at one time or another described as *J. procumbens* and as *J. chinensis* var. *sar-*

gentii, and as always in such circumstances can still be found growing under any name it has once carried. The last mentioned is still recognised by some writers on the coniferae but I follow Van Melle in treating it as a distinct species. It forms a low mat-like, spreading shrub with stout, prostrate main branches and short, ascending branchlets. The leaves are in whorls of three; keeled below, concave with a raised mid-rib above, up to 8 mm long by 2 mm broad and quite free-standing on the main branches, but very much smaller, very colsely set and loosely appressed on the branchlets and growing tips, giving them a cord-like appearance. The foliage has a very unpleasant smell of camphor when bruised, quite unlike any other juniper I know. Zone 4. Colour and habit variations have been selected as follows;

-- **'Compacta'**. *J. chinensis sargentii compacta* Noble, Journ. Calif. Hort. Soc. 1950. Much more compact in growth with juvenile leaves dark green, very glaucous above, adult leaves light green with a dark green margin. I have not found it to be in cultivation.

-- **'Glauca'**. *J. chinensis* var. *sargentii* f. *glauca* Grootn. Mitt. d. d. d. G. **53**: 207. (1940). In this more vigorous form the main branches are ascending at a steep angle and the foliage is blue-green. (See Illus. No. 231)

-- **'Viridis'**. *J. chinensis* var. *sargentii* f. *viridis* Grootn. Mitt. d. d. d. G. **53**: 207. (1940). A form lacking the glaucous bloom and hence appearing greener than the type.

J. saltillensis. (Hall) Silba, Census Conif. 32 (1984). See *J. ashei* var. *saltillensis*.

J. scopulorum. Sargent in Gar. Forest **10**: 420, f. 54 (1897). Western Red Cedar (Grt. Brit.); Rocky Mountain Juniper (Amer.)

Size and habit Rounded bush to 12 m tree, narrowly conical at first, later with rounded crown. Monoecious.

Trunk Frequently divided at base into several trunks. *Bark*: Reddish-brown, shredding into thin strips.

Branches Variously spreading, ascending or drooping; stout; with smooth, peeling bark.

Branchlets Shorter and more densely set than in *J. virginiana*: rounded or ± quadrangular, 1 mm diam.

Winter buds Hidden.

Leaves Scale-like in decussate pairs; tightly appressed but not imbricate; rhombic-ovate, tips acute or acuminate, entire, on the back varying in colour from pale greyish- to dark-green and obscurely glandular, with an umpleasant odour when crushed. Some juvenile (acicular) spiny-pointed, spreading leaves are usually also present.

Fruit A globose berry-like cone 6 mm diam., ripening the 2nd year, dark blue/black but covered with a thick, whitish bloom; the pulp sweetish to the taste.

Seeds Usually 2 ; triangular, reddish-brown, noticeably angular, grooved.

Habitat Rocky Mountains, usually at elevations over 1500 m., from British Columbia and Alberta in Canada to Arizona and Texas, U.S.A.; often in dry & rocky places. Introduced into European cultivation in 1839.

Cultivation This species in its typical form is not outstanding, but, in the past many variants selected for unusual habit or foliage colour have been introduced and named, but of these only a small number have stood the test of time and are still in production. All forms are hardy and tolerant of dry conditions. Zone 3. It is closely allied to *J. virginiana*, but can be distinguished by the somewhat larger fruit, not ripening until the 2nd year, by (in mature trees) the broader, rounded crown with stout, spreading branches, by the shredding bark and (frequently) by a trunk divided near the base. (See Illus. No. 233)

-- **'Admiral'**. *J. scopulorum 'Admiral'* Mount

Figure 232. *Juniperus phoenicea*. The Phoenicean juniper. Detail of foliage.

Arbor Nurs., Cat. 47 **1956**. Broad, pyramidal shrub with grey-green foliage; introduced by the Mount Arbor nurseries, Shenandoah, Iowa, U.S.A., ± 1956.

-- **'Blue Banff'**. A hardy blue form collected near Banff in Alberta, Canada in 5959 by R. Almey and introduced by Lakeshore Tree Farms, Saskatoon, Canada, as 'Blue Banff Juniper' in 1976. It forms a low, spreading plant with prostrate main branches and upright, arching laterals, and carries silvery-blue juvenile foliage. It can be found listed under *J. horizontalis, J. virginiana*, and sometimes carries the simpler name 'Banff'.

-- 'Blue Haven'. A mistake for 'Blue Heaven'.

-- **'Blue Heaven'**. *J. scopulorum 'Blue Heaven'* Plumfield ex Wyman in Amer. Nurs. **117,2**: 12 (1963); *J. scopulorum 'Blue Haven'* Hort. Amer. Neat pyramidal; foliage of a strikingly blue colour in every season; originated at the Plumfield nurseries, Fremont, Nebr., U.S.A. It is one of the most popular clones.

-- **'Chandler's Silver'**. *J. scopulorum 'Chandler's Silver'*. Kumlein, L. L. in Hill's Book of Evergreens 171 (1936.) as the Chandler juniper. Pyramidal, open, loose; branches spreading; branchlets 7–8 cm long, scantily furnished with sprays; leaves awl-shaped, very narrow, nearly appressed, glaucous above, green beneath. According to Kumlien one of the best pyramidal forms, selected in the Black Hills of S. Dakota, U.S.A.

Figure 233. *Juniperus scopulorum*. detail of foliage and fruit.

-- **'Cologreen'**. *J. scopulorum 'Cologreen'* More in the Green Thumb **1956**: 34; *J. scopulorum 'Colorado Green'* Hort. Amer. A medium-sized, columnar tree, regular-shaped with a strong central leader; foliage green all year. Selected in the wild by Marshall, nurseryman at Arlington, Nebr. U.S.A. in 1935. An early selection, but still in production.

-- 'Cupressifolia Erecta'. Now known as 'Greenspire'.

-- **'Cupressifolia Glauca'**. Hetz Nurs. Cat. 6 (1940). Makes a tall, upright-conical tree with ascending branches and dense, somewhat pendulous foliage of rich green with undertones of silvery-blue. It was selected by F. C. Hetz and Sons, Fairview Evergreen Nurs, Fairview, PA., U.S.A. in 1932.

-- **'Dew Drop'**. *J. scopulorum 'Kenyoni'* Monrovia Nurs., Azusa, Calif. 1958. A slow-growing, compact and upright tree, to 4 m, with short branches; sprays crowded, very short; foliage steelblue. Selected from the wild by Kenyon, nurseryman at Dover, Okla., U.S.A.

-- **'Gareei'**. Grootn. Mitt. d. d. d. G. **53**: 209 (1940) is a dwarf, rounded, compact bush reaching to 1.5 m; branches spreading; foliage whitish-blue. Originally selected by a Mr. Garee of Noble, Oklahoma. Wyman records that a 20 year old specimen measured over 3 m across by 1.25 m high.

-- **'Globe'**. Wyman, Amer. Nurs. **117,2**: 13 (1963). A variety of globose habit; to 2 m high in time, with silvery-grey, juvenile foliage.

-- **'Gray Gleam'**. Scott Wilmore, Cat. 5 (1956:) Conical slow-growing small tree; male; foliage distinct silvery grey, becoming more brilliant during the winter and not turning the drab or dull colour so common to other varieties. Originated as a seedling at Wheatridge, Colorado, U.S.A. in 1944. Name registered in 1950 with A.A.N.

-- **'Hillborns Silver Globe'**. *J. scopulorum* 'Hillburn's Silver Globe' Hillier, Man. 49 (1971). A selection that makes a medium sized globose bush with very attractive silvery-grey foliage.

-- **'Kansas Silver'**. Wyman in Amer. Nurs. **117,2**: 13 (1963). Pyramidal form with definite upward growth of the end of the branchlets and with silvery blue foliage. Selected by the Kansas Landscape & Nursery Co., Salina, Kansas, U.S.A.

-- 'Kenyonii'. Now known as 'Dew Drop'.

-- **'Lakewood Globe'**. A selection that lends itself to formal shearing, being naturally compact and somewhat globose. The foliage colour is blue-green.

-- **'Medora'**. A slow-growing form with a neat, dense, fastigiate habit of growth. The foliage is bluish-green. It was named for a town in N. Dakota near which it was found.

-- **'Moonglow'**. A norrowly upright clone, with striking blue foliage, somewhat less severely fastigiate than 'Skyrocket'. A broadly conical form, densely branched; foliage blue. Said to be hardy.

-- **'Pathfinder'**. Kammerer in Bull. Morton Arb. **14**: 39 (1939). This clone forms a small, broadly conical tree. The foliage is dense, but each growing tip is held stiffly outwards, exposing the stomatic bands and giving an attractive silver-blue colour. Selected by L. A. Moffet of the Plumfield Nurseries, Fremont, Nebr., U.S.A. in 1937.

-- **'Repens'**. *J. scopulorum* var. *repens* Hornibrook Dw. Con. Ed. 2, 126 (1939); var. *palmeri* Grootn. Mitt. d. d. d. G. **53**: 209 (1940) 'Palmeri' Den Ouden and Boom, Man. 194 (1965). This is a strong-growing but completely prostrate form found on the hills near Lake Windemere, Kootenay District, British Colombia, Canada. It is a dwarf form with prostrate branches similar to *J. horizontalis* 'Prostrata', branches spreading; branchlets directed forward, prostrate but with tips slightly turned up, leaves awl-shaped, 5 mm long, bluish-green.

-- **'Silver Column'**. A narrow, compact form found in the wild by Dr. M. B. Davis in 1953 in a valley near Smithers in British Columbia, Canada. The combination of colour and shape make it outstanding, and, in the grounds of the Plant Research Station, Ottawa, Ontario, Canada it has proved hardier and more drought resistant than other, similar, clones. Name registered in 1978 with C.O.P.F.

-- **'Skyrocket**. Den Ouden and Boom, Man. 205 (1965); *J. virginiana* 'Skyrocket' Grootn. Nurs. Cat. (1964/65); *J. chinensis* 'Skyrocket' Grootn. Nurs. Cat. 11 (1957); *J. scopulorum* 'Pillaris' Wyman, Amer. Nurs. **117,2**: 13 (1963). This must be one of the most narrowly columnar trees in existence. It can be used with great effect for a number of years in a situation where a strong vertical line is needed in the garden picture. It is widely known and justifiably popular, but does not 'grow old' well. (See Illus. No. 236)

-- **'Springbank'**. A fairly rapid grower with a good, narrowly ascending branch system but a graceful, open texture and extremely thin final branchlets. The foliage is an intense silver-blue. It originated with Mr John Parry of London, Ontario, Canada who gave material to Springbank Park in London, (where the name was suggested, with Mr Parry's consent) and to a nurseryman in Holland, who introduced it to the trade in 1965. See Dendroflora **2**: 48 (1965).

-- **'Tabletop'**. More, Green Thumb, **13,3**: 33 (1956)

Figure 234. *Juniperus squamata* 'Blue Star. A recent arrival that well deserves its popularity.

and picture; 'Table Top Blue' Hort. Amer. 'Table Top' will form a flat-topped shrub to 2 m by 5 m across with wide-spreading main branches, branchlets thin, pale yellow-brown, leaves very small, closely appressed, conspicuously blue. Female plants bear berries which add considerably to the attractiveness of the plant.

-- **'Tolleson's Blue Weeping'**. This remarkable clone, the origin of which I have not been able to trace (Who was Tolleson?) is one of two very similar plants differing only in colour of foliage. This blue form was apparently the earlier selection, so by the law of priority it could claim the name 'Tolleson's Weeping' under which it was first distributed. But the later introduced and similar plant with green foliage began also to be distributed under the same name. In some cases an attempt to lessen the resulting confusion was made by adding the word 'Green' or 'Viridis', but the above name (already in use in the trade) should clear up all uncertainty. This, after all is the only reason for having names, and rules should by our servants, not our masters. 'Tolleson's Blue Weeping', then, is a spectacular weeping tree with arching branches and very thin, thread-like foliage hanging vertically down in bold clusters. Despite the suggested name the colour is rather more silvery-grey than blue.

-- **'Tolleson's Green Weeping'**. This, as the name would suggest, is similar to the last previous cultivar in every respect save colour, which is a soft green. It probably arose as a mutation

Figure 235. *Juniperus virginiana* 'Tripartita'.

Figure 236. *Juniperus scopulorum* 'Skyrocket'. The most columnar form when young, but it tends to lose shape with age.

on the original cultivar.

-- **'Welchii'**. Wyman in Amer. Nurs. **117,2**: 56 (1963). Habit narrow conical to columnar and rather compact; foliage of a greenish-silver colour, more silvery in the young growth. Selected by the Plumfield Nurseries, Fremont, Nebr., U.S.A.

-- **'Wichita Blue'**. Makes a handsome, broadly conical tree of medium size with an ascending branch system and blue foliage which retains its brightness throughout the year.

-- **'Winter Blue'**. A selection made by Clayton Berg, Valley Nurs. Helena, Montana, U.S.A. It is a semi-prostrate, spreading bush with brilliant silver-blue foliage, the colour being maintained well during the winter.

J. semiglobosa. Regel in Acta Hort. Petrop. **6**: 487 (1879). Russian Juniper. A species widespread in central Asia but very rare in cultivation. It makes an upright but pendulous tree 10–12 m high carrying both long-pointed acicular (juvenile) leaves in 3's and scale-like adult leaves, imbricate, closely appressed, glossy green, noticeably glandular on the back. The species is dioecious, the sub-globose fruits 6–7 mm diam. ripening to black, with some glaucous bloom. Afghanistan; U.S.S.R.; China: Kazakhistan to Sinkiang. Zone 4.

J. seravshanica Komarov. Now known as *J. macropoda*.

J. sheppardii (Veitch) Van Melle. Now included in *J. chinensis*.

J. silicicola (Small) Bailey. Now known as *J. virginiana* var. *silicicola*.

J. squamata Buchanan-Hamilton ex Lamb, Genus Pinus **2**: 17 (1824). Flaky Juniper. This is one of a group of junipers from mainland Asia the relationships of which are still imperfectly understood and in need of further study. The long-pointed foliage of var. *fargesii* shews clear affinity with *J. recurva*, but var. *wilsonii* and the popular group of cultivars hitherto included there under (all with shorter, broader and somewhat boat-shaped leaves) are here listed under *J. pingii*. *J. morrisonicola* here preserves its specific independence. For the distinctiveness of *J. procumbens*, see under that species.

Size and habit A spreading shrub or small tree, often lacking a central trunk. *Bark*: Reddish-brown.

Branches Wide-spreading or ascending.

Branchlets Thick, green, often arching over towards the tips.

Winter buds Visible but indistinguishable from the leaves.

Leaves Acicular, in 3s, 4–5 mm long, 1 mm broad, straight, loosely appressed, sharp-pointed; concave above, with a long central groove and two greyish-green stomatic bands; convex above, glaucous green. Old leaves persist as dry brown scales (hence the name *squamata*).

Mature fruits Elliptic, 6–8 mm across, reddish-brown, ripening the 2nd year to purplish black. Each scale carries a small triangular mucro.

Seeds One per cone.

Habitat Afghanistan; India; China; Tibet to Anwhei; Taiwan.

Cultivation The typical form does not appear to be in cultivation, but the following cultivars are widely planted in gardens. Zone 4.

-- 'Blue Alps'. See *J. chinensis* 'Blue Alps'.

-- **'Blue Carpet'**. Van de Laar, in Dendroflora **9**: 47 (1972). This now widely planted cultivar aroseas a mutation on the Jac. Schoemaker Nurs. Boskoop, Holland in about 1962. It forms a large, widely spreading, procumbent plant that is all the better for having its main branches cut back annually for a few years to encourage a well-filled plant, but it should only be planted where it will have plenty of room to develop into a plant several metres across. The foliage is similar in every way, including colour, to 'Meyeri'.

-- **'Blue Spider'**. Van de Laar in Dendroflora **17**: 49 (1980). Another mutation from 'Meyeri', more or less intermediate between 'Blue Carpet' and 'Blue Star'. It was discovered by Joh. Nieuwesteeg, Boskoop, Holland. It is less spreading than 'Blue Carpet' and gradually builds up into a very dense hemi-spherical bush, with elegantly pendulous growth tips. Again, the foliage is similar to that of 'Meyeri', but the blue shade is perhaps a little more intense. The young shoots are orange-brown.

-- **'Blue Star'**. P. de Vogel in Dendroflora **2**: 46 (1965). Although coming third alphabetically, this was the first recorded mutation from the cultivar 'Meyeri'. When shewing me his stock in 1969, Mr. Hoogeveen of Reeuwijk, Holland told me that in a group of stock plants of 'Meyeri', one seemed to be becoming unusually compact and later it produced two very compact shoots. Used as cuttings (this was about 1950) they developed into the new form. The name was suggested by a visitor from Germany as being descriptive of the plant when young. It makes a dense bush, about as wide as high, growth much reduced in every way compared with Meyeri, but of the same lovely blue colour. (See Illus. No. 234)

I have seen several plants shewing some white or yellowish variegation, also said to have arisen as mutations from 'Meyeri'. One such (which I have not seen), has been introduced by the Konijn Nurs. at Reeuwijk, Holland, under the name **'Golden Flame'**. Krüssmann records of it that it has "Yellow spots in the branches" - perhaps falling a little short of a golden flame effect. **'Golden Flash'**, a name listed in New Zealand, presumably originated with a similar mutation.

-- 'Blue Swede' is a superfluous name for 'Hunnetorp'.

-- **'Chinese Silver'**. A very ornamental large shrub or small tree, usually multi-stemmed and of dense habit with loosely held or recurved terminal growths. Leaves are awl-shaped and of an intense silvery-green, and the fruits are ovoid as in the species. It was probably raised from seed collected by T. T. Yu and was for many years listed by Hilliers Nurs, Winchester, U.K. as '*Juniperus sp. Yu 7881*' but it was given this cultivar name by them in about 1977.

-- var. ***fargesii*** Rehder & Wilson ex Sargent, Plant. Wilsonae **2**: 59 (1914); *J. fargesii* Komarov (1924); *J. lemeeana* Leveille & Blinovskij (1914). Tree, 24 m, trunk often divided a few feet above the ground into several erect trunks; bark brown or greyish-brown, fairly smooth, peeling off in longitudinal strips; branches ascending or spreading; branchlets pendulous, tips upright or spreading; leaves awl-shaped, longer and narrower than in the typical form, usually 8 mm long, apex prickly pointed, lurid or bluish-green; cones ovoid, 5–6 mm long, apex blunt, blackish-blue. Grows in woodlands throughout the China-Tibetan borderland. Introduced in 1908 from SW. China into U.S.A. (Arnold Arboretum). In cultivation, but rare. (See Illus. No. 223)

The correct placement of this tree awaits a full investigation based on a wide study of the plants in the wild (and possibly invoking help from other disciplines, such as chemistry) of all the mainland Asian junipers. These species are all widely variable and so, pending clarification at the species level, border-line cases will remain a problem. But since var. *fargesii* has leanings towards *J. recurva*, the misapplication of the name in some collections to plants of *J. pingii* var. *wilsonii* is quite indefensible.

-- **'Filborna'**. This has the same origin as 'Holger' (see below) and was also introduced by the Wansdyke Nursery, Devizes, U.K. It is a much more upright-growing and vigorous plant, and the foliage is green.

-- **'Holger'**. This was at first thought to be a hybrid between *J. squamata* 'Meyeri' and either *Juniperus* × *media* 'Pfitzeriana' or 'Pfitzeriana Aurea', but I treat it here under this species because even if it truly is a hybrid the influence

Figure 237. *Juniperus squamata* 'Holger'. The mother plant at Wansdyke Nursery, devizees, Wilts., U.K. in 1974.

Figure 238. *Juniperus taxifolia* var. *Iuchuensis*. The foliage is soft to the touch, unlike *Juniperus conferta*.

of *J. squamata* is dominant. It was first propagated from cuttings sent by Tage Lundell of Helsingborg to me in October 1968, and was introduced to the trade by Wansdyke Nurs. Devizes, U.K. The name honours Dr. Holger Jensen of Ramlosa Planteskole, Helsingborg, Sweden who made the crosses in 1946 and it was suggested by his son Staffen in a letter to me in January, 1974. 'Holger' makes a low-growing, spreading bush, building up height as it ages. In the winter the blue colour of 'Meyeri' prevails, but the outstanding attraction is that during the spring and summer the young foliage is a clear butter-yellow, giving a beautiful two-tone effect against the glaucous-blue older foliage. 'Yellow Tip' is a superfluous name. (See Illus. No. 237)

-- **'Hunnetorp'**. This also has the same history as 'Holger'. It also is a more vigorous selection, but the foliage is distinctly glaucous, especially at certain times of the year and entirely lacking the yellow-coloured young foliage that is characteristic of 'Holger'. Listed Wansdyke Nurs. Cat. 11 (1976).

-- **'Meyeri'**. *J. squamata* var. *Meyeri* Rehd. Journ. Arn. Arb. **3**: 207 (1922). Meyer Juniper. Bushy, dense shrub, 1–1.5 m ; branches erect or ascending, top slightly decurving; branchlets short, straight; leaves narrow-lenceolate, apex acute, 6–10 mm long, about 1.5 mm broad, striking blue-white above; cones mostly atthe base of branches, very short-stalked, erect, oblong-ovoid, 6 mm long, dark brown, becoming black at maturity, slightly pruinose; seeds ovoid, acute, 2–3 ridged.

Found by F. N. Meyer of the U.S. Department of Agriculture in Tienstin, grown by the Chinese as a potplant and supposed to have come from S. W. Shantung. Although Professor Rehder may have correctly identified this now deservedly poular plant as *J. squamata* (he having the advantage of fruiting material) an attempt to confer botanical status on a plant found, not 'in the wild' but in a Chinese flower pot, is clearly not for us to follow today. 'Meyeri', then, forms a large handsome bush (or if stem-trained as a young plant – a small tree) with strong-growing luxuriant main shoots and short branchlets, gracefully curling over at every growing tip, quite one of the most beautiful of conifers, or indeed of all plants, when well grown. The foliage is dense and of a beautiful blue-grey. Leaves are long, straight and sharp-pointed, 6–10 mm long by 1.5 mm broad.

As this cultivar has the annoying habit found in this species and holds on to its dead leaves for several years it tends to develop a half-dead appearance in time. This can be overcome by regular hard pruning (as is advocated for black currants) to encourage strong young main branches and plenty of new wood.

This name illustrates or throws light on several basic principles of nomenclature that are not always clearly understood. The first is that publishing a new botanical name at any rank that is based only on a description of a plant known only in cultivation is only logically permissible on the assumption that a population exists (or at one time did exist) 'in the wild'.

The second is that should time shew such an assumption to have been unjustified and the plant so described be subsequently considered to be a cultivar, the botanical name falls to the ground, having failed to satisfy the requirements of Article 36 of the *Botanical Code*, a cultivar by definition not a being a taxon.

The third is that, notwithstanding any pronouncement to the contrary, a cultivar can never in real terms be co-extensive with the botanical category under which it is classified. If the essential element of 'selection from within a category' is absent, no cultivar exists: the botanical name suffices for all purposes and any 'cultivar' name is superfluous. The present case *J. squamata* var. *Meyeri* might for many years have been cited to prove the contrary. But the occurrence of mutations giving rise to new cultivars such as 'Blue Carpet', 'Blue Star', etc. (and possibly others to come, all clearly within the botanical category, but fresh cultivars in their own right) shewed otherwise. Even now the possibility of discoveries 'in the wild' of plants within the botanical category but horticulturally inferior to any of the clones in cultivation cannot be ruled out.

-- 'Prostrata'. The transference to *J. pingii* of the one-time *J. squamata* var. *wilsonii*, taking with it the prostrate plant raised from seed by Murray Hornibrook, enables us to clear up another old conundrum. It is most unlikely that either van Melle or Teuscher ever saw Hornibrook's plant and almost equally certain that the plant each of these authorities described and illustrated was a form of *J. procumbens*.

-- var. *wilsonii*. Now known as *J. pingii* var. *wilsonii*.

J. standleyii. Steyermark (1943). A shrubby species from mountainous areas of Mexico and Guatemala. It is not in cultivation.

J. taxifolia. Hooker & Arnott, Bot. Beech. Yoy. 271 (1841). A shrub or small tree 12–15 m. high usually with a twisted trunk. *Bark*: Red-brown. *Branches*: Horizontal, pendulous at the tips, young shoots angular, narrowly winged. *Leaves*: Needle-like, obtuse, bright green, 1–1.3 cm long, 1.3–1.6 mm wide with 2 broad bands of stomata below, keeled above, not decurrent. *Fruit*: Globose, 6–10 mm diam., pale brown with prominent umbos. *Seeds*: 3 per cone. A rare species found only on the Bonin and Ryoku Islands of Japan. Zone 9. It may be in cultivation only as the following:

-- var. ***lutchuensis*** (Koidz.) Satake, Bull. Nat. Sci. Mus. Tokyo **6**: 192 (1962), *J. conferta* var. *maritima* Wils. ex Nakai, Bot. Mag. Tokyo. **36**: 105 (1922) *J. lutchuensis* Koidz. Bot. Mag. Tokyo. **32**: 138 (1918) is a prostrate plant not unlike *J. conferta* but differing from that species by its flatter leaves (less deeply grooved above, and with the mid-rib sometimes evident) its much more prostrate habit and the lack of prickliness to the foliage when brushed, or of pungency when crushed. It is one of the few conifers that will fall down the face of a rock wall like an Aubrieta and with its dense growth and fresh, grass green foliage is a useful ground cover plant and deserves to be much more widely known. (See Illus. No. 238)

J. thurifera L. 1039 (1753); *J. hispanica* Mill. no 13 (1768). Spanish Juniper. Dioecious shrub or tree, 15 m; crown narrowly conical or rounded trunk attaining a large girth; branches widely spreading; branchlets nearly quadrangular, scarcely 1 mm thick, the whole plant having a strong musty scent; leaves greyish-green, on young plants awl-shaped, opposite, with a white band above, green beneath; in adult stage scale-like, opposite, tightly appressed, apex free, back rhomboid-convex, keeled and with a resin gland; cones on short branchlets, upright, solitary, globose, composed of 4–6 mucronate scales, 7–11 mm across, at first blue, afterwards brown to nearly black, glaucous bloomy; seeds 2 or 3, rarely 4, ovate to globose, flat or nearly triangular, brown.

N. Algeria; Spain; S.E. France. Despite its early introduction it is almost unknown in Britain today where it is not reliably hardy. Zone 9.

J. tibetica Komar. in Notul. Syst. Herb. Hort. Petrop. **5**: 27 (1924). Tibetan Juniper. Tree, 30 m; crown very dense; branchlets erect, short; leaves almost triangular, oblong to almost lanceolate, short, yellowish-green, glandular at back; cones to 15 mm long, shining reddish-brown; seeds to 4, globose, 8–10 mm across, tip obtuse, quadrangular, sides pitted and tubercled.

Eastern Tibet and Kansu, W. China, at an elevation of 3500–4000 m. Zone 6.

J. turkestanica Komarov (1924). Now known as *J. pseudsabina var turkestanica.*

J. virginiana L. 1039 (1753); Pencil Cedar (Gr. Brit.); Eastern Red Cedar (Amer.).

Size and habit Slow-growing tree, 12–30 m, usually less in cultivation, with a narrowly conical crown, soon becoming irregular, and broadening in old age.

Trunk Often buttressed and fluted at base. *Bark*: Grey or reddish-brown, peeling off in long, narrow strips spiraling round the trunk.

Branches On young trees, usually ascending; on old trees spreading horizontally or sometimes pendulous.

Branchlets Irregularly arranged round the shoot; pale- or dark-grey; inconspicuously quadrangular, noticeably thin (less than 1 mm diam).

Winter buds Small, visible but hidden by the topmost leaves.

Leaves Predominantly adult (scale-like) in decussate pairs, rhomboic, acute, 1.5 mm long: imbricate, thickened but free at tips; dull dark green with a glandular depression. Some juvenile leaves in pairs (rarely in 3s); 3–6 mm long; convex and grey-green below; concave and glaucous above (except at the margins) are always present, but on old trees only in small tufts at the base and/or tips of the sprays. Mitchell 127 (1972) states that the adult foliage gives off an aroma of soap or paint if crushed.

Flowers Usually, but not invariably, dioecious.

Mature fruits Ovoid, 5–7 mm diam, ripening by the first autumn to brownish-violet with a pronounced glaucous bloom.

Seeds 1–2, ovoid, small, apex blunt-angular.

Habitat S.E. Canada; E. United States: S.W. Maine to N. Florida. Zone 3.

Cultivation The species is almost always represented in cultivation by one of the numerous cultivars that have been selected.

J. virginiana is closely related to *J. scopulorum*, being the equivalent in eastern North America of that species in the west of the continent. It is clearly distinguishable from *J. scopulorum* in its flowering parts and in its habit of ripening its seeds the first year, but in the dwarf forms where we are usually restricted to foliar characteristics, the two species are often difficult to separate. From *J. sabina* forms it is readily distinguishable by the absence of the true savin odour and taste to the foliage. *J. chinensis* is distinct by its much larger 5–7 mm fruit (when present), by the coarser (i.e., thicker) growth, by the awl-like leaves being, usually in threes (seldom in pairs) and more widely occurring over the tree, and by the scent of the crushed foliage, which in *J. chinensis* is sour and resin-like.

-- **'Blue Cloud'** *J.* × *media* 'Blue Cloud' Hillier, Dw. Con. 36 (1964). This is an alleged hybrid raised from seed from *Juniperus virginiana* 'Glauca' with *J.* × *media* 'Pfitzeriana' as the pollen parent. It forms an interesting broad bush of similar parentage to 'Grey Owl' (below) in which the glaucous grey foliage is very dense in the interior of the bush, but long, twisted, whiskery young shoots thrown outwards at all angles give an untidy but picturesque look to the plant. It deserves to be better known than it is.

-- **'Boskoop Purple'**. F. J. Grootend., Cat. 1963: Similar to 'Hillii', but much faster growing; it is a narrow form purplish-brown in winter. Originated in 1960 as a sport of 'Hillii' in the nursery of F. J. Grootendorst & Sons, Boskoop, Netherlands.

-- **'Burkii'**. Bobbink & Atkins ex Slavin 109 (1932). Burk Red Cedar. Narrow-fastigiate: branches ascending, dense; branchlets crowded, regularly set, short; leaves acicular and scale-like, the former gradually finely pointed, sharp, not prickly, with a dull blue band above, margin narrow, green, green beneath, steel-blue in autumn. One of the best blue forms of the species. Common in cultivation.

-- **'Canaertii'** Canaert ex Sénécl. 99 (1868); *J. virginiana* 'Cannarti Beissn. 125 (1891). Compact, pyramidal, slender, dense; branches spreading, stout, the tops protruding out of the compact mass of secundary branches; branchlets stout; leaves on young branchlets scale-like, on older ones awl-shaped, dark green; cones small, grape-like, whitish blue bloomy, usually profusely produced, strongly contrasting with the dark green leaves.

Named after canaert d'Hamale, Mechelen, Belgium. Very common in cultivation.

-- **'Carolina'**. Van der Laar in Dendroflora 20: 71 (1983). A squat shrub with ascending shoots and grey-green adult foliage throughout save at the base of the plant. It was imported into Holland from Denmark by J. Fase & Zoon Nursery, Boskoop, under the inadmissible name 'Globosa'.

-- **'Chamberlaynii'**. *J. virginiana Chamberlainii* Knight, Syn. Conif. 12 (1850); *Chamberlaynii* Carrière, A6 (1855); NON Hillier, Dw. Con. 41 (1964) NEC Welch, Dwarf Conif. 211 (1966). This is a vigorous tree with stout main branches and all terminal growths strongly pendulous; leaves all acicular and mostly appressed, sharp-pointed, glaucous; scale-like leaves closely appressed, dark green, glandular. Hillier misapplied the name to a dwarf plant correctly known as 'Pendula Nana' and I regret having added to the confusion over this cultivar by writing of it in an earlier book as a prostrate dwarf form. I have been unable to trace whose name was thus honoured nor how he himself spelled it, but 'Chamberlaynii' was early described as a tall, pendulous tree. I have seen several such specimens in collections in France. It is a female clone.

-- var. ***crebra***. Fern. & Griscom in Rhodora **37**: 133, t. 332 (1935). Narrowly conical to columnar in growth; branches ascending; leaves narrow ovate, acute, loosely appressed; seeds slightly pitted. It distinguishes the slender type found in north-eastern U.S. from the broader type found in the south Most or all of the cultivated varieties are selections from this variety.

-- **'Cupressifolia'**. Kammerer in Bull. Morton Arb. 7: 32 (1932), not Kumlien (1946). Pyramidal, rather loose; leaves cypresslike, soft yellow-green. The 'Cupressifolia' described by Kumlien is now known as 'Hillspire'.

-- **'Deforest Green'**. Wyman in Amer. Nurs. **117,2**: 60 (1963). A narrowly conical tree like 'Canaertii' but faster growing and deeper green.

-- **'Elegantissima'**. *J. virginiana elegantissima* Hochst., Conif. 92 (1882); *J. virginiana* 'Elegans' Nicholson 214 (1885). Bushy pyramidal; branchlets somewhat pendulous, tips golden yellow; leaves scale-like and awl-shaped, yellowish-green. Common in cultivation.

-- **'Fastigiata'**. Hesse, Cat. 1933/34. Columnar, narrow; branches ascending, at the end many-topped; branchlets erect; sprays very thin, bluish-green, at the base of branchlets crowded, short, cord-like, erect; leaves scale-like, very small, pointed, tightly appressed, bluish-green. Raised and distributed by Herm. A. Hesse, nurseryman, Weener on Ems, Germany.

-- **'Fiore'**. Named in honour of the Fiore Nurs. Ill. U.S.A. who sent scions to Jack Hill, Dendee, Ill. in 1954. These were grafted by Mr. A. Thomsen of Thomsens Planteskole, Skalborg, Denmark, who took two of the grafts back to Denmark and later introduced it under this name. He writes "It is very similar in appearance to 'Fairview' but it shapes better and is much easier to propagate".

-- **'Glauca'**. Carr. 45 (1855). Columnar, narrow, dense; branches spreading; branchlets thin, round, blue-bloomy; leaves scale-like, appressed, small, distinctly glaucous with some awl-shaped leaves inside the plant. Very common in cultivation.

-- **'Globosa'**. Hort. ex Beissner, Handb. Nadelh. 126 (1891). This makes a truly globular plant, being probably one of the most naturally globose conifers we have, even large plants having the appearance of having had some artificial help in keeping their neat shape.

A specimen fifty years old in the Arnold Arboretum, Jamaica Plain, Massachusetts, U.S.A., is only 4.5 m high and was an excellent rounded and densely branched specimen. Except in small plants the foliage is mainly adult with small leaves and fine twigs densely set, the leaves being green. It is one of the most attractive dwarf conifers we have and has been in cultivation many years.

-- **'Grey Owl'**. F. J. Grootendorst & Sons, Cat. 14 (1949/50). In growth similar to *J.* × *media* 'Pfitzeriana', but less obliquely spreading; branches at first spreading, later nearly horizontal with decurving tips; branchlets rather thin; leaves small, appressed, soft silvery grey. Raised from seed of 'Glauca', from a plant standing at a short distance from *J.* × *media* 'Pfitzeriana', and which originated in 1938 in the nursery of Caam Bros, Oudenbosch, Netherlands. Distributed by F. J. Grootendorst & Sons, Boskoop in 1949.

-- **'Hetzii'**. Fairview Nurs. Cat. 4 (1937), as *J. chinensis glauca hetzii*, but it is is nearer to *J. virginiana*, possibly a hybrid. Hetz Blue Juniper. A strong-growing shrub to 1.5 m tall and wide; free branching, sprays thin; leaves predominantly scale-like, conspicuously glaucous. Widely used as an understock. **'Sulphur Spray'** is a colour mutation aptly named.

-- **'Hillii'**. D. Hill, Cat. 8 (1944); D. Hill ex Kuml. 170 (1946); *J. virginiana* pyramidiformis hillii D. Hill, Evergr. 17 (1922). Hill's Dundee or Dundee Juniper. Dense, columnar, slow-growing, 2–4 m; branches and branchlets spreading, short; leaves awl-shaped, 5–8 mm long, with a rather broad, bluish white band above, greenish-blue beneath, the whole foliage conspicuously purplish in winter. Raised in ± 1916 by D. Hill Nursery Co., Dundee, Ill. U.S.A.

-- 'Horizontalis'. Hornb. Dw. Con. ed. 2, 12 (1939); NON horizontalis glauca Kumlien, Friendly Evergreens, 183 (1948) and colour plate. This is an unsatisfactory name; any plants I have seen [see Hillier, Dw. Con. 41 (1964)] or received under this name turning out always to be a strong-growing form of *J. horizontalis*. One thing is certain: Hornibrook's description was not of the cultivar 'Horizontalis Glauca; [Kumlien l.c.], a form found many years ago by Professor Sargent on an exposed cliff site along the coast of Maine and introduced by D. Hill Nursery Co., which has very stiff branches, thick, rugged foliage and a bright bluish colour. The name 'Horizontalis' should be dropped as a case of mistaken identification by Hornibrook. (See also 'Reptans' below).

-- **'Humilis'**. *J. virginica humilis* Loddiges, Nurs. Cat. 42 (1826); *J. virginiana humilis*. Sén. Conif. 100 (1867); 'Repens' Nisbet ex Hillier, Dw. Con. 41 (1964) (?); 'Monstrosa' Welch, Dw. Con. 213 (1966) and Hort. Eng. (in part).] This is an extremely diminutive plant which forms a low, spreading bush with very small, acicular leaves borne in quite dense tufts, from which very thin growing shoots stand out horizontally in every direction. It is dark green in colour, the young growth turning dull mauve-pink in the winter. It deserves to be better known.

-- **'Kobold'**. Den Ouden & Boom, 202 (1965). Dwarf, globular and dense; branches thin, stout; branchlets upward to spreading, 1–1.5 cm long, thin; leaves densely set, 0.5–1 cm long, 0.5 mm broad, bluish above, green beneath. Raised from seed by N. Th. Bosman, nurseryman, Boskoop, Netherlands, in 1952. In 1958 the plant was 15 cm high and 12 cm wide.

-- **'Kosteri'**. Beissn. 466 (1884); M. Koster & Son, Cat. **14**: 23 (1877); *J. virginiana* f. *kosteriana* Beissn. 126 (1891); *J. chinensis kosteriana* Boom. 99 (1942). Dwarf, bushy, 6–7 m in diam; branches procumbent or prostrate, broadly feathered, shoots 0.8–1 m long, nearly upright and spreading in different directions; branchlets on these shoots much shorter; leaves of the second season loosely appressed, slightly convex at back, sharp-pointed, greyish-blue, slightly steel-blue near the top of the branches and shoots, assuming a purplish tint in winter. Sometimes confused in commerce with *J.* × *media* 'Pfitzeriana' but makes a lower and wider-spreading plant. Raised by the firm M. Koster & Son, Boskoop, Netherlands, about 1870.

-- **'Manhattan Blue'**. Scott ex Wyman in Amer. Nurs. **117,2**: 60 (1963). A compact, conical, male tree differing from 'Glauca' by the bluish green foliage. Introduced by R. Scott, nurseryman in Manhattan, Kansas, U.S.A.

-- 'Moonglow'. See under *J. scopulorum*.

-- **'Nana Compacta'**. de Vos, Handb. 242 (1887); Beissn. Handb. ed. 2, 594 (1909); 'Tripartita Nana' Nisbet ex Hillier, Dw. Con. 42 (1964). Dwarf, globose shrub, irregular in outline, bushy, 0.5–0.8 m high, less compact than in 'Globosa' and more irregularly growing; branches spreading; branchlets crowded, short, densely plume-shaped; leaves scale-like and awl-shaped, the latter 2–4 mm long, slightly greenish-blue above, green beneath, purplish in winter. Raised by C. G. Overeynder, nurseryman, Boskoop, Netherlands.

-- **'Pendula'**. Lawson, Fir Tribe 77 (1852), Carrière 45 (1855), Gordon, Pin. Suppl. 36 (1852). A rather unsatisfactory name covering all pendulous forms, of which Gordon distinguished three (including 'Chamberlaynii' and 'Pendula Viridis') in 1852 and there are probably even more clones in cultivation by this time. New cultivar names should be found for any identifiable clone that are in cultivation and worth preserving.

-- **'Pendula Nana'**. Hillier, Nurs. Cat. 194 (1928) and earlier?; 'Chamberlaynii' Hillier, Dw. Con. 41 (1964) and Man. 493 (1971) is a very small, slow-growing form with a strongly pendulous habit so that, as wrongly described by Mr. Hillier under the cultivar name 'Chamberlaynii', it makes an almost prostrate grey-green mat unless trained up in its early days. In the Pygmy Pinetum I had a small plant that was so trained to a height of 30 cm or so and planted where it was flowing down over the face of the rockery.

-- **'Pseudocupressus'**. Kammerer in Bull. Morton Arb. **7**: 30 (1932), den Ouden 182 (1949). Columnar and narrow; branches hugging the stem closely, tapering at the top to form a slender point; leaves awl-shaped appressed at the top of the branchlets, regularly spreading to the base and 6 mm long there, apex sharp-pointed, upper side concave, light green, underside keeled, bluish-green; the colour of the new growth is quite pronounced.

-- 'Pyramidalis'. Carr. 47 (1867). Useful now only as a collective name for the pyramidal-growing cultivars.

-- **'Sherwoodii'** Sherwood Nurs., Cat. 1943: Amer. Nurseryman **88,2**: 41 (1948). Conical, narrow; leaves dark green, similar to cv. '*Canaertii*'; tips of the long slender shoots creamy in spring, solid bright green in summer, very dark plum-coloured in winter. Originated as a seedling in the Sherwood Nursery Company, Portland, Oregon, U.S.A., in 1935.

-- **'Schottii'**. Gord. 157 (1875). Pyramidal and narrow; branches ascending; branchlets very thin, yellowish-green; leaves scale-like, tightly appressed, light green, sometimes slightly yellowish-green. Known in cultivations as early as 1855 in Britain. Very common in cultivation.

-- var. ***silicicola*** (Small) Silba, Census Conif. 37 (1984); *J. salicicola* (Small) Bailey, Cult. Conif. 197 (1933). A variant that is found in coastal areas and swampy dunes at the extreme southerly limit of the species, by some botanists regarded as a separate species, is here (following Silba), listed as a botanical variety. It differs only in its more pendulous growth and slender foliage and in the shape of its cones.

-- **'Silver Spreader'**. *J. virginiana* 'Dammii' Monrovia Nurs. Cat. (1955) *J. virginiana prostrata* 'Silver Spreader' Monrovia Nurs. Cat. 12(1963/64). A widely spreading and rather low form with silvery grey foliage. The Monrovia Nurseries, Azusa, Calif., purchased it from a nursery as a seedling ± 1954. An excellent ground-cover plant where a low-planting effect is required.

-- 'Sky Rocket'. See *J. scopulorum*.

-- **'Triomphe d'Angers'**. de Vos in Sempervirens **3**: 229 (1874); Beissn. 127 (1891). Conical, slow-growing, scanty but dense; branches and branchlets thin, regularly borne; sprays very regular in length, short, white-blotched at tips and with the branchlets themselves as it were sprinkled with white, contrasting well with the dark bluish-green foliage; leaves small, likewise white blotched. Rather tender.

-- **'Tripartita'**. (de Vos) Longone ex Sénéclauze 101 (1868); as *J. tripartita* de Vos 165 (1867). This is a strong-growing form which makes a large, spreading bush 2–2.5 m across by nearly as much high. It usually throws several stiff main branches from near the ground level and bears a mixture of adult and juvenile foliage which is green, turning to dull purple in winter. It is not in my opinion a particularly attractive form, but is useful where a fairly large and wide-spreading plant is required, being roughly the equivalent of *J.* × *media* 'Pfitzeriana', but it is more upright and of a stiffer habit of growth, lacking the graceful, nodding tips of that variety. (See Illus. No. 235)

J. wallichiana. Hooker f. & Thomas ex Parlatore in DC, Prodromus **16** (2); 482 (1868). Tree to 20 m, or sometimes a low, spreading bush. *Bark*: Fibrous, peeling away, fissured, showing orange beneath. *Leaves*: On strong-growing shoots – awl-like (juvenile) in 3s, 5 mm long by 1 mm broad: on weak shoots – scale-like, decussate, ovate, 2–2.5 mm long by 1.25 mm broad. Spring-flowering. Himalayas at high elevations. Although introduced by Sir W. J. Hooker into Britain in 1849 it is rarely seen outside botanical gardens, but it makes an interesting, healthy-looking, medium-size tree which could well be made more use of in gardens. Zone 6.

KETELEERIA

Pinaceae

Keteleeria Carrière in Rev. Hort. 449 1866: A genus, very rare in cultivation, allied to *ABIES*.

Size and habit Evergreen trees, fastigiate when young, later becoming irregular in outline and flat-topped in maturity.

Trunk Often buttressed. *Bark*: On old trees rough, corky and irregularly fissured.

Branches Massive and horizontal.

Branchlets Shoot dull, at first light brown and usually densely pubescent, becoming grey-brown.

Winter buds Ovoid or globose; dark brown, 2–3 mm. *Scales*: Numerous, with a conspicuous, persistent sheath at the base of the shoot; not resinous.

Leaves Spirally set but on side shoots twisted as in *Abies* into two ranks; set at a wide angle to the shoot persisting 5 or more years and leaving a circular scar on falling; sharply pointed 5–6 cm long, 0.5 cm wide on vigorously growing trees, smaller on old trees; hard, green, lustrous, with a prominent raised mid-rib above; pale green with two indistinct bands of stomata and a shiny mid-rib below.

Flowers Monoecious. The female flowers are terminal, on short, leafy shoots; the male flowers in clusters of 5–10, emerging from a scaly bud.

Mature cones Ripening in 1 year; held erect as in *Abies*; not disintegrating on ripening but falling from the tree intact. Scales broad, woody.

Seeds 2 on each scale; large, glandular. *Wing*: Long, hatchet-shaped

Habitat: China, Taiwan and south into Vietnam and Laos. Zone 8.

Cultivation Since the whole genus is only marginally hardy in temperate regions it is very rare in cultivation. It requires a hotter and drier situation than suits *Abies*. Propagation is by seed or by cuttings.

The genus shews considerable variation throughout such an extensive habitat. These variations have not been thoroughly studied, so botanists are not agreed as to the number of species to recognise. The Chinese botanists accept 10, but here I follow Silba [*Census Conif.* 38 (1984)] and the more conservative view and list two species.

K. davidiana. (Bertrand). Beissn. 424 (1891); *Pseudotsuga davidiana* Bertrand in Bull. Soc. Philom. d. Paris **9**: 38 (1872); *K. esquirolii* Léveillé in Fedde, Rep. **8**: 60 (1910); *Abies davidiana* Franchet, Pl. David. **1**: 288 (1884). In this species the cone scales are about 25 mm long, 18–24 mm wide, widest above a claw-like stalk, narrowing to an irregular-shaped top with the apex free or reflexed, exposing the seed-wing to view. The juvenile form of foliage is retained for many years. It is very rare in Britain.

K. fortunei. (Murr.) Carr. in Rev. Hort. 449 **1866**; *Picea fortunei* Murr. in Proc. Hort. Soc. London 421 **1862**. Apart from minor foliar distinctions this differs from *K. davidiana* only in the cone scales, which are smaller and rounded at the tip, and during growth, tightly appressed. The seed and wing also are smaller. No examples are on record in Britain.

LAGAROSTROBUS (Hooker f.) Quinn in Austral. Journ. Bot. **30**: 316 (1982). A new monotypic genus suggested by Quin to replace *Dacrydium franklinii*. Not accepted by De Laubenfels [Flora Malesiana (1984–5)] or Silba 1984.

LARIX

Pinaceae

Larix Miller, Gard, Dict. (Abridg. ed. 4, 2 (1754). Larch.

The Larches form a genus containing about a dozen species of deciduous trees widely distributed throughout the temperate zone in the Northern Hemisphere. They are valuable producers of timber so are of much greater importance in forestry than in horticulture. A great deal of work has been, and is being, done in forestry to improve yields by the development of seed orchards raised from selected ('Plus') mother trees. This work is outside the scope of this book and the genus has not been put to much good use in horticulture. A few clones have been described, but larches have been little planted for ornament. This is surprising, since the young foliage is of an exquisite soft shade of green in early Spring and takes on rich Autumn colouring. It makes an attractive small specimen tree with an interesting branching pattern in winter, and a larch tree is always a favourite with the birds.

A larch is one of the easiest of genera to recognize, but the species are closely related and the presence of hybrids taking after both their parents can make further identification difficult. Foliage is of two kinds; the short, spur-like shoots on second year and older wood bear relatively small, slender leaves in clusters of 12–50 in early Spring. New extension shoots, produced from May onwards have large leaves spirally set. This is characteristic also of *Cedrus* but there need be no uncertainty – is summer the foliage of larch is soft, unlike the harsh feel of cedar leaves and in winter

Figure 239. *Larix decidua*. European larch. A young tree in summer. (5)

Figure 240. *Larix decidua*. European larch. A tree in winter. (10)

Figure 241. *Larix decidua*. Bark on trunk. (10)

the bareness of the deciduous larch is conclusive. Also all larch cones are small and quite unlike the large, barrel-like cones of the cedars.

L. chinensis Beissner 1896. Not now distinguished from *L. potaninii*.

L. dahurica Lawson 1836. Now known as *L. gmelinii*.

L. decidua, Mill. no. 1. (1768); *L. europaea* DC. in Lam. & DC., Fl. France ed. 3 **3**: 277 (1805). Common or European Larch.

Size and habit Monoecious deciduous trees, 25–45 m, the crown on young trees conical. Older trees are broader, with heavy, buttressed branches and often an irregularly shaped crown. Many of the characteristics of growth, habit, cone size, etc, differ according to the provenance, so can only be ascertained from a good forestry manual.

Trunk Straight, 1–1,5 m in girth. *Bark*: On young trees smooth, on old trees grey or brown and fissured, shredding in small plates.

Branches On young trees, dense, short and compact; on old trees spreading, upturned or drooping.

Branchlets Drooping, thin, grey or yellowish, glabrous, furrowed. Short spurs dark brown, nearly black, marked with as many rings as they are old, rings downy.

Winter buds On main shoots ovoid, short-pointed; scales pointed, slightly resinous; lateral buds

Figure 242. *Larix kaempferi*. Japanese larch. (10)

Figure 243. *Larix occidentalis*. Western larch. Foliage and cone. (5)

buds shorter and blunter and those on short spurs small, rounded; golden- to chestnut-brown.

Leaves On growth shoots – single, varying in size up to 6–7 cm long, keeled below, broad and bluntly pointed; on the spurs – in clusters of 30–40, 1.5–3.5 cm long, narrower and blunter. All leaves turn to yellow in Autumn.

Flowers Male strobili stalkless, rounded, yellow, 5–10 mm long, with bract margin fringed, ripening in April and May.

Mature cones Ovoid, to 4 cm long, 2 cm broad, on a twisted stem, light brown ripening to grey the first year. *Scales*: Round, loosely appressed, entire above, striated, margin sometimes wavy, outer surface soft-brown; downy.

Seeds About 4.5 mm long, greyish. *Wing*: mostly 12.5 mm long, 8–9 mm wide, reaching to the top of the scales.

Habitat S.E. France; Switzerland; N. Italy; S. Germany; Austria.

Cultivation The European larch has been in cultivation for timber production over many centuries but has been replaced by the Japanese Larch *L. kaempferi* or the even more profitable hybrid *L.* × *eurolepsis*. For ornamental planting this species will tolerate poor soil conditions but not air pollution of shade. Zone 2. (See Illus. Nos. 239, 240, 241, 249).

Several geographical variants have been recognised botanically, but these are more akin to the 'provenances' of the foresters. Den Ouden and Boom devote two pages to the listing of minor variants to be found 'in the literature', mostly with descriptive Latin-form names. Where these were

Figure 244. *Larix × eurolepsis.* Dunkeld Hybrid larch. Detail of foliage and cone. (5)

not colour variations that turn up from time to time they were peculiarities of habit, leaf or cone, in some cases mere freaks, and none seem to have survived. Re-typification in such circumstances is undesirable: fresh selections based on the same peculiarities should be given new cultivar names.

-- '**Corley**'. Hillier [Man. 494. (1971)] a slow-growing, small, leaderless bush of rounded habit. It was found as a witch's broom by the late R. S. Corley of High Wycombe, Buckinghamshire and introduced to the trade by Hillier's Nursery, Winchester U.K.

-- '**Fastigiata**'. (Sénécl.) K. Koch 259 (1873); *L. europaea fastigiata stricta* Originated with A. Sénéclauze, nurseryman at Bourg-Argental, France.

-- '**Pendula**'. (Laws.) Henk. & Hochst. 134 (1865); *L. europaea pendula* Lawson 386 (1836). Weeping European larch. A tree, becoming tall and

Figure 245. *Larix laricina*. Tamarack. Detail of foliage. (5)

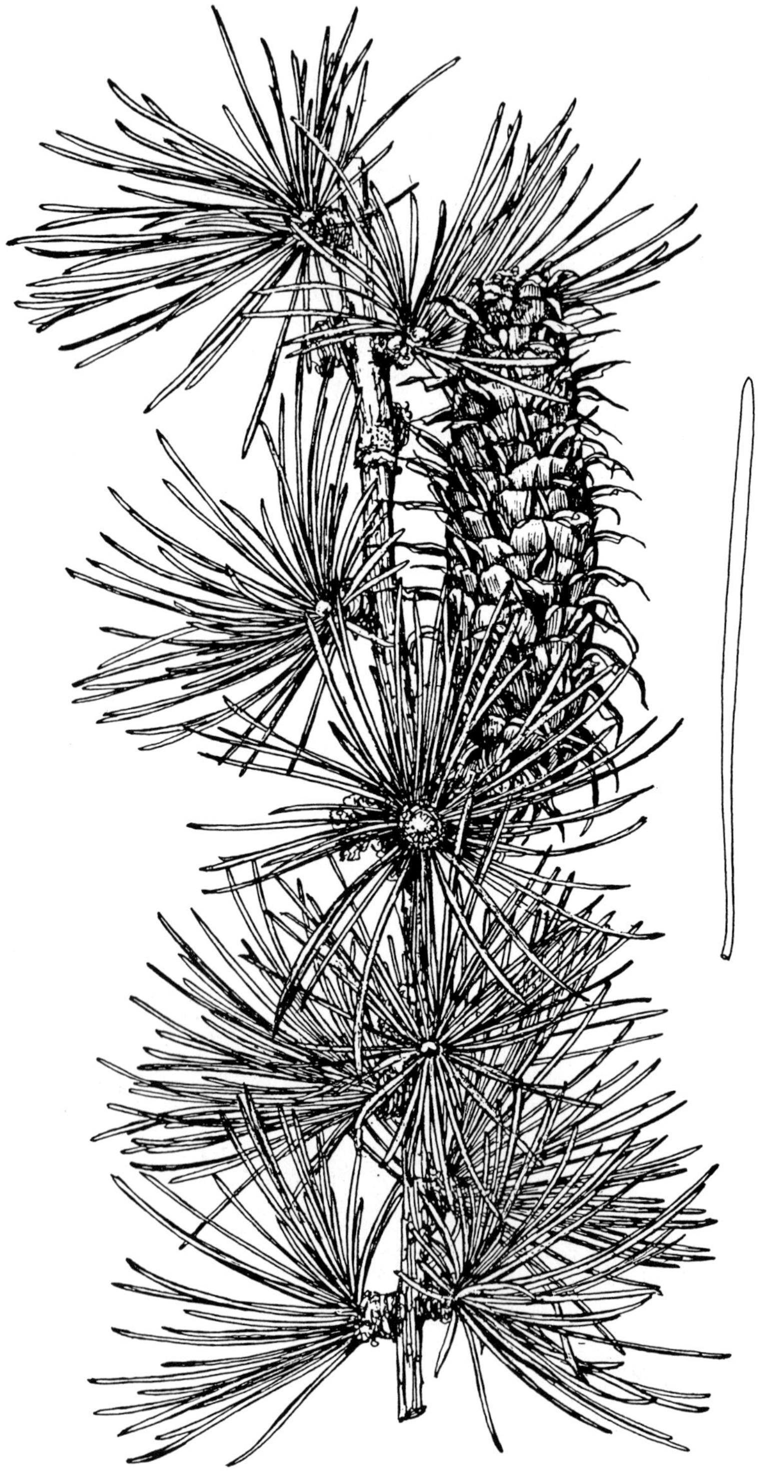

Figure 246. *Larix griffithii*. Sikkim larch. Foliage and cone. (5)

Figure 247. *Larix kaempferi.* Japanese larch. Detail of foliage and cone.(5)

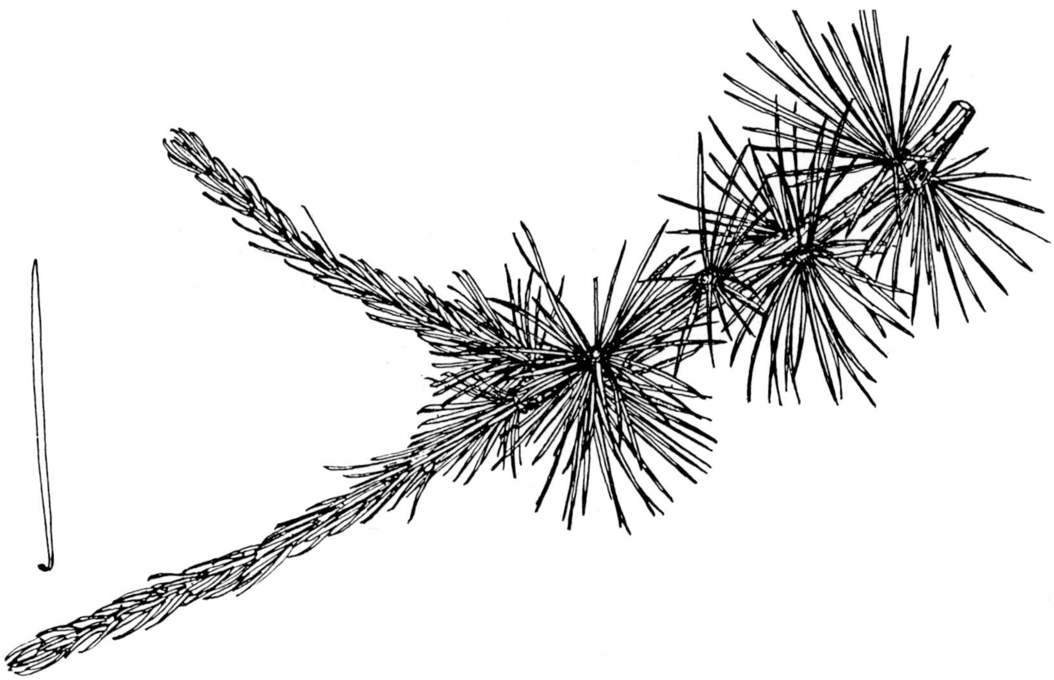

Figure 248. *Larix potaninii.* Chinese larch. Detail of foliage.

irregular in habit, stem with deeply arching branches. Noticeably pendulous specimens of the European larch occasionally turn up, and more than one has found its way into production. Mitchell [134 (1972)] records a specimen of this cultivar in the Botanical Garden at Cambridge, U.K. as being "an irregular umbrella-shaped tree with a tangle of twisting branches above the graft union."

Larches being variable in this respect, more than one species has its own cv. Pendula. There is also the hybrid species *L.* × *pendula*, so confusion can arise.

-- var. ***polinica***'. (Racib.) Ostenf. & Syrach-Larsen in Pflanzenareale **2**: 63 (1930); *L. polonica* Racib in Österr. Bot. Zeit 346, (**1912**). Branchlets slender, pendent; cones truncate like those of *L. sibirica* but not so pointed, light coloured and conspicuously pubescent when young, ripe green; scales more rounded and less emarginate at the free margins than those of *L. decidua*, and often pilose on the outer side. An intermediate form between the European and the West-Sibirian larch, and which appears to be not far from extinction. Only found in small grooves in select forests in Poland and Roumania.

L. × ***eurokurilensis***. Rohm & Dimpflm. in Zeitschr. Forst genetik **1**: 16 (1952) = *L. decidua* × *L. gmelinii* var. *japonica*. Tree, vigorous in growth; lower branches spreading with upward-curving tops, those in upper part of the crown even more ascending; young branchlets yellowish with a reddish tint, glabrous; cones short-stalked, 3 cm long; scales spreading at an angle of 50–70°.

A hybrid which originated spontaneously in the experimental garden at Grafrath, Germany.

L. × ***eurolepis***. Henry in Gard. Chron. 4 (1919) *L. decidua* × *L. kaempferi*. Dunkeld Larch. The name was derived from European larch and *L. leptolepis*, the then accepted name for *L. kaempferi*.

This cross was first induced, probably with intent, on the Duke of Atholl's estate at Dunkeld in Scotland at around the turn of the century and has since been repeated in cultivation many times. It was a big break-through for foresters because the hybrid is more vigorous and satisfactory for them in every respect, but it raises problems of identification because the hybrids are always intermediate between the (not all that dissimilar) parents and each tree may incline to either parent. The hybrid differs from *L. decidua* in its wider (to 2 mm) leaves, dark green above, contrasting with two grey stomatic bands on the underside and in its slightly reflexed cone-scales, and from *L. kaempferi* it differs by its less glaucous, yellow, less pruinose twigs, its shorter leaves with less noticeably white stomatic bands beneath, and in the large size and tall shape of the cones, with the scales upright or only slightly bent outwards. (See Illus. Nos. 244, 251)

L. europaea D.C. ex Lambert (1805). Now known as *L. decidua*.

L. gmelinii. (Rupr.) Kuzeneva in Trav. Mus. Bot. Ac. Sci. Russ. **18**: 41 (1920); *Abies gmelinii* Rupr., Beitr. Pflanzenk. Russ. Reich **2**: 56 (1845). *L. dahurica. (Loud.) Turcz. ex Trautv., Pl. Mag. Fl. Russ. 48 (1846); L. amurensis* Hort. Dahurian Larch. A conical tree 20–30 m, but in cultivation often misshapen and irregular and merely a shrub in the Arctic Circle and on high mountains. *Bark*: Dark reddish-brown, somewhat scaly. *Branches*: Widespreading, with pendulous ultimate growths. *Buds*: Roundish, not resinous, scales free at tips. *Leaves*: 2–4 cm long, thin, slender, very bright green, keeled and with distinct greyish-green lines of stomata below often curved. *Cones*: Often plentiful, ovoid, set on stalks, 20–30 mm long, with 20–40 nearly flat scales, polished, pale brown. Rare in cultivation. *Habitat*: N.E. Asia, from E. Siberia, through Manchuria into Kamchatka, forming large and extensive forests. Introduced in 1827 into Britain. Zone 2.

This E. Asian larch varies considerably in the degree of hairiness of the young shoots, the length of the leaves and the size of the cones. Several distinct forms have been described under specific names, but they are now included as geographical varieties.

-- var. ***japonica***. (Maxim. ex Regel) Pilger 327

(1926); *L. kamtchatika* (Rupr.) Carr. 279 (1855); *L. dahurica japonica* Maxim. ex Regel in Gartenfl. **20**: 105 (1871); *L. kurilensis* Mayr 66 (1890); *L. dahurica* var. *kurilensis* (Mayr) Sarg., Silva N. Am. **12**: 4 (1898). Kurile Larch. Branchlets horizontal spreading, reddish-brown or violet, pubescent when young, often pruinose, spurs very thick; leaves to 2 cm, shorter than in the typical form, and with brighter stomatic bands beneath. Cones also are smaller, to 2.5 cm.

Sachalin, Kurile islands. Cultivated in Japan and found there by Dr. Maxomowicz in a temple-garden in 1861 and introduced by Dr. Mayr into Europe in 1888, but still very rare in Britain.

-- var. ***olgensis***. (Henry) Ostenf. & Syrach-Larsen in Pflanzenareale **2**: 62 (1930); *L. olgensis* Henry in Gard. Chron. **1**: 109 (1915); *L. dahurica* var. *koreana* Nakai, Atlas Ill. geogr. Distr. Korea Pl. **1**: 1 (1939). Olga Bay Larch. Branchlets densely reddish-brown pubescent; leaves more or less sickle-shaped, 10–20 mm long; cones 15–20 mm long, scales 25–30, the upper side covered with short, dense down; bracts producing beyond the scales; seeds about 2,5 mm long.

E. Siberia near Olga and Vladimar Bays about 200 km N.E. of Vladivostock, growing under very cold conditions. Introduced from Olga Bay into Britain by Capt. Clinton Baker in 1911, but Mitchell [136 (1972)] records only one tree in the U.K.

-- var. ***principis-rupprechtii***. (Mayr) Pilg. 327 (1926); *L. principis-rupprechtii* Mayr 309 (1906); *L. dahurica* var. *principis-rupprechtii* (Mayr) R. & W. 21 (1914). Branchlets glabrous, bright reddish-brown; leaves sickle-shaped, 5–10 mm long, rather broad; cones 3–4 cm long; scales 30–40, flat in expanded cones; bracts exserted. Korea, Manchuria and the extreme north-eastern corner of China. Introduced from Korea by Dr. Mayr who found a tree in a temple-garden in Watai-shan in 1903. This, unlike the foregoing varieties, is an improvement on the type since it makes a strong-growing, vigorous open tree, well represented in the best collections in the U.K. Although winter-hardy it comes into leaf early so often suffers from early frosts.

-- var. ***prostrata***'. *L. gmelinii* var. *prostrata* (Regel) Hornibr. 132, (1939); *L. dahurica prostrata* Regel in Gartenfl. **20**.: 105 (1871). Dwarf; branches spreading over the ground. According to Regel the creeping growth is due to its occurrence in cold and swampy valleys and on the highest peaks of the Dahurian alps. Occurs in Dahuria and East Asia but does not appear to be in cultivation in Europe.

L. griffithii. Hooker, f. Himalayan Journ. **2**: 44 (1854); *Abies griffithiana* Hook. f. ex Ldl. & Gord. in Journ. Hort. Sco. **5**: 214 (1850). Himalaya Larch. *Tree*: 20 m; branches widely spreading; *Branchlets*: pendulous, downy, reddish-brown the second year; winter-buds conical, not resinous, downy. *Leaves*: 25–35 mm long, blunt, bright green, keeled below, stomata on both surfaces, but not prominent beneath; *cones*: cylindrical, 7–10 cm long, 2.5–3.5 cm across, violet-purple during growth: *Scales* about 25 mm long and broad; bract-apex abruptly acuminate, exserted, reflexed: *Seeds* oblong, 4–5 mm long; wing about 8.5 mm long.

E. Nepal, Sikkim and Bhutan at elevations of 2500–4800 m, being commonest at 2800–3400 m, and is also found in the Chumbi Valley, Tibet; Zone 6. Introduced form Nepal by Sir Joseph Hooker into England (Kew Gardens) in 1848. In Bhutan it grows either together with *Pinus griffithii* or in pure forests above it, or in smal groups with an undergrowth of Rhododendron. In inhabits deep valleys, but prefers dry, rocky, ancient moraines and grassy slopes where drainage is good. Only hardy in the milder parts of Britain: it does not succeed either at Kew or in the National Pinetum at Bedgebury, but there is a large tree at Coldrennick in Cornwall and another at Strete Raleigh near Exeter. (See Illus. No. 246).

L. kaempferi (Lambert) Carrière Fl. des Serres **11**: 97 (1856); *L. leptolepis* (Siebold & Zuccarini) Gordon, Pin. 128 (1858). Japanese Larch.

Size and habit Tree, to 30 m with a dense crown; habit variable between narrowly conical with

Figure 249. *Larix decidua*. European larch. Cones. (10)

short or ascending branches and a widespreading tree with long, horizontal branches.

Trunk Stout. *Bark*: On young trees reddish, later becoming brown; smooth with shallow furrows, or scaly and peeling off in strips.

Branches Stout, usually horizontal.

Branches Very numerous, clothing the trunk and branches; sometimesw pubescent; yellowish, becoming reddish the second year. Spur shoots stout, dark brown.

Winter buds Rounded, oblong or conical and pointed; resinous; with bright, brown fringed scales.

Leaves On leading shoots – up to 6 cm long, 0.5–1 cm broad; or 3.5–4 cm long in clusters of 40 or more. Flat and soft; glaucous; dark green above and with two wide greenish-grey bands of stomata beneath.

Flowers Male flowers dark red-brown in clusters on weak shoots. Female flowers widely distributed on second-year or older wood.

Mature cones Ovoid or bun-shaped, 3 cm by 3 cm, on short stalks. *Scales*: Numerous, about 10 mm across, rounded, the upper edge rolled back with the tip curled under, giving the ripe cone a rosette-like appearance; ripening to dark brown.

Seeds Ovoid, 3-4 mm, long, 2 mm broad; light brown. *Wing*: Semi-ovate, as long as the scale; glossy dark brown.

Figure 250. *Larix kaempferi*. Cones. (1)

Figure 251. *Larix* × *eurolepsis*. Dunkeld Hybrid larch. Cones. (10)

Habitat Japan, island of Hondo on slopes of volcanic mountains at 1200–1800 m.

Cultivation The Japanese larch is one of the two most important species, not only as being one of the parents of the now widely-planted hybrid *L.* × *eurolepis* but because it is more tolerant than the European larch of unfavourable soil conditions and exposure. It is, however, susceptible to drought. A number of clones have been selected and named, but few have survived in cultivation. (See Illus. Nos. 242, 247, 250).

'Blue Haze'. A selection with attractively coloured foliage, introduced by Hillier & Son, Winchester, U.K.

'Blue Rabbit'. Den Ouden/Boom, 216 (1965). A narrowly conical form with conspicuously glaucous foliage. Found in the Tempelhof nurseries (L. Konijn & Co), Reeuwijk, Boskoop, Netherlands, about 1960.

-- **'Cruwys Morchard'**. A very slow growing form with an interesting columnar habit. The main branches are upright and extend by only 35 mm tips. Leaves are 20 mm long, a glaucous green. This cultivar was propagated from a witch's broom found by Gordon Haddow in a roadside planting near Tiverton, Devon U.K. and was introduced by Kenwith Castle Nurs, Bideford in the same county.

-- **'Dervaes'**. *L. leptolepis pendula Dervaes*, (1949) 190; Dervaes ex den Ouden, 190 (1949);*L. leptolepis pendula* C. de Coninck-Dervaes, Cat. 27. (1946); Branches horizontal; branchlets drooping leaves fresh-green. Raised by C. de Coninck-Dervaes, nurseryman, Maldeghem, Belgium.

-- **'Diane'**. This interesting form was raised from a tree found by J. Bohlje of Westerstede when out hunting, and named by him for the Greek goddess of hunting. It has a normal growth rate, making a small tree in ten years. The twigs are thin, and twisted as in *Corylus avellana* 'Contorta'.

-- **'Kenwith'**. A very slow-growing, globose bush raised from seed on the Kenwith Castle Nursery at Bideford, Devon, which has an ascending branch-system on which all laterals turn upwards at the tip. The leaves are 10 mm long, with a cluster of tiny leaves around the dark brown, globose, 4 mm diam buds.

-- **'Minor'**. *L. leptolepis* var. *minor* Murr. in Proc. Hort. Soc. London **2**: 637 (1862); *L. leptolepis dumosa* Beissn. in Mitt. d. d. d. G. **11**: 73 (1902). Globular bush; branches ascending from the ground. Since these bush-forming seedling variants have appeared more than once, this name will cover them unless and until a form appears that is worthy of a cultivar name. This species is prone to witches' brooms but a number of attempts to propagate these by grafting have not succeeded in fixing the dwarfism. The following are known to me to be in current cultivation. These seem reasonably stable, but no doubt individual plants may revert in the end.

These dwarf forms have the beautiful, grey-green spring foliage characteristic of the species and their dense, picturesque branch pattern and bright red twigs are added attractions during the winter.

-- **'Nana'**. Welch, 247 (1979); *L. decidua* 'Nana' Hillier, List Conif. 8 (1970) (Nom.) This was received by me from the Hillier Nursery, Winchester, U.K., but whether it had earlier appeared in any of that firm's catalogues I cannot say, so the legitimacy of the name is uncertain. The plant is a very dainty little bush, very close jointed and slow-growing, and, since most of the lateral buds develop into shoots, the growth is very congested and compact. The leaves are in dense clusters 8–10 mm long.

-- **'Varley'**. Welch, 247 (1979). This is another attractive form propagated from a witch's broom found by M.J. Varley on a plant of Japanese larch near Ormskirk, Lancashire, in 1968. The tree was about 8 m high and the broom about 2 m form the ground and very near the trunk. It was introduced to the trade by the Wansdyke Nursery, Devizes, Wiltshire.

The annual growth is about half that of 'Nana' (above) and as the leaves are longer

(10–15 mm long–occasionally more) the result is a very dense little bush indeed.

L. laricina. (Duroi) K. Koch. 263, (1873); *Pinus laricina* Duroi, Obs. Bot. 49 (1771); *L. americana* Michx, Fl. Bor. Am. **2**: 203 (1803); *L. alaskensis* Wight in Smiths. Misc. coll. **50**: 174 (1907). Tamarack; Eastern Larch.

Size and habit Tree to 20 m, slender and conical when young, later developing a rounded, often asymmetrical, head. *Bark*: Pinkish-orange to dark red, smooth when young, on older trees silver-grey, flaking off in small circular patches.

Trunk Straight and regular.

Branches Horizontal, short.

Branchlets Shoots numerous, slender, (noticeably so because of the appressed leaves), curling in all directions; at first glabrous with a grey bloom later becoming glossy and pale brown.

Winter buds Rounded, glossy dark red, slightly resinous.

Leaves On growing shoots: – 25–30 mm long, 0.5 mm broad, strongly keeled, with 2 broad pale bands on both sides, giving the foliage a bluish cast; on spur shoots, in bundles of 12–30: – very slender and thin, dark green, with stomata on the underside only.

Flowers Small and usually very numerous. Female flowers red.

Mature cones Ovoid, very small but noticeable because of the bright colour in summer, 13–20 mm long, 8-12 mm broad; violet-red, ripening to yellowish-brown; with 15–20 rounded scales with an erect margin curving inwards. Bracts hidden, narrow and much shorter than the scale.

Seeds Oblong-ovoid, long, brown. Wing 5 mm long, blunt-ended.

Habitat N. America, form the Arctic Circle in Alaska and Canada southwards to N. Pennsylvania, Minnesota and Illinois, reaching an altitude of about 1200 m; Introduced into Britain in 1739.

Cultivation In nature it is frequent in swamps. Being the only larch that tolerates such conditions it should be useful for planting in such situations, but is rarely seen in U.K. Of several variants listed in the past, none has been traced in cultivation. Zone 2. (See Illus. No. 245).

L. leptolepis. Now known as *L. kaempferi*.

L. lyallii. Parlatore in Enum. Sem. Hort. Bot. Florent. 259 (1863). A mountain species native in British Columbia and Alberta in Canada and in Washington, U.S.A. that has never succeeded in cultivation in either America or Europe. Its main distinctive feature is dense grey-or greyish-brown pubescence on the first-year shoots. Zone 3.

L. × ***marschlinsii*** (*L. leptolepis* × *L. russica*) Coaz in Schweiz. Zeitschr. Forstw. **68**: 12 (1917). Differs from *L. leptolepis* in its glabrous branchlets and its larger, more ovoid cones, with not recurved scales and exserted bracts. Originated from seed on *L. leptolepis* probably having been crossed with *L. russica* in the nursery of Tsarnerholz near Moret, Switzerland. According to Coaz, 16 years old trees were 8 m high. In view of the uncertainty of the parentage this name cannot properly be regarded as having priority over *L.* × *eurolepis*.

L. mastersiana Rehd. & Wils. 19, (1914). Masters Larch. Differs from *L. griffithii* by its glabrous branchlets, its pale green, puberulous leaves keeled on both surfaces and its smaller cones (± 3 cm long) with long acuminate bracts. Very local in W. Szechuan, Central China. Introduced by E. H. Wilson into the U.S.A. (Arnold Arboretum) in 1908. Zone 7.

L. occidentalis. Nuttall, N. Amer. Silva **3**: 143 (1849). Western Larch.

Size and habit Narrowly conical tree to 50 m. less in cultivation.

Trunk Columnar, 4–5 m in girth. *Bark*: On young trees thin, scaly, dark grey or brown; on older trees, cinnamon-red and deeply furrowed.

Branches Short, horizontal, noticeably in whorls.

Branchlets Light orange-brown at first with long, scattered hairs in the grooves but soon becoming glabrous; the spurs are chetnut-brown.

Winter buds Round, dark reddish-brown, slightly resinous.

Leaves On growth shoots: – 25–40 mm long; spreading; apex sharp-pointed; stiff; greyish-green with two prominent white stomatic bands beneath; on spurs: – to 3.5 mm long in clusters of 40–50, bright glossy green, sometimes with white stomata on one side.

Mature cones Cylindrical, 25–35 mm long, 25–30 mm broad; purple-brown. Scales numerous 8–12 mm wide, tops reflexed. Bracts long-pointed, projecting well above the scales.

Seeds Obovate, 6–9 mm long, nearly white. *Wing*: 12–18 mm long, pale brown.

Habitat N. America, from Br. Columbia to Montana, between the Rocky Mountains and Blue Mountains of Oregon, at altitude ranging from 750–1800 m.

Cultivation Discovered by D. Douglas in 1826 at the Kittle Falls on the Columbia River and sent by Dr. Sargent in 1881 to the Arnold Arboretum, U.S.A. and R.B.G., Kew, U.K. Several of the original trees still survive at Kew but it is still rare in U.K. Zone 7. (See Illus. No. 243).

L.* × *pendula (*L. decidua* × *L. laricina*) (Sol.) Salisb. in Trans. Linn. Soc. **8**: 314 (1807); *Pinus pendula* Sol. in Ait. **3**: 369, (1789). Weeping Larch. A tree of mysterious origin, differing from both its putative parents in one way or another. It forms a conical tree with bark and foliage resembling that of *L. laricina* but with pendulous terminal shoots resembling a pendulous specimen of *L. decidua*. The large cones also more closely resemble the latter species, but are purple in summer, ripening to grey-brown. The hybrid origin of *L. pendula* is criticized by various authors; the origin is still quite uncertain; it is not known in the wild, but seems to come true from seed. It should not be confused with cultivars of other species having the name 'Pendula'. **'Repens'** is probably a cultivariant, a plant propagated from a pendulous terminal shoot and unable to develop a leader.

L. potaninii Batal. in Act. Hort. Petrop. **8**: 385 (1893). Chinese Larch. Closely related to *L. lyallii* but differing by the less pubescent, shining orange branchlets and the not recurved cone-scales. South northern China to the border of Tibet in altitutes of 2500–3000 m, Zone V. It is a strikingly handsome tree and is the common larch of W. Shechuan. At lower altitudes it is scattered in moist woods, occurring more especially by the sides of streams, with other conifers and broad-leaved trees; it becomes more abundant as it ascends and forms forests in the higher more alpine regions. It is considered as the most valuable coniferous timber in W. China; introduced in 1904 by E. H. Wilson into England (Veitch nurseries). It is still very rare in U.K. (See Illus. No. 248)

L. russica. (Endlicher) Sabine ex Trautvetter in Act. Petrop. **9**: 212 (1884); *L. sibirica* Ledebour. Tree, 40 m, pyramidal, slender; branches on young trees erect, on old trees horizontal; branchlets shining light yellow; winter-buds slightly resinous; leaves 20–40 mm long, thin, soft, apex sharp-pointed, under surface keeled, stomatic lines inconspicuous; cones conical, narrowed near the apex, 3–4 cm long; scales about 30, larger and more widely set than those of *L. decidua*, rather thick, leathery margin undulated, incurved, closely appressed when half mature, slightly breaking up when mature; bracts hidden, those of the lowest rows of scales slightly visible; seeds 6–8 mm long, wing 12–15 mm long.

N.E. Russia and W. Siberia, where it has a very wide distribution; Zone II. Introduced from Archangel (N. Russia) into Britain by the Duke of Atholl in 1806, but it flushes early and regularly suffers from frost so has almost died out.

L. sibirica. Now known as *L. russica*.

L. speciosa Cheng & Fu 1925. Burma Larch. A little known species near *L. griffithii*. Ledebour 1833.

LEPIDOTHAMNUS Philippi in Linnea **30**: 730 (1860). A proposal by Quinn in Austral. Journ. Bot. **30**: 316 (1982) to restore recognition of this old generic name for the Dacrydium species *D. fonkii, D. intermedia* and *D. laxifolium* was not accepted by de Laubenfels [Flora Males. (1984–5)] or Silba (1984), so is not followed here.

LIBOCEDRUS
Cupressaceae

LIBOCEDRUS Endlicher, Synop. Conif. 42 (1847). This genus as established by Endlicher in 1847 comprised a rather assorted collection of 13 species. In 1930 Florin isolated one of them *L. uvifera*, creating for it a new monotypic genus *Pilgerodendron*. In 1953 the botanist Li transferred three more species to his new tropical genus *Papuacedrus*, at the same time transferring three more species to yet another new genus, *Heyderia* (which name, under the inexorable botanical rule of priority later had to give way to *Calocedrus*.) This latest inroad reduced the genus *Libocedrus* to six species, after which it might have expected to be left in peace, but one year later Florin and Boutelje isolated *L. chilensis* in yet another monotypic genus *Austrocedrus*.

This leaves *Libocedrus* to cover five species, two from New Zealand and three from New Caledonia, all with flattened foliage on opposite branchlets in fan-like sprays, of which only *L. bidwillii* and *L. plumosa*, both from New Zealand, are even marginally hardy anywhere in Britain. The taxonomy follows Rushforth 1987.

L. arfakensis (Gibbs) Li. Now known as *Papuacedrus arfakensis*.

L. austro-caledonica Brogniart & Grisebach (1841). Not hardy.

L. bidwillii Hooker f. (1867). Pahautea. A conical tree to 20 m, or a shrub at high altitudes. Adult leaves are triangular, acute, depressed, in decussate pairs, the lateral pair longer and decurrent. Found in both islands in New Zealand, but rare in cultivation. Zone 8.

L. chevalieri Buchholz (1949). Not hardy.

L. chilensis (D. Don) Endlicher 44 (1847). Now known as *Austrocedrus chilensis*

L. decurrens Torrey, Smithson Inst. Contrib. Knowl. 5(1): 7. pl. 3 (1946). Now known as *Calocedrus decurrens.*

L. formosana Florin (1930). Now known as *Calocedrus formosana*.

L. macrolepis Bentham and Hooker (1880). Now known as *Calocedrus macrolepis*.

L. papuana F. Mueller (1889). Now known as *Papuacedrus papuana*.

L. plumosa (D. Don) Sargent (1896). Kawaka. A species close to *L. bidwillii*; differing only in details of foliage and cones. Equally rare in cultivation.

L. torricellensis Schlecter & Lauterbach. (1913). Now known as *Papuacedrus papuana*.

L. uvifera Pilger (1926). Now known as *Pilgerodendron uviferum*.

L. yateensis. Guillamin. (1949). Not hardy.

METASEQUOIA
Taxodiaceae

METASEQUOIA. S. Miki in Jap. Journ. Bot. **9**: 261 (1941). This is a monotypic genus, a monoecious, deciduous tree allied to *Taxodium* and *Glyptostrobus*, differing therefrom by the deciduous shoots, leaf buds, leaves and cone scales being in opposite pairs, not spirally arranged.

Figure 252. *Metasequoia glyptostroboides*. Dawn redwood. A vigorous young tree in summer. (10)

Figure 253. *Metasequoia glyptostroboides*. Dawn redwood. Detail of foliage. (10)

Figure 254. *Metasequoia glyptostroboides*. Detail of buds. (10)

M. glyptostroboides. Hu & Cheng in Bull. Fan. Mem. Inst. of Biol. Peiping (new series), **1** (2): 154 (1948). Dawn Redwood. A tree long known in a fossilised state but only discovered as a living tree in 1941. It makes a narrowly columnar tree, broadening with age, with fresh-green, soft, fern-like foliage and a good autumn colour in most situations and most years. In cultivation it has proved to be hardy and fast-growing, especially in damp, and even boggy, conditions. It varies in the foliage colour, and in habit from broadly conical to almost columnar, and several selections have been made on these grounds. It propagates readily from cuttings and one mutation has been reported in a rooted cutting. (See Illus. Nos. 252–255)

-- **'National'**. Wyman in Arnoldia **23** (7) 1963. This is a clone that was selected in the seed-bed at the U.S. National Arboretum, Washington, DC., in 1958. Apart from having a narrowly-pyramidal and compact habit it is normal to the species.

-- **'Sheridan Spire'** is a narrow, almost columnar form, densely branched, which originated as a sport that appeared in a row of cuttings on the Sherwood Nursery, Oakville, Ontario, Canada in 1960. It was first listed in their 1976 catalogue.

Other names listed by Kruessmann in 1983 are 'Moerheim' and 'Vada' (both stated to have originated in Holland) and 'Emerald Feathers'.

MICROBIOTA

Cupressaceae

MICROBIOTA. Komarov in Not. Syst. Herb. Petrop. **4**: 180 (1923) is a monotypic genus with a restricted habitat in eastern U.S.S.R.

Figure 255. *Metasequoia glyptostroboides*. Bark on main trunk. (10)

Figure 256. *Phyllocladus alpinus*. Detail of foliage. (5)

M. decussata. Komarov l.c. is a quick-growing, dioecious, prostrate shrub that has recently become popular for planting as ground cover. The leaves are in flat, frondose sprays, with the lateral and facing pairs similar as in *Thuja*; acute with incurved but free tips; mid-green in summer but turning an unattractive brown in winter. The cones are very small, globose with 2–4 scales, but carry only a single fertile seed.

MICROCACHRYS

Podocarpaceae

MICROCACHRYS. Hooker f. ex Hooker in Lond. Journ. Bot. **4**: 149 (1845) is another monotypic genus from Tasmania.

M. tetragona. Hooker l.c. A low, spreading bush with long, prostrate branches and scale-like leaves spirally set in 4 ranks (hence the name). The cones are ovoid, 8–10 mm long with rounded scales which become fleshy and bright scarlet as they ripen. It has no garden value but grown in a pot and suitably restrained by pruning it makes an attractive 'show-plant' because of its scarlet fruits.

MICROSTROBUS

Podocarpaceae

MICROSTROBUS. Garden & Johnson in Contrib. N.S.W. Nat. Herb. **1**: 316 (1915). A genus of two shrubby species with restricted habitats, one in Tasmania and one in New South Wales.

It was formerly known as *Pherosphaera*. In each species plants are usually found under shelving rocks at the base of waterfalls, but this seems to be more the result of the ravages of fire than the requirements of the plants themselves.

M. fitzgeraldii. (F. Mueller) Garden & Johnson, l.c. A densely-branched, scrambling shrub. Leaves dark olive-green, imbricated, obtuse, keeled, about 3 mm long, incurved but not tightly appressed. Cones small, with 4–8 scales.

M. niphophilus. Garden & Johnson, l.c.; *Pherosphaera hookeriana* Archer & Hooker, Kew Journal **2**: 52 (1850). A densely branched erect shrub to 2 m. Leaves closely imbricate, decussate, thick, very obtuse, keeled, more densely set than in *M. fitzgeraldii.*

NEOCALLITROPSIS pancheri. (Carrière) De Laubenfels in Flora Nouv. Caled. **4**: 161 (1972). This is the sole occupant of a tropical genus emanating from New Caledonia.

PAPUACEDRUS

Cupressaceae

PAPUACEDRUS. Li in Journ. Arn. Arb. **34**: 25 (1953). A genus of two species allied to *Libocedrus* but separated therefrom on rather minor distinctions in foliage and flowering parts. Neither species is hardy in cool-temperate regions.

P. arfakensis. (Gibbs) Li in Journ. Arn. Arb. **34**: 23 (1953). A tree differing from the following only in minor respects.

P. papuana. (F. Mueller) Li in Trans. Roy. Soc. Vict. new series, **1**: (1) 32 (1889). A tree to 50 m in New Guinea, with foliage similar to *Fokienia*, but doubtfully hardy in cool temperate regions.

PARASITAXUS ustus. (Viellard) De Laubenfels in Flora Nov. Caled. **4**: 44 (1972). The sole occupant of a tropical genus found only in New Caledonia and the only parasitic conifer. It is an endangered species.

PHYLLOCLADUS

Podocarpaceae

PHYLLOCLADUS. L. C. and A. Richard, Comment. Bot. Conif. Cycad. 129 (1826). A curious genus with flattened branchlets (phylloclades) acting as leaves. It contains five species, of which only one is hardy in U.K.

P. alpinus. Hooker f., Flora Novae-Zeland **1**: 235 (1853). Mountain Celery Pine. A tree, usually no more than a bush in cultivation, with the leaf-like cladodes that functions as leaves looking as though made by a small child from modelling clay. It is more interesting than beautiful. (See Illus. No. 256)

Other species: ***P. asplenifolius, P. glaucous, P. hypophyllus*** and ***P. trichomanoides*** are too tender for cool-temperate regions.

Index to Volume One

Key to typography

Names of Genera are printed in ***BOLD ITALICS***, other botanical names and page references to illustrations are printed in *Italics*. Page references preceded by DO refer to *Manual of Dwarf Conifers* by Den Ouden and Boom (1965). Page references preceded by W refer to *Manual of Dwarf Conifers* by Humphrey Welch (1979).

FORESTRY SCIENCES

1. P. Baas (ed.): *New Perspectives in Wood Anatomy*. Published on the Occasion of the 50th Anniversary of the International Association of Wood Anatomists. 1982 ISBN 90-247-2526-7

2. C.F.L. Prins (ed.): *Production, Marketing and Use of Finger-Jointed Sawnwood*. Proceedings of an International Seminar Organized by the Timber Committee of the UNECE (Halmar, Norway, 1980). 1982 ISBN 90-247-2569-0

3. R.A.A. Oldeman (ed.): *Tropical Hardwood Utilization*. Practice and Prospects. 1982 ISBN 90-247-2581-X

4. P. den Ouden, (in collaboration with B.K. Boom): *Manual of Cultivated Conifers*. Hardy in the Cold- and Warm-Temperate Zone. 3rd ed., 1982 ISBN Hb 90-247-2148-2; Pb 90-247-2644-1

5. J.M. Bonga and D.J. Durzan (eds.): *Tissue Culture in Forestry*. 1982 ISBN 90-247-2660-3

6. T. Satoo: *Forest Biomass*. Rev. ed. by H.A.I. Madgwick. 1982 ISBN 90-247-2710-3

7. Tran Van Nao (ed.): *Forest Fire Prevention and Control*. Proceedings of an International Seminar Organized by the Timber Committee of the UNECE (Warsaw, Poland, 1981). 1982 ISBN 90-247-3050-3

8. J.J. Douglas: *A Re-Appraisal of Forestry Development in Developing Countries*. 1983 ISBN 90-247-2830-4

9. J.C. Gordon and C.T. Wheeler (eds.): *Biological Nitrogen Fixation in forest Ecosystems*. Foundations and Applications. 1983 ISBN 90-247-2849-5

10. M. Németh: *Virus, Mycoplasma and Rickettsia Diseases of Fruit Trees*. Rev. (English) ed., 1986 ISBN 90-247-2868-1

11. M.L. Duryea and T.D. Landis (eds.): *Forest Nursery Manual*. Production of Bareroot Seedlings. (1984) 2nd printing 1987 ISBN Hb 90-247-2913-0; Pb 90-247-2914-9

12. F.C. Hummel: *Forest Policy*. A Contribution to Resource Development. 1984 ISBN 90-247-2883-5

13. P.D. Manion (ed.): *Scleroderris Canker of Conifers*. Proceedings of an International Symposium on Scleroderris Canker of Conifers (Syracuse, USA, 1983). 1984 ISBN 90-247-2912-2

14. M.L. Duryea and G.N. Brown (eds.): *Seedling Physiology and Reforestation Success*. Proceedings of the Physiology Working Group Technical Session, Society of American Foresters National Convention (Portland, Oregon, USA, 1983). 1984 ISBN 90-247-2949-1

15. K.A.G. Staaf and N.A. Wiksten (eds.): *Tree Harvesting Techniques*. 1984 ISBN 90-247-2994-7

16. J.D. Boyd: *Biophysical Control of Microfibril Orientation in Plant Cell Walls.* Aquatic and Terrestrial Plants Including Trees. 1985 ISBN 90-247-3101-1

17. W.P.K. Findlay (ed.): *Preservation of Timber in the Tropics.* 1985 ISBN 90-247-3112-7

18. I. Samset: *Winch and Cable Systems.* 1985 ISBN 90-247-3205-0

19. R.A. Leary: *Interaction Theory in Forest Ecology and Management.* 1985 ISBN 90-247-3220-4

20. S.P. Gessel (ed.): *Forest Site and Productivity.* 1986 ISBN 90-247-3284-0

21. T.C. Hennessey, P.M. Dougherty, S.V. Kossuth and J.D. Johnson (eds.): *Stress Physiology and Forest Productivity.* Proceedings of the Physiology Working Group, Technical Session, Society of American Foresters National Convention (Fort Collins, Colorado, USA, 1985). 1986 ISBN 90-247-3359-6

22. K.R. Shepherd: *Plantation Silviculture.* 1986 ISBN 90-247-3379-0

23. S. Sohlberg and V.E. Sokolov (eds.): *Practical Application of Remote Sensing in Forestry.* Proceedings of a Seminar on the Practical Application of Remote Sensing in Forestry (Jönköping, Sweden 1985). 1986 ISBN 90-247-3392-8

24. J.M. Bonga and D.J. Durzan (eds.): *Cell and Tissue Culure in Forestry.* Volume 1: General Principles and Biotechnology. 1987 ISBN 90-247-3430-4

25. J.M. Bonga and D.J. Durzan (eds.): *Cell and Tissue Culure in Forestry.* Volume 2: Specific Principles and Methods: Growth and Development. 1987 ISBN 90-247-3431-2

26. J.M. Bonga and D.J. Durzan (eds.): *Cell and Tissue Culure in Forestry.* Volume 3: Case Histories: Gymnosperms, Angiosperms and Palms. 1987 ISBN 90-247-3432-0

    Set ISBN (Volumes 24-26) 90-247-3433-9

27. E.G. Richards (ed.): *Forestry and the Forest Industries: Past and Future.* Major Developments in the Forest and Forest Industries Sector Since 1947 in Europe, the USSR and North America. In Commemoration of the 40th Anniver-sary of the Timber Committee of the UNECE.1987 ISBN 90-247-3592-0

28. S.V. Kossuth and S.D. Ross (eds.): *Hormonal Control of Tree Growth.* Proceedings of the Physiology Working Group, Technical Session, Society of American Foresters National Convention (Birmingham, Alabama, USA, 1986). 1987 ISBN 90-247-3621-8

29. U. Sundberg and C.R. Silversides: *Operational Efficiency in Forestry.* Vol. 1: Analysis. 1988 ISBN 90-247-3683-8

FORESTRY SCIENCES

30. M.R. Ahuja (ed.): *Somatic Cell Genetics of Woody Plants.* Proceedings of the IUFRO Working Party S2. 04-07 Somatic Cell Genetics (Grosshansdorf, Germany, 1987). 1988 ISBN 90-247-3728-1

31. P.K.R. Nair (ed.): *Agroforestry Systems in the Tropics.* 1989 ISBN 90-247-3790-7

32. C.R. Silversides and U. Sundberg: *Operational Efficiency in Forestry.* Vol 2: Practice. 1989 ISBN 0-7923-0063-7

    Set ISBN (Volumes 29 and 32) 90-247-3684-6

33. T.L. White and G.R. Hodge (eds.): *Predicting Breeding Values with Applications in Forest Tree Improvement.* 1989 ISBN 0-7923-0460-8

34. H.J. Welch: *The Conifer Manual.* Volume 1. 1991 ISBN 0-7923-0616-3

35. P.K.R. Nair, H.L. Gholz, M.L. Duryea (eds.): *Agroforestry Education and Training.* Present and Future. 1990 ISBN 0-7923-0864-6